T5-CSB-043

INTERNATIONAL SERIES IN
NATURAL PHILOSOPHY
General Editor: D. TER HAAR

VOLUME 107

Fluctuations and Non-linear Wave Interactions in Plasmas

Other Pergamon Titles of Interest

Books

KLIMONTOVICH:
Kinetic Theory of Non-ideal Gases and Non-ideal Plasmas

LIFSHITZ and PITAEVSKII: Physical Kinetics
Landau and Lifshitz Curse of Theoretical Physics Volume 10

LOMINADZE:
Cyclotron Waves in Plasme

POZHELA:
Plasmas and Current Instabilities in Semiconductors

*Journals**

Plasma Physics

Journal of Quantitative Spectroscopy
& Radiative Transfer

**Free specimen copy available on request.*

A full list of titles in the International
Series in Natural Philosophy follows the index.

NOTICE TO READERS

Dear Reader

If your library is not already a standing order customer or subscriber to this series, may we recommend that you place a standing or subscription order to receive immediately upon publication all new issues and volumes published in this valuable series. Should you find that these volumes no longer serve your needs your order can be cancelled at any time without notice.

The Editors and the Publisher will be glad to receive suggestions or outlines of suitable titles, reviews or symposia for consideration for rapid publication in this series.

Robert Maxwell
Publisher at Pergamon Press

Fluctuations and Non-linear Wave Interactions in Plasmas

by

A. G. SITENKO

Translated by DR. O. D. KOCHERGA

Institute for Theoretical Physics
Academy of Sciences of the Ukrainian SSR
252130 Kiev—130 USSR

PERGAMON PRESS

OXFORD · NEW YORK · TORONTO · SYDNEY · PARIS · FRANKFURT

U.K. Pergamon Press Ltd., Headington Hill Hall, Oxford OX3 0BW, England

U.S.A. Pergamon Press Inc., Maxwell House, Fairview Park, Elmsford, New York 10523, U.S.A.

CANADA Pergamon Press Canada Ltd., Suite 104, 150 Consumers Rd., Willowdale, Ontario M2J 1P9, Canada

AUSTRALIA Pergamon Press (Aust.) Pty. Ltd., P.O. Box 544. Potts Point, N.S.W. 2011, Australia

FRANCE Pergamon Press SARL, 24 rue des Ecoles, 75240 Paris, Cedex 05, France

FEDERAL REPUBLIC OF GERMANY Pergamon Press GmbH, 6242 Kronberg-Taunus, Hammerweg 6, Federal Republic of Germany

---

First edition 1982

British Library Cataloguing in Publication Data

Sitenko, A.G.
Fluctuations and non-linear wave interactions in plasmas. — (International series in natural philosophy; vol. 107)
1. Plasma (Ionized gases)
2. Nonlinear theories
I. Title II. Series
530.4'4 QC718 80-41990

ISBN 0-08-025051-3

Printed in Hungary by Franklin Printing House.

TO MY DEAR DAUGHTER ALLA

Preface

MODERN plasma theory is founded on a comprehensive application of statistical methods. The most efficient description of plasma properties may be obtained on the basis of the general dynamic approach in Bogoliubov's statistical theory (see Bogoliubov, 1962). The fundamental concept of this treatment involves a sequence of many-particle distribution functions which completely describe all macroscopic properties of the system. The sequence of the many-particle functions satisfy an interconnected chain of equations which are conventionally referred to as the Bogoliubov–Born–Green–Kirkwood–Yvon hierarchy. The approximate solution of this hierarchy of equations by means of a power series expansion in a small plasma parameter makes it possible to express the two-particle distribution function in terms of the one-particle function and thus to derive a kinetic equation with a Landau collision term (the latter being first obtained by Landau, 1937).

The peculiarities that distinguish a plasma from any other macroscopic medium owe to the collective plasma properties which are manifest in the existence of various eigenwaves and eigenoscillations. Vlasov (1938) and Landau (1946) were the first to introduce a self-consistent field approximation, study the plasma eigenoscillations, and thus lay the foundation of kinetic plasma theory. An account of the concepts of this theory is given in the monographs by Klimontovich (1967), Silin (1971), Ecker (1972), and Ichimaru (1973).

A study of the electromagnetic fluctuations in a plasma provides valuable information on plasma properties. It is well known that the fluctuation–dissipation theorem establishes for equilibrium systems a relation between the fluctuations and the energy dissipation in the medium. Therefore, the knowledge of the dielectric permittivity tensor is sufficient to determine completely the fluctuations in an equilibrium plasma. And, vice versa, the dielectric permittivity tensor is easy to find by means of the inversion of the fluctuation–dissipation theorem provided the fluctuations of non-interacting currents in a plasma are given. Thus we have an exhaustive description of the electromagnetic plasma properties (this approach may be extended to a non-equilibrium plasma) without considering any kinetic equations (Sitenko, 1967).

A number of monographs (Silin and Rukhadze, 1961; Stix, 1962; Thompson, 1962; Balescu, 1963; Montgomery and Tidman, 1964; Akhiezer, Akhiezer, Polovin, Sitenko, and Stepanov, 1967, 1975; Ginzburg, 1970, 1975) are devoted to a detailed investigation of the electrodynamic plasma properties. The monograph by Silin and Rukhadze (1961) is one of the first books on the matter; the one by Akhiezer, Akhiezer, Polovin, Sitenko, and Stepanov (1975) is the most detailed study of plasma electrodynamics. The progress of fusion research favoured the development of non-linear plasma electrodynamics. An account of the achievements in the field is given in the monographs by Vedenov

(1965), Galeev and Sagdeev (1973), Kadomtsev (1965), Sagdeev and Galeev (1969), Tsytovich (1970, 1973), Pustovalov and Silin (1972), Davidson (1972), and Karpman (1975). Kadomtsev (1965) constructed the theory of plasma turbulence; non-linear plasma theory was advanced by Sagdeev and Galeev (1969, 1973).

The purpose of this book is to work out a theory of fluctuations in a homogeneous plasma taking into account non-linear wave interactions. The treatment is based on an expansion in the field intensity and on the averaging procedure involving the Bogoliubov correlation weakening principle. A non-linear equation is derived from the microscopic equations and is used to study the dynamics of the electromagnetic fields in a plasma. The electrodynamic plasma properties are completely determined by the coefficients of the non-linear equation which are the plasma dielectric permittivity and the non-linear susceptibilities. A multiple-time-scale perturbation approach is applied to derive from the non-linear equation for the field a hierarchy of equations which describe a variety of processes that may occur in a plasma.

The fluctuation–dissipation theorem is extended to non-linear media. The non-linear plasma susceptibilities are expressed in terms of the correlation functions of various orders for the current density fluctuations in neglect of interactions between particles. Therefore all linear and non-linear electrodynamic plasma properties may be completely described by means of a sequence of the above correlation functions.

The last chapters are devoted to an extensive investigation of fluctuations in a non-equilibrium plasma taking into account the non-linear interactions of waves. A set of non-linear equations is derived, which govern the sequence of correlation functions for plasma fluctuations. The coefficients of the equations are functions of the non-linear susceptibilities, the inhomogeneous parts are written in terms of the correlation functions for the fluctuations in a system of non-interacting particles. The solutions of the set serve to find the spectral distributions of fluctuations in a plasma. It is shown that the non-linear wave interaction removes the singularities in the spectral distributions of the non-equilibrium fluctuations which occur in the linear approximation. Kinetic equations for particles and waves in non-equilibrium plasmas are considered. An extended kinetic equation for waves is derived taking into account the interaction between the waves and the fluctuating plasma fields. Wave scattering and transformation in, and radiation from, a non-equilibrium plasma are studied on the basis of this kinetic equation.

In the English translation of the book a chapter is added in which a kinetic theory of the non-linear wave interaction in a semi-bounded plasma is developed for the specular reflection model. The non-linear equation for the field is derived and used to study the resonant interaction of surface waves that leads to decay and explosive instabilities.

I express my sincere gratitude to A. I. Akhiezer, Yu. L. Klimontovich, and V. P. Silin for most fruitful discussions.

A. G. SITENKO

Contents

List of notations

$$a^2 = \frac{T}{4\pi e^2 n_0}$$

$$\Omega^2 = \frac{4\pi e^2 n_0}{m}$$

$$s^2 = \frac{3T}{m}$$

$$s_i^2 = \frac{3T_i}{m_i}$$

$$z = \sqrt{\frac{3}{2}}\,\frac{\omega}{ks}$$

$$\varphi(z) = 2z\,e^{-z^2}\int_0^z dx\,e^{x^2}$$

$$\omega_B = \frac{eB_0}{mc}$$

$$z_n = \sqrt{\frac{3}{2}}\,\frac{\omega - n\omega_B}{|k_z|\,s}$$

$$\beta = \frac{k_\perp^2 s^2}{3\omega_B^2}$$

$$v_s^2 = \frac{T_e}{m_i}$$

$$v_A^2 = \frac{B_0^2}{4\pi m_i n_0}$$

$$\eta^2 = \frac{k^2 c^2}{\omega^2}$$

$$\mathbf{E}_{\mathbf{k}\omega} = \int dt \int d\mathbf{r}\, e^{-i\mathbf{k}\mathbf{r} + i\omega t}\mathbf{E}(\mathbf{r}, t)$$

$$\mathbf{E}(\mathbf{r}, t) = \sum_{\omega, \mathbf{k}} \mathbf{E}_{\mathbf{k}\omega}\, e^{i(\mathbf{k}\cdot\mathbf{r}) - i\omega t}$$

$$\sum_{\omega, \mathbf{k}} \cdots = \int \frac{d\omega}{2\pi} \int \frac{d\mathbf{k}}{(2\pi)^3} \cdots$$

$$\langle E_i E_j \rangle_{\mathbf{r}t} = \langle E_i(\mathbf{r}_1, t_1)\, E_j(\mathbf{r}_2, t_2) \rangle, \quad \mathbf{r} = \mathbf{r}_1 - \mathbf{r}_2, \quad t = t_1 - t_2$$

$$\langle E_i E_j \rangle_{\mathbf{k}\omega} = \int dt \int d\mathbf{r}\, e^{-i(\mathbf{k}\cdot\mathbf{r}) + i\omega t} \langle E_i E_j \rangle_{\mathbf{r}t}$$

$$\langle E_i E_j \rangle_{\mathbf{k}} = \frac{1}{2\pi} \int d\omega \langle E_i E_j \rangle_{\mathbf{k}\omega}$$

$$\langle E_i E_j \rangle_{\omega} = \frac{1}{(2\pi)^3} \int d\mathbf{k} \langle E_i E_j \rangle_{\mathbf{k}\omega}$$

$$\varepsilon_{ij}(\omega, \mathbf{k}) = \delta_{ij} + \Sigma \varkappa_{ij}^{(1)}(\omega, \mathbf{k})$$

[in an isotropic plasma:

$$\varepsilon_{ij}(\omega, \mathbf{k}) = \frac{k_i k_j}{k^2} \varepsilon_l(\omega, \mathbf{k}) + \left(\delta_{ij} - \frac{k_i k_j}{k^2}\right) \varepsilon_t(\omega, \mathbf{k})\Big]$$

$$\Lambda_{ij}(\omega, \mathbf{k}) = \varepsilon_{ij}(\omega, \mathbf{k}) + \left(\frac{k_i k_j}{k^2} - \delta_{ij}\right)\eta^2$$

$$\Lambda(\omega, \mathbf{k}) = |\Lambda_{ij}(\omega, \mathbf{k})|$$

$$\Lambda(\omega_{\mathbf{k}}, \mathbf{k}) = 0, \quad \Lambda'_{\mathbf{k}} = \frac{\partial \Lambda(\omega_{\mathbf{k}}, \mathbf{k})}{\partial \omega_{\mathbf{k}}}$$

[in an isotropic plasma:

$$\varepsilon_l(\omega_{\mathbf{k}}, \mathbf{k}) = 0, \quad \varepsilon'_{\mathbf{k}} = \frac{\partial \varepsilon_l(\omega_{\mathbf{k}}, \mathbf{k})}{\partial \omega_{\mathbf{k}}}\Big]$$

$$k = \omega, \mathbf{k}$$

$$\varkappa_{ijk}^{(2)}(1, 2) = \varkappa_{ijk}^{(2)}(\omega_1, \mathbf{k}_1; \omega_2, \mathbf{k}_2)$$

$$\varkappa_{ijkl}^{(3)}(1, 2, 3) = \varkappa_{ijkl}^{(3)}(\omega_1, \mathbf{k}_1; \omega_2, \mathbf{k}_2; \omega_3, \mathbf{k}_3)$$

CHAPTER 1

Statistical Description of a Plasma

1.1. Microscopic equations

The microscopic density

A plasma is a fully or partially ionized gas, the mean charge density of which is equal to zero. We shall deal here with the simplest two-component plasma that consists of negatively charged particles—electrons and positive ones—ions. The ion mass is large in comparison with that of an electron, so the ions are assumed to form a fixed neutralizing background and only the dynamics of the electrons are considered.

Let the plasma be enclosed in a volume V and the total number of electrons be N, then the mean electron density is $n_0 \equiv N/V$. Each electron of a mass m carries a charge e.

The states of individual particles are completely determined by giving the coordinates and velocities. (We assume the electrons to be point particles.) In the six-dimensional phase space of the coordinates $\mathbf{r}$ and velocities $\mathbf{v}$ the motion of an electron is associated with a trajectory that is governed by the equations of motion and the initial conditions.

Denote the radius-vector and the velocity of some electron at time t by $\mathbf{r}_\alpha(t)$ and $\mathbf{v}_\alpha(t)$. The equation of motion of the electron may be written as

$$\left.\begin{aligned} \dot{\mathbf{r}}_\alpha(t) &= \mathbf{v}_\alpha(t), \\ \dot{\mathbf{v}}_\alpha(t) &= \frac{e}{m}\left\{\mathbf{E}(\mathbf{r}_\alpha(t),\, t)+\frac{1}{c}\,[\mathbf{v}_\alpha(t)\,\mathbf{B}(\mathbf{r}_\alpha(t),\, t)]\right\}, \end{aligned}\right\} \tag{1.1}$$

where $\mathbf{E}(\mathbf{r}_\alpha(t), t)$ and $\mathbf{B}(\mathbf{r}_\alpha(t), t)$ are the microscopic electric and magnetic fields at the point where the electron is localized at time t. The fields $\mathbf{E}(\mathbf{r}, t)$ and $\mathbf{B}(\mathbf{r}, t)$ are caused by the external sources as well as by the microscopic distribution of charged particles. (We assume that there are no fields of non-electromagnetic nature. Otherwise the right-hand part of the second equation (1.1) would contain a term $\frac{1}{m}\,\mathbf{F}(\mathbf{r}_\alpha(t)t)$, $\mathbf{F}(r, t)$ denoting the force associated with the non-electromagnetic field.) The presence of the proper electric and magnetic fields in a plasma means that the microscopic state of the latter is determined, besides by the coordinates and the velocities of all charged particles, also by the electric and magnetic fields at each point of the plasma volume.

Let us define the microscopic electron distribution function, or the microscopic density, $\mathcal{F}(\mathbf{r}, \mathbf{v}, t)$ by means of the relation

$$\mathcal{F}(\mathbf{r}, \mathbf{v}, t) = \sum_{\alpha=1}^{N} \delta(\mathbf{r}-\mathbf{r}_\alpha(t))\, \delta(\mathbf{v}-\mathbf{v}_\alpha(t)), \tag{1.2}$$

where the summation extends over all electrons. Such a form of the microscopic density is the immediate consequence of the particles being identical—each term in the sum (1.2) is associated with an individual particle. $\mathcal{F}(\mathbf{r}, \mathbf{v}, t)$ describes the microscopic distribution of electrons in the one-particle phase space. Nevertheless it is to be emphasized that according to (1.2) the microscopic density $\mathcal{F}(\mathbf{r}, \mathbf{v}, t)$ depends on the dynamical variables of all plasma particles.

For simplicity of notation we shall write

$$x \equiv (\mathbf{r}, \mathbf{v}). \tag{1.3}$$

Then (1.2) becomes of the form

$$\mathcal{F}(x, t) = \sum_{\alpha=1}^{N} \delta(x-x_\alpha(t)), \tag{1.4}$$

where the six-dimensional delta-function is defined by means of the relation

$$\delta(x-x_\alpha(t)) \equiv \delta(\mathbf{r}-\mathbf{r}_\alpha(t))\, \delta(\mathbf{v}-\mathbf{v}_\alpha(t)). \tag{1.5}$$

The knowledge of the microscopic density $\mathcal{F}(\mathbf{r}, \mathbf{v}, t)$ in every phase point and of the microscopic electric and magnetic fields $\mathbf{E}(\mathbf{r}, t)$ and $\mathbf{B}(\mathbf{r}, t)$ is sufficient for the complete description of the microscopic plasma state at some time t.

The equation of continuity and Maxwell's equations

The microscopic density $\mathcal{F}(\mathbf{r}, \mathbf{v}, t)$ satisfies the equation of continuity in phase space

$$\frac{\partial \mathcal{F}}{\partial t} + \left(\mathbf{v} \cdot \frac{\partial \mathcal{F}}{\partial \mathbf{r}}\right) + \left(\dot{\mathbf{v}} \cdot \frac{\partial \mathcal{F}}{\partial \mathbf{v}}\right) = 0, \tag{1.6}$$

which within the context of the equation of electron motion (1.1) may be rewritten as

$$\frac{\partial \mathcal{F}}{\partial t} + \left(\mathbf{v} \cdot \frac{\partial \mathcal{F}}{\partial \mathbf{r}}\right) + \frac{e}{m}\left(\left\{\mathbf{E} + \frac{1}{c}[\mathbf{vB}]\right\} \cdot \frac{\partial \mathcal{F}}{\partial \mathbf{v}}\right) = 0 \tag{1.7}$$

(if the non-electromagnetic forces would have been taken into account the left-hand part of (1.7) would have contained $\frac{1}{m}\left(\mathbf{F} \cdot \frac{\partial \mathcal{F}}{\partial \mathbf{v}}\right)$). The microscopic fields $\mathbf{E}(\mathbf{r}, t)$ and

$\mathbf{B}(\mathbf{r}, t)$ are governed by Maxwell's equations

$$\left.\begin{aligned}\operatorname{curl} \mathbf{B} &= -\frac{1}{c}\frac{\partial \mathbf{B}}{\partial t},\\ \operatorname{div} \mathbf{B} &= 0,\\ \operatorname{curl} \mathbf{E} &= \frac{1}{c}\frac{\partial \mathbf{E}}{\partial t}+\frac{4\pi}{c}(\mathbf{j}+\mathbf{j}_0),\\ \operatorname{div} \mathbf{E} &= 4\pi(\varrho+\varrho_0),\end{aligned}\right\} \tag{1.8}$$

where ϱ_0 and $\mathbf{j}_0$ are the charge and the current densities associated with the external sources; ϱ and $\mathbf{j}$ are the microscopic charge and current densities caused by the electron motion in the plasma:

$$\varrho = e\left\{\int d\mathbf{v}\mathcal{F}-n_0\right\}, \tag{1.9}$$

$$\mathbf{j} = e\int d\mathbf{v}\,\mathbf{v}\mathcal{F}. \tag{1.10}$$

Equations (1.7) and (1.8), together with the relations (1.9) and (1.10), form a closed set of equations, which is sufficient for the complete microscopic description of the plasma. Equation (1.7) determines the time evolution of the microscopic density $\mathcal{F}(\mathbf{r}, \mathbf{v}, t)$, while (1.8), together with (1.9) and (1.10), serve to find the self-consistent microfields $\mathbf{E}(\mathbf{r}, t)$ and $\mathbf{B}(\mathbf{r}, t)$. It should be noted that the set (1.7)–(1.8) is exact.†

The set (1.7)–(1.8) becomes simplified in the case of a non-relativistic plasma, since the electromagnetic interaction between particles is then of a purely Coulomb nature The equation for the microscopic density (1.7) reduces in such a case to the foll owing

$$\frac{\partial \mathcal{F}}{\partial t}+\left(\mathbf{v}\cdot\frac{\partial \mathcal{F}}{\partial \mathbf{r}}\right)+\frac{e}{m}\left(\left\{\mathbf{E}+\frac{1}{c}[\mathbf{v}\mathbf{B}_0]\right\}\cdot\frac{\partial \mathcal{F}}{\partial \mathbf{v}}\right) = 0, \tag{1.11}$$

where $\mathbf{B}_0$ is a constant external magnetic field and $\mathbf{E}$ is the longitudinal self-consistent electric field that is governed by the last equation (1.8):

$$\operatorname{div} \mathbf{E} = 4\pi(\varrho+\varrho_0). \tag{1.12}$$

In the absence of external fields, the set (1.11) and (1.12) takes the form

$$\frac{\partial \mathcal{F}}{\partial t}+\left(\mathbf{v}\cdot\frac{\partial \mathcal{F}}{\partial \mathbf{r}}\right)+\frac{e}{m}\left(\mathbf{E}\cdot\frac{\partial \mathcal{F}}{\partial \mathbf{v}}\right) = 0, \tag{1.13}$$

$$\operatorname{div} \mathbf{E} = 4\pi\varrho, \tag{1.14}$$

where ϱ is the charge density (1.9).

Besides the microscopic density $\mathcal{F}(\mathbf{r}, \mathbf{v}, t)$, which is responsible for the particle distribution in the six-dimensional phase space $(\mathbf{r}, \mathbf{v})$, we introduce also microscopic densities $\mathcal{F}_2(\mathbf{r}, \mathbf{v}; \mathbf{r}', \mathbf{v}'; t)$, $\mathcal{F}_3(\mathbf{r}, \mathbf{v}; \mathbf{r}', \mathbf{v}'; \mathbf{r}'', \mathbf{v}''; t)$, etc., which describe the probabilities

† Klimontovich (1967) was the first to propose this approach to the description of a plasma.

that two, three, and so on particles be located simultaneously in some definite phase points:†

$$
\begin{aligned}
\mathcal{F}_2(x, x'; t) &\equiv \sum_{\alpha \neq \alpha'} \delta(x - x_\alpha(t))\,\delta(x' - x_{\alpha'}(t)),\\
\mathcal{F}_3(x, x', x''; t) &\equiv \sum_{\alpha \neq \alpha' \neq \alpha''} \delta(x - x_\alpha(t))\,\delta(x' - x_{\alpha'}(t))\,\delta(x'' - x_{\alpha''}(t)),\\
&\dots \\
\mathcal{F}_s(x, x', \dots, x^{(s-1)}; t) &\\
\equiv \sum_{\alpha \neq \alpha' \neq \dots \neq \alpha^{(s-1)}} &\delta(x - x_\alpha(t))\,\delta(x' - x_{\alpha'}(t)) \dots \delta(x^{(s-1)} - x_{\alpha^{(s-1)}}(t)).
\end{aligned}
\tag{1.15}
$$

For the ion motion to be involved in the consideration, the microscopic ion density $\mathcal{F}_i(\mathbf{r}, \mathbf{v}, t)$ must be defined together with (1.2). In the general case of a multicomponent plasma there must be introduced partial microscopic densities

$$\mathcal{F}_a(\mathbf{r}, \mathbf{v}, t) \equiv \sum_{\alpha=1}^{N_a} \delta(\mathbf{r} - \mathbf{r}_\alpha(t))\,\delta(\mathbf{v} - \mathbf{v}_\alpha(t)), \tag{1.16}$$

where a labels different plasma species, the summation extends over all particles of the relevant component. Microscopic densities of the plasma components are governed by equations analogous to (1.7) and (1.11). The self-consistent fields **E** and **B** are described by the same Maxwell set (1.8) or by (1.12), where the sums of the partial densities of the relevant plasma components must be substituted for the microscopic charge and current densities ϱ and $\mathbf{j}$:

$$\varrho = \sum_a e_a \int d\mathbf{v}\,\mathcal{F}_a, \tag{1.17}$$

$$\mathbf{j} = \sum_a e_a \int d\mathbf{v}\,\mathbf{v}\,\mathcal{F}_a. \tag{1.18}$$

To simplify the notation we shall omit the component labels in the analysis that follows.

1.2. The Liouville Distribution

The Liouville equation

The microscopic equations (1.7) and (1.8), that determine the microscopic density and the self-consistent microfields in a plasma, describe all dynamical plasma properties, since their solution is equivalent to the solution of the complete set of dynamical equations for a plasma. However, such a detailed microscopic analysis is not necessary when the macroscopic plasma properties are considered. The appropriate level of description may be provided on the basis of the statistical approach (Landau and Lifshitz, 1969).

The essence of the statistical treatment of many-particle systems is that instead of determining the coordinates and the momenta of all particles of the system at some time

† According to the definition (1.15), the microscopic density (1.2) should be denoted by $\mathcal{F}_1(\mathbf{r}, \mathbf{v}, t)$. However, we shall always suppress the subscripts of the one-particle functions.

by their given initial values, an ensemble of such systems, which differ only in the particle states, is considered, and the distribution of the systems in different states is studied (Gibbs, 1902; Boltzmann, 1964). These distributions are conventionally described by means of the Liouville distribution function in the phase space of the relevant number of particles. The N-particle Liouville distribution $D_N(x_1, x_2, \ldots, x_N; t)$ is a function of the coordinates and velocities of N particles. If the latter are identical, then the Liouville function is symmetric relative to the permutations of particle coordinates and velocities. The Liouville distribution function satisfies the normalization condition

$$\int \prod_{\alpha=1}^{N} dx_\alpha \, D_N(x_1, x_2, \ldots, x_N; t) = 1 \tag{1.19}$$

and the continuity condition in the N-particle phase space

$$\frac{\partial D_N}{\partial t} + \sum_{\alpha=1}^{N} \left(\dot{\mathbf{r}}_\alpha \frac{\partial D_N}{\partial \mathbf{r}_\alpha} + \dot{\mathbf{v}}_\alpha \frac{\partial D_N}{\partial \mathbf{v}_\alpha} \right) = 0. \tag{1.20}$$

Using the equations of motion one can easily derive from (1.20) the Liouville equation

$$\frac{\partial D_N}{\partial t} + \{D_N, H\} = 0, \tag{1.21}$$

where $\{\ ,\ \}$ is a Poisson bracket and H is the Hamiltonian of the system taking into account the particle interactions. The latter is for a non-relativistic system of N particles interacting through a Coulomb law of force (in the absence of external fields) of the form

$$H = \sum_{\alpha=1}^{N} \frac{1}{2} m^{(\alpha)} v_\alpha^2 + \sum_{\alpha < \alpha'} \frac{e^{(\alpha)} e^{(\alpha')}}{|\mathbf{r}_\alpha - \mathbf{r}_{\alpha'}|}. \tag{1.22}$$

The Liouville equation governs the time evolution of the distribution function and is suitable for finding the distribution function at any time $D_N(x_1, x_2, \ldots, x_N; t)$ from the given initial distribution $D_N(x_1, x_2, \ldots, x_N; 0)$. It follows from (1.21) that the Liouville distribution is conserved along the trajectory in the N-particle phase space

$$D_N(x_1, x_2, \ldots, x_N; t) = D_N(x_1(0), x_2(0), \ldots, x_N(0); 0). \tag{1.23}$$

Statistical averaging

Making use of the Liouville distribution function we can carry out the averaging of any microscopic quantity. Let some microscopic quantity $Q(x, x', \ldots, t)$ be given in the points $x, x', \ldots,$ of the six-dimensional phase space. Q depends on time through particle coordinates and velocities, i.e.

$$Q(x, x', \ldots; t) \equiv Q(x, x', \ldots; \{x_\alpha(t)\}). \tag{1.24}$$

Let us define the average of Q according to

$$\langle Q(x, x', \ldots; t \rangle = \int \prod_{\alpha=1}^{N} dx_\alpha \, D_N(x_1, x_2, \ldots, x_N; t) \, Q(x, x', \ldots; \{x_\alpha\}). \tag{1.25}$$

As the Liouville distribution is conserved along the trajectory, the average over the distribution at time t in (1.25) may be replaced by that over the initial distribution. The particle coordinates and velocities at time t are governed by the initial conditions $\{x_\alpha(t)\} \equiv \{x_\alpha(\{x_\alpha(0)\}, t)\}$, so within the context of the relation (1.23) the average (1.25) reduces to

$$\langle Q(x, x', \dots; t)\rangle$$
$$= \int \prod_{\alpha=1}^{N} dx_\alpha(0)\, D_N(x_1(0), x_2(0), \dots, x_N(0); 0)\, Q\big(x, x', \dots; \{x_\alpha(x_\alpha\{(0)\}, t)\}\big). \quad (1.26)$$

The average of any microscopic quantity may be found by means of (1.25) or (1.26) provided the Liouville distribution is given. However, it is in fact impossible to find D_N explicitly for the ensembles of very large numbers of particles, for to do this we must exactly integrate the equations of motion. The exhaustive knowledge of D_N is in fact not needed in case we are interested in the macroscopic quantities which either are independent of particle correlations or depend only on the correlations between a few particles. Examples of such quantities are the mean particle density (the average number of particles per unit volume), the mean particle density in phase space, the mean velocity of an individual particle, the mean total particle energy, and so on. All we have to know to calculate these quantities is the one-particle distribution function or the distribution functions of a few particles.

1.3. The One-particle Distribution Function and Correlation Functions

Sequence of distribution functions

Let us define the one-particle distribution function $f(x, t)$ as the average of the microscopic density (1.4) over the Liouville distribution:

$$f(x, t) \equiv \langle \mathcal{F}(x, t)\rangle. \quad (1.27)$$

Making use of the definition (1.4) and having in mind that the distribution $D_N(x_1, x_2, \dots, x_N; t)$ is symmetric with respect to its arguments, we easily find the following expression for the one-particle distribution function:

$$f(x, t) = N \int dx_2 \dots dx_N D_N(x, x_2, \dots, x_N; t). \quad (1.28)$$

According to (1.19), $f(x, t)$ is normalized to the total number of particles N:

$$\int dx f(x, t) = N. \quad (1.29)$$

Let us define the two-particle distribution function $f_2(x, x'; t)$ to be the average of the two-particle microscopic density (1.15) over the Liouville distribution:

$$f_2(x, x'; t) \equiv \langle \mathcal{F}_2(x, x'; t)\rangle. \quad (1.30)$$

Within the context of the explicit form of $\mathcal{F}_2(x, x'; t)$ and of the symmetry properties of the Liouville distribution the two-particle distribution function is of the following form:

$$f_2(x, x'; t) = N(N-1) \int dx_3 \dots dx_N D_N(x, x', x_3, \dots, x_N; t). \tag{1.31}$$

Similarly we may define the s-particle distribution function $f_s(x, x', \dots, x^{(s-1)}, t)$ as

$$f_s(x, x', \dots, x^{(s-1)}; t) \equiv \langle \mathcal{F}_s(x, x', \dots, x^{(s-1)}; t) \rangle, \quad s \leqslant N, \tag{1.32}$$

which may be expressed in terms of the Liouville function as

$$f_s(x_1, x_2, \dots, x_s; t) = \frac{N!}{(N-s)!} \int dx_{s+1} \dots dx_N D_N(x_1, x_2, \dots, x_N; t). \tag{1.33}$$

The knowledge of the few simplest functions from the introduced sequence is sufficient for the description of the macroscopic properties of the system.

By virtue of (1.19) the many-particle distribution functions $f_s(x_1, \dots, x_s; t)$ satisfy the normalization conditions

$$\int dx_1 \dots dx_s f_s(x_1, \dots, x_s; t) = \frac{N!}{(N-s)!}. \tag{1.34}$$

It should be pointed out that some authors use an alternative normalization of the many-particle functions $\tilde{f}_s(x_1, \dots, x_s; t)$:

$$\int dx_1 \dots dx_s \tilde{f}_s(x_1, \dots x_s; t) = N^s. \tag{1.35}$$

(This normalization is associated with the following definition of the many-particle distribution functions:

$$\tilde{f}_s(x_1 \dots x_s; t) = N^s \int dx_{s+1} \dots dx_N D_N(x_1, \dots x_N; t).) \tag{1.36}$$

It is clear that f_s and $\tilde{f}_s$ are related according to

$$f_s(x_1 \dots x_s; t) = C_s \tilde{f}_s(x_1, \dots x_s; t), \tag{1.37}$$

with the proportionality coefficient

$$C_s = \frac{N!}{N^s(N-s)!} \tag{1.38}$$

C_s turns into unity in the limiting case $N \to \infty$ and for the finite values of s. So the difference between f_s and $\tilde{f}_s$ may be neglected for systems of large numbers of particles.

Connection between one-particle and many-particle distribution functions

Let us find the relation between the mean product of the microscopic densities in two different phase points and the two-particle distribution function. Within the context of the definition (1.4) the product of the microscopic densities associated with two phase points may be written as

$$\mathcal{F}(x,t)\mathcal{F}(x',t) = \delta(x-x')\sum_{\alpha}\delta(x-x_\alpha(t)) + \sum_{\alpha\neq\alpha'}\delta(x-x_\alpha(t))\,\delta(x'-x_{\alpha'}(t)).$$

The mean value of the first term in the right-hand part is the one-particle distribution function; the average of the second term yields the two-particle distribution function:

$$\langle\mathcal{F}(x,t)\mathcal{F}(x',t)\rangle = \delta(x-x')f(x,t)+f_2(x,x';t). \tag{1.39}$$

It is suitable to divide the product of the microscopic densities in three different phase points $\mathcal{F}(x,t)$, $\mathcal{F}(x',t)$, and $\mathcal{F}(x'',t)$ in three groups of terms: the first group includes the terms with all identical subscripts, the second one contains those with two identical subscripts, and the third one consists of those with all different subscripts:

$$\begin{aligned}\mathcal{F}(x,t)\mathcal{F}(x',t)\mathcal{F}(x'',t) &= \delta(x-x')\,\delta(x-x'')\sum_{\alpha}\delta(x-x_\alpha(t))\\ &+[\delta(x-x')+\delta(x-x'')]\sum_{\alpha\neq\alpha'}\delta(x'-x_\alpha(t))\,\delta(x''-x_{\alpha'}(t))\\ &+\delta(x'-x'')\sum_{\alpha\neq\alpha'}\delta(x-x_\alpha(t))\,\delta(x'-x_{\alpha'}(t))\\ &+\sum_{\alpha\neq\alpha'\neq\alpha''}\delta(x-x_\alpha(t))\,\delta(x'-x_{\alpha'}(t))\,\delta(x''-x_{\alpha''}(t)).\end{aligned}$$

The average of this expression is

$$\begin{aligned}\langle\mathcal{F}(x,t)\mathcal{F}(x',t)\mathcal{F}(x'',t)\rangle &= \delta(x-x')\,\delta(x-x'')f(x,t)\\ &+[\delta(x-x')+\delta(x-x'')]f_2(x',x'';t)\\ &+\delta(x'-x'')f_2(x,x';t)+f_3(x,x',x'';t).\end{aligned} \tag{1.40}$$

Analogous arguments lead to the conclusion that the mean value of the product of microscopic densities in s different phase points may be expressed in terms of a sequence of functions $f, f_2, \ldots, f_{s-1}$, and f_s.

In the absence of particle interactions the many-particle distribution function becomes a product of s one-particle functions. If the particles do interact, then the many-particle distribution function cannot be reduced to such a product; in particular, the two-particle distribution cannot be presented by a product of two one-particle functions. The interdependence of the particle motions may be described by means of the correlation functions. For example, the two-particle distribution function may be written in the form

$$f_2(x,x';t) = f(x,t)f(x',t)+g(x,x';t). \tag{1.41}$$

The first term in the right-hand part of (1.41) is the two-particle distribution function where we neglect particle interactions; the second one describes the two-particle correlations and is conventionally referred to as the two-particle correlation function.† Similarly,

† In fact, the two-particle correlation function should be defined by means of the equality

$$\tilde{f}_2(x,x';t) = f(x,t)f(x',t)+\tilde{g}(x,x';t).$$

However, the difference between $\tilde{g}(x,x';t)$ and $g(x,x';t)$ may be neglected as $N\to\infty$. The treatment of the higher-order correlation functions implies analogous arguments.

the three-particle distribution function may be presented as

$$f_3(x, x', x''; t) = f(x, t) f(x', t) f(x'', t) + f(x, t) g(x', x''; t) + f(x', t) g(x, x''; t) + f(x'', t) g(x, x'; t) + h(x, x', x''; t). \quad (1.42)$$

The first term in the right-hand part of (1.42) is the distribution function of three non-interacting particles; the next three terms take into account binary correlations in the system; the last term describes the simultaneous three-particle correlations and is called the ternary correlation function. The higher-order correlation functions, which are associated with the effects that are produced by the correlations of four and more particles, may be introduced in a similar manner. The knowledge of the sequence of correlation functions provides a complete description of all effects which may occur by virtue of the many-particle properties of the system. It is clear that the one-particle distribution function and the set of correlation functions $g(x, x'; t)$, $h(x, x', x''; t)$, etc., may serve to study the macroscopic properties of the system as well as the set of distribution functions $f(x, t)$, $f_2(x, x'; t)$, $\ldots$, $f_s(x, x', \ldots, x^{(s-1)}; t)$.

1.4. The Bogoliubov–Born–Green–Kirkwood–Yvon Hierarchy

The microscopic distribution functions of the succession $f(x, t)$, $f_2(x, x'; t)$, $\ldots$, $f_s(x, x', \ldots, x^{(s-1)}; t)$ satisfy a set of coupled equations, the so-called Bogoliubov–Born–Green–Kirkwood–Yvon (BBGKY) hierarchy. One of the ways to derive the latter is to carry out the statistical averaging of the set of equations for the microscopic distribution function and the microscopic field.†

Suppose for simplicity that there are no external fields $\mathbf{E}_0$ and $\mathbf{B}_0$. The set of (1.13) and (1.14) reduces then to a single equation. Indeed, the solution of (1.14) within the context of (1.9) may be written as

$$\mathbf{E}(\mathbf{r}, t) = -e' \frac{\partial}{\partial \mathbf{r}} \int d\mathbf{r}'\, d\mathbf{v}' \frac{\mathcal{F}(\mathbf{r}', \mathbf{v}', t)}{|\mathbf{r}-\mathbf{r}'|}. \quad (1.43)$$

Substituting (1.43) into (1.13) we obtain the following equation for the microscopic function $\mathcal{F}(x, t)$:

$$\left(\frac{\partial}{\partial t} + \left(\mathbf{v} \cdot \frac{\partial}{\partial \mathbf{r}}\right)\right) \mathcal{F}(x, t) = \int dx' \mathcal{V}(x, x') \mathcal{F}(x, t) \mathcal{F}(x', t), \quad (1.44)$$

where $\mathcal{V}(x, x')$ is the two-particle Coulomb operator

$$\mathcal{V}(x, x') \equiv \frac{ee'}{m} \left(\frac{\partial}{\partial \mathbf{r}} \frac{1}{|\mathbf{r}-\mathbf{r}'|}\right) \frac{\partial}{\partial \mathbf{v}}. \quad (1.45)$$

Let us average (1.44) over the Liouville distribution. Multiply the left- and right-hand parts of equation (1.44) by D_N and carry out the integration over the N-particle phase

† Since the distribution functions $f, f_2, \ldots, f_N$ may be expressed in terms of D_N (within the context of (1.33), the BBGKY hierarchy may be also obtained by means of a direct integration of the Liouville equation (1.21) (Uhlenbeck and Ford, 1963; Gurov, 1966).

space. It becomes evident that by virtue of (1.26) the averaging and the differentiation in the left-hand part of the resulting equality commute. Then making use of (1.39) in the right-hand part of the equality we obtain

$$\left(\frac{\partial}{\partial t}+\left(\mathbf{v}\cdot\frac{\partial}{\partial \mathbf{r}}\right)\right)f(x, t) = \int dx'\mathcal{V}(x, x')\{\delta(x-x')f(x, t)+f_2(x, x'; t)\}. \tag{1.46}$$

The first term in the right-hand part of (1.46) is associated with the proper Coulomb energies of the charged particles and may be omitted. (When deriving (1.44) one must remember that the microscopic electric field (1.43) in a point that is occupied by some charge is governed only by the charges of the rest of the particles.) Thus we obtain the following relation between the one- and two-particle distribution functions $f(x, t)$ and $f_2(x, x'; t)$:

$$\left(\frac{\partial}{\partial t}+\left(\mathbf{v}\cdot\frac{\partial}{\partial \mathbf{r}}\right)\right)f(x, t) = \int dx'\mathcal{V}(x, x')f_2(x, x'; t). \tag{1.47}$$

Equation (1.47) is the first equation of the BBGKY hierarchy. To solve (1.47) we have to derive an equation for $f_2(x, x'; t)$. To find this let us multiply the equation for the microscopic distribution function $\mathcal{F}(x, t)$ by the microscopic distribution function $\mathcal{F}(x', t)$, that for $\mathcal{F}(x', t)$ by $\mathcal{F}(x, t)$, and add the resulting equalities. We obtain

$$\left(\frac{\partial}{\partial t}+\left(\mathbf{v}\cdot\frac{\partial}{\partial \mathbf{r}}\right)+\left(\mathbf{v}'\cdot\frac{\partial}{\partial \mathbf{r}'}\right)\right)\mathcal{F}(x, t)\mathcal{F}(x', t)$$
$$= \int dx''[\mathcal{V}(x, x'')+\mathcal{V}(x', x'')]\mathcal{F}(x, t)\mathcal{F}(x', t)\mathcal{F}(x'', t). \tag{1.48}$$

Averaging (1.48) over the Liouville distribution within the context of (1.40) we derive an equation which contains the distribution functions f, f_2, and f_3. After the terms that are produced by the proper Coulomb energy are neglected and use of (1.47) is made, we have:

$$\left\{\frac{\partial}{\partial t}+\left(\mathbf{v}\cdot\frac{\partial}{\partial \mathbf{r}}\right)+\left(\mathbf{v}'\cdot\frac{\partial}{\partial \mathbf{r}'}\right)-[\mathcal{V}(x, x')+\mathcal{V}(x', x)]\right\}f_2(x, x'; t)$$
$$= \int dx''[\mathcal{V}(x, x'')+\mathcal{V}(x', x'')]f_3(x, x', x''; t). \tag{1.49}$$

This equation connects the two- and three-particle distribution functions and is the second in the BBGKY hierarchy.

The similar procedure yields the whole BBGKY hierarchy. In the general case the equation for the s-particle distribution function f_s contains the $(s+1)$-particle distribution function f_{s+1}:

$$\left\{\frac{\partial}{\partial t}+\sum_{i=1}^{s}\left(\mathbf{v}_i\cdot\frac{\partial}{\partial \mathbf{r}_i}\right)-\sum_{i\neq j}^{s}\mathcal{V}(x_i, x_j)\right\}f_s(x_1, x_2, \ldots, x_s; t)$$
$$= \sum_{i=1}^{s}\int dx_{s+1}\mathcal{V}(x_i, x_{s+1})f_{s+1}(x_1, x_2, \ldots, x_{s+1}; t). \tag{1.50}$$

Note that when deriving (1.47), (1.49), and (1.50) we have carried out integrations within an infinite range. This corresponds to the limiting case $V \to \infty$ and $N \to \infty$ while $n_0 = N/V = \text{const}$. If the ensemble is enclosed in a finite volume, then the equations contain additional terms which are due to the boundary conditions.

The BBGKY hierarchy of coupled equations is basic in the kinetic plasma theory. Note that the set of the hierarchy equations is not closed. The equation for the one-particle distribution function contains the two-particle one; the equation for the latter involves the three-particle distribution function, and so on. If the total number of N particles is very large, then the equations make in fact an infinite hierarchy which must be cut off in order to obtain a closed set of equations. The various kinetic approximations are associated with different methods of a meaningful closure of the hierarchy at the appropriate level of description. The simplest treatment is Vlasov's approach or the self-consistent field approximation which implies that all particle correlations be neglected.

1.5. Self-consistent Field Approximation

The Vlasov kinetic equation

In case the particles move independently, all correlation functions must be set equal to zero: $g(x, x'; t), h(x, x', x''; t) = 0$ and so on. Then all many-particle distribution functions reduce to products of one-particle functions

$$f_s(x_1, x_2, \ldots, x_s; t) = \prod_{i=1}^{s} f(x_i, t). \tag{1.51}$$

Substituting the product of two one-particle functions $f(x, t)$ and $f(x', t)$ for the two-particle distribution function $f_2(x, x'; t)$ in the first equation of the hierarchy (1.47), we obtain

$$\left\{\frac{\partial}{\partial t} + \left(\mathbf{v} \cdot \frac{\partial}{\partial \mathbf{r}}\right) - \int dx' \mathcal{V}(x, x') f(x', t)\right\} f(x, t) = 0. \tag{1.52}$$

This equation governs the one-particle distribution function of uncorrelated particles and is referred to as the Vlasov kinetic equation. It is peculiar for this approximation that the particle interactions are suggested to be manifested through the average self-consistent field (Vlasov, 1950).

Indeed, the average of (1.43) over the Liouville distribution yields the following expression for the mean electric field in a plasma:

$$\overline{\mathbf{E}(\mathbf{r}, t)} = -e' \frac{\partial}{\partial \mathbf{r}} \int d\mathbf{r}' \, d\mathbf{v}' \frac{f(r', \mathbf{v}', t)}{|\mathbf{r} - \mathbf{r}'|}, \tag{1.53}$$

and (1.52) reduces to:

$$\frac{\partial f}{\partial t} + \left(\mathbf{v} \cdot \frac{\partial f}{\partial \mathbf{r}}\right) + \frac{e}{m}\left(\overline{\mathbf{E}} \cdot \frac{\partial f}{\partial \mathbf{v}}\right) = 0. \tag{1.54}$$

(If an external magnetic field $\mathbf{B}_0$ is present, the Vlasov equation should be written as

$$\frac{\partial f}{\partial t}+\left(\mathbf{v}\cdot\frac{\partial f}{\partial \mathbf{r}}\right)+\frac{e}{m}\left(\overline{\mathbf{E}}\cdot\frac{\partial f}{\partial \mathbf{v}}\right)+\frac{e}{mc}\left([\mathbf{v}_1\mathbf{B}_0]\cdot\frac{\partial f}{\partial \mathbf{v}}\right)=0. \tag{1.55}$$

It may be easily verified that the electric field (1.53) is the solution of the following equation for the self-consistent field:

$$\operatorname{div}\overline{\mathbf{E}} = 4\pi\bar{\varrho}, \tag{1.56}$$

where $\overline{\varrho(\mathbf{r}, t)}$ is the mean charge density:

$$\overline{\varrho(\mathbf{r}, t)} = \Sigma e \int d\mathbf{v} f(\mathbf{r}, \mathbf{v}, t). \tag{1.57}$$

(To derive the microscopic equation (1.56) one has to average the microscopic equation (1.14) over the Liouville distribution.)

The closed set of (1.54) and (1.56) are the basis of the Vlasov approximation, which treats the macroscopic plasma properties neglecting particle correlations.

It is of considerable interest to point out the seeming identity of the Vlasov kinetic equation (1.54) and the equation for the microscopic density (1.13). The set of (1.54) and (1.56) is formally similar to (1.13) and (1.14). However, it is very important to observe the deep physical distinction between these two sets. Equation (1.13) governs the microscopic distribution function which provides a complete account of the random motion of individual particles, while the Vlasov kinetic equation (1.54) determines only the macrodistribution, i.e. the statistical average of the microscopic distribution function. The fluctuations due to the individual particle motion and the fluctuating variations of the fields are not described by the Vlasov equation. Similarly, the macroscopic field that is the solution of (1.56) may be obtained as the statistical average of the microscopic field governed by (1.14). The set of (1.13) and (1.14) is exact; that of the kinetic Vlasov equation (1.54) and the equation for self-consistent field (1.56) is approximate and has a restricted range of validity.

Effects of electromagnetic interactions between particles

It follows from definition (1.53) that equations (1.54) and (1.56) take into account only the Coulomb interaction between charged particles. However, these equations may be extended to include the electromagnetic particle interactions into our consideration too. To achieve this we must introduce the average self-consistent (both potential and rotational) electric and magnetic fields, which are caused by the random charge and current distributions, and take into account the influence of these fields on the distribution function in the kinetic equation.

The kinetic Vlasov equation, taking into account the electromagnetic interactions between charged particles in a plasma, is in the general case as follows:

$$\frac{\partial f}{\partial t}+\left(\mathbf{v}\cdot\frac{\partial f}{\partial \mathbf{r}}\right)+\frac{e}{m}\left(\left\{\overline{\mathbf{E}}+\frac{1}{c}[\mathbf{v}\overline{\mathbf{B}}]\right\}\cdot\frac{\partial f}{\partial \mathbf{v}}\right)=0. \tag{1.58}$$

The self-consistent electric and magnetic fields $\overline{\mathbf{E}}$ and $\overline{\mathbf{B}}$ are governed by the Maxwell equations

$$\left.\begin{aligned}\operatorname{curl}\overline{\mathbf{E}} &= -\frac{1}{c}\frac{\partial\overline{\mathbf{B}}}{\partial t},\\ \operatorname{div}\overline{\mathbf{B}} &= 0,\end{aligned}\right\}\qquad \left.\begin{aligned}\operatorname{curl}\overline{\mathbf{B}} &= \frac{1}{c}\frac{\partial\overline{\mathbf{E}}}{\partial t}+\frac{4\pi}{c}\overline{\mathbf{j}},\\ \operatorname{div}\overline{\mathbf{E}} &= 4\pi\overline{\varrho},\end{aligned}\right\} \tag{1.59}$$

where the mean charge and current distributions $\overline{\varrho(\mathbf{r}, t)}$ and $\overline{\mathbf{j}(\mathbf{r}, t)}$ depend on their arguments through the distribution function $f(\mathbf{r}, \mathbf{v}, t)$ according to

$$\overline{\varrho} = \Sigma e \int d\mathbf{v} f, \tag{1.60}$$

$$\overline{\mathbf{j}} = \Sigma e \int d\mathbf{v}\mathbf{v} f. \tag{1.61}$$

(If external charges and currents ϱ_0 and $\mathbf{j}_0$ are present, then the relevant terms must be added to the the right-hand parts of the third and the fourth equations of (1.59). If the plasma is in a constant external magnetic field $\mathbf{B}_0$, then $\mathbf{B}+\mathbf{B}_0$ must be substituted for the magnetic field $\mathbf{B}$ in (1.58).) The set of (1.58) and (1.59) provides a complete description of the electromagnetic plasma properties when particle correlations are neglected and may be taken as the basis of macroscopic plasma electrodynamics.

It may be easily verified that the kinetic Vlasov equation (1.54) (or (1.58)) is invariant under time inversion. Thus its solutions are time-reversible and the equation itself does not describe the relaxation of a system towards thermal equilibrium. (The distributions which satisfy the Vlasov equation lead to entropy conservation.) It may be easily shown that any distribution which is homogeneous in space and stationary in time satisfies (1.54) (the mean electric field is equal to zero in a homogeneous medium by virtue of the neutrality of the latter). To describe the relaxation of a plasma to a state of thermal equilibrium we must necessarily take into account the correlations between charged particles.

1.6. Influence of Binary Correlations and the Collision Term

The kinetic equation with collisions

The Vlasov approximation ignores all particle correlations. Let us take into account the binary correlations in an ensemble of charged particles and neglect the correlations between three and more particles. In other words, we shall assume that the two-particle correlation function $g(x, x'; t)$ does not vanish and set all correlation functions of higher orders equal to zero. Substituting (1.41) and (1.42) into the two first equations of the BBGKY hierarchy we obtain within the context of the assumption $h(x, x', x''; t) = 0$ a closed set of equations for $f(x, t)$ and $g(x, x'', t)$. The first equation is of the form

$$\left\{\frac{\partial}{\partial t}+\left(\mathbf{v}\cdot\frac{\partial}{\partial \mathbf{r}}\right)+\frac{e}{m}\left(\overline{\mathbf{E}(\mathbf{r}, t)}\cdot\frac{\partial}{\partial \mathbf{v}}\right)\right\} f(x, t) = \int dx' \mathcal{V}(x, x')\, g(x, x'; t) \tag{1.62}$$

where $\overline{\mathbf{E}(\mathbf{r}, t)}$ is the self-consistent field governed by (1.56) and $\mathcal{V}(x, x')$ is the two-particle Coulomb operator (1.45). The left-hand part of (1.62) is the same as that of the kinetic

Vlasov equation (1.54), and the right-hand part contains a term that depends on the binary correlation function. This term describes the binary particle encounters and is conventionally referred to as the collision term.

After some straightforward algebra with the use of (1.62) the second equation of the set reduces to

$$\left\{\frac{\partial}{\partial t}+\left(\mathbf{v}\cdot\frac{\partial}{\partial \mathbf{r}}\right)+\left(\mathbf{v}\cdot\frac{\partial}{\partial \mathbf{r}'}\right)+\frac{e}{m}\left(\overline{\mathbf{E}(\mathbf{r},t)}\cdot\frac{\partial}{\partial \mathbf{v}}\right)+\frac{e'}{m'}\left(\overline{\mathbf{E}(\mathbf{r}',t)}\cdot\frac{\partial}{\partial \mathbf{v}'}\right)-\mathcal{V}(x,x')-\mathcal{V}(x',x)\right\}$$

$$\times g(x,x';t)-\int dx''\mathcal{V}(x,x'')f(x,t)\,g(x',x'';t)$$

$$-\int dx''\mathcal{V}(x',x'')f(x',t)\,g(x,x'';t)$$

$$=[\mathcal{V}(x,x')+\mathcal{V}(x',x)]f(x,t)f(x',t). \tag{1.63}$$

The left-hand part of this equation is homogeneous in $g(x, x'; t)$, so the right-hand part of (1.63) may be treated as an inducing force which by virtue of the Coulomb interaction causes the correlations of the particles.

The set of coupled equations (1.62) and (1.63) becomes rather simplified in the limiting case of weak particle interactions. First, we may neglect the influence of the self-consistent field, then the third term in the left-hand part of (1.62) and the fourth and fifth terms in the left-hand part of (1.63) vanish. It should be observed that these terms are exactly zero if the medium is spatially homogeneous. We shall suggest our system to be slightly inhomogeneous, i.e. we assume the relevant self-consistent field weak as compared to the mean correlation field. Second, since the interaction is weak, we may neglect also the sixth and the seventh terms in the left-hand part of (1.63). (This implies $|g(x,x';t)| \ll f(x,t)f(x',t)$.) The same reason allows us to omit the eighth and the ninth terms in the left-hand part of (1.63). (The interaction between two particles may be neglected if one of those is close to the third particle.) As a result the set of equations governing the distribution function $f(x, t)$ a nd the correlation function $g(x, x'; t)$ reduces to

$$\left(\frac{\partial}{\partial t}+\left(\mathbf{v}\cdot\frac{\partial}{\partial \mathbf{r}}\right)\right)f(x,t)=\frac{1}{m}\frac{\partial}{\partial \mathbf{r}}\int dx'\,U(\mathbf{r}-\mathbf{r}')\frac{\partial}{\partial \mathbf{v}}g(x,x';t), \tag{1.64}$$

$$\left(\frac{\partial}{\partial t}+\left(\mathbf{v}\cdot\frac{\partial}{\partial \mathbf{r}}\right)+\left(\mathbf{v}'\cdot\frac{\partial}{\partial \mathbf{r}'}\right)\right)g(x,x';t)$$

$$=\left(\frac{\partial}{\partial \mathbf{r}}\cdot U(\mathbf{r}-\mathbf{r}')\left\{\frac{1}{m}\frac{\partial f(x,t)}{m\,\partial \mathbf{v}}f(x',t)-\frac{1}{m}\frac{\partial f(x',t)}{\partial \mathbf{v}'}f(x,t)\right\}\right), \tag{1.65}$$

$$U(\mathbf{r}-\mathbf{r}')\equiv\frac{ee'}{|\mathbf{r}-\mathbf{r}'|}.$$

To derive from this set a kinetic equation, i.e. an equation for the one-particle distribution, we eliminate the correlation function $g(x, x'; t)$ from (1.64) within the context of (1.65).

There is no difficulty to find the general solution of (1.65) if the distribution is assumed to be almost homogeneous. Indeed, the integration in this approximation may be carried out along straight-line trajectories. Using the notation $g_0(x, x')$ for the correlation function at the initial time t_0, we may write the general solution of (1.65) in the form

$$g(x, x'; t) = g_0(\mathbf{r}-\mathbf{v}(t-t_0), \mathbf{v}; \mathbf{r}'-\mathbf{v}'(t-t_0), \mathbf{v}')$$
$$+\int_{t_0}^{t} dt' \left(\frac{\partial}{\partial \mathbf{r}} \cdot U(\mathbf{r}-\mathbf{r}'-(\mathbf{v}-\mathbf{v}')(t-t')) \cdot \left\{ \frac{1}{m} \frac{\partial f(x, t')}{\partial \mathbf{v}} f(x', t') - \frac{1}{m'} \frac{\partial f(x', t')}{\partial \mathbf{v}'} f(x, t') \right\} \right). \tag{1.66}$$

If there are no appreciable changes in the distribution function during the interaction time, we may neglect the retardation (then the expression within the braces in (1.66) may be put before the integration symbol at $t = t'$).

Let us assume the time segment between the initial time t_0 and the time under consideration t to be very long. Then the initial correlations in the system may be neglected (Bogoliubov, 1962):

$$\lim_{t_0 \to -\infty} g_0(\mathbf{r}-\mathbf{v}(t-t_0), \mathbf{v}; \mathbf{r}'-\mathbf{v}'(t-t_0), \mathbf{v}') = 0. \tag{1.67}$$

This requirement is sometimes called the Bogoliubov correlation weakening principle.†

Some simple algebra yields correlation function $g(x, x'; t)$ in the form

$$g(x, x'; t) = \frac{1}{2} \left(\left\{ \frac{1}{m} \frac{\partial f(\mathbf{v}, t)}{\partial \mathbf{v}} f(\mathbf{v}', t) - \frac{1}{m'} \frac{\partial f(\mathbf{v}', t)}{\partial \mathbf{v}'} f(\mathbf{v}, t) \right\} \right.$$
$$\left. \times \int_{-\infty}^{\infty} dt' \frac{\partial}{\partial \mathbf{r}} U(\mathbf{r}-\mathbf{r}'-(\mathbf{v}-\mathbf{v}')t') \right). \tag{1.68}$$

By virtue of spatial homogeneity, the correlation function (1.68) depends on the coordinates through the relative distances between the particles.

Substituting the correlation function (1.68) into (1.64) we obtain the kinetic equation

$$\frac{\partial f}{\partial t} + \mathbf{v} \frac{\partial f}{\partial \mathbf{r}} = J\{f\} \tag{1.69}$$

with the collision term

$$J\{f\} \equiv \frac{\partial}{\partial v_i} \sum \int d\mathbf{v}' I_{ij}(\mathbf{v}-\mathbf{v}') \left\{ \frac{1}{m} \frac{\partial f(\mathbf{v}, t)}{\partial v_j} f(\mathbf{v}', t) - \frac{1}{m'} \frac{\partial f(\mathbf{v}', t)}{\partial v_j'} f(\mathbf{v}, t) \right\} \tag{1.70}$$

the kernel of which is determined by the expression

$$I_{ij}(\mathbf{v}) \equiv \frac{1}{2m} \int d\mathbf{r} \frac{\partial U(\mathbf{r})}{\partial r_i} \int_{-\infty}^{\infty} dt \frac{\partial U(\mathbf{r}-\mathbf{v}t)}{\partial r_j}. \tag{1.71}$$

† It is to be observed that the requirement (1.67) is in accordance with Boltzmann's assumption of initial molecular chaos (Boltzmann, 1964).

Coulomb collisions

To carry out the integration in (1.71) we make use of the Fourier representation of the potential energy $U(\mathbf{r})$:

$$U(\mathbf{r}) = \int \frac{d\mathbf{k}}{(2\pi)^3} U(\mathbf{k})\, e^{i(\mathbf{k}\cdot\mathbf{r})}, \qquad U(\mathbf{k}) = \frac{4\pi ee'}{k^2}. \tag{1.72}$$

Substituting (1.72) into (1.71) we find

$$I_{ij}(\mathbf{v}) = \frac{1}{8\pi^2 m} \int d\mathbf{k}\, k_i k_j U^2(\mathbf{k})\, \delta(\mathbf{k}\mathbf{v}). \tag{1.73}$$

The tensor nature of $I_{ij}(\mathbf{v})$ allows us to reduce the right-hand part of (1.73) to

$$I_{ij}(\mathbf{v}) = \frac{1}{16\pi^2 m} \left(\delta_{ij} - \frac{v_i v_j}{v^2}\right) \int d\mathbf{k} k^2 U^2(\mathbf{k})\, \delta((\mathbf{k}\cdot\mathbf{v})). \tag{1.74}$$

Let us introduce spherical coordinates in the **k**-space, then the integration over angles may be easily performed by virtue of the δ-function. The k-integration must be carried out within a finite range from $k_{\min}$ up to $k_{\max}$ $\Big(k_{\max}$ is determined by the mean distance of the closest encounter of two particles, $r_{\min} = ee'/T$, where T is the plasma temperature in energy units and $k_{\min}$ depends on the Debye length $a = \sqrt{\dfrac{T}{4\pi e^2 n_0}}\Big)$. Thus we obtain the following result which is correct up to the logarithm:

$$I_{ij}(\mathbf{v}) = \frac{2\pi e^2 e'^2}{m} \frac{v^2\delta_{ij} - v_i v_j}{v^3} \ln \Lambda, \tag{1.75}$$

where $\ln \Lambda$ is the so-called Coulomb logarithm ($\Lambda = a/r_{\min}$). Formula (1.75) is valid only within the context of the condition $\Lambda \gg 1$.

The collision term (1.70) with the kernel (1.75) was first derived by Landau (1937) and is conventionally called the Landau collision term. The relevant kinetic equation is referred to as the Landau kinetic equation. It is clear from the derivation that the Landau equation (1.69) is adequate if the particle interaction is weak. Besides, the Landau collision term was obtained neglecting screening effects.

With the screening taken into consideration, a more exact expression for the collision term in the kinetic equation (1.69) may be derived. This was done by Balescu (1960) and Lénard (1960). In order to obtain the Balescu–Lénard collision term from the set of (1.62) and (1.63), one must retain in the left-hand part of (1.63) the eighth and the ninth terms which give rise to the screening of the particle interactions. The solution of the refined equation (1.65) results in a (1.70)-like collision term but with a kernel:

$$I_{ij}(\mathbf{v}, \mathbf{v}') = \frac{2e^2e'^2}{m} \int d\mathbf{k} k_i k_j \frac{\delta[\mathbf{k}(\mathbf{v}-\mathbf{v}')]}{k^4\, |\varepsilon(\mathbf{k}\mathbf{v}, \mathbf{k})|^2} \tag{1.76}$$

($\varepsilon(\omega, \mathbf{k})$ denotes the plasma dielectric permittivity). This expression may be formally derived from (1.73) provided the screening due to the plasma polarization is taken into account when calculating the two-particle potential energy. It is to be pointed out that by virtue of the screening the integral in (1.75) becomes convergent at small k. (For a detailed derivation of the Balescu–Lénard collision term, see Chapter 9.)

Correlation weakening principle and irreversibility

In contrast to the Vlasov equation (1.54) the kinetic equation with the collision term (1.70) leads to an entropy increase with time. It may be easily shown within the context of the explicit expression (1.70) that systems, which are described by the kinetic equation (1.69), satisfy the requirements of the H-theorem (Sommerfeld, 1956):

$$\frac{\partial S}{\partial t} = -\varkappa \int dx(1+\ln f)\frac{\partial f}{\partial t} \geqslant 0, \tag{1.77}$$

where S is the entropy and $\varkappa$ is the Boltzmann constant.

The entropy growth with time indicates that eq. (1.69) describes irreversible processes. There exists a direct connection between irreversibility and the correlation weakening principle. The requirement (1.67), use of which was made to derive (1.70), distinguishes the direction of time. The time direction being definite implies the irreversibility of the processes that are determined by (1.69) and that are called kinetic processes.

There is no difficulty to prove the following properties of the solutions of the kinetic equation (1.69). (i) If the initial distribution function is positive definite, it remains positive definite at any time. (ii) The Maxwell distribution is a stationary solution of (1.69) and is associated with the maximum entropy. (iii) Any solution of (1.69) tends to the Maxwell distribution as $t \to \infty$.

The time, which is needed for the equilibrium to be established in the system, is usually called the relaxation time. When considering a many-particle system we may introduce a hierarchy of relaxation times which are characteristic for different dynamical processes (Bogoliubov, 1962). The time evolution of the one-particle distribution function is governed by the kinetic processes which occur in the system and is described by the kinetic equation (1.69). The time needed for a one-particle distribution function to relax to a local equilibrium distribution may be estimated as to order of magnitude as the ratio of the mean free path l to the mean particle velocity s:

$$\tau_1 \simeq \frac{l}{s}. \tag{1.78}$$

It is clear from (1.70) and (1.75) that the mean free path l is governed by the Coulomb encounters.

The relaxation time of the binary correlation function may be treated similarly. It may be estimated as to order of magnitude as the ratio of the correlation distance, which is the plasma Debye length a, to the mean particle velocity s:

$$\tau_2 \simeq \frac{a}{s}. \tag{1.79}$$

3

The Coulomb mean free path l in an almost equilibrium plasma is much greater than the Debye length a:

$$l \gg a. \tag{1.80}$$

Therefore

$$\tau_1 \gg \tau_2, \tag{1.81}$$

i.e. the distribution function f relaxes very slowly in comparison with the correlation function g.

It becomes evident that the kinetic processes may be treated without taking into account the initial perturbation of the correlation function, since within the context of (1.81) the correlation function rapidly relaxes to the value that does not depend on the initial perturbation. (This is in accordance with the above-mentioned correlation weakening principle.) Besides, since the correlation function g may be taken to be quasi-stationary during the evolution of the distribution function f, it depends on time only through f (Bogoliubov, 1962).

The neglect of correlations between three and more particles is reasonable in a plasma near thermal equilibrium. In such a case the closure of the correlation functions hierarchy is associated with the expansion in series in the plasma parameter. A suitable parameter is the inverse number of particles in the Debye sphere

$$\varepsilon = \frac{1}{a^3 n_0}. \tag{1.82}$$

It may be easily verified that the plasma parameter (1.82) is proportional to the ratio of the mean densities of the potential to the kinetic energy of a plasma. (This ratio is very small in a plasma in thermal equilibrium.)

The case $\varepsilon = 0$ is associated with the neglect of correlations. This value of the plasma parameter may be realized in the so-called fluid model ($e \to 0$, $m \to 0$, $en_0 = \text{const}$, $mn_0 = \text{const}$), in which the plasma is treated as a continuous medium (Rostoker and Rosenbluth, 1960). The BBGKY hierarchy in the fluid limit reduces to a single equation —the kinetic Vlasov equation.

If the plasma parameter is different from zero but very small, then the interconnected chain of equations (1.50) may be cut off by means of the expansion in series in ε, thus reducing to a finite set of equations. Taking into account the linear terms of the expansion corresponds to binary correlations and results in the kinetic equation with collisions. The successive terms of the ε-expansion give rise to higher-order particle correlations.

CHAPTER 2

Non-linear Electrodynamic Equations

2.1. The Non-linear Equation for the Potential Field in a Plasma

Solution of the kinetic equation

The electrodynamic properties of any material medium are described by the electric susceptibility. The latter determines the general non-linear relationship between the polarization and the field intensity that may be reduced to a series expansion provided the fields be weak. The linear approximation implies that only the leading term is retained and thus describes the electrodynamic properties of the medium in terms of the linear susceptibility which in the case of a space–time dispersive medium depends on the frequency and the wave vector. To consider the non-linear effects we must take into account also the higher-order terms in the expansion of the polarization in the field strength. The expansion coefficients represent the higher-order non-linear susceptibilities. Those depend, naturally, on the frequencies and the wave vectors of all interacting waves. The knowledge of the sequence of susceptibilities is sufficient to describe all linear and non-linear electrodynamic properties of the medium.†

In order to find the relation between the polarization and the field intensity in a plasma of uncorrelated particles, we start from the kinetic Vlasov equation. It is a non-linear equation, so the resulting relation must also be non-linear. It has not still been written down for the general case, but it is not difficult to derive an expansion for it and thus to determine the linear and non-linear susceptibilities.

Consider a homogeneous stationary plasma. The appropriate kinetic equation may be solved by means of a Fourier analysis. Let us denote the unperturbed distribution function by $f_0(\mathbf{v})$ and its perturbation due to the self-consistent field by $f(\mathbf{r}, \mathbf{v}, t)$. Consider first the potential electric field. The Fourier transformed kinetic equation (1.54) becomes

$$-i(\omega-\mathbf{k}\cdot\mathbf{v})f_{\mathbf{k}\omega}+\frac{e}{m}\left(\mathbf{E}_{\mathbf{k}\omega}\cdot\frac{\partial f_0}{\partial \mathbf{v}}\right)+\frac{e}{m}\sum_{\omega',\mathbf{k}'}\left(\mathbf{E}_{\mathbf{k}'\omega'}\cdot\frac{\partial f_{\mathbf{k}-\mathbf{k}',\,\omega-\omega'}}{\partial \mathbf{v}}\right)=0, \qquad (2.1)$$

† The non-linear plasma susceptibilities were introduced by Galeev, Karpman, and Sagdeev (1965), Gorbunov, Pustovalov, and Silin (1965), Kadomtsev (1965), and Coppi, Rosenbluth, and Sudan (1969). For a detailed consideration of the non-linear electrodynamic plasma properties, see Pustovalov and Silin (1972a).

where $\mathbf{E}_{\mathbf{k}\omega}$ and $f_{\mathbf{k}\omega}$ are respectively the space–time Fourier components of the field strength and of the distribution function perturbation:

$$\mathbf{E}_{\mathbf{k}\omega} = \int dt \int d\mathbf{r}\, e^{-i\mathbf{k}\cdot\mathbf{r}+i\omega t}\mathbf{E}(\mathbf{r}, t), \tag{2.2}$$

$$f_{\mathbf{k}\omega}(v) = \int dt \int d\mathbf{r}\, e^{-i\mathbf{k}\cdot\mathbf{r}+i\omega t} f(\mathbf{r}, \mathbf{v}, t) \tag{2.3}$$

($\mathbf{k}$ and ω denote the wave vector and the frequency). The summation symbol in (2.1) implies

$$\sum_{\omega, \mathbf{k}} \dots \rightarrow \int \frac{d\omega}{2\pi} \int \frac{d\mathbf{k}}{(2\pi)^3} \dots \tag{2.4}$$

The Fourier transformed field equation (1.56) reduces to

$$i(\mathbf{k}\cdot\mathbf{E}_{\mathbf{k}\omega}) = 4\pi(\varrho_{\mathbf{k}\omega} + \varrho^0_{\mathbf{k}\omega}), \tag{2.5}$$

where $\varrho_{\mathbf{k}\omega}$ and $\varrho^0_{\mathbf{k}\omega}$ are the space–time Fourier components of the induced and of the external charge density. The induced charge density $\varrho_{\mathbf{k}\omega}$ is directly related to the distribution function perturbation $f_{\mathbf{k}\omega}$:

$$\varrho_{\mathbf{k}\omega} = \int d\mathbf{v} f_{\mathbf{k}\omega}(\mathbf{v}). \tag{2.6}$$

Introducing the polarization $\mathbf{P}_{\mathbf{k}\omega}$ by means of the relation

$$i(\mathbf{k}\cdot\mathbf{P}_{\mathbf{k}\omega}) = -4\pi\varrho_{\mathbf{k}\omega}, \tag{2.7}$$

we can rewrite (2.5) in the form

$$i(\mathbf{k}\cdot\mathbf{D}_{\mathbf{k}\omega}) \equiv i(\mathbf{k}\cdot\{\mathbf{E}_{\mathbf{k}\omega} + \mathbf{P}_{\mathbf{k}\omega}\}) = 4\pi\varrho^0_{\mathbf{k}\omega}. \tag{2.8}$$

($\mathbf{D}_{\mathbf{k}\omega}$ is the electric field induction.)

Let us iterate the non-linear equation (2.1). The distribution function perturbation $f_{\mathbf{k}\omega}$ may be expanded according to

$$f_{\mathbf{k}\omega} = \sum_{n=1}^{\infty} f^{(n)}_{\mathbf{k}\omega}. \tag{2.9}$$

Separate terms of (2.9) correspond to ascending powers of the electric field strength $\mathbf{E}_{\mathbf{k}\omega}$. In the linear approximation we find from (2.1):

$$f^{(1)}_{\mathbf{k}\omega} = -i\frac{e}{m}\frac{1}{\omega - \mathbf{k}\cdot\mathbf{v}}\mathbf{E}_{\mathbf{k}\omega}\frac{\partial f_0}{\partial \mathbf{v}}. \tag{2.10}$$

The higher-order iterations satisfy the following recurrence relation

$$f^{(n)}_{\mathbf{k}\omega} = -i\frac{e}{m}\frac{1}{\omega - (\mathbf{k}\cdot\mathbf{v})}\sum_{\omega', \mathbf{k}'}\left(\mathbf{E}_{\mathbf{k}'\omega'}\cdot\frac{\partial f^{(n-1)}_{\mathbf{k}-\mathbf{k}', \omega-\omega'}}{\partial \mathbf{v}}\right). \tag{2.11}$$

Relation between the polarization and the electric field strength

Making use of (2.6) and (2.7) we obtain the following expression for the polarization $\mathbf{P}_{\mathbf{k}\omega}$ in terms of the potential electric field strength:

$$P_{\mathbf{k}\omega} = \varkappa^{(1)}(\omega, \mathbf{k})E_{\mathbf{k}\omega} + \sum_{\substack{\omega_1+\omega_2=\omega \\ \mathbf{k}_1+\mathbf{k}_2=\mathbf{k}}} \varkappa^{(2)}(\omega_1, \mathbf{k}_1; \omega_2, \mathbf{k}_2)E_{\mathbf{k}_1\omega_1}E_{\mathbf{k}_2\omega_2}$$
$$+ \sum_{\substack{\omega_1+\omega_2+\omega_3=\omega \\ \mathbf{k}_1+\mathbf{k}_2+\mathbf{k}_3=\mathbf{k}}} \varkappa^{(3)}(\omega_1, \mathbf{k}_1; \omega_2, \mathbf{k}_2; \omega_3, \mathbf{k}_3)E_{\mathbf{k}_1\omega_1}E_{\mathbf{k}_2\omega_2}E_{\mathbf{k}_3\omega_3} + \ldots$$
$$+ \sum_{\substack{\omega_1+\omega_2+\ldots+\omega_n=\omega \\ \mathbf{k}_1+\mathbf{k}_2+\ldots+\mathbf{k}_n=\mathbf{k}}} \varkappa^{(n)}(\omega_1, \mathbf{k}_1; \omega_2, \mathbf{k}_2; \ldots; \omega_n, k_n)E_{\mathbf{k}_1\omega_1}E_{\mathbf{k}_2\omega_2}\ldots E_{\mathbf{k}_n\omega_n} + \ldots, \quad (2.12)$$

where $\varkappa^{(1)}(\omega, \mathbf{k})$ is the linear electric susceptibility of a plasma, $\varkappa^{(2)}(\omega_1, \mathbf{k}_1; \omega_2, \mathbf{k}_2)$, $\varkappa^{(3)}(\omega_1, \mathbf{k}_1; \omega_2, \mathbf{k}_2; \omega_3, \mathbf{k}_3)$, etc., are the non-linear ones. The summation in (2.12) extends over all intermediate frequencies and wave vectors which satisfy the relations:

$$\left.\begin{aligned} \omega_1+\omega_2+\ \ldots\ +\omega_n &= \omega, \\ \mathbf{k}_1+\mathbf{k}_2+\ \ldots\ +\mathbf{k}_n &= \mathbf{k}. \end{aligned}\right\} \quad (2.13)$$

Since the polarization and the field intensity are longitudinal vectors in the potential case, the susceptibilities, i.e. the coefficients in the expansion (2.12), are scalars. They are called the longitudinal electric susceptibilities.

In contrast to $\varkappa^{(1)}(\omega, \mathbf{k})$, the non-linear plasma susceptibilities $\varkappa^{(n)}(\omega_1, \mathbf{k}_1; \omega_2, \mathbf{k}_2; \ldots; \omega_n, \mathbf{k}_n)$ $(n \geqslant 2)$ are dimensional quantities. The dimension of the nth order susceptibility $\varkappa^{(n)}(\omega_1, \mathbf{k}_1; \omega_2, \mathbf{k}_2; \ldots; \omega_n, \mathbf{k}_n)$ is equal to the inverse dimension of the field strength to the $(n-1)$st power.

Substituting (2.12) in (2.7) we obtain the basic non-linear equation for the potential macroscopic field in a plasma:

$$ik\Bigg\{\varepsilon(\omega, \mathbf{k})E_{\mathbf{k}\omega} + \sum_{\substack{\omega_1+\omega_2=\omega \\ \mathbf{k}_1+\mathbf{k}_2=\mathbf{k}}} \varkappa^{(2)}(\omega_1, \mathbf{k}_1; \omega_2, \mathbf{k}_2)E_{\mathbf{k}_1\omega_1}E_{\mathbf{k}_2\omega_2}$$
$$+ \sum_{\substack{\omega_1+\omega_2+\omega_3=\omega \\ \mathbf{k}_1+\mathbf{k}_2+\mathbf{k}_3=\mathbf{k}}} \varkappa^{(3)}(\omega_1, \mathbf{k}_1; \omega_2, \mathbf{k}_2; \omega_3, \mathbf{k}_3)E_{\mathbf{k}_1\omega_1}E_{\mathbf{k}_2\omega_2}E_{\mathbf{k}_3\omega_3} + \ldots\Bigg\} = 4\pi\varrho^0_{\mathbf{k}\omega}, \quad (2.14)$$

where $\varepsilon(\omega, \mathbf{k})$ is the longitudinal dielectric permittivity:

$$\varepsilon(\omega, \mathbf{k}) = 1+\varkappa^{(1)}(\omega, \mathbf{k}).$$

Here and below we shall restrict ourselves to taking into account in (2.14) only the linear, quadratic, and cubic terms in the field intensity.

The non-linear equation (2.14) provides a complete description of the macroscopic field dynamics in a plasma. Its solution uniquely determines the microscopic field $E_{\mathbf{k}\omega}$ in terms of the external charge distribution $\varrho^0_{\mathbf{k}\omega}$. As the non-linear interaction is assumed to be weak, eq. (2.14) may be solved by means of a multiple-time-scale perturbation analysis, which allows us to remove, order by order, any time secularities that may occur in the solutions due to resonances in the relevant approximations.

2.2. Longitudinal Dielectric Permittivity and Non-linear Electric Susceptibilities of a Plasma

General relations

Making use of the iteration solutions of the kinetic equation (2.1) we can derive explicit expressions for the longitudinal dielectric permittivity and the non-linear longitudinal susceptibilities of a plasma. The one for the plasma dielectric permittivity $\varepsilon(\omega, \mathbf{k})$ is well known:

$$\varepsilon(\omega, \mathbf{k}) = 1+\sum \frac{4\pi e^2}{m} \frac{1}{k^2} \int d\mathbf{v} \frac{1}{\omega-(\mathbf{k}\cdot\mathbf{v})+i0}\left(\mathbf{k}\cdot \frac{\partial f_0(\mathbf{v})}{\partial \mathbf{v}}\right). \tag{2.15}$$

The second-order electric susceptibility $\varkappa^{(2)}(\omega_1, \mathbf{k}_1; \omega_2, \mathbf{k}_2)$ is of the form

$$\varkappa^{(2)}(\omega_1, \mathbf{k}_1; \omega_2, \mathbf{k}_2) = \sum \frac{(-i)}{2} \frac{4\pi e^2}{m} \frac{e}{m} \frac{1}{k_1 k_2 |\mathbf{k}_1+\mathbf{k}_2|} \int d\mathbf{v} \frac{1}{\omega_1+\omega_2-(\{\mathbf{k}_1+\mathbf{k}_2\}\cdot\mathbf{v})+i0}$$
$$\times\left[\left(\mathbf{k}_1\cdot \frac{\partial}{\partial \mathbf{v}}\left\{\frac{1}{\omega_2-(\mathbf{k}_2\cdot\mathbf{v})+i0}\left(\mathbf{k}_2\cdot\frac{\partial}{\partial \mathbf{v}}\right)\right\}\right)+\left(\mathbf{k}_2\cdot \frac{\partial}{\partial \mathbf{v}}\left\{\frac{1}{\omega_1-\mathbf{k}_1\mathbf{v}+i0}\left(\mathbf{k}_1\cdot\frac{\partial}{\partial \mathbf{v}}\right)\right\}\right)\right] f_0(\mathbf{v}), \tag{2.16}$$

and the third-order susceptibility $\varkappa^{(3)}(\omega_1, \mathbf{k}_1; \omega_2, \mathbf{k}_2; \omega_3, \mathbf{k}_3)$ may be written as

$$\varkappa^{(3)}(\omega_1, \mathbf{k}_1; \omega_2, \mathbf{k}_2; \omega_3, \mathbf{k}_3) = \tfrac{1}{3}\{\bar{\varkappa}^{(3)}(1, 2, 3)+\bar{\varkappa}^{(3)}(2, 1, 3)+\bar{\varkappa}^{(3)}(3, 2, 1)\}, \tag{2.17}$$

where

$$\bar{\varkappa}^{(3)}(\omega_1, \mathbf{k}_1; \omega_2, \mathbf{k}_2; \omega_3, \mathbf{k}_3)$$
$$= \sum \frac{(-i)^2}{2} \frac{4\pi e^2}{m}\left(\frac{e}{m}\right)^2 \frac{1}{k_1 k_2 k_3 |\mathbf{k}_1+\mathbf{k}_2+\mathbf{k}_3|} \int d\mathbf{v} \frac{1}{\omega_1+\omega_2+\omega_3-(\{\mathbf{k}_1+\mathbf{k}_2+\mathbf{k}_3\}\cdot\mathbf{v})+i0}$$
$$\times\left(\mathbf{k}_1\cdot\frac{\partial}{\partial \mathbf{v}}\right)\left\{\frac{1}{\omega_2+\omega_3-(\{\mathbf{k}_2+\mathbf{k}_3\}\cdot\mathbf{v})+i0}\left[\left(\mathbf{k}_2\cdot\frac{\partial}{\partial \mathbf{v}}\right)\left\{\frac{1}{\omega_3-(\mathbf{k}_3\cdot\mathbf{v})+i0}\left(\mathbf{k}_3\cdot\frac{\partial}{\partial \mathbf{v}}\right)\right\}\right.\right.$$
$$\left.\left.+\left(\mathbf{k}_3\cdot\frac{\partial}{\partial \mathbf{v}}\right)\left\{\frac{1}{\omega_2-\mathbf{k}_2\mathbf{v}+i0}\left(\mathbf{k}_2\cdot\frac{\partial}{\partial \mathbf{v}}\right)\right\}\right]\right\} f_0(\mathbf{v}). \tag{2.18}$$

Expressions (2.16) and (2.17) are the $n = 2$ and $n=3$ realizations of the following general formula for the nth order electric susceptibility:

$$\varkappa^{(n)}(\omega_1, \mathbf{k}_1; \omega_2, \mathbf{k}_2; \ldots; \omega_n, \mathbf{k}_n)$$
$$= \sum \frac{(-i)^{n-1}}{n!} \frac{4\pi e^2}{m}\left(\frac{e}{m}\right)^{n-1} \mathcal{D} \frac{1}{k_1 k_2 \ldots k_n |\mathbf{k}_1+\mathbf{k}_2+\ldots+\mathbf{k}_n|}$$
$$\times \int d\mathbf{v} \frac{1}{\omega_1+\omega_2+\ldots+\omega_n-(\{\mathbf{k}_1+\mathbf{k}_2+\ldots+\mathbf{k}_n\}\cdot\mathbf{v})+i0}$$

$$\times\left(\mathbf{k}_1\cdot\frac{\partial}{\partial\mathbf{v}}\right)\left\{\frac{1}{\omega_2+\omega_3+\ldots+\omega_n-(\{k_1+\mathbf{k}_2+\ldots+\mathbf{k}_n\}\cdot\mathbf{v})+i0}\right.$$
$$\times\left(\mathbf{k}_2\cdot\frac{\partial}{\partial\mathbf{v}}\right)\left[\frac{1}{\omega_3+\ldots+\omega_n-(\{\mathbf{k}_3+\ldots+\mathbf{k}_n\}\cdot\mathbf{v})+i0}\right.$$
$$\left.\left.\times\left(\mathbf{k}_3\cdot\frac{\partial}{\partial\mathbf{v}}\right)\ldots\left\{\frac{1}{\omega_n-(\mathbf{k}_n\cdot\mathbf{v})+i0}\left(\mathbf{k}_n\cdot\frac{\partial f_0(\mathbf{v})}{\partial\mathbf{v}}\right)\right\}\ldots\right]\right\}, \tag{2.19}$$

where the symbol $\mathcal{P}$ denotes all possible permutations of the arguments $\omega_1, \mathbf{k}_1$; $\omega_2, \mathbf{k}_2; \ldots; \omega_n, \mathbf{k}_n$. The small additions $i0$ in the denominators of the integrands in (2.15), (2.16), and so on, stand for the singularities to be bypassed from below when carrying out the integrations over the relevant velocity components. It is not difficult to derive the bypass rules. We introduce the collision term $\nu f_{\mathbf{k}\omega}$, where ν is the effective collision frequency, in the right-hand part of (2.1) and let ν tend to zero. The summations in (2.15), (2.16), etc., extend over the different plasma components.

Symmetry properties

The polarization $\mathbf{P}(\mathbf{r}, t)$ and the field strength $\mathbf{E}(\mathbf{r}, t)$ are real, so the plasma susceptibilities satisfy the following conditions:

$$\left.\begin{aligned}\varkappa^{(1)}(\omega,\mathbf{k}) &= \varkappa^{(1)*}(-\omega, -\mathbf{k}),\\ \varkappa^{(2)}(\omega_1,\mathbf{k}_1;\omega_2,\mathbf{k}_2) &= -\varkappa^{(2)*}(-\omega_1, -\mathbf{k}_1; -\omega_2, -\mathbf{k}_2),\\ \varkappa^{(3)}(\omega_1,\mathbf{k}_1;\omega_2,\mathbf{k}_2;\omega_3,\mathbf{k}_3) &= \varkappa^{(3)*}(-\omega_1, -\mathbf{k}_1; -\omega_2, -\mathbf{k}_2; -\omega_3, -\mathbf{k}_3).\end{aligned}\right\} \tag{2.20}$$

The non-linear plasma susceptibilities are symmetric relative to all the arguments:

$$\left.\begin{aligned}\varkappa^{(2)}(\omega_1,\mathbf{k}_1;\omega_2,\mathbf{k}_2) &= \varkappa^{(2)}(\omega_2,\mathbf{k}_2;\omega_1,\mathbf{k}_1),\\ \varkappa^{(3)}(\omega_1,\mathbf{k}_1;\omega_2,\mathbf{k}_2;\omega_3,\mathbf{k}_3) &= \varkappa^{(3)}(\omega_2,\mathbf{k}_2;\omega_1,\mathbf{k}_1;\omega_3,\mathbf{k}_3)\\ &= \varkappa^{(3)}(\omega_3,\mathbf{k}_3;\omega_2,\mathbf{k}_2;\mathbf{k}_2;\omega_1,\mathbf{k}_1).\end{aligned}\right\} \tag{2.21}$$

This symmetry follows immediately from the expansion (2.12). Note that $\bar{\varkappa}^{(3)}(\omega_1, \mathbf{k}_1; \omega_2, \mathbf{k}_2; \omega_3, \mathbf{k}_3)$, which was defined in (2.18), is symmetric relative to the permutation $\omega_2, \mathbf{k}_2 \rightleftarrows \omega_3, \mathbf{k}_3$, but non-symmetric with respect to the interchanges $\omega_1, \mathbf{k}_1 \rightleftarrows \omega_2, \mathbf{k}_2$ and $\omega_1, \mathbf{k}_1 \rightleftarrows \omega_3, \mathbf{k}_3$. Besides, the second-order susceptibilities possess the following useful symmetry property:

$$\varkappa^{(2)}(\omega_1,\mathbf{k}_1;\omega_2,\mathbf{k}_2) = \varkappa^{(2)}(\omega_1+\omega_2,\mathbf{k}_1+\mathbf{k}_2;-\omega_2,-\mathbf{k}_2). \tag{2.22}$$

The long-wave limiting case

Integration by parts reduces the dielectric permittivity and non-linear susceptibilities to the following:

$$\varepsilon(\omega,\mathbf{k}) = 1-\sum\frac{4\pi e^2}{m}\int d\mathbf{v}\,\frac{1}{(\omega-(\mathbf{k}\cdot\mathbf{v})+i0)^2}\,f_0(\mathbf{v}); \tag{2.23}$$

$$\varkappa^{(2)}(\omega_1, \mathbf{k}_1; \omega_2, \mathbf{k}_2)$$

$$= \sum \frac{(-i)}{2} \frac{4\pi e^2}{m} \frac{e}{m} \frac{1}{k_1 k_2 k} \int d\mathbf{v} \frac{1}{(\omega_1 - (\mathbf{k}_1 \cdot \mathbf{v}) + i0)(\omega_2 - (\mathbf{k}_2 \cdot \mathbf{v}) + i0)(\omega - (\mathbf{k} \cdot \mathbf{v}) + i0)}$$

$$\times \left(\frac{k_1^2 \cdot (\mathbf{k} \cdot \mathbf{k}_2)}{\omega_1 - (\mathbf{k}_1 \cdot \mathbf{v}) + i0} + \frac{k_2^2 \cdot (\mathbf{k} \cdot \mathbf{k}_1)}{\omega_2 - (\mathbf{k}_2 \cdot \mathbf{v}) + i0} - \frac{k^2 \cdot (\mathbf{k}_1 \cdot \mathbf{k}_2)}{\omega - (\mathbf{k} \cdot \mathbf{v}) + i0} \right) f_0(\mathbf{v}),$$

where

$$\omega = \omega_1 + \omega_2, \quad \mathbf{k} = \mathbf{k}_1 + \mathbf{k}_2; \tag{2.24}$$

$$\bar{\varkappa}^{(3)}(\omega_1, \mathbf{k}_1; \omega_2, \mathbf{k}_2; \omega_3, \mathbf{k}_3)$$

$$= \sum \frac{1}{2} \frac{4\pi e^2}{m} \left(\frac{e}{m} \right)^2 \frac{(\mathbf{k} \cdot \mathbf{k}_1)}{k_1 k_2 k_3 k} \int d\mathbf{v} \frac{1}{(\omega_2 - (\mathbf{k}_2 \cdot \mathbf{v}) + i0)(\omega_3 - (\mathbf{k}_3 \cdot \mathbf{v}) + i0)(\omega - (\mathbf{k} \cdot \mathbf{v}) + i0)^2}$$

$$\times \left\{ \frac{2}{\omega - (\mathbf{k} \cdot \mathbf{v}) + i0} \left(3 \frac{(\mathbf{k} \cdot \mathbf{k}) \cdot (\mathbf{k}_2 \cdot \mathbf{k}_3)}{\omega - (\mathbf{k} \cdot \mathbf{v}) + i0} + \frac{k_2^2 \cdot (\mathbf{k} \cdot \mathbf{k}_3)}{\omega_2 - (k_2 \cdot \mathbf{v}) + i0} + \frac{k_3^2 \cdot (k \cdot k_2)}{\omega_3 - (\mathbf{k}_3 \cdot \mathbf{v}) + i0} \right) \right.$$

$$+ \frac{1}{\omega' - (\mathbf{k} \cdot \mathbf{v}) + i0} \left(2 \frac{(\mathbf{k} \cdot \mathbf{k})' \cdot (\mathbf{k}_2 \cdot \mathbf{k}_3)}{\omega - (\mathbf{k} \cdot \mathbf{v}) + i0} + \frac{k_2^2 \cdot (\mathbf{k}' \cdot \mathbf{k}_3)}{\omega_2 - (\mathbf{k}_2 \cdot \mathbf{v}) + i0} \right.$$

$$\left. \left. + \frac{k_3^2 \cdot (\mathbf{k}' \cdot \mathbf{k}_2)}{\omega_3 - (\mathbf{k}_3 \cdot \mathbf{v}) + i0} + \frac{k'^2 \cdot (\mathbf{k}_2 \cdot \mathbf{k}_3)}{\omega' - (\mathbf{k}' \cdot \mathbf{v}) + i0} \right) \right\} f_0(\mathbf{v}), \tag{2.25}$$

where

$$\omega = \omega_1 + \omega_2 + \omega_3, \quad \mathbf{k} = \mathbf{k}_1 + \mathbf{k}_2 + \mathbf{k}_3, \quad \omega' = \omega_2 + \omega_3, \quad \mathbf{k}' = \mathbf{k}_2 + \mathbf{k}_3.$$

In the long-wavelength limiting case ($k \to 0$) we obtain from (2.23):

$$\varepsilon(\omega, 0) = 1 - \sum \frac{\Omega^2}{\omega^2}, \tag{2.26}$$

which is similar to the dielectric permittivity of a cold plasma ($T = 0$). (It should be pointed out that the latter does not depend on the wave vector, i.e. there is no spatial dispersion in a cold plasma.)

If the plasma is in a state of thermal equilibrium, then the distribution function $f_0(\mathbf{v})$ is Maxwellian and the longitudinal plasma dielectric permittivity takes the form (Sitenko, 1967):

$$\varepsilon(\omega, \mathbf{k}) = 1 + \sum \frac{1}{a^2 k^2} \{1 - \varphi(z) + i \sqrt{\pi} z e^{-z^2}\}, \tag{2.27}$$

where $z = \sqrt{\frac{3}{2}} \frac{\omega}{ks}$ is the dimensionless frequency, $\varphi(z)$ is the following real function:

$$\varphi(z) \equiv 2z e^{-z^2} \int_0^z dx\, e^{x^2}. \tag{2.28}$$

It follows from (2.24) and (2.25) that the non-linear plasma susceptibilities $\varkappa^{(2)}(\omega_1, \mathbf{k}_1; \omega_2, \mathbf{k}_2)$ and $\varkappa^{(3)}(\omega_1, \mathbf{k}_1; \omega_2, \mathbf{k}_2; \omega_3, \mathbf{k}_3)$ vanish in the long-wavelength limiting case $k_1 = k_2 \to 0$ and $k_1 = k_2 = k_3 \to 0$:

$$\varkappa^{(2)}(\omega_1, 0; \omega_2, 0) = 0, \quad \varkappa^{(3)}(\omega_1, 0; \omega_2, 0; \omega_3, 0) = 0. \tag{2.29}$$

It is to be observed that, in contrast to the dielectric permittivity, the long-wavelength non-linear susceptibilities differ from those of a cold plasma.

Non-linear susceptibilities of a cold plasma

To find the cold plasma non-linear susceptibilities we expand the factors of the distribution function $f_0(\mathbf{v})$ in (2.24) and (2.25) in a series in the velocity. Thus we obtain the following expression for the second-order susceptibility:

$$\varkappa^{(2)}(\omega_1, \mathbf{k}_1; \omega_2, \mathbf{k}_2) = \sum \frac{(-i)}{2} \frac{e}{m} \frac{\Omega^2}{\omega_1\omega_2\omega} \frac{1}{k_1k_2k} \left(\frac{k_1^2}{\omega_1}(\mathbf{k}\cdot\mathbf{k}_2) + \frac{k_2^2}{\omega_2}(\mathbf{k}\cdot\mathbf{k}_1) + \frac{k^2}{\omega}(\mathbf{k}_1\cdot\mathbf{k}_2) \right), \tag{2.30}$$

where $\omega = \omega_1+\omega_2$, $\mathbf{k} = \mathbf{k}_1+\mathbf{k}_2$, which is valid within the context of the conditions

$$k_1 s \ll \omega_1, \quad k_2 s \ll \omega_2, \quad ks \ll \omega. \tag{2.31}$$

Exactly the same expression may be derived within the framework of the hydrodynamic model. The latter approach deals with macroscopic electron and ion densities $n(\mathbf{r}, t)$ and velocities $\mathbf{v}(\mathbf{r}, t)$, which satisfy the hydrodynamic equations of continuity and motion:

$$\frac{\partial n}{\partial t} + \operatorname{div}(n\mathbf{v}) = 0, \quad \frac{\partial \mathbf{v}}{\partial t} + (\mathbf{v}\cdot\nabla)\mathbf{v} = \frac{e}{m}\mathbf{E}. \tag{2.32}$$

The self-consistent electric field $\mathbf{E}$ is governed by (1.56), where the charge density is to be written in terms of the partial particle densities $\varrho = \Sigma en$.

The thermal correction to (2.30), involving the assumption (2.31), is of the following form:

$$\begin{aligned} &\Delta\varkappa^{(2)}(\omega_1, \mathbf{k}_1; \omega_2, \mathbf{k}_2) \\ &= \sum \frac{(-i)}{2} \frac{e}{m} \frac{\Omega^2}{\omega_1\omega_2\omega} \frac{1}{k_1k_2k} \frac{T}{m} \left\{ \frac{k^2}{\omega} \left(3\frac{k^2}{\omega^2} + \frac{k_1^2}{\omega_1^2} + \frac{k_2^2}{\omega_2^2} + 2\frac{(\mathbf{k}\cdot\mathbf{k}_1)}{\omega\omega_1} + 2\frac{(\mathbf{k}\cdot\mathbf{k}_2)}{\omega\omega_2} \right.\right. \\ &\left.\left. + 2\frac{(\mathbf{k}_1\cdot\mathbf{k}_2)}{\omega_1\omega_2} \right)(\mathbf{k}_1\cdot\mathbf{k}_2) + (\omega, \mathbf{k} \rightleftarrows \omega_1, \mathbf{k}_1) + (\omega, \mathbf{k} \rightleftarrows \omega_2, \mathbf{k}_2) \right\}. \end{aligned} \tag{2.33}$$

Formula (2.30) is inadequate at low frequencies that do not satisfy (2.31). In case one of the frequencies entering $\varkappa^{(2)}(\omega_1, \mathbf{k}_1; \omega_2, \mathbf{k}_2)$ (e.g. ω_1) and the total frequency ω are high enough, so that

$$k_1 s \ll \omega_1 \quad \text{and} \quad ks \ll \omega$$

(ω_2 is arbitrary), we can easily derive from (2.24) the following approximate relation:

$$\begin{aligned} \varkappa^{(2)}(\omega_1, \mathbf{k}_1; \omega_2, \mathbf{k}_2) = \sum \frac{(-i)}{2} \frac{e}{m} \frac{\Omega^2}{\omega^2} \frac{1}{k_1k_2k} \left\{ \left[2\frac{(\mathbf{k}\cdot\mathbf{k}_2)}{\omega\omega_2}(1 - a^2k_2^2\varkappa^{(1)}(\omega_2, \mathbf{k}_2)) \right.\right. \\ \left.\left. - \frac{k_2^2}{\Omega^2}\varkappa^{(1)}(\omega_2, \mathbf{k}_2) \right](\mathbf{k}\cdot\mathbf{k}_1) + \frac{(\mathbf{k}\cdot\mathbf{k}_1)}{\omega_1}\left(\frac{k_1^2}{\omega_1} + 2\frac{(\mathbf{k}\cdot\mathbf{k}_1)}{\omega} \right) \right\}, \end{aligned} \tag{2.34}$$

where $\varkappa^{(1)}(\omega_2, \mathbf{k}_2)$ is the partial linear susceptibility. In the $\omega_2 = 0$ limiting case (2.34) reduces to

$$\varkappa^{(2)}(\omega_1, \mathbf{k}; 0, \mathbf{k}_2) = \sum \frac{(-i)}{2} \frac{e}{m} \frac{\Omega^2}{\omega^2} \frac{1}{k_1k_2k} \left\{ -3\frac{(\mathbf{k}\cdot\mathbf{k}_1)}{s^2} + \frac{(\mathbf{k}\cdot\mathbf{k}_1)}{\omega^2}(k_1^2 + 2(\mathbf{k}\cdot\mathbf{k}_1)) \right\}. \tag{2.35}$$

If $k_2 s \ll \omega_2$, then (2.34) reduces to the hydrodynamical expression (2.30).

According to (2.25) the non-symmetrized non-linear susceptibility $\bar{\varkappa}^{(3)}(\omega_1, \mathbf{k}_1; \omega_2, \mathbf{k}_2; \omega_3, \mathbf{k}_3)$ at $T = 0$ is equal to

$$\bar{\varkappa}^{(3)}(\omega_1, \mathbf{k}_1; \omega_2, \mathbf{k}_2; \omega_3, \mathbf{k}_3)$$
$$= \sum \frac{1}{2} \frac{e^2}{m^2} \frac{\Omega^2}{\omega_2\omega_3\omega^2} \frac{1}{k_1k_2k_3k} \left\{ \frac{1}{\omega'} \left(2\frac{(\mathbf{k}\cdot\mathbf{k}')}{\omega} + \frac{k_2^2}{\omega_2} + \frac{k_3^2}{\omega_3} + \frac{k'^2}{\omega'} \right) \mathbf{k}_2\cdot\mathbf{k}_3 + \frac{k_2^2k_3^2}{\omega_2\omega_3} \right.$$
$$\left. + \frac{2}{\omega} \left(3\frac{(\mathbf{k}\cdot\mathbf{k}_2)\cdot(\mathbf{k}\cdot\mathbf{k}_3)}{\omega} + \frac{k_2^2}{\omega_2}(\mathbf{k}\cdot\mathbf{k}_3) + \frac{k_3^2}{\omega_3}(\mathbf{k}\cdot\mathbf{k}_2) \right) \right\} (\mathbf{k}\cdot\mathbf{k}_1). \tag{2.36}$$

At the same time, it follows from the hydrodynamical equation (2.32) that

$$\bar{\varkappa}^{(3)}(\omega_1, \mathbf{k}_1; \omega_2, \mathbf{k}_2; \omega_3, \mathbf{k}_3)$$
$$= \sum \frac{1}{2} \frac{e^2}{m^2} \frac{\Omega^2}{\omega_1\omega_2\omega_3\omega} \frac{1}{k_1k_2k_3k} \left\{ \frac{1}{\omega'} \left(\frac{k^2}{\omega} - \frac{k_1^2}{\omega_1} + \frac{k_2^2}{\omega_2} + \frac{k_3^2}{\omega_3} + \frac{k'^2}{\omega'} \right) (\mathbf{k}\cdot\mathbf{k}_1)\cdot(\mathbf{k}_2\cdot\mathbf{k}_3) \right.$$
$$\left. + \frac{1}{\omega\omega_1} k^2k_1^2\cdot(\mathbf{k}_2\cdot\mathbf{k}_3) + \frac{1}{\omega_2\omega_3} k_2^2k_3^2\cdot(\mathbf{k}\cdot\mathbf{k}_1) \right\}. \tag{2.37}$$

The difference between (2.36) and (2.37) arises by virtue of the ambiguous definition of the non-symmetrized susceptibility $\bar{\varkappa}^{(3)}(\omega_1, \mathbf{k}_1; \omega_2, \mathbf{k}_2; \omega_3, \mathbf{k}_3)$ (which is symmetric only relative to the interchange $\omega_2, \mathbf{k}_2 \rightleftarrows \omega_3, \mathbf{k}_3$). But it may be easily verified that the symmetrization of (2.36) and (2.37) with respect to the permutations $\omega_1, \mathbf{k}_1 \rightleftarrows \omega_2, \mathbf{k}_2$ and $\omega_1, \mathbf{k}_1 \rightleftarrows \omega_3, \mathbf{k}_3$ yields the same symmetrized susceptibility of a cold plasma:

$$\varkappa^{(3)}(\omega_1, \mathbf{k}_1; \omega_2, \mathbf{k}_2; \omega_3, \mathbf{k}_3)$$
$$= \sum \frac{1}{6} \frac{e^2}{m^2} \frac{\Omega^2}{\omega_1\omega_2\omega_3\omega} \frac{1}{k_1k_2k_3k} \left\{ \left(\frac{1}{\omega'} \left[\frac{k^2}{\omega} - \frac{k_1^2}{\omega_1} + \frac{k_2^2}{\omega_2} + \frac{k_3^2}{\omega_3} + \frac{k'^2}{\omega'} \right] (\mathbf{k}\cdot\mathbf{k}_1)\cdot(\mathbf{k}_2\cdot\mathbf{k}_3) \right. \right.$$
$$\left. \left. + \frac{1}{\omega\omega_1} k^2k_1^2\cdot(\mathbf{k}_2\cdot\mathbf{k}_3) + \frac{1}{\omega_2\omega_3} k_2^2k_3^2\cdot(\mathbf{k}\cdot\mathbf{k}_1) \right) + (\omega_1\mathbf{k}_1 \rightleftarrows \omega_2, \mathbf{k}_2) + (\omega_1, \mathbf{k}_1 \rightleftarrows \omega_3, \mathbf{k}_3) \right\}. \tag{2.38}$$

An alternative equivalent presentation may be written as:

$$\varkappa^{(3)}(\omega_1, \mathbf{k}_1; \omega_2, \mathbf{k}_2; \omega_3, \mathbf{k}_3)$$
$$= \sum \frac{1}{6} \frac{e^2}{m^2} \frac{\Omega^2}{k_1k_2k_3k} \left\{ \frac{(\mathbf{k}\cdot\mathbf{k}_1)\cdot(\mathbf{k}\cdot\mathbf{k}_2)\cdot(\mathbf{k}\cdot\mathbf{k}_3)}{\omega_1^2\omega_2^2\omega_3^2} + \left(\frac{(\mathbf{k}_1\cdot\mathbf{k}_2)\cdot(\mathbf{k}_1\cdot\mathbf{k}_3)}{\omega_2^2\omega_3^2\omega^2} + \frac{(\mathbf{k}_1\cdot\mathbf{k}_2)\cdot(\mathbf{k}_2\cdot\mathbf{k}_3)}{\omega_3^2\omega^2\omega'^2} \right. \right.$$
$$\left. + \frac{(\mathbf{k}_1\cdot\mathbf{k}_3)\cdot(\mathbf{k}_2\cdot\mathbf{k}_3)}{\omega_2^2\omega^2\omega'^2} - \frac{(\mathbf{k}\cdot\mathbf{k}_2)\cdot(\mathbf{k}_2\cdot\mathbf{k}_3)}{\omega_1^2\omega_3^2\omega'^2} - \frac{(\mathbf{k}\cdot\mathbf{k}_3)\cdot(\mathbf{k}_2\cdot\mathbf{k}_3)}{\omega_1^2\omega_2^2\omega'^2} \right) (\mathbf{k}\cdot\mathbf{k}_1)$$
$$\left. + (\omega_1, \mathbf{k}_1 \rightleftarrows \omega_2, \mathbf{k}_2) + (\omega_1, \mathbf{k}_1 \rightleftarrows \omega_3, \mathbf{k}_3) \right\}. \tag{2.39}$$

Such a form is very suitable when there is a need to compare the susceptibilities of plasmas with and without external magnetic fields.

It should be mentioned that (2.36) and (2.37) have similar singularities at $\omega' \equiv \omega_2 + \omega_3 = 0$. Consequently, (2.36) differs from (2.37) by the ambiguous non-singular terms. The symmetrization of the non-singular portions of (2.36) or (2.37) with respect to all possible permutations yields:

$$\bar{\varkappa}'^{(3)}(\omega_1, \mathbf{k}_1; \omega_2, \mathbf{k}_2; \omega_3, \mathbf{k}_3)$$
$$= \sum \frac{1}{2} \frac{e^2}{m^2} \frac{\Omega^2}{\omega_1\omega_2\omega_3\omega} \frac{1}{k_1k_2k_3k} \left\{ \frac{1}{\omega'} \left(\frac{k^2}{\omega} - \frac{k_1^2}{\omega_1} + \frac{k_2^2}{\omega_2} + \frac{k_3^2}{\omega_3} + \frac{k'^2}{\omega'} \right) (\mathbf{k}\cdot\mathbf{k}_1)\cdot(\mathbf{k}_2\cdot\mathbf{k}_3) \right.$$
$$+ \frac{1}{3} \left(\frac{k_1^2k_2^2}{\omega_1\omega_2}(\mathbf{k}\cdot\mathbf{k}_3) + \frac{k_1^2k_3^2}{\omega_1\omega_3}(\mathbf{k}\cdot\mathbf{k}_2) + \frac{k_2^2k_3^2}{\omega_2\omega_3}(\mathbf{k}\cdot\mathbf{k}_1) + \frac{k_1^2k^2}{\omega_1\omega}(\mathbf{k}_2\cdot\mathbf{k}_3) \right.$$
$$\left.\left. + \frac{k_2^2k^2}{\omega_2\omega}(\mathbf{k}_1\cdot\mathbf{k}_3) + \frac{k_3^2k^2}{\omega_3\omega}(\mathbf{k}_1\cdot\mathbf{k}_2) \right) \right\}. \tag{2.40}$$

It should be kept in mind that (2.40), like (2.36) and (2.37), is valid only within the context of the conditions $k_1s \ll \omega_1$, $k_2s \ll \omega_2$, $k_3s \ll \omega_3$, and $k's \ll \omega'$. If the last requirement is not satisfied, then ω'^{-1} and ω'^{-2} must be replaced by

$$\omega'^{-n} \to \frac{1}{n_0} \int d\mathbf{v} \frac{f_0(\mathbf{v})}{(\omega' - (\mathbf{k}\cdot\mathbf{v}') + i0)^n}, \qquad n = 1, 2.$$

The symmetrized susceptibility $\varkappa^{(3)}(\omega_1, \mathbf{k}_1; \omega_2, \mathbf{k}_2; \omega_3, \mathbf{k}_3)$ remains related to (2.40) according to (2.17).

To find the thermal correction to (2.38) we make use of the non-symmetrized susceptibility in the following suitable form:

$$\bar{\varkappa}^{(3)}(\omega_1, \mathbf{k}_1; \omega_2, \mathbf{k}_2; \omega_3, \mathbf{k}_3) = \sum \frac{1}{2} \frac{4\pi e^2}{m}$$
$$\times \left(\frac{e}{m}\right)^2 \frac{1}{k_1k_2k_3k} \int d\mathbf{v} \frac{1}{(\omega_1 - (\mathbf{k}_1\cdot\mathbf{v}) + i0)(\omega_2 - (\mathbf{k}_2\cdot\mathbf{v}) + i0)(\omega_3 - (\mathbf{k}_3\cdot\mathbf{v}) + i0)(\omega' - (\mathbf{k}'\cdot\mathbf{v}) + i0)}$$
$$\times \frac{1}{(\omega - (\mathbf{k}\cdot\mathbf{v}) + i0)} \left(\frac{\mathbf{k}_1^2\cdot(\mathbf{k}_2\cdot\mathbf{k}_3)\cdot(\mathbf{k}\cdot\mathbf{k})}{\omega_1 - (\mathbf{k}_1\cdot\mathbf{v}) + i0} + \frac{\mathbf{k}_2^2\cdot(\mathbf{k}_1\cdot\mathbf{k})\cdot(\mathbf{k}_3\cdot\mathbf{k}')}{\omega_2 - (\mathbf{k}_2\cdot\mathbf{v}) + i0} + \frac{\mathbf{k}_3^2\cdot(\mathbf{k}_1\cdot\mathbf{k})\cdot(\mathbf{k}_2\cdot\mathbf{k}')}{\omega_3 - (\mathbf{k}_3\cdot\mathbf{v}) + i0} \right.$$
$$\left. + \frac{k'^2\cdot(\mathbf{k}_1\cdot\mathbf{k})\cdot(\mathbf{k}_2\cdot\mathbf{k}_3)}{\omega' - (\mathbf{k}\cdot\mathbf{v}) + i0} + \frac{k^2\cdot(\mathbf{k}_1\cdot\mathbf{k})\cdot(\mathbf{k}_2\cdot\mathbf{k}_3)}{\omega - (\mathbf{k}\cdot\mathbf{v}) + i0} \right) f_0(\mathbf{v}). \tag{2.41}$$

The definition of the non-symmetrized susceptibility $\bar{\varkappa}^{(3)}(\omega_1, \mathbf{k}_1; \omega_2, \mathbf{k}_2; \omega_3, \mathbf{k}_3)$ is unique but for the additive terms which are antisymmetric with respect to the interchanges $(\omega_1, \mathbf{k}_1 \rightleftarrows \omega_2, \mathbf{k}_2)$ and $(\omega_1, \mathbf{k}_1 \rightleftarrows \omega_3, \mathbf{k}_3)$. It may be easily verified that (2.41) differs from (2.25) just by such antisymmetric terms.

At $T = 0$ the non-symmetrized susceptibility (2.41) immediately reduces to the hydrodynamical expression (2.37). Substituting the Maxwellian function for $f_0(\mathbf{v})$ and expanding the integrand of (2.41) in a series in v, we find the following thermal correction to (2.37):

$$\Delta\bar{\varkappa}^{(3)}(\omega_1, \mathbf{k}_1; \omega_2, \mathbf{k}_2; \omega_3, \mathbf{k}_3)$$

$$= \sum \frac{1}{2}\frac{e^2}{m^2}\frac{\Omega^2}{\omega_1\omega_2\omega_3\omega\omega'}\frac{1}{k_1k_2k_3k}\frac{T}{m}\left\{\frac{k'^2}{\omega'}\left(\frac{k_1^2}{\omega_1^2}+\frac{k_2^2}{\omega_2^2}+\frac{k_3^2}{\omega_3^2}+\frac{k^2}{\omega^2}\right.\right.$$

$$+3\frac{k'^2}{\omega'^2}+\frac{(\mathbf{k}_1\cdot\mathbf{k}_2)}{\omega_1\omega_2}+\frac{(\mathbf{k}_1\cdot\mathbf{k}_3)}{\omega_1\omega_3}+\frac{(\mathbf{k}_2\cdot\mathbf{k}_3)}{\omega_2\omega_3}+\frac{(\mathbf{k}_1\cdot\mathbf{k})}{\omega_1\omega}+\frac{(\mathbf{k}_2\cdot\mathbf{k})}{\omega_2\omega}+\frac{(\mathbf{k}_3\cdot\mathbf{k})}{\omega_3\omega}$$

$$\left.+2\frac{(\mathbf{k}_1\cdot\mathbf{k}')}{\omega_1\omega'}+2\frac{(\mathbf{k}_2\cdot\mathbf{k}')}{\omega_2\omega'}+2\frac{(\mathbf{k}_3\cdot\mathbf{k}')}{\omega_3\omega'}+2\frac{(\mathbf{k}\cdot\mathbf{k}')}{\omega\omega'}\right)(\mathbf{k}\cdot\mathbf{k}_1)\cdot(\mathbf{k}_2\cdot\mathbf{k}_3)$$

$$\left.+(\omega_1, \mathbf{k}_1 \rightleftarrows \omega', \mathbf{k}')+(\omega_2, \mathbf{k}_2 \rightleftarrows \omega', \mathbf{k}')+(\omega_3, \mathbf{k}_3 \rightleftarrows \omega', \mathbf{k}')+(\omega, \mathbf{k} \rightleftarrows \omega', \mathbf{k}')\right\}. \tag{2.42}$$

The thermal correction to (2.38) may be obtained by symmetrizing (2.42) relative to the permutations $\omega_1, \mathbf{k}_1 \rightleftarrows \omega_2, \mathbf{k}_2$ and $\omega_1, \mathbf{k}_1 \rightleftarrows \omega_3, \mathbf{k}_3$.

An observation is to be made that the non-linear plasma susceptibilities are always functions of the wave vectors of interacting fields (in contrast to the linear susceptibility that is **k**-independent if the thermal motion is not involved in the consideration). This dependence may be interpreted as the non-linear spatial dispersion of the plasma, since the relationship between the induced charges and the field is non-local by virtue of the particle transport caused by the oscillations under the influence of the interacting fields, but not as a result of the thermal motion (Pustovalov and Silin, 1972a).

2.3. Multiple Time-scale Perturbation Analysis of the Non-linear Equation for the Longitudinal Field

Hierarchy of times and expansion in the non-linearity parameter series

Let us solve the non-linear equation for the macroscopic plasma field (2.14). When deriving it we expanded the polarization in the electric field strength series. However, we cannot solve it by an iterative procedure, since the standard perturbation analysis may yield secularities in the solutions by virtue of the resonances in the relevant equations. This difficulty may be overcome if we solve the non-linear equation (2.14) by means of the multiple time-scale perturbation approach (Frieman, 1963; Sandri, 1963, 1965; Davidson, 1972), which is a modification of the general Bogoliubov–Krylov method (Krylov and Bogoliubov, 1947).

Let us denote the dimensionless small parameter of the non-linear interaction by α and extend the number of time variables $t, t', t'', \ldots$, with characteristic periods that increase successively by a factor α^{-1}. We expand the solution of (2.14) according to

$$\mathbf{E}_\mathbf{k}(t) \to \alpha\mathbf{E}_\mathbf{k}^{(1)}(t, t', t'', \ldots)+\alpha^2\mathbf{E}_\mathbf{k}^{(2)}(t, t', t'', \ldots)+\alpha^3\mathbf{E}_\mathbf{k}^{(3)}(t, t', t'', \ldots)+ \ldots \quad (\varrho_\mathbf{k}^0 \to \alpha\varrho_\mathbf{k}^0) \tag{2.43}$$

and suppose the separate terms of the expansion to be functions of the independent variables t, t', t'', ..., etc. The dielectric permittivity $\varepsilon(\omega, \mathbf{k})$ and the non-linear electric susceptibilities $\varkappa^{(2)}(\omega_1, \mathbf{k}_1; \omega_2, \mathbf{k}_2)$, $\varkappa^{(3)}(\omega_1, \mathbf{k}_1; \omega_2, \mathbf{k}_2; \omega_3, \mathbf{k}_3)$, etc., are to be regarded as differential operators $\left(\omega \to i\frac{\partial}{\partial t}\right)$, the time derivatives being expanded according to

$$\frac{\partial}{\partial t} \to \frac{\partial}{\partial t}+\alpha\frac{\partial}{\partial t'}+\alpha^2\frac{\partial}{\partial t''}+ \dots . \tag{2.44}$$

Substituting (2.43) and (2.44) into (2.14) and equating to zero the coefficients of successive powers of α, we obtain (Sitenko, 1973*b*) a hierarchy of equations for the terms of the expansion (2.43):

$$\varepsilon(\omega, \mathbf{k})E^{(1)}_{\mathbf{k}\omega} = -\frac{4\pi i}{k}\overset{0}{\varrho}_{\mathbf{k}\omega}, \tag{2.45}$$

$$\varepsilon(\omega, \mathbf{k})E^{(2)}_{\mathbf{k}\omega} = -\sum_{\substack{\omega_1+\omega_2=\omega \\ \mathbf{k}_1+\mathbf{k}_2=\mathbf{k}}} \varkappa^{(2)}(\omega_1, \mathbf{k}_1; \omega_2, \mathbf{k}_2)E^{(1)}_{\mathbf{k}_1\omega_1}E^{(1)}_{\mathbf{k}_2\omega_2} - i\frac{\partial\varepsilon(\omega, \mathbf{k})}{\partial\omega}\frac{\partial}{\partial t'}E^{(1)}_{\mathbf{k}\omega}, \tag{2.46}$$

$$\begin{aligned}\varepsilon(\omega, \mathbf{k})E^{(3)}_{\mathbf{k}\omega} = &-\sum_{\substack{\omega_1+\omega_2+\omega_3=\omega \\ \mathbf{k}_1+\mathbf{k}_2+\mathbf{k}_3=\mathbf{k}}} \varkappa^{(3)}(\omega_1, \mathbf{k}; \omega_2, \mathbf{k}_2; \omega_3, \mathbf{k}_3)E^{(1)}_{\mathbf{k}_1\omega_1}E^{(1)}_{\mathbf{k}_2\omega_2}E^{(1)}_{\mathbf{k}_3\omega_3} \\ &-\sum_{\substack{\omega_1+\omega_2=\omega \\ \mathbf{k}_1+\mathbf{k}_2=\mathbf{k}}} \varkappa^{(2)}(\omega_1, \mathbf{k}_1; \omega_2, \mathbf{k}_2)\left(E^{(1)}_{\mathbf{k}_1\omega_1}E^{(2)}_{\mathbf{k}_2\omega_2}+E^{(2)}_{\mathbf{k}_1\omega_1}E^{(1)}_{\mathbf{k}_2\omega_2}\right) \\ &-\sum_{\substack{\omega_1+\omega_2=\omega \\ \mathbf{k}_1+\mathbf{k}_2=\mathbf{k}}} \left\{\frac{\partial\varkappa^{(2)}(\omega_1, \mathbf{k}_1; \omega_2, \mathbf{k}_2)}{\partial\omega_1}E^{(1)}_{\mathbf{k}_2\omega_2}\frac{\partial}{\partial t'}E^{(1)}_{\mathbf{k}_1\omega_1}+\frac{\partial\varkappa^{(2)}(\omega_1, \mathbf{k}_1; \omega_2, \mathbf{k}_2)}{\partial\omega_2}E^{(1)}_{\mathbf{k}_1\omega_1}\frac{\partial}{\partial t'}E^{(1)}_{\mathbf{k}_2\omega_2}\right\} \\ &-i\frac{\partial\varepsilon(\omega, \mathbf{k})}{\partial\omega}\frac{\partial}{\partial t'}E^{(2)}_{\mathbf{k}\omega}+\frac{1}{2}\frac{\partial^2\varepsilon(\omega, \mathbf{k})}{\partial\omega^2}\frac{\partial}{\partial t'^2}E^{(1)}_{\mathbf{k}\omega}-i\frac{\partial\varepsilon(\omega, \mathbf{k})}{\partial\omega}\frac{\partial}{\partial t''}E^{(1)}_{\mathbf{k}\omega}.\end{aligned} \tag{2.47}$$

The times t, t', t'' are assumed to be independent. The arbitrariness in the definitions of the time variables may be removed within the context of the requirement that the solutions of the set (2.45)–(2.47) contain no secular terms.

Linear approximation

Let the linearized equation (2.45) be the first approximation. Within the context of the requirement that the right-hand part has to vanish we obtain an equation that describes the free oscillations:

$$\varepsilon(\omega, \mathbf{k})E^{(1)}_{\mathbf{k}\omega} = 0. \tag{2.48}$$

For this equation to have non-zero solutions, the frequencies and the wave vectors must satisfy the dispersion relation

$$\varepsilon(\omega, \mathbf{k}) = 0. \tag{2.49}$$

The roots of (2.49) are associated with the plasma eigenmodes. If we fix some real wave vector $\mathbf{k}$, the root of (2.49) determines the frequency and the damping rate of the relevant mode $\omega^\alpha_{\mathbf{k}}$ and $\gamma^\alpha_{\mathbf{k}}$ (α is introduced to label different eigenoscillations).

Let the solution of (2.48), which corresponds to a certain root of the dispersion relation (2.49), be written in the form

$$E_{\mathbf{k}}^{\alpha}(t) = E_{\mathbf{k}}^{\alpha} \cos(\omega_{\mathbf{k}}^{\alpha} t + \phi_{\mathbf{k}}^{\alpha}), \quad E_{\mathbf{k}}^{\alpha} \equiv E_{\mathbf{k}0}^{\alpha} e^{-\gamma_{\mathbf{k}}^{\alpha} t}, \tag{2.50}$$

where $E_{\mathbf{k}0}^{\alpha}$ and $\phi_{\mathbf{k}}^{\alpha}$ are the initial amplitude and phase of the oscillation. The dielectric permittivity $\varepsilon(\omega, \mathbf{k})$ for the case of an isotropic distribution $f_0(\mathbf{v})$ is a function of the wave-vector modulus:

$$\varepsilon(\omega, \mathbf{k}) \equiv \varepsilon(\omega, k).$$

In such a case $\omega_{\mathbf{k}}^{\alpha}$, $\gamma_{\mathbf{k}}^{\alpha}$, and $\phi_{\mathbf{k}}^{\alpha}$ are real functions of k and by virtue of the realness of the field strength

$$\mathbf{E}_{\mathbf{k}}^{*} = \mathbf{E}_{-\mathbf{k}}. \tag{2.51}$$

The Fourier time-component of the solution of (2.50) in neglect of damping is given by the expression

$$E_{\mathbf{k}}^{\alpha} = \pi E_{\mathbf{k}}^{\alpha} \left\{ e^{-i\phi_{\mathbf{k}}^{\alpha}} \delta(\omega - \omega_{\mathbf{k}}^{\alpha}) + e^{i\phi_{\mathbf{k}}^{\alpha}} \delta(\omega + \omega_{\mathbf{k}}^{\alpha}) \right\}. \tag{2.52}$$

Quadratic interaction and three-wave resonances

Let us consider now the non-linear effects that are due to the quadratic terms in the field amplitude in (2.14). (The wave interaction is described in this approximation by the second-order non-linear susceptibilities.) The field intensity of the eigenoscillations taking into account the non-linear wave interaction is governed in the first approximation by (2.50) or (2.52) where $E_{\mathbf{k}}^{\alpha}$ and $\phi_{\mathbf{k}}^{\alpha}$ must be regarded as functions of the time t':

$$E_{\mathbf{k}\omega}^{\alpha}(t') = \pi E_{\mathbf{k}}^{\alpha}(t') \left\{ e^{-i\phi_{\mathbf{k}}^{\alpha}(t')} \delta(\omega - \omega_{\mathbf{k}}^{\alpha}) + e^{i\phi_{\mathbf{k}}^{\alpha}(t')} \delta(\omega + \omega_{\mathbf{k}}^{\alpha}) \right\}. \tag{2.53}$$

The correction to the first approximation $E_{\mathbf{k}\omega}^{(2)}$ is determined by (2.46), the right-hand part of which contains the resonant terms. For example, the last term in the right-hand part of (2.46) (if it does not vanish) is always in resonance with the left-hand part.

Let the fields $E_{\mathbf{k}_1\omega_1}$, $E_{\mathbf{k}_2\omega_2}$, and $E_{\mathbf{k}\omega}$ be associated with the eigenoscillations $\omega_{\mathbf{k}_1}^{\alpha}$, $\omega_{\mathbf{k}_2}^{\beta}$, and $\omega_{\mathbf{k}}^{\gamma}$. If the frequencies satisfy the condition

$$\omega_{\mathbf{k}_1}^{\alpha} + \omega_{\mathbf{k}_2}^{\beta} = \omega_{\mathbf{k}}^{\gamma}, \tag{2.54}$$

then the non-linear term in the right-hand part of (2.46) is also resonant (three-wave resonance). From the requirement that the resonant right-hand part of (2.46) be equal to zero we derive an equation which is responsible for the slow time-development of the linear amplitude $E_{\mathbf{k}}^{\gamma}$ and the phase $\phi_{\mathbf{k}}^{\gamma}$:

$$\frac{\partial}{\partial t'} E_{\mathbf{k}}^{\gamma} e^{-i\phi_{\mathbf{k}}^{\gamma}} = \frac{i}{2} \frac{1}{\dfrac{\partial \varepsilon(\omega_{\mathbf{k}}^{\gamma}, \mathbf{k})}{\partial \omega_{\mathbf{k}}^{\gamma}}} \sum_{\mathbf{k}_1 + \mathbf{k}_2 = \mathbf{k}} \varkappa^{(2)}(\omega_{\mathbf{k}_1}^{\alpha}, \mathbf{k}_1; \omega_{\mathbf{k}_2}^{\beta}, \mathbf{k}_2) E_{\mathbf{k}_1}^{\alpha} E_{\mathbf{k}_2}^{\beta} e^{-i(\phi_{\mathbf{k}_1}^{\alpha} + \phi_{\mathbf{k}_2}^{\beta})}. \tag{2.55}$$

The extended condition of the three-wave resonance may be written in the form

$$\left| \omega_{\mathbf{k}_1}^{\alpha} \pm \omega_{\mathbf{k}_2}^{\beta} \right| = \omega_{\mathbf{k}}^{\gamma}. \tag{2.56}$$

Equation (2.55) holds only if the two waves with frequencies $\omega_{\mathbf{k}_1}^{\alpha}$ and $\omega_{\mathbf{k}_2}^{\beta}$ are in resonance with the third one $\omega_{\mathbf{k}}^{\gamma}$, i.e. within the context of the three-wave resonance condition (2.56); otherwise the amplitudes are t'-independent:

$$\frac{\partial}{\partial t'} E_{\mathbf{k}}^{\gamma} = 0. \tag{2.57}$$

In the latter case the second-order correction to the field intensity may be expressed in terms of the linear field strengths:

$$E_{\mathbf{k}\omega}^{(2)} = -\frac{1}{\varepsilon(\omega, \mathbf{k})} \sum_{\substack{\omega_1+\omega_2=\omega \\ \mathbf{k}_1+\mathbf{k}_2=\mathbf{k}}} \varkappa^{(2)}(\omega_1, \mathbf{k}_1; \omega_2, \mathbf{k}_2) E_{\mathbf{k}_1\omega_1}^{(1)} E_{\mathbf{k}_2\omega_2}^{(1)}. \tag{2.58}$$

Note that in the absence of resonances the frequency of the second-order correction to the field strength differs from the eigenfrequency in spite of the fact that the linear fields correspond to the eigenoscillations.

Cubic interaction and four-wave resonances

Suppose there are no three-wave resonances. Let us consider the next correction to the field strength associated with the cubic terms in (2.14). Since the linear amplitude is t'-independent in the absence of three-wave resonances, the second-order amplitude also does not depend on t' by virtue of (2.58). Therefore (2.47) for the third-order correction to the field intensity becomes considerably simplified and may be rewritten as:

$$\begin{aligned} \varepsilon(\omega, \mathbf{k}) E_{\mathbf{k}\omega}^{(3)} = & \sum_{\substack{\omega_1+\omega_2+\omega_3=\omega \\ \mathbf{k}_1+\mathbf{k}_2+\mathbf{k}_3=\mathbf{k}}} \frac{1}{3}\left\{\left(2\frac{\varkappa^{(2)}(\omega_1, \mathbf{k}_1; \omega_2+\omega_3, \mathbf{k}_2+\mathbf{k}_3)\varkappa^{(2)}(\omega_2, \mathbf{k}_2; \omega_3, \mathbf{k}_3)}{\varepsilon(\omega_2+\omega_3, \mathbf{k}_2+\mathbf{k}_3)}\right.\right. \\ & \left.\left. -\varkappa^{(3)}(\omega_1, \mathbf{k}_1; \omega_2, \mathbf{k}_2; \omega_3, \mathbf{k}_3)\right) + (\omega_1, \mathbf{k}_1 \rightleftarrows \omega_2, \mathbf{k}_2) + (\omega_1, \mathbf{k}_1 \rightleftarrows \omega_3, \mathbf{k}_3)\right\} \\ & \times E_{\mathbf{k}_1\omega_1}^{(1)} E_{\mathbf{k}_2\omega_2}^{(1)} E_{\mathbf{k}_3\omega_3}^{(1)} - i\frac{\partial\varepsilon(\omega, \mathbf{k})}{\partial\omega} \frac{\partial}{\partial t''} E_{\mathbf{k}\omega}^{(1)}. \end{aligned} \tag{2.59}$$

(Equation (2.59) has been written taking account of the symmetry of the susceptibility $\varkappa^{(2)}(\omega_1, \mathbf{k}_1; \omega_2, \mathbf{k}_2)$.) The last term in the right-hand part of (2.59) is resonant; the non-linear terms may be also of a resonance nature. The requirement for the resonant right-hand part of (2.59) to vanish yields an equation for the t''-dependence of the linear amplitude.

Substituting the field intensity (2.53) in the right-hand part of (2.59) we see that each term of the sum in (2.59) is associated with an inducing force with some of the combination frequencies $|\omega_{\mathbf{k}_1} \pm \omega_{\mathbf{k}_2} \pm \omega_{\mathbf{k}_3}|$. It is clear that the resonance can occur if at least one of the combination frequencies coincides with $\omega_{\mathbf{k}}$. So, the following four-wave resonance condition may be written:

$$|\omega_{\mathbf{k}_1} \pm \omega_{\mathbf{k}_2} \pm \omega_{\mathbf{k}_3}| = \omega_{\mathbf{k}}. \tag{2.60}$$

Assuming (2.60) to be satisfied we obtain, from the requirement that the resonant right-hand part of (2.59) be equal to zero, the following equation for the t''-evolution of the amplitude $E_{\mathbf{k}}$:

$$\frac{\partial}{\partial t''} E_{\mathbf{k}} e^{-i\phi_{\mathbf{k}}}$$

$$= -\frac{i}{12} \frac{1}{\dfrac{\partial \varepsilon(\omega_{\mathbf{k}}, \mathbf{k})}{\partial \omega_{\mathbf{k}}}} \sum_{\mathbf{k}_1+\mathbf{k}_2+\mathbf{k}_3=\mathbf{k}}' \left\{ \left(2 \frac{\varkappa^{(2)}(\omega_{\mathbf{k}_1}, \mathbf{k}_1; \omega_{\mathbf{k}_2}+\omega_{\mathbf{k}_3}, \mathbf{k}_2+\mathbf{k}_3)\, \varkappa^{(2)}(\omega_{\mathbf{k}_2}, \mathbf{k}_2; \omega_{\mathbf{k}_3}, \mathbf{k}_2)}{\varepsilon(\omega_{\mathbf{k}_2}+\omega_{\mathbf{k}_3}, \mathbf{k}_2+\mathbf{k}_3)} \right.\right.$$

$$\left. - \bar{\varkappa}^{(3)}(\omega_{\mathbf{k}_1}, \mathbf{k}_1; \omega_{\mathbf{k}_2}, \mathbf{k}_2; \omega_{\mathbf{k}_3}, \mathbf{k}_3) \right) + (\omega_{\mathbf{k}_1}, \mathbf{k}_1 \rightleftarrows \omega_{\mathbf{k}_2}, \mathbf{k}_2)$$

$$\left. + (\omega_{\mathbf{k}_1}, \mathbf{k}_1 \rightleftarrows \omega_{\mathbf{k}_3}, \mathbf{k}_3) \right\} E_{\mathbf{k}_1} E_{\mathbf{k}_2} E_{\mathbf{k}_3} e^{-i(\phi_{\mathbf{k}_1}+\phi_{\mathbf{k}_2}+\phi_{\mathbf{k}_3})}. \tag{2.61}$$

The prime at the sum symbol reminds us of the necessity of taking into account all combinations of waves, the frequencies of which fit the resonance condition (2.60).

2.4. The General Non-linear Equation for the Field in a Plasma

Maxwell's equations and the non-linear material relation

In the previous sections we have considered the potential interactions in plasmas. Let us study the general case now giving up the assumption of vanishing rotational fields. The general electromagnetic field in neglect of correlations obeys the Maxwell set (1.59) with the induced charge and current densities being governed by the kinetic equation (1.58).

Eliminating the magnetic field $\mathbf{B}$ from (1.59), we reduce the Maxwell set to the following equation for the electric field strength $\mathbf{E}$:

$$\operatorname{curl}\operatorname{curl}\mathbf{E} + \frac{1}{c^2}\frac{\partial^2 \mathbf{E}}{\partial t^2} = -\frac{4\pi}{c^2}\frac{\partial}{\partial t}(\mathbf{j}+\mathbf{j}_0). \tag{2.62}$$

Here $\mathbf{j}$ is the induced current density (1.61) and $\mathbf{j}_0$ is the external current density. Introducing the polarization $\mathbf{P}$ instead of the induced current density by means of the relation

$$\dot{\mathbf{P}} = 4\pi\mathbf{j}, \tag{2.63}$$

and making use of (1.61), we may find the relation between $\mathbf{P}$ and the distribution function perturbation due to the self-consistent field. Note that the polarization contains in the general case both a potential (longitudinal) and a rotational (transverse) component.

Let us Fourier analyze (1.58). We carry out the space–time Fourier decomposition of the distribution function perturbation and of the fields $\mathbf{E}$ and $\mathbf{B}$ and express the magnetic field $\mathbf{B}$ in terms of the electric field $\mathbf{E}$ within the context of the first equation (1.59)

The result is:

$$-i(\omega-(\mathbf{k}\cdot\mathbf{v}))f_{\mathbf{k}\omega}+\frac{e}{m}\left(\left\{E_{\mathbf{k}\omega}+\frac{1}{\omega}[\mathbf{v}[\mathbf{k}\mathbf{E}_{\mathbf{k}\omega}]]\right\}\cdot\frac{\partial f_0}{\partial \mathbf{v}}\right)$$

$$+\frac{e}{m}\sum_{\omega',\mathbf{k}'\neq\omega,\mathbf{k}}\left(\left\{\mathbf{E}_{\mathbf{k}'\omega'}+\frac{1}{\omega'}[\mathbf{v}[\mathbf{k}'\mathbf{E}_{\mathbf{k}'\omega'}]]\cdot\frac{\partial}{\partial\mathbf{v}}\right)f_{\mathbf{k}-\mathbf{k}',\,\omega-\omega'}=0. \quad (2.64)$$

To iterate this equation we expand $f_{\mathbf{k}\omega}$ in a series in the electric field $\mathbf{E}_{\mathbf{k}\omega}$. Substituting this expansions in the expression for the induced current density (1.62) and making use of (2.63) we obtain the polarization expansion in the electric field strength series:

$$P_i(\omega, k) = \varkappa_{ij}^{(1)}(\omega,\mathbf{k})E_j(\omega,\mathbf{k})+\sum_{\substack{\omega_1+\omega_2=\omega\\ \mathbf{k}_1+\mathbf{k}_2=\mathbf{k}}}\varkappa_{ijk}^{(2)}(\omega_1,\mathbf{k}_1;\omega_2,\mathbf{k}_2)E_j(\omega_1,\mathbf{k}_1)E_k(\omega_2,\mathbf{k}_2)$$

$$+\sum_{\substack{\omega_1+\omega_2+\omega_3=\omega\\ \mathbf{k}_1+\mathbf{k}_2+\mathbf{k}_3=\mathbf{k}}}\varkappa_{ijkl}^{(3)}(\omega_1,\mathbf{k}_1;\omega_2,\mathbf{k}_2;\omega_3,\mathbf{k}_3)E_j(\omega_1,\mathbf{k}_1)E_k(\omega_2,\mathbf{k}_2)E_l(\omega_3,\mathbf{k}_3)+\ldots, \quad (2.65)$$

where $\varkappa_{ij}^{(1)}(\omega,\mathbf{k})$ denotes the linear susceptibility tensor, $\varkappa_{ijk}^{(2)}(\omega_1,\mathbf{k}_1;\omega_2,\mathbf{k}_2)$, $\varkappa_{ijkl}^{(3)}(\omega_1,\mathbf{k}_1;\omega_2,\mathbf{k}_2;\omega_3,\mathbf{k}_3)$, etc., are the non-linear tensor plasma susceptibilities. Linear and non-linear plasma susceptibilities are scalars in the potential case. In the general case under consideration the linear coefficients $\varkappa_{ij}^{(1)}(\omega,\mathbf{k})$ of the relation between $\mathbf{P}$ and $\mathbf{E}$ form a second rank tensor, and the non-linear susceptibilities $\varkappa_{ijk}^{(2)}(\omega_1,\mathbf{k}_1;\omega_2,\mathbf{k}_2)$, $\varkappa_{ijkl}^{(3)}(\omega_1,\mathbf{k}_1;\omega_2,\mathbf{k}_2;\omega_3,\mathbf{k}_3)$, etc., are tensors of the third, fourth, and so on rank.

The non-linear equation for the electric field in a plasma

Fourier transformation of (2.62) within the context of (2.63) and (2.65) yields the following non-linear equation for the macroscopic field in a plasma:

$$\Lambda_{ij}(\omega,\mathbf{k})E_j(\omega,\mathbf{k})+\sum_{\substack{\omega_1+\omega_2=\omega\\ \mathbf{k}_1+\mathbf{k}_2=\mathbf{k}}}\varkappa_{ijk}^{(2)}(\omega_1,\mathbf{k}_1;\omega_2,\mathbf{k}_2)E_j(\omega_1,\mathbf{k}_1)E_k(\omega_2,\mathbf{k}_2)$$

$$+\sum_{\substack{\omega_1+\omega_2+\omega_3=\omega\\ \mathbf{k}_1+\mathbf{k}_2+\mathbf{k}_3=\mathbf{k}}}\varkappa_{ijkl}^{(3)}(\omega_1,\mathbf{k}_1;\omega_2,\mathbf{k}_2;\omega_3,\mathbf{k}_3)E_j(\omega_1,\mathbf{k}_1)E_k(\omega_2,\mathbf{k}_2)E_l(\omega_3,\mathbf{k}_3)+\ldots$$

$$=-\frac{4\pi i}{\omega}j_i^0(\omega,\mathbf{k}), \quad (2.66)$$

where we have denoted

$$\Lambda_{ij}(\omega,\mathbf{k})\equiv\varepsilon_{ij}(\omega,\mathbf{k})-\left(\delta_{ij}-\frac{k_ik_j}{k^2}\right)\eta^2,\quad \eta^2\equiv\frac{k^2c^2}{\omega^2}, \quad (2.67)$$

$\varepsilon_{ij}(\omega,\mathbf{k})$ is the dielectric permittivity tensor

$$\varepsilon_{ij}(\omega,\mathbf{k})=\delta_{ij}+\varkappa_{ij}^{(1)}(\omega,\mathbf{k}). \quad (2.68)$$

Similarly to (2.14) for the longitudinal field, the non-linear equation (2.66) describes completely the macroscopic field dynamics in the plasma in the general case. The linear and the non-linear plasma properties are reflected in the dielectric permittivity tensor and the non-linear tensor susceptibilities. The solution of (2.66) for a given distribution of the external currents $\mathbf{j}_0$ uniquely determines the macroscopic field in the plasma.

2.5. The Dielectric Permittivity Tensor and Non-linear Tensor Susceptibilities of a Plasma

General formulas

In the absence of external fields the dielectric permittivity tensor and the non-linear tensor susceptibilities of a plasma may be written as follows:

$$\varepsilon_{ij}(\omega, \mathbf{k}) = \delta_{ij}+\sum \frac{4\pi e^2}{m}\frac{1}{\omega}\int d\mathbf{v}\frac{v_i}{\omega-(\mathbf{k}\cdot\mathbf{v})+i0}\left[\left(1-\frac{(\mathbf{k}\cdot\mathbf{v})}{\omega}\right)\delta_{jk}+\frac{k_k v_j}{\omega}\right]\frac{\partial}{\partial v_k}f_0(v); \quad (2.69)$$

$$\varkappa^{(2)}_{ijk}(\omega_1, \mathbf{k}_1; \omega_2, \mathbf{k}_2) = \sum \frac{(-i)}{2}\frac{4\pi e}{m}\frac{e}{m}\frac{1}{\omega}\int d\mathbf{v}\frac{v_i}{\omega-(\mathbf{k}\cdot\mathbf{v})+i0}$$

$$\times\Bigg(\left[\left(1-\frac{(\mathbf{k}_1\cdot\mathbf{v})}{\omega_1}\right)\delta_{jl}+\frac{k_{1l}v_j}{\omega_1}\right]\frac{\partial}{\partial v_l}\frac{1}{\omega_2-(\mathbf{k}_2\cdot\mathbf{v})+i0}\left[\left(1-\frac{(\mathbf{k}_2\mathbf{v})}{\omega_2}\right)\delta_{km}+\frac{k_{1m}v_k}{\omega_2}\right]\frac{\partial}{\partial v_m}$$

$$+\left[\left(1-\frac{(\mathbf{k}_2\cdot\mathbf{v})}{\omega_2}\right)\delta_{kl}+\frac{k_{2l}v_k}{\omega_2}\right]\frac{\partial}{\partial v_l}\frac{1}{\omega_1-(\mathbf{k}_1\cdot\mathbf{v})+i0}\left[\left(1-\frac{(\mathbf{k}_1\cdot\mathbf{v})}{\omega_1}\right)\delta_{jm}+\frac{k_{1m}v_j}{\omega_1}\right]\frac{\partial}{\partial v_m}\Bigg)f_0(\mathbf{v}), \quad (2.70)$$

where $\omega = \omega_1+\omega_2$, $\mathbf{k} = \mathbf{k}_1+\mathbf{k}_2$;

$$\varkappa^{(3)}_{ijkl}(\omega_1, \mathbf{k}_1; \omega_2, \mathbf{k}_2; \omega_3, \mathbf{k}_3) \equiv \tfrac{1}{3}\{\bar{\varkappa}^{(3)}_{ijkl}(\omega_1, \mathbf{k}_1; \omega_2, \mathbf{k}_2; \omega_3, \mathbf{k}_3)+\bar{\varkappa}^{(3)}_{ikjl}(\omega_2, \mathbf{k}_2; \omega_1, \mathbf{k}_1; \omega_3, \mathbf{k}_3)+\bar{\varkappa}^{(3)}_{ilkj}(\omega_3, \mathbf{k}_3; \omega_2, \mathbf{k}_2; \omega_1, \mathbf{k}_1)\}, \quad (2.71)$$

where

$$\bar{\varkappa}^{(3)}_{ijkl}(\omega_1, \mathbf{k}_1; \omega_2, \mathbf{k}_2; \omega_3, \mathbf{k}_3)$$

$$= \sum \frac{(-i)^2}{2}\frac{4\pi e^2}{m}\left(\frac{e}{m}\right)^2\frac{1}{\omega}\int d\mathbf{v}\frac{v_i}{\omega-(\mathbf{k}\cdot\mathbf{v})+i0}\left[\left(1-\frac{(\mathbf{k}_1\cdot\mathbf{v})}{\omega_1}\right)\delta_{jm}+\frac{k_{1m}v_j}{\omega_1}\right]\frac{\partial}{\partial v_m}$$

$$\times\Bigg\{\frac{1}{\omega'-(\mathbf{k}'\cdot\mathbf{v})+i0}\Bigg(\left[\left(1-\frac{(\mathbf{k}_2\cdot\mathbf{v})}{\omega_2}\right)\delta_{kn}+\frac{k_{2n}v_k}{\omega_2}\right]\frac{\partial}{\partial v_n}\frac{1}{\omega_3-(\mathbf{k}_3\cdot\mathbf{v})+i0}$$

$$\times\left[\left(1-\frac{(\mathbf{k}_3\cdot\mathbf{v})}{\omega_3}\right)\delta_{lp}+\frac{k_{3p}v_l}{\omega_3}\right]\frac{\partial}{\partial v_p}+\left[\left(1-\frac{\mathbf{k}_3\cdot\mathbf{v}}{\omega_3}\right)\delta_{ln}+\frac{k_{3n}v_l}{\omega_3}\right]\frac{\partial}{\partial v_n}\frac{1}{\omega_2-(\mathbf{k}_2\cdot\mathbf{v})+i0}$$

$$\times\left[\left(1-\frac{(\mathbf{k}_2\cdot\mathbf{v})}{\omega_2}\right)\delta_{kp}+\frac{k_{2p}v_k}{\omega_2}\right]\frac{\partial}{\partial v_p}\Bigg)\Bigg\}f_0(\mathbf{v}), \quad (2.72)$$

where $\omega = \omega_1+\omega_2+\omega_3$, $\mathbf{k} = \mathbf{k}_1+\mathbf{k}_2+\mathbf{k}_3$, $\omega' = \omega_2+\omega_3$, $\mathbf{k}' = \mathbf{k}_2+\mathbf{k}_3$.

$$\varkappa^{(n)}_{ij_1j_2\ldots j_n}(\omega_1, \mathbf{k}_1; \omega_2, \mathbf{k}_2; \ldots; \omega_n, \mathbf{k}_n) = \sum \frac{(-i)^{n-1}}{n!} \frac{4\pi e^2}{m} \left(\frac{e}{m}\right)^{n-1} \frac{1}{\omega} \mathcal{P} \int d\mathbf{v}\, v_i$$

$$\times \frac{1}{\omega_1+\omega_2+\ldots+\omega_n-(\{\mathbf{k}_1+\mathbf{k}_2+\ldots+\mathbf{k}_n\}\cdot\mathbf{v})+i0} \left[\left(1-\frac{(\mathbf{k}_1\cdot\mathbf{v})}{\omega_1}\right)\delta_{j_1k_1}+\frac{k_{1k_1}v_{j_1}}{\omega_1}\right]\frac{\partial}{\partial v_{k_1}}$$

$$\times \left\{\frac{1}{\omega_2+\omega_3+\ldots+\omega_n-(\{\mathbf{k}_2+\mathbf{k}_3+\ldots+\mathbf{k}_n\}\cdot\mathbf{v})+i0} \left[\left(1-\frac{(\mathbf{k}_2\cdot\mathbf{v})}{\omega_2}\right)\delta_{j_2k_2}+\frac{k_{2k_2}v_{j_2}}{\omega_2}\right]\right.$$

$$\left.\times \frac{\partial}{\partial v_{k_2}} \cdot \ldots \cdot \left(\frac{1}{\omega_n-(\mathbf{k}_n\cdot\mathbf{v})+i0}\left[\left(1-\frac{(\mathbf{k}_n\cdot\mathbf{v})}{\omega_n}\right)\delta_{j_nk_n}+\frac{k_{nk_n}v_{j_n}}{\omega_n}\right]\frac{\partial f_0(\mathbf{v})}{\partial v_{k_n}}\right)\ldots\right\}. \tag{2.73}$$

(The symbol $\mathcal{P}$ denotes all possible permutations of the frequencies, the wave vectors, and the relevant subscripts.)

Symmetries of the dielectric permittivity tensor

The knowledge of the dielectric permittivity tensor (2.69) is sufficient for a complete description of the linear electrodynamic plasma properties. This tensor satisfies the following symmetry conditions:

$$\varepsilon_{ij}(\omega, \mathbf{k}) = \varepsilon^*_{ij}(-\omega, -\mathbf{k}), \quad \varepsilon_{ij}(\omega, \mathbf{k}) = \varepsilon_{ji}(\omega, -\mathbf{k}). \tag{2.74}$$

Note that it is impossible to separate the longitudinal component of the electromagnetic field from the transverse ones in a plasma with an anisotropic distribution of particle velocities. These components are distinguishable only in a plasma of particles, the velocity distribution of which is isotropic. Such a plasma will be henceforth called an isotropic plasma.

The dielectric permittivity tensor of an isotropic plasma is of the form

$$\varepsilon_{ij}(\omega, \mathbf{k}) = \frac{k_ik_j}{k^2}\varepsilon_l(\omega, \mathbf{k})+\left(\delta_{ij}-\frac{k_ik_j}{k^2}\right)\varepsilon_t(\omega, \mathbf{k}), \tag{2.75}$$

where ε_l and ε_t denote the longitudinal and the transverse dielectric permittivities. ε_l and ε_t are scalar functions of the frequency and of the wave-vector modulus.

The explicit expressions both for the longitudinal dielectric permittivity and for the transverse one may be easily obtained within the context of (2.69):

$$\varepsilon_l(\omega, \mathbf{k}) = 1+\sum\frac{4\pi\varepsilon^2}{mk^2}\int \delta\mathbf{v}\frac{1}{\omega-(\mathbf{k}\cdot\mathbf{v})+i0}\left(\mathbf{k}\cdot\frac{\partial f_0(\mathbf{v})}{\partial\mathbf{v}}\right), \tag{2.76}$$

$$\varepsilon_t(\omega, \mathbf{k}) = 1+\sum\frac{2\pi e^2}{m\omega k^2}\int d\mathbf{v}\frac{1}{\omega-(\mathbf{k}\cdot\mathbf{v})+i0}\left([\mathbf{k}[\mathbf{v}\mathbf{k}]]\cdot\frac{\partial f_0(\mathbf{v})}{\partial\mathbf{v}}\right). \tag{2.77}$$

It is natural that $\varepsilon_l(\omega, \mathbf{k})$ is equal to the dielectric permittivity $\varepsilon(\omega, \mathbf{k})$ which was introduced when considering the potential case.

In the cold plasma ε_l and ε_t coincide and are equal to the limiting value

$$\varepsilon(\omega) = 1-\sum \frac{\Omega^2}{\omega^2}. \tag{2.78}$$

Thus neither the longitudinal field nor the transverse one undergo spatial dispersion in the cold plasma.

Symmetries of the non-linear susceptibilities

The symmetry properties of the non-linear tensor plasma susceptibilities follow immediately from (2.65):

$$\varkappa_{ijk}^{(2)}(\omega_1, \mathbf{k}_1; \omega_2, \mathbf{k}_2) = \varkappa_{ikj}^{(2)}(\omega_2, \mathbf{k}_2; \omega_1, \mathbf{k}_1), \tag{2.79}$$

$$\begin{aligned}\varkappa_{ijkl}^{(3)}(\omega_1, \mathbf{k}_1; \omega_2, \mathbf{k}_2; \omega_3, \mathbf{k}_3) &= \varkappa_{ijlk}^{(3)}(\omega_1, \mathbf{k}_1; \omega_3, \mathbf{k}_3; \omega_2, \mathbf{k}_2)\\ = \varkappa_{ikjl}^{(3)}(\omega_2, \mathbf{k}_2; \omega_1, \mathbf{k}_1; \omega_3, \mathbf{k}_3) &= \varkappa_{iklj}^{(3)}(\omega_2, \mathbf{k}_2; \omega_3, \mathbf{k}_3; \omega_1, \mathbf{k}_1)\\ = \varkappa_{iljk}^{(3)}(\omega_3, \mathbf{k}_3; \omega_1, \mathbf{k}_1; \omega_2, \mathbf{k}_2) &= \varkappa_{ilkj}^{(3)}(\omega_3, \mathbf{k}_3; \omega_2, \mathbf{k}_2; \omega_1, \mathbf{k}_1).\end{aligned} \tag{2.80}$$

(The tensor $\bar{\varkappa}_{ijkl}^{(3)}(\omega_1, \mathbf{k}_1; \omega_2, \mathbf{k}_2; \omega_3, \mathbf{k}_3)$, which was defined by (2.72), is symmetric only relative to the interchange $k \rightleftarrows l$ and $\omega_2, \mathbf{k}_2 \rightleftarrows \omega_3, \mathbf{k}_3$.) Besides, the non-linear tensor susceptibilities satisfy the conditions

$$\varkappa_{ijk}^{(2)}(\omega_1, \mathbf{k}_1; \omega_2, \mathbf{k}_2) = \varkappa_{ijk}^{(2)*}(-\omega_1, -\mathbf{k}_1; -\omega_2, -\mathbf{k}_2), \tag{2.81}$$

$$\varkappa_{ijkl}^{(3)}(\omega_1, \mathbf{k}_1; \omega_2, \mathbf{k}_2; \omega_3, \mathbf{k}_3) = \varkappa_{ijkl}^{(3)*}(-\omega_1, -\mathbf{k}_1; -\omega_2, -\mathbf{k}_2; -\omega_3, -\mathbf{k}_3), \tag{2.82}$$

which follow from the requirement that the fields be real.

Non-linear tensor susceptibilities of a cold plasma

The integration of (2.70) and (2.72) by parts yields

$$\begin{aligned}&\varkappa_{ijk}^{(2)}(\omega_1, \mathbf{k}_1; \omega_2, \mathbf{k}_2)\\ &= \sum \frac{(-i)}{2}\frac{4\pi e^2}{m}\frac{e}{m}\frac{1}{\omega}\int d\mathbf{v} f_0(\mathbf{v})\left(\left[\left(1-\frac{(\mathbf{k}_1\cdot\mathbf{v})}{\omega_1}\right)\delta_{jl}+\frac{k_{1l}v_j}{\omega_1}\right]\frac{\partial}{\partial v_l}\frac{1}{\omega_1-(\mathbf{k}_1\cdot\mathbf{v})+i0}\right.\\ &\times\left[\left(1-\frac{(\mathbf{k}_2\cdot\mathbf{v})}{\omega_2}\right)\delta_{km}+\frac{k_{2m}v_k}{\omega}\right]\frac{\partial}{\partial v_m}+\left[\left(1-\frac{(\mathbf{k}_2\cdot\mathbf{v})}{\omega_2}\right)\delta_{kl}+\frac{k_{2l}v_k}{\omega_2}\right]\frac{\partial}{\partial v_l}\frac{1}{\omega_2-(\mathbf{k}_2\cdot\mathbf{v})+i0}\\ &\times\left.\left[\left(1-\frac{(\mathbf{k}_1\cdot\mathbf{v})}{\omega_1}\right)\delta_{jm}+\frac{k_{1m}v_j}{\omega_1}\right]\frac{\partial}{\partial v_m}\right)\frac{v_i}{\omega-(\mathbf{k}\cdot\mathbf{v})+i0}.\end{aligned} \tag{2.83}$$

where $\omega = \omega_1+\omega_2$, $\mathbf{k} = \mathbf{k}_1+\mathbf{k}_2$;

$$\bar{\varkappa}^{(3)}_{ijkl}(\omega_1, \mathbf{k}_1; \omega_2, \mathbf{k}_2; \omega_3, \mathbf{k}_3) = \sum \frac{1}{2}\frac{4\pi e^2}{m}\left(\frac{e}{m}\right)^2 \frac{1}{\omega}\int d\mathbf{v} f_0(\mathbf{v})$$

$$\times\left(\left[\left(1-\frac{(\mathbf{k}_2\cdot\mathbf{v})}{\omega_2}\right)\delta_{km}+\frac{k_{2m}v_k}{\omega_2}\right]\frac{\partial}{\partial v_m}\frac{1}{\omega_2-(\mathbf{k}_2\cdot\mathbf{v})+i0}\left[\left(1-\frac{(\mathbf{k}_3\cdot\mathbf{v})}{\omega_3}\right)\delta_{ln}+\frac{k_{3n}v_l}{\omega_3}\right]\frac{\partial}{\partial v_n}\right.$$

$$\left.+\left[\left(1-\frac{(\mathbf{k}_3\cdot\mathbf{v})}{\omega_3}\right)\delta_{lm}+\frac{k_{3m}v_l}{\omega_3}\right]\frac{\partial}{\partial v_m}\frac{1}{\omega_3-(\mathbf{k}_3\cdot\mathbf{v})+i0}\left[\left(1-\frac{(\mathbf{k}_2\cdot\mathbf{v})}{\omega_2}\right)\delta_{kn}+\frac{k_{2n}v_k}{\omega_2}\right]\frac{\partial}{\partial v_n}\right)$$

$$\times\frac{1}{\omega'-(\mathbf{k}'\cdot\mathbf{v})+i0}\left[\left(1-\frac{(\mathbf{k}_1\cdot\mathbf{v})}{\omega_1}\right)\delta_{jp}+\frac{k_{1p}v_j}{\omega_1}\right]\frac{\partial}{\partial v_p}\frac{v_i}{\omega-(\mathbf{k}\cdot\mathbf{v})+i0}, \tag{2.84}$$

where $\omega = \omega_1+\omega_2+\omega_3$, $\mathbf{k} = \mathbf{k}_1+\mathbf{k}_2+\mathbf{k}_3$, $\omega' = \omega_2+\omega_3$, $\mathbf{k}' = \mathbf{k}_2+\mathbf{k}_3$.

In a cold plasma, (2.83) and (2.84) reduce to

$$\varkappa^{(2)}_{ijk}(\omega_1, \mathbf{k}_1; \omega_2, \mathbf{k}_2) = \sum\frac{(-i)}{2}\frac{e}{m}\frac{\Omega^2}{\omega_1\omega_2\omega}\left(\frac{k_i}{\omega}\delta_{jk}+\frac{k_{1j}}{\omega_1}\delta_{ik}+\frac{k_{2k}}{\omega_2}\delta_{ij}\right), \tag{2.85}$$

$$\varkappa^{(3)}_{ijkl}(\omega_1, \mathbf{k}_1; \omega_2, \mathbf{k}_2; \omega_3, \mathbf{k}_3)$$

$$= \sum\frac{1}{6}\frac{e^2}{m^2}\frac{\Omega^2}{\omega_1\omega_2\omega_3\omega}\left\{\left(\frac{1}{\omega'}\left[\frac{k'^2}{\omega'}\delta_{ij}\delta_{kl}+\left(\frac{k_ik_j}{\omega}-\frac{k_{1i}k_{1j}}{\omega_1}\right)\delta_{kl}+\left(\frac{k_{2k}k_{2l}}{\omega_2}+\frac{k_{3k}k_{3l}}{\omega_3}\right)\delta_{ij}\right]\right.\right.$$

$$\left.\left.+\frac{k_ik_{1j}}{\omega\omega_1}\delta_{kl}+\frac{k_{2k}k_{3l}}{\omega_2\omega_3}\delta_{ij}\right)+(1\rightleftarrows 2)+(1\rightleftarrows 3)\right\}. \tag{2.86}$$

These expressions for the symmetrized non-linear susceptibilities $\varkappa^{(2)}_{ijk}(\omega_1, \mathbf{k}_1; \omega_2, \mathbf{k}_2)$ and $\varkappa^{(3)}_{ijkl}(\omega_1, \mathbf{k}_1; \omega_2, \mathbf{k}_2; \omega_3, \mathbf{k}_3)$ may also be derived starting from the hydrodynamical equations

$$\left.\begin{aligned}&\frac{\partial n}{\partial t}+\operatorname{div}(n\mathbf{v}) = 0,\\&\frac{\partial \mathbf{v}}{\partial t}+(\mathbf{v}\cdot\nabla)\mathbf{v} = \frac{e}{m}\left(\mathbf{E}+\frac{1}{c}[\mathbf{vB}]\right),\end{aligned}\right\} \tag{2.87}$$

where charge and current densities are the following functions of the macroscopic densities and the particle velocities:

$$\varrho = \sum en, \quad \mathbf{j} = \sum en\mathbf{v}. \tag{2.88}$$

We present here also an explicit form of the non-symmetrized non-linear susceptibility $\bar{\varkappa}'^{(3)}_{ijkl}(\omega_1, \mathbf{k}_1; \omega_2, \mathbf{k}_2; \omega_3, \mathbf{k}_3)$ of a cold plasma. This is by definition a sum of the portion of (2.84) that is singular at $\omega' \equiv \omega_2+\omega_3 = 0$ as $T = 0$ and the rest of (2.84), symmetrized with respect to all possible permutations:

$$\bar{\varkappa}'^{(3)}_{ijkl}(\omega_1, \mathbf{k}_1; \omega_2, \mathbf{k}_2; \omega_3, \mathbf{k}_3)$$

$$= \sum\frac{1}{2}\frac{e^2}{m^2}\frac{\Omega^2}{\omega_1\omega_2\omega_3\omega}\left(\frac{1}{\omega'}\left[\frac{k'^2}{\omega'}\delta_{ij}\delta_{kl}+\left(\frac{k_ik_j}{\omega}-\frac{k_{1i}k_{1j}}{\omega_1}\right)\delta_{kl}+\left(\frac{k_{2k}k_{2l}}{\omega_2}+\frac{k_{3k}k_{3l}}{\omega_3}\right)\delta_{ij}\right]\right.$$

$$\left.+\frac{1}{3}\left[\frac{k_i}{\omega}\left(\frac{k_{1j}}{\omega_1}+\frac{k_{2j}}{\omega_2}+\frac{k_{3j}}{\omega_3}\right)\delta_{kl}+\left(\frac{k_{1k}k_{3l}}{\omega_1\omega_3}+\frac{k_{2k}k_{1l}}{\omega_1\omega_2}+\frac{k_{2k}k_{3l}}{\omega_2\omega_3}\right)\delta_{ij}\right]\right). \tag{2.89}$$

According to (2.85), (2.86), and (2.89), the non-linear tensor susceptibilities of a cold plasma depend on the wave vectors of the interacting fields analogously to the longitudinal ones.

2.6. Hierarchy of Multiple Time-scale Approximate Equations for the Electromagnetic Field in a Plasma

Basic set of equations

Let us try to solve the non-linear equation (2.66) for the macroscopic electric field in a plasma by means of the multiple time-scale perturbation analysis. In a manner analogous to the consideration of the longitudinal case, we expand the electric field strength (which includes now both the potential and the rotational contributions) in a power series in the parameter of the non-linear interaction α

$$\mathbf{E}_\mathbf{k}(t) \rightarrow \alpha \mathbf{E}_\mathbf{k}^{(1)}(t, t', t'', \ldots)+\alpha^2 \mathbf{E}_\mathbf{k}^{(2)}(t, t', t'', \ldots)+\alpha^3 \mathbf{E}_\mathbf{k}(t, t', t'' \ldots)+ \ldots, \tag{2.90}$$

and take the dielectric permittivity $\varepsilon_{ij}(\omega, \mathbf{k})$ and the non-linear susceptibilities $\varkappa_{ijk}^{(2)}(\omega_1, \mathbf{k}_1; \omega_2, \mathbf{k}_2)$ and $\varkappa_{ijkl}^{(3)}(\omega_1, \mathbf{k}_1; \omega_2, \mathbf{k}_2; \omega_3, \mathbf{k}_3)$ to be differential operators. The dependence of those on α may be defined by means of the formal replacement

$$\omega \rightarrow i\left(\frac{\partial}{\partial t}+\alpha \frac{\partial}{\partial t'}+\alpha^2 \frac{\partial}{\partial t''}+ \ldots\right).$$

Substituting the above expressions in the non-linear equation (2.66) and equating to zero the coefficients of the successive powers of α, we obtain the following hierarchy of equations that determine the terms of the expansion (2.90):

$$\Lambda_{ij}(\omega, \mathbf{k})\, E_j^{(1)}(\omega, \mathbf{k}) = -\frac{4\pi i}{\omega} j_i^{(0)}(\omega, \mathbf{k}), \tag{2.91}$$

$$\begin{aligned}\Lambda_{ij}(\omega, \mathbf{k})\, E_j^{(2)}(\omega, \mathbf{k}) = &-\sum_{\substack{\omega_1+\omega_2=\omega \\ \mathbf{k}_1+\mathbf{k}_2=\mathbf{k}}} \varkappa_{ijk}^{(2)}(\omega_1, \mathbf{k}_1; \omega_2, \mathbf{k}_2)\, E_j^{(1)}(\omega_1, \mathbf{k}_1)\, E_k^{(1)}(\omega_2, \mathbf{k}_2) \\ &- i\frac{\partial}{\partial \omega}\Lambda_{ij}(\omega, \mathbf{k}) \frac{\partial}{\partial t'} E_j^{(1)}(\omega, \mathbf{k}),\end{aligned} \tag{2.92}$$

$$\begin{aligned}&\Lambda_{ij}(\omega, \mathbf{k})\, E_j^{(3)}(\omega, \mathbf{k}) \\ &= -\sum_{\substack{\omega_1+\omega_2+\omega_3=\omega \\ \mathbf{k}_1+\mathbf{k}_2+\mathbf{k}_3=\mathbf{k}}} \varkappa_{ijkl}^{(3)}(\omega_1, \mathbf{k}_1; \omega_2, \mathbf{k}_2; \omega_3, \mathbf{k}_3)\, E_j^{(1)}(\omega_1, \mathbf{k}_1)\, E_k^{(1)}(\omega_2, \mathbf{k}_2)\, E_l^{(1)}(\omega_3, \mathbf{k}_3) \\ &\quad - \sum_{\substack{\omega_1+\omega_2=\omega \\ \mathbf{k}_1+\mathbf{k}_2=\mathbf{k}}} \varkappa_{ijk}^{(2)}(\omega_1, \mathbf{k}_1; \omega_2, \mathbf{k}_2)\left(E_j^{(1)}(\omega_1, \mathbf{k}_1)\, E_k^{(2)}(\omega_2, \mathbf{k}_2)+E_j^{(2)}(\omega_1, \mathbf{k}_1)\, E_k^{(1)}(\omega_2, \mathbf{k}_2)\right) \\ &\quad - i\sum_{\substack{\omega_1+\omega_2=\omega \\ \mathbf{k}_1+\mathbf{k}_2=\mathbf{k}}} \left\{\frac{\partial \varkappa_{ijk}^{(2)}(\omega_1, \mathbf{k}_1; \omega_2, \mathbf{k}_2)}{\partial \omega_1} E_k^{(1)}(\omega_2, \mathbf{k}_2) \frac{\partial}{\partial t'} E_j^{(1)}(\omega_1, \mathbf{k}_1)\right. \\ &\quad \left. + \frac{\partial \varkappa_{ijk}^{(2)}(\omega_1, \mathbf{k}_1; \omega_2, \mathbf{k}_2)}{\partial \omega_2} E_j^{(1)}(\omega_1, \mathbf{k}_1) \frac{\partial}{\partial t'} E_k^{(1)}(\omega_2, \mathbf{k}_2)\right\} - i\frac{\partial}{\partial \omega}\Lambda_{ij}(\omega, \mathbf{k}) \frac{\partial}{\partial t'} E_j^{(2)}(\omega, \mathbf{k}) \\ &\quad + \frac{1}{2}\frac{\partial^2}{\partial \omega^2}\Lambda_{ij}(\omega, \mathbf{k}) \frac{\partial}{\partial t'^2} E_j^{(1)}(\omega, \mathbf{k}) - i\frac{\partial}{\partial \omega}\Lambda_{ij}(\omega, \mathbf{k}) \frac{\partial}{\partial t''} E_j^{(1)}(\omega, \mathbf{k}).\end{aligned} \tag{2.93}$$

Since the times $t, t', t'', \ldots$, are supposed to be independent variables, this set of equations is in fact indefinite. To remove this uncertainty we require the equations to contain no secular terms. The conditions that the secular terms of each equation vanish make a supplementary set of equations. The latter and the set (2.91)–(2.93) determine uniquely the dependence of the terms of the expansion (2.90) on the times $t, t', t'', \ldots$.

Linear approximation

Consider the linearized equation (2.91). In the case where there are no external charges and currents in the plasma, we obtain

$$\Lambda_{ij}(\omega, \mathbf{k})\, E_j(\omega, \mathbf{k}) = 0, \tag{2.94}$$

where

$$\Lambda_{ij}(\omega, \mathbf{k}) \equiv \varepsilon_{ij}(\omega, \mathbf{k}) + \left(\frac{k_i k_j}{k^2} - \delta_{ij}\right)\eta^2, \quad \eta^2 \equiv \frac{k^2 c^2}{\omega^2}.$$

Equation (2.94) is conventionally referred to as the wave equation, for it determines the electromagnetic field in the absence of external charges and fields, i.e. the electromagnetic waves. The wave equation has non-zero solutions only within the context of the dispersion relation; for the latter to be satisfied the determinant of $\Lambda_{ij}(\omega, \mathbf{k})$ must vanish:

$$\Lambda(\omega, \mathbf{k}) \equiv |\Lambda_{ij}(\omega, \mathbf{k})| = 0. \tag{2.95}$$

The roots of the dispersion equation (2.95) determine the interdependence between the frequencies and the wave vectors of the electromagnetic waves that propagate in the plasma.

We denote the frequency and the damping rate associated with some root of (2.95) that corresponds to a fixed real value of the wave vector $\mathbf{k}$ by $\omega_{\mathbf{k}}^{\alpha}$ and $\gamma_{\mathbf{k}}^{\alpha}$ (α labels different roots of the dispersion equation that give rise to the electromagnetic eigenwaves in the plasma). The relevant solution of (2.94) may be written as

$$\mathbf{E}_{\mathbf{k}}^{\alpha}(t) = \mathbf{E}_{\mathbf{k}}^{\alpha} \cos(\omega_{\mathbf{k}}^{\alpha} t + \phi_{\mathbf{k}}^{\alpha}), \quad \mathbf{E}_{\mathbf{k}}^{\alpha} \equiv \mathbf{E}_{\mathbf{k}0}^{\alpha}\, e^{-\gamma_{\mathbf{k}}^{\alpha} t}, \tag{2.96}$$

where $\mathbf{E}_{\mathbf{k}0}$ and $\phi_{\mathbf{k}}^{\alpha}$ are the initial amplitude and phase. Suppose $\omega_{\mathbf{k}}^{\alpha}$, $\gamma_{\mathbf{k}}^{\alpha}$, and $\phi_{\mathbf{k}}^{\alpha}$ to be real functions of the wave-vector modulus. The field (2.96) is real, so that

$$\mathbf{E}_{\mathbf{k}}^{*} = \mathbf{E}_{-\mathbf{k}}. \tag{2.97}$$

The Fourier time-component of (2.96) in neglect of damping is of the form

$$\mathbf{E}_{\mathbf{k}\omega}^{\alpha} = \pi \mathbf{E}_{\mathbf{k}}^{\alpha} \left\{ e^{-i\phi_{\mathbf{k}}^{\alpha}} \delta(\omega - \omega_{\mathbf{k}}^{\alpha}) + e^{i\phi_{\mathbf{k}}^{\alpha}} \delta(\omega + \omega_{\mathbf{k}}^{\alpha}) \right\}. \tag{2.98}$$

Three-wave resonances

Let us consider now the non-linear effects that are due to the quadratic terms in the electric field strength in (2.66). The non-linear wave interaction is reflected in this approximation in $\varkappa_{ijk}^{(2)}(\omega_1, \mathbf{k}_1; \omega_2, \mathbf{k}_2)$. The eigenwave electric field strength may be written as before in the form (2.98), where a slowly varying function of t' must be substituted for $\mathbf{E}^{\alpha}_{\mathbf{k}}$:

$$\mathbf{E}^{\alpha}_{\mathbf{k}\omega}(t') = \pi\mathbf{E}^{\alpha}_{\mathbf{k}}(t')\left\{e^{-i\phi^{\alpha}_{\mathbf{k}}}\delta(\omega-\omega^{\alpha}_{\mathbf{k}})+e^{i\phi^{\alpha}_{\mathbf{k}}}\delta(\omega+\omega^{\alpha}_{\mathbf{k}})\right\}. \tag{2.99}$$

The linear field amplitude varies in time due to the non-linear wave interaction and only within the context of the resonance conditions.

To derive an equation to describe the t'-dependence of the amplitude $\mathbf{E}^{\alpha}_{\mathbf{k}}$, we start from (2.92) which determines the second-order correction $\mathbf{E}^{\alpha(2)}_{\mathbf{k}\omega}$. Let the resonance conditions be satisfied, i.e. suppose that

$$\omega^{\alpha}_{\mathbf{k}_1}+\omega^{\beta}_{\mathbf{k}_2} = \omega^{\gamma}_{\mathbf{k}_3}. \tag{2.100}$$

From the requirement that the resonant terms in the right-hand part of (2.92) be equal to zero, we obtain

$$\frac{\partial}{\partial\omega^{\gamma}_{\mathbf{k}}}\Lambda_{ij}(\omega^{\gamma}_{\mathbf{k}},\mathbf{k})\frac{\partial}{\partial t'}E^{\gamma}_{j\mathbf{k}}\,e^{-i\phi^{\gamma}_{\mathbf{k}}} = \frac{i}{2}\sum_{\mathbf{k}_1+\mathbf{k}_2=\mathbf{k}}\varkappa^{(2)}_{ijk}(\omega^{\alpha}_{\mathbf{k}_1},\mathbf{k}_1;\omega^{\beta}_{\mathbf{k}_2},\mathbf{k}_2)E_{j\mathbf{k}_1}E_{k\mathbf{k}_2}\,e^{-i(\phi^{\alpha}_{\mathbf{k}_1}+\phi^{\beta}_{\mathbf{k}_2})}. \tag{2.101}$$

The solution of (2.101) determines the time evolution of the linear amplitude $E^{\gamma}_{\mathbf{k}}$ due to the three-wave resonant coupling.

If (2.100) is not satisfied, then the linear amplitude $\mathbf{E}^{\gamma}_{\mathbf{k}}$ is time-independent:

$$\frac{\partial}{\partial t'}\mathbf{E}^{\gamma}_{\mathbf{k}} = 0. \tag{2.102}$$

The non-linear correction to the field strength (2.99) is then equal to

$$E^{(2)}_i(\omega,\mathbf{k}) = -\Lambda^{-1}_{ij}(\omega,\mathbf{k})\sum_{\substack{\omega_1+\omega_2=\omega\\ \mathbf{k}_1+\mathbf{k}_2=\mathbf{k}}}\varkappa^{(2)}_{jkl}(\omega_1,\mathbf{k}_1;\omega_2,\mathbf{k}_2)\,E^{(1)}_k(\omega_1,\mathbf{k}_1)\,E^{(1)}_l(\omega_2,\mathbf{k}_2), \tag{2.103}$$

where the tensor $\Lambda^{-1}_{ij}(\omega,\mathbf{k})$ is the inverse of (2.67). The frequency of the correction (2.103) may be considerably different from the eigenfrequency.

Four-wave resonances

Suppose the frequencies do not satisfy (2.100) and there are no three-wave resonances in the system. Then the four-wave or higher-order resonances become of importance. Consider the next approximation for the field strength $\mathbf{E}^{\gamma(3)}_{\mathbf{k}\omega}$, which is due to the cubic terms in (2.66). The linear amplitude $\mathbf{E}^{\gamma}_{\mathbf{k}}$ is t'-independent, so the first correction (2.103) also does not depend on t'. Then (2.93) for the second correction to (2.99) becomes much simplified:

$$\begin{aligned}\Lambda_{ij}(\omega,\mathbf{k})\,E^{(3)}_j(\omega,\mathbf{k}) &= \sum_{\substack{\omega_1+\omega_2+\omega_3=\omega\\ \mathbf{k}_1+\mathbf{k}_2+\mathbf{k}_3=\mathbf{k}}}\{2\varkappa^{(2)}_{ijj'}(\omega_1,\mathbf{k}_1;\omega_2+\omega_3,\mathbf{k}_2+\mathbf{k}_3)\,\Lambda^{-1}_{j'i'}(\omega_2+\omega_3,\mathbf{k}_2+\mathbf{k}_3)\,\varkappa^{(2)}_{i'kl}(\omega_2,\mathbf{k}_2;\omega_3,\mathbf{k}_3)\\ &\quad-\varkappa^{(3)}_{ijkl}(\omega_1,\mathbf{k}_1;\omega_2,\mathbf{k}_2;\omega_3,\mathbf{k}_3)\}\,E^{(1)}_j(\omega_1,\mathbf{k}_1)\,E^{(1)}_k(\omega_2,\mathbf{k}_2)\,E^{(1)}_l(\omega_3,\mathbf{k}_3)\\ &\quad-i\frac{\partial}{\partial\omega}\Lambda_{ij}(\omega,\mathbf{k})\frac{\partial}{\partial t''}E^{(1)}_j(\omega,\mathbf{k}).\end{aligned} \tag{2.104}$$

We substitute the t''-dependent linear amplitudes (2.99) in (2.104) and suppose that the frequencies fit the resonance condition

$$|\omega_{\mathbf{k}_1} \pm \omega_{\mathbf{k}_2} \pm \omega_{\mathbf{k}_3}| = \omega_{\mathbf{k}_3}. \tag{2.105}$$

Then, equating to zero the resonant right-hand part of (2.104), we obtain an equation that determines the time-evolution of the amplitude $\mathbf{E}_{\mathbf{k}}(t'')$:

$$\frac{\partial}{\partial \omega_k} \Lambda_{ij}(\omega_{\mathbf{k}}, \mathbf{k}) \frac{\partial}{\partial t''} E_{j\mathbf{k}}\, e^{-i\phi_{\mathbf{k}}}$$

$$= -\frac{i}{4} \sum_{\mathbf{k}_1+\mathbf{k}_2+\mathbf{k}_3=\mathbf{k}}{}' \{2\varkappa^{(2)}_{ijj'}(\omega_{\mathbf{k}_1}, \mathbf{k}_1; \omega_{\mathbf{k}_2}+\omega_{\mathbf{k}_3}, \mathbf{k}_2+\mathbf{k}_3)\, \Lambda^{-1}_{j'i'}(\omega_{\mathbf{k}_2}+\omega_{\mathbf{k}_3}, \mathbf{k}_2+\mathbf{k}_3)\, \varkappa^{(2)}_{i'kl}(\omega_{\mathbf{k}_2}, \mathbf{k}_2; \omega_{\mathbf{k}_3}, \mathbf{k}_3)$$

$$- \varkappa^{(3)}_{ijkl}(\omega_{\mathbf{k}_1}, \mathbf{k}_1; \omega_{\mathbf{k}_2}, \mathbf{k}_2; \omega_{\mathbf{k}_3}, \mathbf{k}_3)\} E_{j\mathbf{k}_1} E_{k\mathbf{k}_2} E_{l\mathbf{k}_3}\, e^{-i(\phi_{k_1}+\phi_{k_2}+\phi_{k_3})}. \tag{2.106}$$

The prime at the sum symbol indicates that we must take into account all wave combinations which satisfy (2.105).

In the absence of four-wave resonances we must take into account the higher-order non-linear wave interactions. In a manner similar to the above analysis we may derive equations that determine the time-evolution of the field amplitude for the multiwave resonances. Such equations, together with (2.101) and (2.106), make a hierarchy that determines the time-development of the field amplitude on the whole hierarchy of times. Later on we shall make use of these equations to describe a variety of effects which are associated with the non-linear wave interactions in plasmas.

CHAPTER 3

Electrodynamic Properties of a Magneto-active Plasma

3.1. Longitudinal Dielectric Permittivity and Non-linear Susceptibilities of a Plasma in the Presence of a Magnetic Field

Solution of the kinetic equation

In the previous chapter we derived the non-linear equations which provide an adequate description of plasmas both with and without external fields. The influence of the latter on the electrodynamic plasma properties is very significant, being reflected, for instance, in the modified dielectric permittivity and non-linear plasma susceptibilities.

Especially peculiar is the behaviour of a plasma in an external magnetic field. The presence of the magnetic field makes the plasma properties anisotropic and causes the gyrotropy which manifests itself in its optical activity (for this reason a plasma with an external magnetic field is called magneto-active). In view of future applications we shall consider only the simplest and the most interesting case: we shall study the electrodynamic properties of a plasma in an external constant and uniform magnetic field.†

First of all let us find the dielectric permittivity and the non-linear electric susceptibilities of a magneto-active plasma with a potential self-consistent field. All considerations are analogous to the analysis of the field-free plasma. Starting from the kinetic equation (1.55) we obtain instead of (2.1) the following equation, which governs the distribution function perturbation due to the self-consistent field in the presence of the magnetic field:

$$-i[\omega-(\mathbf{k}\cdot\mathbf{v})]f_{\mathbf{k}\omega}-\omega_B\frac{\partial f_{\mathbf{k}\omega}}{\partial\varphi}+\frac{e}{m}\left(\mathbf{E}_{\mathbf{k}\omega}\cdot\frac{\partial f_0}{\partial\mathbf{v}}\right)+\frac{e}{m}\sum_{\omega',\mathbf{k}'}\left(\mathbf{E}_{\mathbf{k}'\omega'}\cdot\frac{\partial f_{\mathbf{k}-\mathbf{k}',\,\omega-\omega'}}{\partial\mathbf{v}}\right)=0. \quad (3.1)$$

Here $\omega_B=\dfrac{eB_0}{mc}$ is the cyclotron frequency and φ is the azimuth of the vector $\mathbf{v}$ in the plane normal to the external magnetic field $\mathbf{B}_0$. (Equation (3.1) involves the assumption of an isotropic initial distribution function f_0 with respect to the velocity directions in

† The linear electrodynamic properties of magneto-active plasmas have been investigated in detail by many authors (Silin and Rukhadze, 1961; Allis, Buchsbaum, and Bers, 1963; Shafranov, 1963; Sitenko and Kirochkin, 1966; Sitenko, 1967; Ginzburg and Rukhadze, 1970; Akhiezer, Akhiezer, Polovin, Sitenko, and Stepanov, 1975). The non-linear case is discussed by Pustovalov and Silin (1972a).

the plane, perpendicular to the external field, i.e. that f_0 is a function of the longitudinal ($v_{||}$) and the transverse ($v_\perp$) velocity components relative to the external field $\mathbf{B}_0$ and does not depend on the azimuth in velocity space.)

Neglecting the non-linear terms in (3.1) we obtain

$$-i[\omega-(\mathbf{k}\cdot\mathbf{v})]f_{\mathbf{k}\omega}-\omega_B\frac{\partial f_{\mathbf{k}\omega}}{\partial\varphi}+\frac{e}{m}\left(\mathbf{E}_{\mathbf{k}\omega}\cdot\frac{\partial f_0}{\partial\mathbf{v}}\right)=0. \tag{3.2}$$

By contrast to the linear part of (2.1) that reduces to an algebraic equation for $f_{\mathbf{k}\omega}$, (3.2) is a differential equation in velocity space. Integration of (3.2) yields

$$f^{(1)}_{\mathbf{k}\omega}(\mathbf{v})=\frac{e}{m\omega_B}\exp\left[-\frac{i}{\omega_B}\int_0^{\varphi}d\varphi\{\omega-(\mathbf{k}\cdot\mathbf{v})\}\right]\cdot\int^{\varphi}d\varphi\exp\left[\frac{i}{\omega_B}\int_0^{\varphi}d\varphi\{\omega-(\mathbf{k}\cdot\mathbf{v})\}\right]\left(\mathbf{E}_{\mathbf{k}\omega}\cdot\frac{\partial f_0}{\partial\mathbf{v}}\right), \tag{3.3}$$

where the value of the integral must be taken only at the upper limit. It may be easily verified that $f_{\mathbf{k}\omega}(\varphi+2\pi)=f_{\mathbf{k}\omega}(\varphi)$.

Let us choose the coordinate system so that $k_y=0$. Then

$$\int_0^{\varphi}d\varphi[\omega-(\mathbf{k}\cdot\mathbf{v})]=(\omega-k_{||}v_{||})\varphi-k_\perp v_\perp\sin\varphi.$$

Substituting this into (3.3) and expanding the integrand according to

$$e^{ia\sin\varphi}=\sum_{n=-\infty}^{\infty}J_n(a)\,e^{-in\varphi}, \tag{3.4}$$

$\left(J_n(a)\text{ is the }n\text{th order Bessel function, }a=\dfrac{k_\perp v_\perp}{\omega_B}\right)$, we find

$$f^{(1)}_{k\omega}(v)=-i\frac{e}{mk}\sum_n J_n(a)\frac{e^{ia\sin\varphi-in\varphi}}{\omega-k_{||}v_{||}-n\omega_B+i0}\left(\frac{n\omega_B}{v_\perp}\frac{\partial}{\partial v_\perp}+k_{||}\frac{\partial}{\partial v_{||}}\right)f_0(v_\perp,v_{||})E_{\mathbf{k}\omega}. \tag{3.5}$$

Evidently, if we had considered an arbitrary coordinate system, φ would denote the difference between the azimuths of the vectors $\mathbf{v}$ and $\mathbf{k}$.

Making use of (3.5) we can find the induced charge density (2.8) and, hence, the linear plasma susceptibility $\varkappa^{(1)}(\omega,\mathbf{k})$.

Thus within the context of (2.15) we obtain the following expression for the longitudinal dielectric permittivity of a magneto-active plasma:

$$\varepsilon(\omega,\mathbf{k})=1+\sum\frac{4\pi e^2}{mk^2}\int d\mathbf{v}\sum_n\frac{J_n^2(a)}{\omega-k_{||}v_{||}-n\omega_B+i0}\left(\frac{n\omega_B}{v_\perp}\frac{\partial}{\partial v_\perp}+k_{||}\frac{\partial}{\partial v_{||}}\right)f_0(v_\perp,v_{||}), \tag{3.6}$$

where $a=\dfrac{k_\perp v_\perp}{\omega_B}$ and $k^2=k_\perp^2+k_{||}^2$.

The second and the third iterations of (3.1),

$$f^{(2)}_{\mathbf{k}\omega}(v) = \frac{e}{m\omega_B} \exp\left[-\frac{i}{\omega_B}\int_0^{\varphi} d\varphi\{\omega-(\mathbf{k}\cdot\mathbf{v})\}\right]\cdot\int^{\varphi} d\varphi \exp\left[\frac{i}{\omega_B}\int_0^{\varphi} d\varphi\{\omega-(\mathbf{k}\cdot\mathbf{v})\}\right]$$

$$\times \sum_{\substack{\omega_1+\omega_2=\omega \\ \mathbf{k}_1+\mathbf{k}_2=\mathbf{k}}} \left(\mathbf{E}_{\mathbf{k}_2\omega_1}\cdot\frac{\partial f^{(1)}_{\mathbf{k}_1\omega_2}(v)}{\partial\mathbf{v}}\right),$$

and so on, serve to derive the following non-linear longitudinal electric susceptibilities of a plasma in an external magnetic field:

$$\varkappa^{(2)}(\omega_1,\mathbf{k}_1;\omega_2,\mathbf{k}_2) = \sum\frac{(-i)}{2}\frac{4\pi e^2}{m}\frac{e}{m}\frac{1}{k_1k_2k}\int d\mathbf{v}\sum_n\frac{e^{in\varphi-ia\sin\varphi}}{\omega-k_{\|}v_{\|}-n\omega_B+i0}J_n(a)$$

$$\times\left\{\left(\mathbf{k}_1\cdot\frac{\partial}{\partial\mathbf{v}}\right)\sum_{n_2}\frac{e^{-in_2(\varphi-\varphi_2)+ia_2\sin(\varphi-\varphi_2)}}{\omega_2-k_{2\|}v_{\|}-n_2\omega_B+i0}J_{n_2}(a_2)\left(\frac{n_2\omega_B}{v_\perp}\frac{\partial}{\partial v_\perp}+k_{2\|}\frac{\partial}{\partial v_{\|}}\right)\right.$$

$$\left.+\left(\mathbf{k}_2\cdot\frac{\partial}{\partial\mathbf{v}}\right)\sum_{n_1}\frac{e^{-in_1(\varphi-\varphi_1)+ia_1\sin(\varphi-\varphi_1)}}{\omega_1-k_{1\|}v_{\|}-n_1\omega_B+i0}J_{n_1}(a_1)\left(\frac{n_1\omega_B}{v_\perp}\frac{\partial}{\partial v_\perp}+k_{1\|}\frac{\partial}{\partial v_{\|}}\right)\right\}f_0(v_\perp,v_{\|}), \tag{3.7}$$

where $\omega = \omega_1+\omega_2$, $\mathbf{k} = \mathbf{k}_1+\mathbf{k}_2$;

$$\varkappa^{(3)}(\omega_1,\mathbf{k}_1;\omega_2,\mathbf{k}_2;\omega_3,\mathbf{k}_3) = \tfrac{1}{3}\{\bar{\varkappa}^{(3)}(\omega_1,\mathbf{k}_1;\omega_2,\mathbf{k}_2;\omega_3,\mathbf{k}_3)$$
$$+\bar{\varkappa}^{(3)}(\omega_2,\mathbf{k}_2;\omega_1,\mathbf{k}_1;\omega_3,\mathbf{k}_3)+\bar{\varkappa}^{(3)}(\omega_3,\mathbf{k}_3;\omega_2,\mathbf{k}_2;\omega_1,\mathbf{k}_1)\}, \tag{3.8}$$

$$\bar{\varkappa}^{(3)}(\omega_1,\mathbf{k}_1;\omega_2,\mathbf{k}_2;\omega_3,\mathbf{k}_3) = \sum\frac{(-i)^2}{2}\frac{4\pi e^2}{m}\left(\frac{e}{m}\right)\frac{1}{k_1k_2k_3k}\int d\mathbf{v}\frac{e^{in\varphi-ia\sin\varphi}}{\omega-k_{\|}v_{\|}-n\omega_B+i0}J_n(a)$$

$$\times\mathbf{k}_1\frac{\partial}{\partial\mathbf{v}}\sum_{n'}\frac{e^{-in'(\varphi-\varphi')+ia'\sin(\varphi-\varphi')}}{\omega'-k'_{\|}v_{\|}-n'\omega_B+i0}\int_0^{2\pi}\frac{d\varphi}{2\pi}e^{in'(\varphi-\varphi')-ia'\sin(\varphi-\varphi')}$$

$$\times\left\{\mathbf{k}_2\frac{\partial}{\partial\mathbf{v}}\sum_{n_3}\frac{e^{-in_3(\varphi-\varphi_3)+ia_3\sin(\varphi-\varphi_3)}}{\omega_3-k_{3\|}v_{\|}-n_3\omega_B+i0}J_{n_3}(a_3)\left(\frac{n_3\omega_B}{v_\perp}\frac{\partial}{\partial v_\perp}+k_{3\|}\frac{\partial}{\partial v_{\|}}\right)\right.$$

$$\left.+\mathbf{k}_3\frac{\partial}{\partial\mathbf{v}}\sum_{n_2}\frac{e^{-in_2(\varphi-\varphi_2)+ia_2\sin(\varphi-\varphi_2)}}{\omega_2-k_{2\|}v_{\|}-n_2\omega_B+i0}J_{n_2}(a_2)\left(\frac{n_2\omega_B}{v_\perp}\frac{\partial}{\partial v_\perp}+k_{2\|}\frac{\partial}{\partial v_{\|}}\right)f_0(v_\perp,v_{\|}),\right\} \tag{3.9}$$

where $\omega = \omega_1+\omega_2+\omega_3$, $\mathbf{k} = \mathbf{k}_1+\mathbf{k}_2+\mathbf{k}_3$, $\omega' = \omega_2+\omega_3$, $\mathbf{k}' = \mathbf{k}_2+\mathbf{k}_3$.

Here φ is the azimuth of the vector $\mathbf{v}$ and $\varphi_1, \varphi_2, \varphi_3, \varphi'$ are those of the vectors $\mathbf{k}_1, \mathbf{k}_2, \mathbf{k}_3$, and $\mathbf{k}'$ in the coordinate system with the z-axis along the magnetic field $\mathbf{B}_0$ and the x-axis in the $(\mathbf{k}, \mathbf{B}_0)$ plane.

The dielectric permittivity (3.6) and the non-linear susceptibilities (3.7) and (3.8) completely describe all linear and non-linear electrodynamic properties of a magneto-active plasma similarly to (2.15)–(2.18) for a plasma without external fields. It may be immediately verified that the expressions (3.6), (3.7), and (3.9) reduce as $B_0 \to 0$ to (2.15), (2.16), and (2.18) respectively.

The dielectric permittivity of an equilibrium plasma

Integration by parts reduces (3.6) to the following:

$$\varepsilon(\omega, \mathbf{k}) = 1 - \sum \frac{4\pi e^2}{mk^2} \sum_n \int d\mathbf{v} f_0(v_\perp, v_{||}) \left(\frac{n\omega_B}{v_\perp} \frac{\partial}{\partial v_\perp} + k_{||} \frac{\partial}{\partial v_{||}} \right) \frac{J_n^2(a)}{\omega - k_{||}v_{||} - n\omega_B + i0} \quad (3.10)$$

and for a cold plasma:

$$\varepsilon(\omega, \mathbf{k}) = 1 - \sum \Omega^2 \left(\frac{\sin^2 \vartheta}{\omega^2 - \omega_B^2} + \frac{\cos^2 \vartheta}{\omega^2} \right), \quad (3.11)$$

where ϑ is the angle between the wave vector $\mathbf{k}$ and the magnetic field $\mathbf{B}_0$. As in the field-free case, the dielectric permittivity of a cold plasma (3.11) does not depend on the wave vector, i.e. there is no spatial dispersion in cold plasmas either with or without the magnetic field. However, the dielectric permittivity (3.11) depends on the angle between the wave vector and the magnetic field, so a magneto-active plasma is anisotropic even when the thermal motion of particles is neglected.

In the case of an equilibrium plasma we carry out in (3.10) the integration procedure with respect to the velocity components transverse to the magnetic field. Taking a Maxwellian distribution for $f_0(v)$ and making use of the relation

$$\int_0^\infty dt\, t\, e^{-p^2t^2} J_n^2(at) = \frac{1}{2p^2} e^{-\frac{a^2}{2p^2}} I_n\left(\frac{a^2}{2p^2}\right), \quad (3.12)$$

where $I_n(x)$ is a modified Bessel function, we obtain the dielectric permittivity of the equilibrium magneto-active plasma in the following form (Sitenko and Stepanov, 1957):

$$\varepsilon(\omega, \mathbf{k}) = 1 + \sum \frac{1}{a^2k^2} \left\{ 1 - e^{-\beta} \sum_n I_n(\beta) \frac{z_0}{z_n} \left[\varphi(z_n) - i\sqrt{\pi} z_n e^{-z_n^2} \right] \right\}, \quad (3.13)$$

where

$$\beta = \frac{k_\perp^2 s^2}{3\omega_B^2}, \quad \omega_B = \frac{eB_0}{mc}, \quad s^2 = \frac{3T}{m},$$

$$a^2 = \frac{T}{4\pi e^2 n_0}, \quad z_n = \sqrt{\frac{3}{2}} \frac{\omega - n\omega_B}{|k_{||}| s}. \quad (3.14)$$

The dispersion function $\varphi(z)$ was defined by (2.28). Note, that (3.13) reduces to (2.27) as $k_\perp = 0$.

The dielectric permittivity of a plasma with an anisotropic velocity distribution

Consider the dielectric permittivity of a magneto-active plasma with an anisotropic velocity distribution. Let the temperatures of the transverse and longitudinal velocity distributions $T_\perp$ and $T_{||}$ be different:

$$f_0(v_\perp, v_{||}) = n_0 \left(\frac{m}{2\pi T_\perp} \right)^{3/2} \sqrt{\frac{T_\perp}{T_{||}}}\, e^{-\frac{mv_\perp^2}{2T_\perp} - \frac{mv_{||}^2}{2T_{||}}}. \quad (3.15)$$

Then by virtue of (3.6) we have

$$\varepsilon(\omega, \mathbf{k}) = 1+\sum \frac{1}{a^2k^2}\left\{1-e^{-\beta_\perp}\sum_n I_n(\beta_\perp)\left[1-\frac{n\omega_B}{\omega}\left(1-\frac{T_{||}}{T_\perp}\right)\right]\frac{z_0^{||}}{z_n^{||}}\left[\varphi(z_n^{||})-i\sqrt{\pi}z_n^{||}e^{-z_n^{||2}}\right]\right\}, \tag{3.16}$$

where

$$\beta_\perp = \frac{k_\perp^2 s_\perp^2}{2\omega_B}, \quad s_\perp^2 = \frac{2T_\perp}{m}, \quad s_{||}^2 = \frac{T_{||}}{m},$$

$$a_{||}^2 = \frac{T_{||}}{4\pi e^2 n_0}, \quad z_n^{||} = \frac{1}{\sqrt{2}}\frac{\omega-n\omega_B}{|k_{||}|\, s_{||}}. \tag{3.17}$$

Non-linear susceptibilities of a cold plasma

The susceptibilities (3.7) and (3.9) may be integrated by parts. The result is as follows:

$$\varkappa^{(2)}(\omega_1, \mathbf{k}_1; \omega_2, \mathbf{k}_2) = \sum \frac{(-i)}{2}\frac{4\pi e^2}{m}\frac{e}{m}\frac{1}{k_1k_2k}$$

$$\times\left\{\left(\sum_{n, n_2}\int d\mathbf{v} f_0(v_\perp, v_{||})\left(\frac{n_2\omega_B}{v_\perp}\frac{\partial}{\partial v_\perp}+k_{2||}\frac{\partial}{\partial v_{||}}\right)\right.\right.$$

$$\times\left.\left.\frac{e^{-in_2(\varphi-\varphi_2)+ia_2\sin(\varphi-\varphi_2)}}{\omega_2-k_{2||}v_{||}-n_2\omega_B+i0}J_{n_2}(a_2)\left(\mathbf{k}_1\cdot\frac{\partial}{\partial \mathbf{v}}\right)\frac{e^{in\varphi-ia\sin\varphi}}{\omega-k_{||}v_{||}-n\omega_B+i0}J_n(a)\right)+(1\rightleftarrows 2)\right\}. \tag{3.18}$$

$$\bar{\varkappa}^{(3)}(\omega_1, \mathbf{k}_1; \omega_2, \mathbf{k}_2; \omega_3, \mathbf{k}_3) = \sum\frac{1}{2}\frac{4\pi e^2}{m}\left(\frac{e}{m}\right)^2\frac{1}{k_1k_2k_3k}\left\{\left(\sum_{n, n', n_3}\int d\mathbf{v} f_0(v_\perp, v_{||})\right.\right.$$

$$\times\left(\frac{n_3\omega_B}{v_\perp}\frac{\partial}{\partial v_\perp}+k_{3||}\frac{\partial}{\partial v_{||}}\right)\cdot\frac{e^{-in_3(\varphi-\varphi_3)+ia_3\sin(\varphi-\varphi_3)}}{\omega_3-k_{3||}v_{||}-n_3\omega_B+i0}J_{n_3}(a_3)\left(\mathbf{k}_2\cdot\frac{\partial}{\partial \mathbf{v}}\right)e^{in'(\varphi-\varphi')-ia'\sin(\varphi-\varphi')}$$

$$\times\left.\left.\int_0^{2\pi}\frac{d\varphi}{2\pi}\frac{e^{-in'(\varphi-\varphi')+ia\sin(\varphi-\varphi')}}{\omega'-k'_{||}v_{||}-n'\omega_B+i0}\left(\mathbf{k}_1\cdot\frac{\partial}{\partial \mathbf{v}}\right)\frac{e^{in\varphi-ia\sin\varphi}}{\omega-k_{||}v_{||}-n\omega_B+i0}J_n(a)\right)+(2\rightleftarrows 3)\right\}. \tag{3.19}$$

In a cold plasma (3.18) reduces to

$$\varkappa^{(2)}(\omega_1, \mathbf{k}_1; \omega_2, \mathbf{k}_2) = \sum\frac{(-i)}{2}\frac{e}{m}\frac{\Omega^2}{\omega_1\omega_2\omega}\frac{1}{k_1k_2k}\{\omega(\mathbf{k}\Gamma(\omega_1)\mathbf{k}_1)(\mathbf{k}\Gamma(\omega_2)\mathbf{k}_2)$$

$$-\omega_1(\mathbf{k}\Gamma(\omega)\mathbf{k}_1)(\mathbf{k}_1\Gamma(\omega_2)\mathbf{k}_2)-\omega_2(\mathbf{k}\Gamma(\omega)\mathbf{k}_2)(\mathbf{k}_2\Gamma(\omega_1)\mathbf{k}_1)\}, \tag{3.20}$$

where

$$(\mathbf{k}_1\Gamma(\omega)\mathbf{k}_2) \equiv k_{1i}\Gamma_{ij}(\omega)k_{2j} \equiv \frac{1}{2}\left(\frac{1}{\omega-\omega_B}+\frac{1}{\omega+\omega_B}\right)k_{1\perp}k_{2\perp}\cos(\varphi_2-\varphi_1)$$

$$+\frac{i}{2}\left(\frac{1}{\omega-\omega_B}-\frac{1}{\omega+\omega_B}\right)k_{1\perp}k_{2\perp}\sin(\varphi_2-\varphi_1)+\frac{1}{\omega}k_{1||}k_{2||}. \tag{3.21}$$

The tensor $\Gamma_{ij}(\omega)$ may be written through the unit vector $\mathbf{b}$ directed along the magnetic field, the symmetric tensor δ_{ij}, and the fully antisymmetric tensor ε_{ijk}:

$$\Gamma_{ij}(\omega) = \frac{1}{2}\left(\frac{1}{\omega-\omega_B}+\frac{1}{\omega+\omega_B}\right)(\delta_{ij}-b_ib_j)-\frac{i}{2}\left(\frac{1}{\omega-\omega_B}-\frac{1}{\omega+\omega_B}\right)\varepsilon_{ijk}b_k+\frac{1}{\omega}b_ib_j. \quad (3.22)$$

Making use of (3.19) we obtain the symmetrized third-order susceptibility of a cold magneto-active plasma:

$$\begin{aligned}
&\varkappa^{(3)}(\omega_1, \mathbf{k}_1; \omega_2, \mathbf{k}_2; \omega_3, \mathbf{k}_3) \\
&\quad = \sum \frac{1}{6}\frac{e^2}{m^2}\frac{\Omega^2}{\omega_1\omega_2\omega_3\omega}\frac{1}{k_1k_2k_3k}\Bigg\{\omega(\mathbf{k}\Gamma(\omega_1)\mathbf{k}_1)(\mathbf{k}\Gamma(\omega_2)\mathbf{k}_2)(\mathbf{k}\Gamma(\omega_3)\mathbf{k}_3) \\
&\quad +\Bigg(\omega_1\Bigg[(\mathbf{k}_1\Gamma(\omega_2)\mathbf{k}_2)(\mathbf{k}_1\Gamma(\omega_3)\mathbf{k}_3)+\frac{\omega_2}{\omega'}(\mathbf{k}_1\Gamma(\omega')\mathbf{k}_2)(\mathbf{k}_2\Gamma(\omega_3)\mathbf{k}_3) \\
&\quad +\frac{\omega_3}{\omega'}(\mathbf{k}_1\Gamma(\omega')\mathbf{k}_3)(\mathbf{k}_3\Gamma(\omega_2)\mathbf{k}_2)\Bigg](\mathbf{k}\Gamma(\omega)\mathbf{k}_1)-\frac{\omega}{\omega'}[\omega_2(\mathbf{k}\Gamma(\omega')\mathbf{k}_2)(\mathbf{k}_2\Gamma(\omega_3)\mathbf{k}_3) \\
&\quad +\omega_3(\mathbf{k}\Gamma(\omega')\mathbf{k}_3)(\mathbf{k}_3\Gamma(\omega_2)\mathbf{k}_2)](\mathbf{k}\Gamma(\omega_1)\mathbf{k}_1)\Bigg)+(1 \rightleftarrows 2)+(1 \rightleftarrows 3)\Bigg\}. \qquad (3.23)
\end{aligned}$$

The non-linear susceptibilities (3.20) and (3.23) do not depend on the wave vectors of the interacting fields at $T = 0$. As $\mathbf{B}_0 \to 0$ (3.20) and (3.23) reduce to (2.30) and (2.39), respectively.

The non-symmetrized susceptibility $\bar{\varkappa}'^{(3)}(\omega_1, \mathbf{k}_1; \omega_2, \mathbf{k}_2; \omega_3, \mathbf{k}_3)$ of a cold magneto-active plasma may be written as:

$$\begin{aligned}
&\bar{\varkappa}'^{(3)}(\omega_1, \mathbf{k}_1; \omega_2, \mathbf{k}_2; \omega_3, \mathbf{k}_3) = \sum \frac{1}{2}\frac{e^2}{m^2}\frac{\Omega^2}{\omega_1\omega_2\omega_3\omega}\frac{1}{k_1k_2k_3k} \\
&\times\Bigg\{\Bigg(\frac{\omega_1\omega_2}{\omega'}\Bigg[(\mathbf{k}[\Gamma(\omega_1)-\Gamma(\omega)]\mathbf{k}_1)\Big(\mathbf{k}_1[\Gamma(\omega_2)-\Gamma(\omega')]\frac{1}{2}\mathbf{k}_1+\mathbf{k}_2\Big) \\
&+\omega'(\mathbf{k}\Gamma(\omega)\mathbf{k}_1)(\mathbf{k}\Gamma(\omega)[\Gamma(\omega_2)-\Gamma(\omega')]\mathbf{k}_2)+\frac{\omega'}{\omega_1}(\mathbf{k}_1\Gamma(\omega_1)\mathbf{k}_1)\Big(\mathbf{k}[\Gamma(\omega_2)-\Gamma(\omega')]\mathbf{k}_2-\frac{1}{2}\mathbf{k}\Big) \\
&+\Big((\mathbf{k}'[\Gamma(\omega_2)-\Gamma(\omega')]\mathbf{k}_2)+\frac{\omega_3}{\omega_2}(\mathbf{k}_2\Gamma(\omega_2)\mathbf{k}_2)\Big)(\mathbf{k}\Gamma(\omega_1)\mathbf{k}_1)\Bigg](\mathbf{k}_2\Gamma(\omega_3)\mathbf{k}_3)\Bigg)+(2 \rightleftarrows 3) \\
&+\frac{1}{3}\Bigg(\Bigg[\Bigg\{\omega_2\Big(\frac{1}{2}(\mathbf{k}_1\Gamma(\omega_1)\mathbf{k}_1)(\mathbf{k}[\Gamma(\omega_2)-\Gamma(\omega')]\mathbf{k}) \\
&-\frac{1}{2}\frac{\omega_1}{\omega'}(\mathbf{k}[\Gamma(\omega_1)-\Gamma(\omega)]\mathbf{k}_1)(\mathbf{k}_1[\Gamma(\omega_2)-\Gamma(\omega')]\mathbf{k}_1) \\
&-\omega_1(\mathbf{k}_1\Gamma(\omega_1)\mathbf{k}_1)(\mathbf{k}\Gamma(\omega)[\Gamma(\omega_2)-\Gamma(\omega')]\mathbf{k}_2)\Big)(\mathbf{k}_2\Gamma(\omega_3)\mathbf{k}_3)\Bigg\}+\{2 \rightleftarrows 3\} \\
&+\omega_1(\mathbf{k}\Gamma(\omega_1)\mathbf{k}_1)(\mathbf{k}_2\Gamma(\omega_2)\mathbf{k}_2)(\mathbf{k}_3\Gamma(\omega_3)\mathbf{k}_3)\Bigg]+[1 \rightleftarrows 2]+[1 \rightleftarrows 3]\Bigg)\Bigg\}. \qquad (3.24)
\end{aligned}$$

This expression differs from (3.19) at $T = 0$ by the symmetrization (with respect to the permutations $(1 \rightleftarrows 2)$ and $(1 \rightleftarrows 3)$) of the part of (3.19) that is non-singular at $\omega' \equiv \omega_2+\omega_3 = 0$.

Expressions (3.20) and (3.23) may be derived as well from the hydrodynamical equations with an external magnetic field. The hydrodynamical expression for the non-symmetrized susceptibility $\bar{\varkappa}^{(3)}(\omega_1, \mathbf{k}_1; \omega_2, \mathbf{k}_2; \omega_3, \mathbf{k}_3)$ is the following:

$$\bar{\varkappa}^{(3)}(\omega_1, \mathbf{k}_1; \omega_2, \mathbf{k}_2; \omega_3, \mathbf{k}_3) = \sum \frac{1}{2} \frac{e^2}{m^2} \frac{\Omega^2}{\omega_1\omega_2\omega_3\omega} \frac{1}{k_1k_2k_3k}$$

$$\times \Bigg\{ \Bigg(\omega_2 \Bigg[\frac{\omega_1}{\omega'} (\mathbf{k}\Gamma(\omega_1)\mathbf{k}_1)(\mathbf{k}'\Gamma(\omega_2)\mathbf{k}_2)(\mathbf{k}_3\Gamma(\omega_3)\mathbf{k}_3) + \Bigg\{ \frac{\omega_1}{\omega'} ((\mathbf{k}\Gamma(\omega_1)\mathbf{k}_1)(\mathbf{k}[\Gamma(\omega)-\Gamma(\omega')]\mathbf{k}_2)$$

$$-(\mathbf{k}\Gamma(\omega)\mathbf{k}_1)(\mathbf{k}_1[\Gamma(\omega_2)-\Gamma(\omega')]\mathbf{k}_2)) + (\mathbf{k}_1\Gamma(\omega_1)\mathbf{k}_1)(\mathbf{k}[\Gamma(\omega_2)-\Gamma(\omega')]\mathbf{k}_2)$$

$$+\omega_1(\mathbf{k}'\Gamma(\omega_1)\mathbf{k}_1)(\mathbf{k}\Gamma(\omega)[\Gamma(\omega_2)-\Gamma(\omega')]\mathbf{k}_2) \Bigg\} (\mathbf{k}_2\Gamma(\omega_3)\mathbf{k}_3) \Bigg] \Bigg) + (2 \rightleftarrows 3) \Bigg\}. \tag{3.25}$$

The symmetrization relative to the permutations $(1 \rightleftarrows 2)$ and $(1 \rightleftarrows 3)$ reduces (3.25) to (3.23). To find the thermal corrections to (3.20) and (3.23), the general formulas (3.18) and (3.19) are to be made use of.

3.2. The Dielectric Permittivity Tensor of a Magneto-active Plasma

Solution of the linearized general kinetic equation

Let us consider the electrodynamic properties of a magneto-active plasma making no special assumption of the potential nature of the self-consistent field (Sitenko and Stepanov, 1957). To do this we must introduce the dielectric permittivity tensor and the non-linear tensor susceptibilities in the same manner as we have done when considering the plasma in the absence of external fields.

The dielectric permittivity tensor and the non-linear tensor susceptibilities of a magneto-active plasma depend in an essential way on the magnetic field. In particular, the dielectric permittivity tensor of an isotropic plasma in an external magnetic field cannot be divided into longitudinal and transverse components, in contrast to an isotropic plasma when there are no external fields.

To find the dielectric permittivity of a plasma that is exposed to an external magnetic field we start from the kinetic equation (1.58) and the Maxwell set (1.59).

In the presence of an external field $\mathbf{B}_0$ the distribution function perturbation, which is caused by the self-consistent electromagnetic field, is by virtue of (1.58) governed by the following equation:

$$-i[\omega-(\mathbf{k}\cdot\mathbf{v})]f_{\mathbf{k}\omega} - \omega_B \frac{\partial f_{\mathbf{k}\omega}}{\partial\varphi} + \frac{e}{m}\left(\left\{E_{\mathbf{k}\omega} + \frac{1}{\omega}[\mathbf{v}[\mathbf{k}\mathbf{E}_{\mathbf{k}\omega}]]\right\}\cdot\frac{\partial f_0}{\partial\mathbf{v}}\right)$$

$$+\frac{e}{m}\sum_{\omega'\mathbf{k}' \neq \omega, \mathbf{k}} \left(\left\{\mathbf{E}_{\mathbf{k}'\omega'} + \frac{1}{\omega'}[\mathbf{v}[\mathbf{k}'\mathbf{E}_{\mathbf{k}'\omega'}]]\right\}\cdot\frac{\partial}{\partial\mathbf{v}}\right) f_{\mathbf{k}-\mathbf{k}', \omega-\omega'} = 0. \tag{3.26}$$

We shall solve this equation by means of successive approximations, similarly to the analysis of the longitudinal case. It is not difficult to expand the field polarization in a series in the electric field strength. Thus, the coefficients of this expansion will determine the tensor susceptibilities of the plasma.

The linear solution of (3.26) is of the form

$$f_{\mathbf{k}\omega}^{(1)}(\mathbf{v}) = \frac{e}{m\omega_B} \exp\left[-\frac{i}{\omega_B}\int_0^{\varphi} d\varphi\{\omega-(\mathbf{k}\cdot\mathbf{v})\}\right] \times \int^{\varphi} d\varphi \exp\left[\frac{i}{\omega_B}\int_0^{\varphi} d\varphi\{\omega-(\mathbf{k}\cdot\mathbf{v})\}\right]\left(\left\{\mathbf{E}_{\mathbf{k}\omega}+\frac{1}{\omega}[\mathbf{v}[\mathbf{k}\mathbf{E}_{\mathbf{k}\omega}]]\right\}\cdot\frac{\partial f_0}{\partial\mathbf{v}}\right). \tag{3.27}$$

After an expansion in a Bessel series (as in (3.4)) and integration, we have:

$$f_{\mathbf{k}\omega}^{(1)}(\mathbf{v}) = -i\frac{e}{m\omega}e^{ia\sin(\varphi-\varphi_{\mathbf{k}})}\sum_n \frac{e^{-in(\varphi-\varphi_k)}}{\omega-k_{||}v_{||}-n\omega_B+i0}\left(\mathbf{E}_{\mathbf{k}\omega}\cdot Z^{(n)}\right)f_0, \tag{3.28}$$

where φ and $\varphi_{\mathbf{k}}$ are the azimuths of the vectors $\mathbf{v}$ and $\mathbf{k}$ (relative to an arbitrary reference point) in the plane perpendicular to the magnetic field $\mathbf{B}_0$, $\omega_B = \frac{eB_0}{mc}$ and $a = \frac{k_\perp v_\perp}{\omega_B}$. $\mathbf{Z}^{(n)}$ denotes the following differential operator vector:

$$Z_i^{(n)}(\omega, \mathbf{k}; \mathbf{v}) \equiv X_i^{(n)*}\left(\frac{\omega-k_{||}v_{||}}{v_\perp}\frac{\partial}{\partial v_\perp}+k_{||}\frac{\partial}{\partial v_{||}}\right) - \delta_{iz}J_n(a)(\omega-k_{||}v_{||}-n\omega_B)\left(\frac{v_{||}}{v_\perp}\frac{\partial}{\partial v_\perp}-\frac{\partial}{\partial v_{||}}\right), \tag{3.29}$$

where the components $X_i^{(n)}$ make the following column:

$$X_i^{(n)}(\mathbf{k}_\perp; \mathbf{v}) \equiv \begin{pmatrix} \frac{n\omega_B}{k_\perp}J_n(a)\cos\varphi_{\mathbf{k}}+iv_\perp J_n'(a)\sin\varphi_{\mathbf{k}} \\ \frac{n\omega_B}{k_\perp}J_n(a)\sin\varphi_{\mathbf{k}}-iv_\perp J_n'(a)\cos\varphi_{\mathbf{k}} \\ v_{||}J_n(a) \end{pmatrix}. \tag{3.30}$$

An alternative form of $Z_i^{(n)}$ may be written as well:

$$Z_i^{(n)}(\omega, \mathbf{k}; \mathbf{v}) \equiv X_i^{(n)*}\left(\frac{n\omega_B}{v_\perp}\frac{\partial}{\partial v_\perp}+k_{||}\frac{\partial}{\partial v_{||}}\right) + \frac{\omega-k_{||}v_{||}-n\omega_B}{v_\perp}\begin{pmatrix} \left(\frac{n\omega_B}{k_\perp}J_n(a)\cos\varphi_{\mathbf{k}}-iv_\perp J_n'(a)\sin\varphi_{\mathbf{k}}\right)\frac{\partial}{\partial v_\perp} \\ \left(\frac{n\omega_B}{k_\perp}J_n(a)\sin\varphi_{\mathbf{k}}+iv_\perp J_n'(a)\cos\varphi_{\mathbf{k}}\right)\frac{\partial}{\partial v_\perp} \\ v_\perp J_n(a)\frac{\partial}{\partial v_{||}} \end{pmatrix}, \tag{3.31}$$

which is very suitable for the calculations.

The general form of the dielectric permittivity tensor

Making use of solution (3.28) we obtain the following general expression for the dielectric permittivity of a plasma in a constant uniform external magnetic field:

$$\varepsilon_{ij}(\omega, \mathbf{k}) = \left(1-\sum \frac{\Omega^2}{\omega^2}\right)\delta_{ij}$$

$$+\sum \frac{4\pi e^2}{m\omega^2}\sum_n \int d\mathbf{v} \frac{\Pi_{ij}^{(n)}(v)}{\omega-k_{\|}v_{\|}-n\omega_B+i0}\left(\frac{n\omega_B}{v_\perp}\frac{\partial}{\partial v_\perp}+k_{\|}\frac{\partial}{\partial v_{\|}}\right) f_0(v_\perp, v_{\|}), \quad (3.32)$$

where the tensor $\Pi_{ij}^{(n)}(\mathbf{v})$ is the following dyad:

$$\Pi_{ij}^{(n)}(\mathbf{v}) \equiv X_i^{(n)} X_j^{(n)*}. \quad (3.33)$$

We can easily find explicit expressions for the components of $\Pi_{ij}^{(n)}(\mathbf{v})$ within the context of the definition (3.30). It should be remembered here that the vector $X_i^{(n)}$ was introduced in a coordinate system with its z-axis directed along the external magnetic field $\mathbf{B}_0$, so that the tensor (3.33) is given in the same system. The explicit form of the tensor $\Pi_{ij}^{(n)}(\mathbf{v})$ becomes considerably simplified when written in a coordinate system with its x-axis in the $(\mathbf{k}, \mathbf{B}_0)$ plane. In such a case $\varphi_{\mathbf{k}} = 0$ and (3.33) reduces to:

$$\Pi_{ij}^{(n)}(\mathbf{v}) = \begin{pmatrix} \frac{n\omega_B}{k_\perp} J_n(a) \\ -iv_\perp J_n'(a) \\ v_{\|} J_n(a) \end{pmatrix}_i \begin{pmatrix} \frac{n\omega_B}{k_\perp} J_n(a) \\ iv_\perp J_n'(a) \\ v_{\|} J_n(a) \end{pmatrix}_j . \quad (3.34)$$

This means that $\Pi_{ij}^{(n)}(\mathbf{v})$ has the following matrix form:

$$\Pi_{ij}^{(n)}(\mathbf{v}) = \begin{pmatrix} \frac{n^2\omega_B^2}{k_\perp^2} J_n^2(a) & i\frac{n\omega_B}{k} v_\perp J_n(a) J_n'(a) & \frac{n\omega_B}{k_\perp} v_{\|} J_n^2(a) \\ -i\frac{n\omega_B}{k_\perp} v_\perp J_n(a) J_n'(a) & v_\perp^2 J_n'^2(a) & -iv_\perp v_{\|} J_n(a) J_n'(a) \\ \frac{n\omega_B}{k_\perp} v_{\|} J_n^2(a) & iv_\perp v_{\|} J_n(a) J_n'(a) & v_{\|}^2 J_n^2(a) \end{pmatrix}. \quad (3.35)$$

Thus, in a coordinate system, in which z is directed along the external magnetic field $\mathbf{B}_0$ and x lies in the $(\mathbf{k}, \mathbf{B}_0)$ plane, the dielectric permittivity tensor $\varepsilon_{ij}(\omega, \mathbf{k})$ is of the general form (3.32) with the matrix (3.35) for $\Pi_{ij}^{(n)}(\mathbf{v})$. The summation in (3.32) extends over all plasma components, the distributions of which depend only on the transverse and the longitudinal velocity components relative to the external magnetic field.

It is straightforward to show that the dielectric permittivity (3.32) of a magneto-active plasma satisfies the following symmetry condition:

$$\varepsilon_{ij}(\omega, \mathbf{k}; \mathbf{B}_0) = \varepsilon_{ji}(\omega, -\mathbf{k}; -\mathbf{B}_0). \quad (3.36)$$

An equilibrium plasma

The dielectric permittivity of an equilibrium plasma or a non-isothermal one with different temperatures of the Maxwellian particle distributions may be written as follows (Sitenko and Stepanov, 1957):

$$\varepsilon_{ij}(\omega, \mathbf{k}) = \delta_{ij} - \sum \frac{\Omega^2}{\omega^2}\left\{e^{-\beta}\sum_n \frac{z_0}{z_n}\pi_{ij}(z_n)\left[\varphi(z_n) - i\sqrt{\pi}z_n e^{-z_n^2}\right] - 2z_0^2 b_i b_j\right\}, \quad (3.37)$$

where $\mathbf{b}$ is a unit vector directed along $\mathbf{B}_0$,

$$\pi_{ij}(z_n) = \begin{pmatrix} \frac{n^2}{\beta} I_n & -in(I_n - I_n') & \frac{k_{||}}{|k_{||}|}\sqrt{\frac{2}{\beta}}\, nz_n I_n \\ in(I_n - I_n') & \left(\frac{n^2}{\beta} + 2\beta\right) I_n - 2\beta I_n' & i\frac{k_{||}}{|k_{||}|}\sqrt{2\beta}\, z_n(I_n - I_n') \\ \frac{k_{||}}{|k_{||}|}\sqrt{\frac{2}{\beta}}\, nz_n I_n & -i\frac{k_{||}}{|k_{||}|}\sqrt{2\beta}\, z_n(I_n - I_n') & 2z_n^2 I_n \end{pmatrix} \quad (3.38)$$

$I_n = I_n(\beta)$ and $I_n' = \frac{\partial I_n(\beta)}{\partial \beta}$ are, respectively, a modified Bessel function and its derivative $\left(\sqrt{\beta} = \frac{k_\perp s}{\sqrt{3}\omega_B},\ \omega_B = \frac{eB_0}{mc},\ \Omega^2 = \frac{4\pi e^2 n_0}{m},\ s^2 = \frac{3T}{m}\right)$, the definition of z_n is given by (3.14):

$$z_n = \sqrt{\frac{3}{2}}\,\frac{\omega - n\omega_B}{|k_{||}|s}.$$

A plasma with an anisotropic velocity distribution

The dielectric permittivity tensor of a plasma with different temperatures of the longitudinal component and the transverse one ($T_\perp \neq T_{||}$) may be written as:

$$\varepsilon_{ij}(\omega, \mathbf{k}) = \delta_{ij} - \sum \frac{\Omega^2}{\omega^2}\left\{e^{-\beta_\perp}\sum_n \pi_{ij}(z_n^{||})\left[1 - \frac{T_\perp}{T_{||}} + \left(\frac{z_0^{||}}{z_n^{||}} - 1 + \frac{T_\perp}{T_{||}}\right)\right.\right.$$
$$\left.\left.\times\left[\varphi(z_n^{||}) - i\sqrt{\pi}z_n^{||} e^{-z_n^{||2}}\right]\right] - 2\frac{T_{||}}{T_\perp} z_0^{||2} b_i b_j\right\}, \quad (3.39)$$

where

$$\pi_{ij}(z_n) = \begin{pmatrix} \frac{n^2}{\beta_\perp} I_n & -in(I_n - I_n') & \frac{k_{||}}{|k_{||}|}\sqrt{\frac{2}{\beta_\perp}}\sqrt{\frac{T_{||}}{T_\perp}}\, nz_n I_n \\ in(I_n - I_n') & \left(\frac{n^2}{\beta_\perp} + 2\beta_\perp\right) I_n - 2\beta_\perp I_n' & i\frac{k_{||}}{|k_{||}|}\sqrt{2\beta_\perp}\sqrt{\frac{T_{||}}{T_\perp}} \times z_n(I_n - I_n') \\ \frac{k_{||}}{|k_{||}|}\sqrt{\frac{2}{\beta_\perp}}\sqrt{\frac{T_{||}}{T_\perp}} \times nz_n I_n & -i\frac{k_{||}}{|k_{||}|}\sqrt{2\beta_\perp}\sqrt{\frac{T_{||}}{T_\perp}} \times z_n(I_n - I_n') & 2\frac{T_{||}}{T_\perp} z_n^2 I_n \end{pmatrix} \quad (3.40)$$

$\beta_\perp$, $s_\perp$, $s_{||}$, and z_n'' were defined in (3.17). It should be observed that longitudinal contraction reduces (3.39) to the longitudinal dielectric permittivity (3.16) of a plasma.

A cold magneto-active plasma

The dielectric permittivity of a magneto-active plasma in neglect of the thermal motion of particles (cold plasma) does not depend on the wave vector and is determined by the following well-known expression:

$$\varepsilon_{ij}(\omega) = \begin{pmatrix} \varepsilon_1 & -i\varepsilon_2 & 0 \\ i\varepsilon_2 & \varepsilon_1 & 0 \\ 0 & 0 & \varepsilon_3 \end{pmatrix}, \tag{3.41}$$

where

$$\varepsilon_1 = 1 - \sum \frac{\Omega^2}{\omega^2 - \omega_B^2}, \quad \varepsilon_2 = \sum \frac{\omega_B}{\omega} \frac{\Omega^2}{\omega^2 - \omega_B^2}, \quad \varepsilon_3 = 1 - \sum \frac{\Omega^2}{\omega^2}. \tag{3.42}$$

Again, the neglect of the thermal motion leads to the absence of spatial dispersion. Thus, (3.41) may be treated also as the long wave ($k \to 0$) limiting value of the plasma dielectric permittivity tensor (3.37). It becomes evident within the context of (3.41) that the magneto-active plasma remains anisotropic and gyrotropic even in the absence of spatial dispersion.

3.3. Non-linear Tensor Susceptibilities of a Magneto-active Plasma

The general case

To find the non-linear tensor susceptibilities of a magneto-active plasma we make use of (3.37) and the next few iterations of (3.26). The second iteration of (3.26) may be written as

$$f_{k\omega}^{(2)}(\mathbf{v}) = \frac{e}{m\omega_B} e^{ia \sin(\varphi - \varphi_{\mathbf{k}}) - i \frac{\omega - k_{||}v_{||}}{\omega_B} \varphi}$$
$$\times \int^{\varphi} d\varphi\, e^{-ia \sin(\varphi - \varphi_{\mathbf{k}}) + i \frac{\omega - k_{||}v_{||}}{\omega_B} \varphi} \sum_{\substack{\omega_1 + \omega_2 = \omega \\ \mathbf{k}_1 + \mathbf{k}_2 = \mathbf{k}}} \frac{1}{\omega_1} (\mathbf{E}_{\mathbf{k}_1\omega_1} \cdot \mathbf{Y}(\omega_1, \mathbf{k}_1; \mathbf{v})) f_{\mathbf{k}_2\omega_2}^{(1)}(\mathbf{v}), \tag{3.43}$$

where

$$Y_i(\omega_1, \mathbf{k}_1; \mathbf{v}) \equiv [\{\omega_1 - (\mathbf{k}_1 \cdot \mathbf{v})\}\delta_{ij} + k_{1j}v_i] \frac{\partial}{\partial v_j}; \tag{3.44}$$

relevant expressions may be written also for the third iteration and so on. The intermediate algebra is simple but awkward, so we give here only the results:

$$\varkappa_{ijk}^{(2)}(\omega_1, \mathbf{k}_1; \omega_2, \mathbf{k}_2) = \sum \frac{(-i)}{2} \frac{4\pi e^2}{m} \frac{e}{m} \frac{1}{\omega_1\omega_2\omega} \int d\mathbf{v} \sum_n \frac{e^{in\varphi - ia\sin\varphi}}{\omega - k_{\parallel}v_{\parallel} - n\omega_B + i0} X_i^{(n)}(\mathbf{k}_\perp; \mathbf{v}).$$
$$\times \Bigg\{ Y_j(\omega_1, \mathbf{k}_1; \mathbf{v}) \sum_{n_2} \frac{e^{-in_2(\varphi-\varphi_2)+ia_2\sin(\varphi-\varphi_2)}}{\omega_2 - k_{2\parallel}v_{\parallel} - n_2\omega_B + i0} Z_k^{(n_2)}(\omega_2, \mathbf{k}_2; \mathbf{v})$$
$$+ Y_k(\omega_2, \mathbf{k}_2; \mathbf{v}) \sum_{n_1} \frac{e^{-in_1(\varphi-\varphi_1)+ia_1\sin(\varphi-\varphi_1)}}{\omega_1 - k_{1\parallel}v_{\parallel} - n_1\omega_B + i0} Z_j^{(n_1)}(\omega_1, \mathbf{k}_1; \mathbf{v}) \Bigg\} f_0(v_\perp, v_\parallel), \tag{3.45}$$

where $\omega = \omega_1 + \omega_2$, $\mathbf{k} = \mathbf{k}_1 + \mathbf{k}_2$;

$$\varkappa_{ijkl}^{(3)}(\omega_1, \mathbf{k}_1; \omega_2, \mathbf{k}_2; \omega_3, \mathbf{k}_3) = \tfrac{1}{3}\{\bar{\varkappa}_{ijkl}^{(3)}(\omega_1, \mathbf{k}_1; \omega_2, \mathbf{k}_2; \omega_3, \mathbf{k}_3)$$
$$+ \bar{\varkappa}_{ikjl}^{(3)}(\omega_2, \mathbf{k}_2; \omega_1, \mathbf{k}_1; \omega_3, \mathbf{k}_3)$$
$$+ \bar{\varkappa}_{ilkj}^{(3)}(\omega_3, \mathbf{k}_3; \omega_2, \mathbf{k}_2; \omega_1, \mathbf{k}_1)\}, \tag{3.46}$$

$$\bar{\varkappa}_{ijkl}^{(3)}(\omega_1, \mathbf{k}_1; \omega_2, \mathbf{k}_2; \omega_3, \mathbf{k}_3)$$
$$= \sum \frac{(-i)^2}{2} \frac{4\pi e^2}{m} \left(\frac{e}{m}\right)^2 \frac{1}{\omega_1\omega_2\omega_3\omega} \int d\mathbf{v} \sum_n \frac{e^{in\varphi - ia\sin\varphi}}{\omega - k_{\parallel}v_{\parallel} - n\omega_B + i0} X_i^{(n)}(\mathbf{k}_\perp, \mathbf{v})$$
$$\times Y_j(\omega', \mathbf{k}'; \mathbf{v}) \sum_{n'} \frac{e^{in'(\varphi-\varphi') + ia\sin(\varphi-\varphi')}}{\omega' - k'v - n'\omega_B + i0} \int_0^{2\pi} \frac{d\varphi}{2\pi} e^{in'(\varphi-\varphi') - ia'\sin(\varphi-\varphi')}$$
$$\times \Bigg\{ Y_k(\omega_2, \mathbf{k}_2; \mathbf{v}) \sum_{n_3} \frac{e^{-in_3(\varphi-\varphi_3)+ia_3\sin(\varphi-\varphi_3)}}{\omega_3 - k_{3\parallel}v_{\parallel} - n_3\omega_B + i0} Z_l^{(n_3)}(\omega_3, \mathbf{k}_3; \mathbf{v})$$
$$+ Y_l(\omega_3, \mathbf{k}_3; \mathbf{v}) \sum_{n_2} \frac{e^{-in_2(\varphi-\varphi_2)+ia_2\sin(\varphi-\varphi_2)}}{\omega_2 - k_{2\parallel}v_{\parallel} - n_2\omega_B + i0} Z_k^{(n_2)}(\omega_2, \mathbf{k}_2; \mathbf{v}) \Bigg\} f_0(v_\perp, v_\parallel), \tag{3.47}$$

where $\omega = \omega_1 + \omega_2 + \omega_3$, $\mathbf{k} = \mathbf{k}_1 + \mathbf{k}_2 + \mathbf{k}_3$, $\omega' = \omega_2 + \omega_3$, $\mathbf{k}' = \mathbf{k}_2 + \mathbf{k}_3$.

The notation is the same as in the potential case. $X_i^{(n)}(\mathbf{k}_\perp, \mathbf{v})$ and $Z_i^{(n)}(\omega, \mathbf{k}; \mathbf{v})$ were defined in (3.30) and (3.31).

The symmetries of the susceptibilities of a magneto-active plasma $\varkappa_{ijk}^{(2)}(\omega_1, \mathbf{k}_1; \omega_2, \mathbf{k}_2)$ and $\varkappa_{ijkl}^{(3)}(\omega_1, \mathbf{k}_1; \omega_2, \mathbf{k}_2; \omega_3, \mathbf{k}_3)$ are similar to (2.79) and (2.80). Besides, the non-linear tensor susceptibilities (3.45) and (3.47), as well as the dielectric permittivity tensor (3.32), satisfy the conditions (2.81) and (2.82) for the field being real.

It may be easily shown that (3.45) and (3.47) reduce to (2.70) and (2.72) as $\mathbf{B}_0 \to 0$. Contracting (3.45) and (3.47) with the relevant wave vectors we obtain the longitudinal formulas (3.7) and (3.9).

A cold plasma

In the limiting case of a cold magneto-active plasma ($T = 0$) we obtain from (3.45), (3.46), and (3.47) the following simplified expressions for the symmetrized non-linear susceptibilities:

$$\varkappa^{(2)}_{ijk}(\omega_1, \mathbf{k}_1; \omega_2, \mathbf{k}_2) = \sum \frac{(-i)}{2} \frac{e}{m} \frac{\Omega^2}{\omega_1\omega_2\omega} \{\omega_1 \Gamma_{ij}(\omega_1) \Gamma_{lk}(\omega_2) k_l + \omega_2 \Gamma_{ik}(\omega_2) \Gamma_{lj}(\omega_1) k_l$$
$$+\omega_1 \Gamma_{kj}(\omega_1) \Gamma_{il}(\omega) k_{2l} - \omega \Gamma_{ik}(\omega) \Gamma_{lj}(\omega_1) k_{2l}$$
$$+\omega_2 \Gamma_{jk}(\omega_2) \Gamma_{il}(\omega) k_{1l} - \omega \Gamma_{ij}(\omega) \Gamma_{lk}(\omega_2) k_{1l}\}. \tag{3.48}$$

$$\varkappa^{(3)}_{ijkl}(\omega_1, \mathbf{k}_1; \omega_2, \mathbf{k}_2; \omega_3, \mathbf{k}_3)$$
$$= \sum \frac{1}{6} \frac{e^2}{m^2} \frac{\Omega^2}{\omega_1\omega_2\omega_3\omega} \left\{ \left(\frac{\omega_1}{\omega'} (\omega_2 k'_{j'} k_{3k'} + \omega_3 k_{2j'} k'_{k'}) \Gamma_{ij}(\omega_1) \Gamma_{j'k}(\omega_2) \Gamma_{k'l}(\omega_3) \right.\right.$$
$$+ \left[\frac{\omega_1}{\omega'} k_{k'} \Gamma_{ij}(\omega_1) - \frac{\omega}{\omega'} k_{1k'} \Gamma_{ij}(\omega) + k_{1j'}(\delta_{ik'} \Gamma_{jj'}(\omega_1) \times \delta_{jk'} \Gamma_{ij'}(\omega)) \right.$$
$$\left. + \omega_1 \Gamma_{ik'}(\omega) k'_{j'} \Gamma_{jj'}(\omega_1) \right] [\omega_2 k_{3l'} \Gamma_{k'l'}(\omega') \Gamma_{lk}(\omega_2) + \omega_3 k_{3l'} \Gamma_{l'k}(\omega_2) \Gamma_{k'l}(\omega_3)$$
$$- \omega' k_{3l'} \Gamma_{l'k}(\omega_2) \Gamma_{k'l}(\omega') + \omega_3 k_{2l'} \Gamma_{k'l'}(\omega') \Gamma_{kl}(\omega_3) + \omega_2 k_{2l'} \Gamma_{k'k}(\omega_2) \Gamma_{l'l}(\omega_3)$$
$$\left.\left. - \omega' k_{2l'} \Gamma_{l'l}(\omega_3) \Gamma_{k'k}(\omega')] \right) + (1 \rightleftarrows 2) + (1 \rightleftarrows 3) \right\}. \tag{3.49}$$

The same expressions for $\varkappa^{(2)}_{ijk}(\omega_1, \mathbf{k}_1; \omega_2, \mathbf{k}_2)$ and $\varkappa^{(3)}_{ijkl}(\omega_1, \mathbf{k}_1; \omega_2, \mathbf{k}_2; \omega_3, \mathbf{k}_3)$ may be derived starting from the hydrodynamic equations.

The third order non-symmetrized tensor susceptibility for the magneto-active plasma reduces in the hydrodynamical approximation to the following:

$$\bar{\varkappa}^{(3)}_{ijkl}(\omega_1, \mathbf{k}_1; \omega_2, \mathbf{k}_2; \omega_3, \mathbf{k}_3)$$
$$= \sum \frac{1}{2} \frac{e^2}{m^2} \frac{\Omega^2}{\omega_1\omega_2\omega_3\omega} \{(\omega_1\omega_2 \Gamma_{ij}(\omega_1) k'_{j'} \Gamma_{j'k}(\omega_2) k_{3k'} \Gamma_{k'l}(\omega_3))$$
$$+ [\omega_1 k_{k'} \Gamma_{ij}(\omega_1) - \omega k_{1k'} \Gamma_{ij}(\omega) + \omega' k_{1j'}(\delta_{ik'} \Gamma_{j'j}(\omega_1) + \delta_{jk'} \Gamma_{ij'}(\omega))$$
$$+ \omega_1 \omega \Gamma_{ik'}(\omega) k'_{j'} \Gamma_{jj'}(\omega_1)] [\omega_2 k_{3l'} \Gamma_{k'l'}(\omega') \Gamma_{lk}(\omega_2) + \omega_3 k_{3l'} \Gamma_{l'k}(\omega_2) \Gamma_{k'l}(\omega_3)$$
$$- \omega' k_{3l'} \Gamma_{l'k}(\omega_2) \Gamma_{k'l}(\omega')] + (2 \rightleftarrows 3)\}. \tag{3.50}$$

Various expressions for the dielectric permittivity and the non-linear susceptibilities that we have derived in this chapter will be helpful later on for the study of the wave processes that occur in a homogeneous plasma in the presence of an external magnetic field.

CHAPTER 4

Waves in Plasmas

4.1. Eigenoscillations and Eigenwaves in Space–Time Dispersive Media (Linear Approximation)

The dispersion equation

Before we begin a systematic study of non-linear wave interactions let us pay some more attention to the eigenoscillations and eigenwaves in dispersive media.† We have pointed out already that the linear eigenoscillations and eigenwaves are determined by the solutions of the wave equation, the latter for a spatially homogeneous stationary plasma being written in the form

$$\Lambda_{ij}(\omega, \mathbf{k}) E_j(\omega, \mathbf{k}) = 0, \tag{4.1}$$

where

$$\Lambda_{ij}(\omega, \mathbf{k}) \equiv \varepsilon_{ij}(\omega, \mathbf{k}) + \left(\frac{k_i k_j}{k^2} - \delta_{ij}\right)\eta^2 \tag{4.2}$$

and $\eta^2 \equiv \dfrac{k^2c^2}{\omega^2}$ is the refractive index. For (4.1) to have non-zero solution, the determinant of the matrix (4.2) must be equal to zero:

$$\Lambda(\omega, \mathbf{k}) \equiv |\Lambda_{ij}(\omega, \mathbf{k})| = 0. \tag{4.3}$$

The condition (4.3) is conventionally referred to as the dispersion equation. The roots of this equation are associated with the eigenwaves and determine the frequency dependence of the refractive index $\eta^\alpha(\omega)$ or that of the frequency on the wave vector $\omega^\alpha(\mathbf{k})$ (α labels different eigenwaves).

It is convenient to introduce one more matrix λ_{ij} which elements are the cofactors of Λ_{ij}

$$\Lambda_{ij}\lambda_{jk} = \Lambda\delta_{ik} \tag{4.4}$$

and the following interrelation occurs:

$$\lambda_{ij} = \tfrac{1}{2}\varepsilon_{ikl}\varepsilon_{jmn}\Lambda_{mk}\Lambda_{ni}, \tag{4.5}$$

where ε_{ikl} is the completely antisymmetric unit tensor.

† An account of the linear theory of electromagnetic waves in a plasma is given by Stix (1962), Shafranov (1963), Allis, Buchsbaum, and Bers (1963), Ginzburg (1970), Ginzburg and Rukhadze (1970), and Akhiezer, Akhiezer, Polovin, Sitenko, and Stepanov (1975). Waves in bounded plasmas were studied in detail by Vandenplas (1968).

The tensor (4.2) consists in the general case of a Hermitian and an anti-Hermitian part:

$$\Lambda_{ij}(\omega, \mathbf{k}) = \Lambda_{ij}^{H}(\omega, \mathbf{k}) + \Lambda_{ij}^{A}(\omega, \mathbf{k}), \tag{4.6}$$

$$\Lambda_{ij}^{H}(\omega, \mathbf{k}) = \Lambda_{ji}^{H*}(\omega, \mathbf{k}), \quad \Lambda_{ij}^{A}(\omega, \mathbf{k}) = -\Lambda_{ji}^{A*}(\omega, \mathbf{k}), \tag{4.7}$$

so that the determinant $\Lambda(\omega, \mathbf{k})$ is a complex function of ω and $\mathbf{k}$, and the condition (4.3) reduces to two requirements that the real and the imaginary parts of $\Lambda(\omega, \mathbf{k})$ be zero:

$$\operatorname{Re} \Lambda(\omega, \mathbf{k}) = 0 \quad \text{and} \quad \operatorname{Im} \Lambda(\omega, \mathbf{k}) = 0.$$

In the transparency domain (which is associated with such values of ω and $\mathbf{k}$ that the anti-Hermitian part of the dielectric permittivity is small compared with the Hermitian one) the real part of $\Lambda(\omega, \mathbf{k})$ is much greater than the imaginary one. Thus the dispersion equation (4.3) in neglect of wave damping may be approximately written as

$$\operatorname{Re} \Lambda(\omega, \mathbf{k}) = 0. \tag{4.8}$$

Suppose the wave vector to be real. Substituting the solution of (4.8) for the eigenfrequency $\omega_{\mathbf{k}}^{\alpha}$ ($\omega_{\mathbf{k}}^{\alpha} > 0$) and suggesting the damping be weak, we obtain the following approximate damping rate:

$$\gamma_{\mathbf{k}}^{\alpha} \simeq \frac{\operatorname{Im} \Lambda(\omega_{\mathbf{k}}^{\alpha}, \mathbf{k})}{\dfrac{\partial \operatorname{Re} \Lambda(\omega_{\mathbf{k}}^{\alpha}, \mathbf{k})}{\partial \omega_{\mathbf{k}}^{\alpha}}}. \tag{4.9}$$

It is straightforward to verify that in the transparency domain $\operatorname{Im} \Lambda(\omega, \mathbf{k})$ may be written in terms of the Hermitian part of the matrix $\lambda_{ij}(\omega, \mathbf{k})$ and the anti-Hermitian part of the dielectric permittivity $\varepsilon_{ij}(\omega, \mathbf{k})$:

$$\operatorname{Im} \Lambda(\omega, \mathbf{k}) = \frac{1}{4i} (\varepsilon_{ij}(\omega, \mathbf{k}) - \varepsilon_{ji}^{*}(\omega, \mathbf{k}))(\lambda_{ji}(\omega, \mathbf{k}) + \lambda_{ij}^{*}(\omega, \mathbf{k})). \tag{4.10}$$

The quantities $\gamma_{\mathbf{k}}^{\alpha}$ from (4.9) are always positive in an equilibrium plasma ($\gamma_{\mathbf{k}}^{\alpha} > 0$), i.e. the oscillations are damped. In a non-equilibrium plasma $\gamma_{\mathbf{k}}^{\alpha}$ may be as well negative ($\gamma_{\mathbf{k}}^{\alpha} < 0$) and then the eigenoscillation amplitude grows exponentially and the plasma becomes driven unstable (in such a case $|\gamma_{\mathbf{k}}^{\alpha}|$ is called the growth rate).†

Wave polarization

The time-dependence of an eigenwave with a fixed wave vector $\mathbf{k}$ is described by the following expression:

$$\mathbf{E}_{\mathbf{k}}^{\alpha}(t) = \mathbf{E}_{\mathbf{k}}^{\alpha} \cos(\omega_{\mathbf{k}}^{\alpha} t + \phi_{\mathbf{k}}^{\alpha}), \quad \mathbf{E}_{\mathbf{k}}^{\alpha} \equiv \mathbf{E}_{\mathbf{k}0}^{\alpha}\, e^{-\gamma_{\mathbf{k}}^{\alpha} t}, \tag{4.11}$$

where $\mathbf{E}_{\mathbf{k}0}^{\alpha}$ and $\phi_{\mathbf{k}}^{\alpha}$ are the initial amplitude and phase ($\omega_{\mathbf{k}}^{\alpha}$, $\gamma_{\mathbf{k}}^{\alpha}$ and $\phi_{\mathbf{k}}^{\alpha}$ are real functions of $\mathbf{k}$). The field amplitude satisfies the reality condition

$$\mathbf{E}_{\mathbf{k}}^{\alpha*} = \mathbf{E}_{-\mathbf{k}}^{\alpha}. \tag{4.12}$$

† The theory of plasma instabilities is discussed in detail by Briggs (1964) and Mikhailovskii (1974).

The linear amplitude of the wave field is time-independent in neglect of damping $\mathbf{E}_{\mathbf{k}}^{\alpha} = \mathbf{E}_{\mathbf{k}0}^{\alpha}$.

Let us introduce a unit vector **e** directed along the field strength $\mathbf{E}_{\mathbf{k}}$. (Here and below **e** will be referred to as the wave polarization.)

$$\mathbf{e} \equiv \mathbf{e}(\mathbf{k}) = \frac{\mathbf{E}_{\mathbf{k}}}{|\mathbf{E}_{\mathbf{k}}|}. \tag{4.13}$$

According to (4.1) the polarization $\mathbf{e}^{\alpha}$ of the wave α satisfies the equation

$$\Lambda_{ij}(\omega_{\mathbf{k}}^{\alpha}, \mathbf{k}) e_j^{\alpha} = 0, \tag{4.14}$$

the general solution of which may be written as

$$e_i^{\alpha} = C\lambda_{ij}(\omega_{\mathbf{k}}^{\alpha}, \mathbf{k}) a_j, \tag{4.15}$$

where **a** is an arbitrary vector and C is a constant which is governed by the normalization condition

$$\mathbf{e}^{\alpha}\mathbf{e}^{\alpha *} = 1. \tag{4.16}$$

Let us show that the relation (4.15) determines the polarization $\mathbf{e}^{\alpha}$ but for a phase factor (Sitenko and Kirochkin, 1966). First of all an observation is to be made that irrespective of frequency the matrixes λ_{ij} and Λ_{ij} are related according to

$$\lambda_{ij}\lambda_{kl} = \lambda_{il}\lambda_{kj} + \Lambda\varepsilon_{ikm}\varepsilon_{jln}\Lambda_{nm}. \tag{4.17}$$

To prove (4.17) it is sufficient to multiply the left-hand and the right-hand parts of the equality

$$\Lambda\varepsilon_{abc} = \varepsilon_{mnp}\Lambda_{ma}\Lambda_{nb}\Lambda_{pc}$$

by $\frac{1}{\Lambda}\lambda_{aj}\lambda_{bl}\varepsilon_{ikc}$. The eigenfrequencies satisfy the dispersion equation $\Lambda = 0$ so that (4.17) becomes simplified:

$$\lambda_{ij}\lambda_{kl} = \lambda_{il}\lambda_{kj}. \tag{4.18}$$

Neglecting the anti-Hermitian part of λ_{ij} in the plasma transparency domain, we obtain from (4.18):

$$\frac{\lambda_{il}a_l\lambda_{jk}^{*}a_k}{\lambda_{mn}a_m a_n} = \frac{\lambda_{il}a_l'\lambda_{jk}^{*}a_k'}{\lambda_{mn}a_m' a_n'}, \tag{4.19}$$

where **a** and **a′** are arbitrary real vectors. Note that the scalar products $(\mathbf{a}\lambda\mathbf{a})$ and $(\mathbf{a}'\lambda\mathbf{a}')$ have the same sign:

$$(\mathbf{a}\lambda\mathbf{a})(\mathbf{a}'\lambda\mathbf{a}') = |(\mathbf{a}\lambda\mathbf{a}')|^2. \tag{4.20}$$

In particular, all diagonal elements of the Hermitian part of λ_{ij} have the same sign.

Within the context of (4.18) the normalization constant in (4.15) may be written as

$$C = [(\mathbf{a}\lambda\mathbf{a})\,\mathrm{Tr}\,\lambda]^{-1/2}. \tag{4.21}$$

Then the normalized polarization of the wave with dispersion (4.3) becomes

$$\mathbf{e}^{\alpha} = \left.\frac{\lambda \mathbf{a}}{\sqrt{(\mathbf{a}\lambda\mathbf{a})\,\mathrm{Tr}\,\lambda}}\right|_{\omega=\omega_{\mathbf{k}}^{\alpha}}. \tag{4.22}$$

According to (4.19), the product $e_i e_j^*$ is invariant under any changes of $\mathbf{a}$, hence an arbitrary rotation of the latter can change only the phase factor in (4.22).

Density and flux of energy

The energy conservation law for a plasma follows immediately from the Maxwell equations (1.59) and the linear material equation (2.63):

$$\frac{1}{4\pi}\left[\left(\mathbf{E}\cdot\frac{\partial \mathbf{D}}{\partial t}\right)+\left(\mathbf{B}\cdot\frac{\partial \mathbf{B}}{\partial t}\right)\right]+\frac{c}{4\pi}\,\mathrm{div}[\mathbf{EB}] = 0. \tag{4.23}$$

Consider the field of a nearly monochromatic wave with frequency $\omega_{\mathbf{k}}^{\alpha}$ and wave vector $\mathbf{k}$ (the wave amplitude $\mathbf{E}_{\mathbf{k}}^{\alpha}$ is a slowly varying function of coordinates and time). Averaging (4.23) over a time interval that is much longer than the period of the eigenoscillations and upon noting that the anti-Hermitian part of the dielectric permittivity is small in comparison with the Hermitian part, we obtain the following equation that describes energy transfer in a space–time dispersive plasma:

$$\frac{\partial W_{\mathbf{k}}^{\alpha}}{\partial t}+\mathrm{div}\,\mathbf{S}_{\mathbf{k}}^{\alpha}+Q_{\mathbf{k}}^{\alpha} = 0, \tag{4.24}$$

where $W_{\mathbf{k}}^{\alpha}$ is the mean energy density.†

$$W_{\mathbf{k}}^{\alpha} = \frac{1}{16\pi}\left\{\frac{\partial}{\partial \omega_{\mathbf{k}}^{\alpha}}\left[\omega_{\mathbf{k}}^{\alpha}\varepsilon_{ij}^{H}(\omega_{\mathbf{k}}^{\alpha},\mathbf{k})\right]+\eta^2\left(\delta_{ij}-\frac{k_i k_j}{k^2}\right)\right\}E_i^{\alpha *}(\mathbf{k})\,E_j^{\alpha}(\mathbf{k}), \tag{4.25}$$

$\mathbf{S}_{\mathbf{k}}^{\alpha}$ is the mean density of the energy flux along the direction of the wave propagation

$$\mathbf{S}_{\mathbf{k}i}^{\alpha} = \frac{1}{16\pi}\left\{(2k_i\delta_{jk}-\delta_{ij}k_k-\delta_{ik}k_j)\,\frac{c^2}{\omega_{\mathbf{k}}^{\alpha}}-\frac{\partial}{\partial k_i}\,[\omega_{\mathbf{k}}^{\alpha}\varepsilon_{jk}^{H}(\omega_{\mathbf{k}}^{\alpha},\mathbf{k})]\right\}E_j^{\alpha *}(\mathbf{k})\,E_{\mathbf{k}}^{\alpha}(\mathbf{k}), \tag{4.26}$$

and $Q_{\mathbf{k}}^{\alpha}$ denotes the energy losses due to the linear dissipation

$$Q_{\mathbf{k}}^{\alpha} = -\frac{i\omega_{\mathbf{k}}^{\alpha}}{16\pi}\,[\varepsilon_{ij}(\omega,\mathbf{k})-\varepsilon_{ji}^{*}(\omega,\mathbf{k})]\,E_i^{\alpha *}(\mathbf{k})\,E_j^{\alpha}(\mathbf{k}). \tag{4.27}$$

$W_{\mathbf{k}}^{\alpha}$ involves both the wave-field energy and the kinetic energy of the particles that oscillate under the influence of the latter. Note that $W_{\mathbf{k}}^{\alpha}$ is always positive for eigenwaves in an equilibrium plasma. In non-equilibrium plasmas $W_{\mathbf{k}}^{\alpha}$ can be sometimes negative

† The mean energy density of a wave field in a space–time dispersive medium was calculated by Landau and Lifshitz (1960).

(negative-energy waves). The sign of $W_{\mathbf{k}}^{\alpha}$ is governed by the dispersion properties of the plasma.†

The mean density of the energy flux $\mathbf{S}_{\mathbf{k}}^{\alpha}$ is related to $W_{\mathbf{k}}^{\alpha}$ according to

$$\mathbf{S}_{\mathbf{k}}^{\alpha} = \mathbf{v}_{\mathbf{k}}^{\alpha} W_{\mathbf{k}}^{\alpha}, \tag{4.28}$$

where $\mathbf{v}_{\mathbf{k}}^{\alpha}$ is the group velocity of the wave

$$\mathbf{v}_{\mathbf{k}}^{\alpha} = \frac{\partial \omega_{\mathbf{k}}^{\alpha}}{\partial \mathbf{k}}. \tag{4.29}$$

It is clear from (4.27) that the energy losses are associated with the anti-Hermitian part of $\varepsilon_{ij}(\omega, \mathbf{k})$. Hence we can connect $Q_{\mathbf{k}}^{\alpha}$ with the imaginary part of the determinant (4.3). Indeed, making use of (4.10) and (4.22) we obtain

$$\operatorname{Im} \Lambda(\omega_{\mathbf{k}}^{\alpha}, \mathbf{k}) = \frac{\operatorname{Tr} \Lambda(\omega_{\mathbf{k}}^{\alpha}, \mathbf{k})}{2i} [\varepsilon_{ij}(\omega_{\mathbf{k}}^{\alpha}, \mathbf{k}) - \varepsilon_{ji}^{*}(\omega_{\mathbf{k}}^{\alpha}, \mathbf{k})] e_i^{\alpha *} e_j^{\alpha}. \tag{4.30}$$

The entropy of thermally stable systems can only increase, so that $Q_{\mathbf{k}}^{\alpha} > 0$ and the following helpful relation may be obtained by comparing (4.30) to (4.27):

$$\frac{\operatorname{Tr} \lambda(\omega_{\mathbf{k}}^{\alpha}, \mathbf{k})}{\omega_{\mathbf{k}}^{\alpha}} \operatorname{Im} \Lambda(\omega_{\mathbf{k}}^{\alpha}, \mathbf{k}) > 0. \tag{4.31}$$

Fluctuations and averaging over initial phases

The Fourier time-component of (4.11) in neglect of damping may be written in the form

$$\mathbf{E}_{\mathbf{k}\omega}^{\alpha} = \pi \mathbf{E}_{\mathbf{k}}^{\alpha} \left\{ e^{-i\phi_{\mathbf{k}}^{\alpha}} \delta(\omega - \omega_{\mathbf{k}}^{\alpha}) + e^{i\phi_{\mathbf{k}}^{\alpha}} \delta(\omega + \omega_{\mathbf{k}}^{\alpha}) \right\}. \tag{4.32}$$

It is evident that (4.32) describes also the fluctuation oscillations of the plasma electric field if the initial phase $\phi_{\mathbf{k}}^{\alpha}$ be treated as a random parameter and the amplitude $\mathbf{E}_{\mathbf{k}}^{\alpha}$ be governed by the state of the system. By virtue of plasma homogeneity and stationarity the mean product of the electric field fluctuations is equal to

$$\langle \mathbf{E}_{\mathbf{k}\omega} \mathbf{E}_{\mathbf{k}\omega'}^{*} \rangle = 2\pi V \delta(\omega - \omega') \langle E^2 \rangle_{\mathbf{k}\omega}, \tag{4.33}$$

where V is the plasma volume and $\langle E^2 \rangle_{\mathbf{k}\omega}$ is the spectral correlation function. The substitution of (4.32) in the left-hand part of (4.33) and the averaging over the initial phases yield

$$\langle E^2 \rangle_{\mathbf{k}\omega} = \pi I_{\mathbf{k}}^{\alpha} \{ \delta(\omega - \omega_{\mathbf{k}}^{\alpha}) + \delta(\omega + \omega_{\mathbf{k}}^{\alpha}) \}, \tag{4.34}$$

where

$$I_{\mathbf{k}}^{\alpha} = \frac{1}{2} \frac{|E_{\mathbf{k}}^{\alpha}|^2}{V}. \tag{4.35}$$

The quantity $I_{\mathbf{k}}^{\alpha}$ characterizes the total intensity of the fluctuation oscillations with eigenfrequencies $\omega_{\mathbf{k}}^{\alpha}$ and $-\omega_{\mathbf{k}}^{\alpha}$.

† The possibility that negative-energy waves may exist in a plasma was predicted by Kadomtsev, Mikhailovskii, and Timofeev (1965).

4.2. Waves in an Isotropic Plasma

Dispersion relations for longitudinal and transverse waves

In the homogeneous isotropic plasma with dielectric permittivity (2.75) the dispersion equation (4.3) becomes transformed into the following relation:

$$\Lambda(\omega, \mathbf{k}) \equiv \varepsilon_l(\omega, k)\,[\varepsilon_t(\omega, k) - \eta^2] = 0. \tag{4.36}$$

which decouples into two equations describing the longitudinal and the transverse waves. Indeed, for the longitudinal waves we have $(\mathbf{k} \cdot \mathbf{E}) \neq 0$, $\left(\delta_{ij} - \frac{k_i k_j}{k^2}\right) E_j = 0$ and (4.1) takes the form

$$\varepsilon_l(\omega, k)\mathbf{E}_{\mathbf{k}\omega} = 0. \tag{4.37}$$

Non-zero solutions of this equation exist only if

$$\varepsilon_l(\omega, \mathbf{k}) = 0. \tag{4.38}$$

This requirement is the dispersion equation for the longitudinal waves.

For transverse waves $(\mathbf{k} \cdot \mathbf{E}) = 0$ and $\left(\delta_{ij} - \frac{k_i k_j}{k^2}\right) E_j \neq 0$. The wave equation (4.1) then reduces to

$$[\varepsilon_t(\omega, k) - \eta^2]\mathbf{E}_{\mathbf{k}\omega} = 0. \tag{4.39}$$

The solutions of (4.39) do not vanish provided

$$\varepsilon_t(\omega, k) = \eta^2, \tag{4.40}$$

which is the dispersion equation for the transverse electromagnetic waves in a plasma.

In the general case, when the particle velocity distribution is anisotropic, or external fields are present, the dielectric permittivity cannot be reduced to a (2.75)-like form, and we cannot separate the longitudinal waves from the transverse ones.

Longitudinal Langmuir and ion-sound waves

There exist slowly damped longitudinal waves of two kinds in the homogenous isotropic plasma, namely high-frequency electron Langmuir waves and ion-sound waves with low frequencies. The frequencies and the damping rates of those may be easily found from the dispersion equation (4.38), where (2.27) is to be substituted for the dielectric permittivity.

The frequency and the damping rate of the Langmuir waves are the following:

$$\omega_k = \sqrt{\Omega^2 + k^2 s^2}, \quad \gamma_k = \sqrt{\frac{\pi}{8}}\,\frac{\Omega}{a^3 k^3}\,e^{-\frac{3}{2}\frac{\omega_k^2}{k^2 s^2}}, \tag{4.41}$$

where $\Omega^2 = \dfrac{4\pi e^2 n_0}{m}$ is the Langmuir frequency, $s^2 = \dfrac{3T}{m}$ is the mean square of the electron velocity. The expressions (4.41) are valid if $ak \ll 1$. This condition must be satisfied for the damping to be weak and to be governed by the resonant coupling between the electrons and the wave field (Landau damping). If $ak \gg 1$, then the Langmuir waves damp rapidly ($\gamma_k \gtrsim \omega_k$). The polarization of the Langmuir waves is directed along the wave propagation

$$\mathbf{e} = \frac{\mathbf{k}}{k}. \tag{4.42}$$

The ion-sound frequency is equal to

$$\omega_{\mathbf{k}} = \frac{kv_s}{\sqrt{1+a^2k^2}}, \tag{4.43}$$

where $v_s = \sqrt{\dfrac{T_e}{m_i}}$ is the so-called non-isothermal sound velocity. The ion-sound waves are weakly damped only in strongly non-isothermal plasmas with $T_e \gg T_i$. The frequency and the damping rate of those at $ak \ll I$ are

$$\omega_k = kv_s, \quad \gamma_k = \sqrt{\frac{\pi}{8}\frac{m_e}{m_i}}\, kv_s. \tag{4.44}$$

The ion-sound frequency is close to the ion Langmuir frequency when $ak \gg 1$,

$$\omega_k \simeq \Omega_i. \tag{4.45}$$

If $T_e \simeq T_i$, the ion-sound waves are strongly damped ($\gamma_{\mathbf{k}} \simeq \omega_{\mathbf{k}}$). The ion-sound polarization is determined by (4.42) similarly to that of the Langmuir waves.

Transverse electromagnetic waves

The refractive index of transverse electromagnetic waves in an isotropic plasma is equal to

$$\eta^2(\omega) = \varepsilon_t(\omega, k). \tag{4.46}$$

It is clear that an electromagnetic wave can propagate only in a plasma with $\eta^2(\omega) > 0$.

The frequencies of the propagating waves are given by

$$\omega_{\mathbf{k}} = \sqrt{\Omega^2 + k^2c^2}. \tag{4.47}$$

Taking into account the thermal motion gives rise to only small corrections in (4.47) of the order of magnitude T/mc^2. The electromagnetic waves have much greater phase velocities than the light velocity and are not damped in a collisionless plasma.

The condition $\eta^2(\omega) = 0$ determines the threshold of the wave propagation. Making use of (2.77) we find that the threshold frequency ω_0 is equal to the Langmuir frequency in an isotropic plasma:

$$\omega_0 = \Omega. \tag{4.48}$$

(The subscript 0 implies that ω_0 is the limiting value of the eigenfrequency (4.47) associated with $k \to 0$.)

Each of the frequencies (4.47) may be associated either with two electromagnetic waves, the polarizations of which $\mathbf{e}_1$ and $\mathbf{e}_2$ are perpendicular to each other (and, of course, each of those is orthogonal to the wave vector **k**), or with two circularly polarized waves with right- and left-handed polarizations:

$$\mathbf{e}_r = \frac{1}{\sqrt{2}}(\mathbf{e}_1 - i\mathbf{e}_2), \quad \mathbf{e}_l = -\frac{i}{\sqrt{2}}(\mathbf{e}_1 + i\mathbf{e}_2). \tag{4.49}$$

In Fig. 1 we show the dispersion curves for the eigenwaves of an isotropic plasma (i.e. the frequencies as functions of the wave vectors).

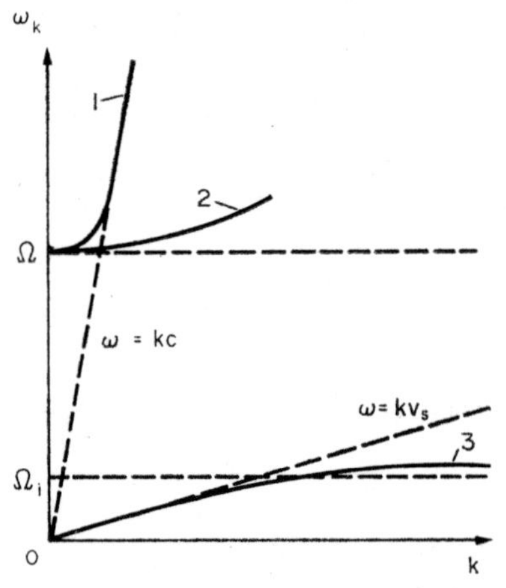

FIG. 1. Frequencies of various eigenwaves in an isotropic plasma as functions of the wave vector: 1, transverse electromagnetic wave; 2, Langmuir wave; 3, ion-sound wave.

4.3. Waves in a Magneto-active Plasma

The dispersion equation for waves in a magneto-plasma

The dispersion equations for the waves that propagate in a magneto-active plasma are of the form

$$A\eta^4 + B\eta^2 + C = 0, \tag{4.50}$$

where

$$\left.\begin{aligned} A &= \varepsilon_{11}\sin^2\vartheta + \varepsilon_{33}\cos^2\vartheta + 2\varepsilon_{13}\sin\vartheta\cos\vartheta, \\ B &= 2(\varepsilon_{12}\varepsilon_{23} - \varepsilon_{22}\varepsilon_{13})\sin\vartheta\cos\vartheta - (\varepsilon_{11}\varepsilon_{22} + \varepsilon_{12}^2)\sin^2\vartheta \\ &\quad - (\varepsilon_{22}\varepsilon_{33} + \varepsilon_{23}^2)\cos^2\vartheta - \varepsilon_{11}\varepsilon_{33} + \varepsilon_{13}^2, \\ C &= \varepsilon_{11}\varepsilon_{22}\varepsilon_{33} + \varepsilon_{11}\varepsilon_{23}^2 - \varepsilon_{22}\varepsilon_{13}^2 + \varepsilon_{33}\varepsilon_{12}^2 + 2\varepsilon_{12}\varepsilon_{23}\varepsilon_{31}, \end{aligned}\right\} \tag{4.51}$$

ϑ is the angle between the direction of the wave vector **k** and the magnetic field $\mathbf{B}_0$. Equation (4.50) determines the connection between the frequencies and the wave vectors of the eigenwaves. If we fix the wave vector **k**, then we can find the eigenfrequencies from (4.50):

$$\omega_{\mathbf{k}}^{\alpha} = \omega^{\alpha}(k, \vartheta). \tag{4.52}$$

The number of the roots of (4.52) and the roots themselves are governed by the ω- and **k**-dependence of the dielectric permittivity.

In neglect of the thermal motion the dielectric permittivity tensor of a cold magneto-active plasma is of the form (3.41); the coefficients A, B, and C in the dispersion equation turn out to depend only on the frequency and the direction of the wave vector

$$\left.\begin{aligned} A_0 &= \varepsilon_1 \sin^2 \vartheta + \varepsilon_2 \cos^2 \vartheta, \\ B_0 &= -[(\varepsilon_1^2 - \varepsilon_2^2) \sin^2 \vartheta + \varepsilon_1 \varepsilon_2 (1 + \cos^2 \vartheta)], \\ C_0 &= (\varepsilon_1^2 - \varepsilon_2^2)\varepsilon_3. \end{aligned}\right\} \tag{4.53}$$

(ε_1, ε_2, and ε_3 were defined in (3.42).) Hence we can find the solution of (4.50) in terms of η^2 and thus determine the dependence of the refractive indexes of the waves on the frequencies and the wave vectors. Equation (4.50) with the coefficients (4.53) has two different solutions,

$$\eta_{\pm}^2 = \frac{-B_0 \pm \sqrt{B_0^2 - 4A_0 C_0}}{2A_0}, \tag{4.54}$$

which are associated with the refractive indexes of the so-called ordinary and extra-ordinary electromagnetic waves.

By virtue of (4.54) there may occur in a magneto-plasma two waves with a given frequency and different indexes of refraction. In case a wave propagates in an arbitrary direction (with respect to the magnetic field), its polarization is elliptic. The polarization vectors may be written as:

$$e = \left\{ \cos \varphi - i \frac{\varepsilon_2}{\eta^2 - \varepsilon_1} \sin \varphi, \quad \sin \varphi + i \frac{\varepsilon_2}{\eta^2 - \varepsilon_1} \cos \varphi, \quad \frac{\eta^2 \sin \vartheta \cos \vartheta}{\eta^2 \sin^2 \vartheta - \varepsilon_1} \right\}, \tag{4.55}$$

where φ is the azimuth of the vector **k**. The electromagnetic waves that propagate along the magnetic field are transverse; the ordinary wave is characterized by the right-handed circular polarization, the extraordinary waves are left-handedly polarized.

Plasma resonances

Electromagnetic waves in a magneto-plasma cannot be separated into longitudinal and transverse components. However, if $A \equiv \frac{k_i k_j}{k^2} \varepsilon_{ij}(\omega, \mathbf{k}) \to 0$, then the longitudinal component of the electric field is considerably greater than the transverse one. To verify this we multiply the wave equation (4.50) by k:

$$E^l = -\frac{k_i \varepsilon_{ij} E_j^t}{kA}. \tag{4.56}$$

It is clear that $\Lambda(\omega, \mathbf{k}) = 0$ and $A(\omega, \mathbf{k}) = 0$ for purely longitudinal oscillations ($E^t = 0$). In fact, these two conditions cannot be satisfied together. However, in the band of frequencies where η^2 is great, only the leading term in η^2 must be taken into account in (4.50) and the dispersion equation reduces to the requirement

$$A(\omega, \mathbf{k}) = 0. \tag{4.57}$$

The frequencies that satisfy (4.57) are called the plasma resonance frequencies. Note that one of the refractive indexes (4.54) tends to infinity as the frequency approaches the resonance value

$$\eta_+^2 = -\frac{B}{A}, \tag{4.58}$$

while the second one remains finite

$$\eta_-^2 = -\frac{C}{B}. \tag{4.59}$$

Making use of (4.53) and (3.42) we obtain the following equation for plasma resonances in neglect of thermal effects:

$$1-\sum \frac{\Omega^2}{\omega^2} \cos^2 \vartheta - \sum \frac{\Omega^2}{\omega^2-\omega_B^2} \sin^2 \vartheta = 0 \tag{4.60}$$

(the summation extends over the electrons and the ions). This equation is of the third power in ω^2 and hence it determines three resonance frequencies:

$$\omega_\infty^{(\alpha)}(\vartheta) \quad (\alpha = 1, 2, 3).$$

The subscript ∞ of the resonance frequency $\omega_\infty(\vartheta)$ implies that $\omega_\infty(\vartheta)$ is the eigenfrequency associated with the infinitely large wave vector $k \to \infty$.

Equation (4.60) reduces in neglect of ion contributions to a quadratic equation for ω^2. The solutions of the latter are:

$$\omega_\infty^{(1,2)}(\vartheta) = \left[\tfrac{1}{2}(\Omega^2+\omega_B^2) \pm \sqrt{(\Omega^2+\omega_B^2)^2 - 4\Omega^2\omega_B^2 \cos^2 \vartheta}\right]^{1/2} \tag{4.61}$$

and as $\vartheta = 0$

$$\omega_\infty^{(1)}(0) = \max(\Omega, \omega_B), \quad \omega_\infty^{(2)}(0) = \min(\Omega, \omega_B). \tag{4.62}$$

The resonance frequency $\omega_\infty^{(1)}(\vartheta)$ grows, and $\omega_\infty^{(2)}(\vartheta)$ decreases with increasing ϑ. According to (4.61), $\omega_\infty^{(2)}(\vartheta)$ vanishes at $\vartheta = \pi/2$. However, this formula is valid for $\omega_\infty^{(2)}(\vartheta)$ only provided $\cos^2 \vartheta \gg m_e/m_i$ and yields an inadequate result if the angles are close to $\pi/2$, since the ion motion cannot be neglected when $\omega_\infty^{(2)}(\vartheta)$ competes in the same order with ω_B.

The additional resonance frequency $\omega_\infty^{(3)}(\vartheta)$, which is also a solution of (4.60), depends in an essential way on the ion motion. In case $\cos^2 \vartheta \gg m_e/m_i$ we obtain the following approximate expression:

$$\omega_\infty^{(3)}(\vartheta) = \left(1 - \frac{1}{2} \frac{m_e}{m_i} \tan^2 \vartheta\right)\omega_{Bi}. \tag{4.63}$$

Refractive indexes and eigenwave frequencies in a cold magneto-plasma

To reveal the frequency dependence of the refractive indexes in a cold magneto-plasma we must find the zeros of $\eta_{\pm}^2(\omega)$, which determine the boundaries between the domains of propagation for different waves. According to (4.54), $\eta_{\pm}^2(\omega) = 0$, if the coefficient C is equal to zero. Neglecting the ion contribution we find

$$\left.\begin{aligned} \omega_0^{(1)} &= \sqrt{\Omega^2+\tfrac{1}{2}\omega_B^2}+\tfrac{1}{2}\omega_B, \\ \omega_0^{(2)} &= \Omega, \\ \omega_0^{(3)} &= \sqrt{\Omega^2+\tfrac{1}{2}\omega_B^2}-\tfrac{1}{2}\omega_B. \end{aligned}\right\} \tag{4.64}$$

Notice that the threshold frequencies $\omega_0^{(\alpha)}$ are the limiting values of the eigenfrequencies as $k \to 0$. It may be easily verified that

$$\omega_0^{(1)} > \omega_\infty^{(1)} > \omega_0^{(2)} > \omega_\infty^{(2)}; \qquad \omega_0^{(3)} > \omega_\infty^{(3)}.$$

It depends on the angle ϑ which of the frequencies $\omega_\infty^{(2)}$ and $\omega_0^{(3)}$ is the higher.

It follows from (4.53) and (4.54) that as $\omega \to 0$:

$$\eta_+^2 = \varepsilon_1(0), \quad \eta_-^2 = \frac{\varepsilon_1(0)}{\cos^2\vartheta}\,. \tag{4.65}$$

If $\omega \to \infty$, then $\eta_{\pm}^2 \to 1$.

Once we know the zeros and the poles, as well as the values of $\eta_+(\omega)$ and $\eta_-(\omega)$ at $\omega = 0$ and $\omega = \infty$, we can draw the plots of these functions. In Fig. 2 we show the frequency dependence of the refractive indexes for $0 < \vartheta < \pi/2$. The propagation

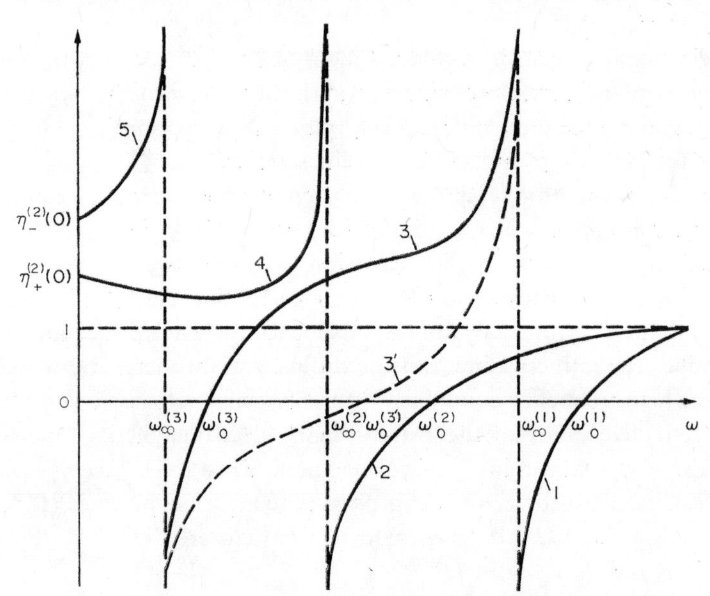

FIG. 2. Squares of refractive indexes for various eigenwaves in a magneto-active plasma as functions of the frequency ($0 < \vartheta < \pi/2$). 1, fast extraordinary wave; 2, ordinary wave; 3, 3′, slow extraordinary wave; 4, maneto-sound wave; 5, Alfvén wave.

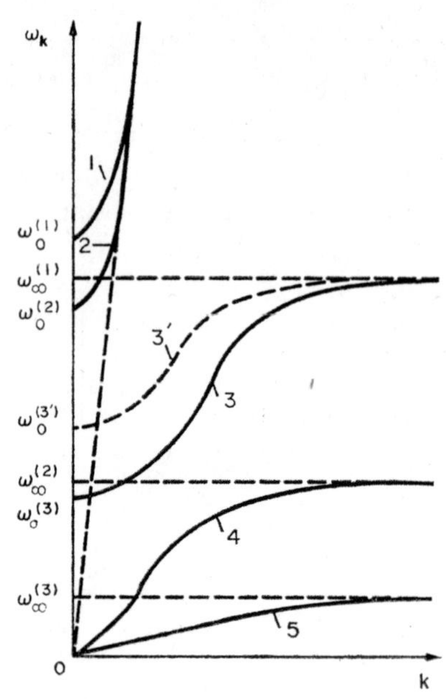

FIG. 3. Frequencies of various eigenwaves in a magneto-active plasma as functions of the wave vector: 1, fast extraordinary wave; 2, ordinary wave; 3, 3′; slow extraordinary wave; 4, magneto-sound wave; 5, Alfvén wave.

bands ($\eta^2(\omega) > 0$) are associated with the eigenfrequencies $\omega_{\mathbf{k}}^{\alpha}$. The latter are shown in Fig. 3 as functions of the wave vector. The eigenfrequencies $\omega_{\mathbf{k}}^{\alpha}$ are determined by (4.50). Since in neglect of the thermal motion (4.50) is a fifth-power equation in ω^2, we can find five eigenfrequencies $\omega_{\mathbf{k}}^{\alpha}$.

Thus five eigenmodes occur in a cold magneto-plasma: the Alfvén, the fast magneto-sonic, the slow extraordinary, the ordinary, and the fast extraordinary waves (Shafranov, 1963). In Figs. 2 and 3 the plots $\eta^2(\omega)$ and $\omega(k)$ for a slow extraordinary wave are shown by solid lines for $\omega_0^{(3)} < \omega_\infty^{(2)}$ and by dashed lines for $\omega_0^{(3)} \to \omega_\infty^{(2)}$.

The frequencies of the propagating waves $\omega_{\mathbf{k}}^{\alpha}$ grow when the wave vector $\mathbf{k}$ increases, i.e. the cold magneto-plasma is a medium with normal dispersion.

Let us consider in more detail the dispersion of various waves in a magneto-active plasma. In the high-frequency band ($\omega^2 \gg \omega_{B_e}\omega_{B_i}$) the ion contribution to (3.42) may be neglected and the electrons only are responsible for the wave dispersion. The quantity $\eta_+(\omega)$ determines the refractive index of the ordinary electromagnetic wave for $\omega > \Omega$ and that of the high-frequency component of the fast magneto-sonic wave for $\omega < \omega_\infty^{(2)}$. $\eta_-(\omega)$ is the refractive index of the extraordinary electromagnetic wave for $\omega > \omega_0^{(1)}$ and that of the slow extraordinary electromagnetic wave for $\omega_\infty^{(1)} > \omega > \omega_0^{(3)}$.

Only the fast magneto-sonic wave can propagate in dense plasmas ($\Omega \gg \omega_{B_e}$) if the frequencies are low ($\omega < \Omega$) and the angles are not close to $\pi/2$. The refractive index of the latter is of the form

$$\eta^2 = \frac{\Omega^2}{\omega(\omega_B \cos\vartheta - \omega)}, \qquad \omega < \omega_B \cos\vartheta. \tag{4.66}$$

The frequency depends on the wave vector as follows:

$$\omega_k = \omega_B \frac{k^2c^2}{\Omega^2 + k^2c^2} \cos\vartheta, \tag{4.67}$$

and for $\omega < \omega_B$

$$\omega_k = \omega_B \frac{k^2c^2}{\Omega^2} \cos\vartheta. \tag{4.68}$$

The fast magneto-sonic wave, the frequency of which is proportional to the square of the wave vector, is conventionally called a helicon or a whistler. The polarization of this wave is circular.

In the low-frequency band ($\omega \ll \omega_{B_e}$, $\omega \ll \Omega$) there may occur Alfvén and fast magneto-sonic waves. The dispersion of those is essentially governed by the ions. When $\Omega_i \gg \omega_{B_i}$ the frequencies of these two waves are the following:

$$\omega_k = kv_A(b \mp \sqrt{b^2 - \cos^2\vartheta})^{1/2}, \quad b \equiv \frac{1}{2}\left[1 + \left(1 + \frac{k^2c^2}{\omega^2}\right)\cos^2\vartheta\right], \tag{4.69}$$

where $v_A = \dfrac{B_0}{\sqrt{4\pi m_i n_0}}$ is the Alfvén velocity ($v_A \ll c$). As the frequencies are very low ($\omega \ll \omega_{B_i}$), (4.69) reduces to

$$\omega_{\mathbf{k}} = kv_A \cos\vartheta, \quad \omega_{\mathbf{k}} = kv_A. \tag{4.70}$$

If the frequency of the Alfvén wave approaches the ion cyclotron frequency (to be more exact, the resonance frequency given by (4.63)), the refractive index of the wave tends to infinity. Such Alfvén waves with $\omega \simeq \omega_{B_i}$ are referred to as ion cyclotron waves. The refractive index and the frequency of the latter are given by the following expressions:

$$\eta^2(\omega) = \frac{1+\cos^2\vartheta}{2\cos^2\vartheta} \frac{\omega}{\omega_{B_i} - \omega} \frac{c^2}{v_A^2}, \quad \omega_{\mathbf{k}} = \left(1 - \frac{1+\cos^2\vartheta}{2\cos^2\vartheta} \frac{\Omega_i}{k^2c^2}\right)\omega_{B_i}. \tag{4.71}$$

Note that the refractive index of the fast magneto-sonic wave is finite at $\omega \simeq \omega_{B_i}$:

$$\eta^2(\omega) = \frac{1}{1+\cos^2\vartheta} \frac{c^2}{v_A^2}. \tag{4.72}$$

Thermal waves in a magneto-plasma

The thermal particle motion makes a significant effect on the waves that propagate in a magneto-active plasma. First, by virtue of the thermal motion there occur some new weakly damped waves (in the absence of a magnetic field the thermal motion of electrons in a plasma with cold ions gives rise to the ion-sound waves). Second, as in the absence of a magnetic field, resonant wave-particle interactions cause energy dissipation (Landau damping). Besides, there exists a cyclotron mechanism of dissipation in a magneto-active plasma, since the charged particles, which move along spiral trajectories under the influence of a magnetic field, can radiate and absorb electromagnetic waves at the cyclotron frequency and its multiples.†

The dielectric permittivity of a thermal plasma evidently depends on **k**, hence the order of the dispersion equation (4.50) is higher than in a cold medium and there appear new roots which are associated with additional oscillations that may occur in a hot magneto-plasma. Taking into account the thermal corrections in (3.37) we obtain an equation of the third order in η^2. If the thermal corrections are small, then two roots of this equation are close to (4.54) and the third one is associated with the condition

$$A \equiv A_0 - \frac{s^2}{c^2} A_1 \eta^2 = 0, \tag{4.73}$$

where

$$A_1 \equiv \left(\cos^4\vartheta + \frac{1}{3}\,\frac{6\omega^6 - 3\omega_B^2\omega^4 + \omega_B^4\omega^2}{(\omega^2-\omega_B^2)^3}\sin^2\vartheta\cos^2\vartheta - \frac{\omega^4}{(\omega^2-\omega_B^2)(\omega^2-4\omega_B^2)}\sin^4\vartheta\right)\frac{\Omega^2}{\omega^2},$$

and determines the refractive index of the longitudinal plasma wave

$$\eta^2(\omega) \equiv \frac{c^2}{s^2}\,\frac{A_0}{A_1}. \tag{4.74}$$

There exist two longitudinal low-frequency modes in a strongly non-isothermal magneto-active plasma—the slow magneto-sonic and the cyclotron sound waves—instead of the single ion-sound wave that can propagate in a non-isothermal plasma without external fields. The frequencies of those two waves are equal to

$$\omega_{\mathbf{k}}^2 = \tfrac{1}{2}(\omega_s^2+\omega_{B_i}^2) \mp \sqrt{(\omega_s^2+\omega_{B_i}^2)^2 - 4\omega_s^2\omega_{B_i}^2\cos^2\vartheta}, \tag{4.75}$$

where ω_s is the ion-sound frequency in the absence of a magnetic field. As $k \to 0$:

$$\omega_{\mathbf{k}} = kv_s\cos\vartheta, \qquad \omega_{\mathbf{k}} = \left(1+\frac{1}{2}\,\frac{k^2v_s^2}{\omega_{B_i}^2}\sin^2\vartheta\right)\omega_{B_i}. \tag{4.76}$$

The frequency of a cyclotron sound wave propagating in a dense plasma ($\Omega_i \gg \omega_{B_i}$) for $ak \ll 1$, $kv_s > \omega_{B_i}$, is equal to the ion-sound frequency in the absence of a magnetic field.

It is suitable to describe the thermal particle motion in the presence of a magnetic field in terms of the Larmor radius $\varrho_\alpha \equiv s_\alpha/\omega_{B\alpha}$. If the latter is finite, then there occur cyclotron eigenwaves that propagate along $\vartheta = \pi/2$. The frequencies of such waves are multiples of the electron or the ion cyclotron frequencies as $k \to 0$ and $k \to \infty$:

$$\omega_{\mathbf{k}} = n\,|\omega_{B\alpha}|, \qquad n = 1, 2, 3, \ldots \tag{4.77}$$

† For a detailed consideration of wave dispersion and absorption in a magneto-plasma taking into account kinetic effects see Sitenko (1967), Ginzburg and Rukhadze (1970), Lominadze (1975), and Akhiezer, Akhiezer, Polovin, Sitenko, and Stepanov (1975).

In the case of quasi-transverse wave propagation (the phase velocity along the magnetic field is very small as compared to the electron thermal velocity but much greater than the ion thermal velocity $s_e > \omega/k_{||} \gg s_i$) there exist weakly damped ion cyclotron waves and non-potential ion cyclotron waves. In the case of transverse propagation (the phase velocity along the magnetic field is much greater than the electron thermal velocity $\omega/k_{||} > s_e$), there may occur longitudinal, ordinary, and extraordinary electron and ion cyclotron waves; the damping of these is very weak (Sitenko and Stepanov, 1957; Bernstein, 1958; Drummond, 1958; Dnestrovskii and Kostomarov, 1961).

CHAPTER 5

Non-linear Wave Interactions

5.1. Non-resonant Wave Interaction and Plasma Echoes

Resonant and non-resonant wave interactions

Non-linear wave interactions manifest themselves both under resonance conditions and far from resonance. In the linear approximation any perturbation in a plasma may be taken to be a superposition of independent eigenoscillations. The time evolution of the perturbation is governed by the linear dissipation which is caused by the resonant wave-particle interactions. For the latter to be significant the particle velocity must be close to the wave phase velocity, i.e.

$$\omega_{\mathbf{k}} = (\mathbf{k} \cdot \mathbf{v}), \tag{5.1}$$

where $\omega_{\mathbf{k}}$ is the eigenfrequency, $\mathbf{k}$ is the wave vector, and $\mathbf{v}$ is the particle velocity. The resonant wave-particle interaction is responsible for the damping of the eigenoscillations as well as for the occurrence of various linear instabilities in a non-equilibrium plasma.

Suppose there are no linear instabilities. The resonant wave-particle interaction in the plasma transparency domain is weak and may be ignored in many cases. Then non-linear wave–wave interactions become important, particularly if the frequencies and the wave vectors of the coupled waves satisfy the matching conditions

$$\omega_{\mathbf{k}} = \sum_{i=1}^{n} \omega_{\mathbf{k}_i}, \quad \mathbf{k} = \sum_{i=1}^{n} \mathbf{k}_i. \tag{5.2}$$

The number of terms on the right-hand side of (5.2) determines the order of the non-linear coupling. The total number of the waves involved into the interaction is larger by one. By virtue of the non-linear wave coupling the coefficients of the perturbation decomposition in eigenoscillations are no longer constants but vary slowly with time. After a sufficiently long time they may be considerably modified from the values predicted by the linear theory.†

The non-linear wave interactions may also be of importance under non-resonance conditions and cause a variety of peculiar effects. The plasma echo is one of the simplest among those.

† A brilliant account of the theory of non-linear wave interactions is given by Sagdeev and Galeev (1969) (see also Galeev and Sagdeev, 1973).

Plasma echoes

The echo generation mechanism in a plasma is directly associated with the non-linear wave interactions. The linear oscillations of the macroscopic quantities (e.g. field, charge density, etc.) are damped exponentially with time even in a collisionless plasma, while the oscillations of the non-equilibrium perturbations of the distribution function may remain undamped (Landau, 1946). The existence of such undamped oscillations of the distribution function is caused by the fact that a collisionless plasma cannot be driven to equilibrium by the self-consistent field alone, since the latter has no effect on the entropy of the system. The presence of the undamped oscillations in the distribution function is that very reason that makes plasma echoes possible.†

Indeed, when macroscopic oscillations of the electric field are induced in a plasma, the interaction between them may be in fact neglected after a period of time which is long compared to the characteristic damping time (which is governed by Landau damping). However, the non-linear coupling of the undamped oscillations of the distribution function (which appear together with the induced electric field oscillations) turns out to be very significant and may excite secondary macroscopic oscillations of the electric field—the echo. The formation times and the shapes of the echoes in a collisionless plasma are governed by the retardation time and the nature of the external perturbations, as well as by the nature of the echo oscillations. Since the plasma echo is by its very nature a non-linear effect, there may exist higher-order echoes. There also may occur a spatial echo in a plasma. The continuous excitement of the electric field oscillations with several frequencies in some two points, spaced at a much greater distance than the damping length, leads to echo oscillations with combination frequencies in some definite regions of the space occupied by the plasma.

Longitudinal echo oscillations

Let us study in detail the temporal plasma echo, specializing for simplicity to the longitudinal echo oscillations associated with perturbations of the distribution function which are caused by external charges. Let the external charge density be of the form

$$\varrho_0(\mathbf{r}, t) = \varrho_1 e^{i(\mathbf{k}_1\cdot\mathbf{r})}\delta(\omega_0 t)+\varrho_2 e^{i(\mathbf{k}_2\cdot\mathbf{r})}\delta[\omega_0(t-\tau)], \tag{5.3}$$

i.e. we assume the external perturbations to be applied to the plasma at $t = 0$ and $t = \tau$ ($\tau \gg \gamma^{-1}$, where γ is the Landau damping rate of the relevant oscillations) and assume a plane wave spatial dependence of the perturbations (ω_0 is a quantity with the dimensions of a frequency, ϱ_1 and ϱ_2 are the amplitudes of the external perturbations). The

† Gould, O'Neil, and Malmberg were the first who discovered the echo phenomena in plasmas (Gould, O'Neil, and Malmberg, 1967; O'Neil and Gould, 1968). They considered Langmuir echo oscillations resulting from two succeeding longitudinal plane wave perturbations with anti-parallel wave vectors. Plasma echoes caused by three succeeding pulses, as well as those associated with the transformation of waves in a plasma, were investigated by Sitenko, Nguen van Chong, and Pavlenko (1970a, b, c). A survey of various echo effects is given by Bachmann, Sauer, and Wallis (1972).

space–time Fourier component of the charge density is given by the following expression:

$$\varrho^0_{\mathbf{k}\omega} = \frac{(2\pi)^3}{\omega_0}\{\varrho_1\delta(\mathbf{k}-\mathbf{k}_1)+\varrho_2\, e^{i\omega\tau}\delta(\mathbf{k}-\mathbf{k}_2)\}. \tag{5.4}$$

Then in the linear approximation

$$\mathbf{E}^{(1)}_{\mathbf{k}\omega} = -4\pi i\frac{\mathbf{k}}{k^2}\,\frac{\varrho^0_{\mathbf{k}\omega}}{\varepsilon(\omega,\mathbf{k})}\,. \tag{5.5}$$

The time asymptotics of the field (5.5) may be easily found by means of an inverse Fourier transformation. The perturbation (5.3) induces damped oscillations of the electric field $\mathbf{E}^{(1)}_{\mathbf{k}}(t)$ at plasma eigenfrequencies. The perturbations ϱ_1 and ϱ_2 cause field oscillations at $t = 0$ and $t = \tau$, respectively. Provided $\tau \gg \gamma^{-1}$ we can neglect the direct coupling between the oscillations associated with ϱ_1 and ϱ_2.

The second iteration of the non-linear field equation (2.14) yields

$$E^{(2)}_{\mathbf{k}\omega} = -\frac{1}{\varepsilon(\omega,\mathbf{k})}\sum_{\omega',\mathbf{k}'}\varkappa^{(2)}(\omega-\omega',\mathbf{k}-\mathbf{k}';\omega',\mathbf{k}')E^{(1)}_{\mathbf{k}-\mathbf{k}',\,\omega-\omega'}E^{(1)}_{\mathbf{k}'\omega'}, \tag{5.6}$$

where $\varkappa^{(2)}(\omega-\omega',\mathbf{k}-\mathbf{k}';\omega',\mathbf{k}')$ is the second-order non-linear electric susceptibility (2.16). We substitute (5.5) for the fields $E^{(1)}_{\mathbf{k}',\omega'}$ and $E^{(1)}_{\mathbf{k}-\mathbf{k}',\,\omega-\omega'}$ and the explicit expressions for $\varrho^0_{\mathbf{k}',\omega'}$ and $\varrho^0_{\mathbf{k}-\mathbf{k}',\,\omega-\omega'}$ in (5.6). Since the plasma echo results from the interference of the two perturbations we must retain only the cross terms in the product of the external charge density components, i.e.

$$\varrho^0_{\mathbf{k}-\mathbf{k}',\,\omega-\omega'}\varrho^0_{\mathbf{k}',\,\omega'} \to \frac{(2\pi)^6}{\omega_0^2}\varrho_1\varrho_2\delta(\mathbf{k}-\mathbf{k}_1-\mathbf{k}_2)\left\{\delta(\mathbf{k}'-\mathbf{k}_2)\,e^{i\omega'\tau}+\delta(\mathbf{k}'-\mathbf{k}_1)\,e^{i(\omega-\omega')\tau}\right\}. \tag{5.7}$$

According to (5.7), the wave vector of the quadratic signal is equal to the sum of the wave vectors of the perturbations

$$\mathbf{k} = \mathbf{k}_1+\mathbf{k}_2. \tag{5.8}$$

After (2.17) is substituted for the non-linear susceptibility and the inverse Fourier transformation is carried out, (5.6) takes the form

$$\mathbf{E}^{(2)}_{\mathbf{k}}(t) = 8(2\pi)^4\, i\mathbf{k}\,\frac{e^3\varrho_1\varrho_2}{m^2\omega_0^2k_1^2k_2^2k^2}\,\delta(\mathbf{k}-\mathbf{k}_1-\mathbf{k}_2)\int d\mathbf{v}\int_{-\infty}^{\infty} d\omega\,\frac{e^{-i\omega t}}{(\omega-(\mathbf{k}\cdot\mathbf{v})+i0)^2\,\varepsilon(\omega,\mathbf{k})}$$

$$\times\int_{-\infty}^{\infty} d\omega'\left\{\frac{(\mathbf{k}\cdot\mathbf{k}_1)\,e^{i\omega'\tau}}{(\omega'-\mathbf{k}_2\mathbf{v}+i0)\,\varepsilon(\omega-\omega',\mathbf{k}_1)\,\varepsilon(\omega',\mathbf{k}_2)}\left(\mathbf{k}_2\cdot\frac{\partial f_0}{\partial\mathbf{v}}\right)\right.$$

$$\left.-\frac{(\mathbf{k}\cdot\mathbf{k}_2)\,e^{i(\omega-\omega')\tau}}{(\omega'-(\mathbf{k}_1\cdot\mathbf{v})+i0)\,\varepsilon(\omega-\omega',\mathbf{k}_2)\,\varepsilon(\omega',\mathbf{k}_1)}\left(\mathbf{k}_1\cdot\frac{\partial f_0}{\partial\mathbf{v}}\right)\right\}. \tag{5.9}$$

The ω' and ω integrations may be carried out using the Cauchy theorem, with the integration contours closed in the relevant complex planes by infinite semicircles in the upper or the lower half-plane, depending on the sign of the factor of ω' or ω in the exponent. The first term within the braces in (5.9) vanishes when integrated over ω'

along the semicircle in the upper half-plane. The semicircle for the ω'-integration of the second term has to lie in the lower half-plane. If τ and the time segment between the beginning of the echo and the second perturbation are long in comparison with γ^{-1}, then only the pole $\omega' = (\mathbf{k}_1 \cdot \mathbf{v})$ contributes in the ω'-integration. (The contributions of the poles in the points where the dielectric permittivity is equal to zero may be neglected since $\gamma\tau \gg 1$.) The ω-integration has to be carried out in a similar manner, taking into account only the second-order pole $\omega = (\mathbf{k}\cdot\mathbf{v})$. The result is

$$\mathbf{E}_{\mathbf{k}}^{(2)}(t) = -8(2\pi)^6 i\mathbf{k}\frac{e^3\varrho_1\varrho_2}{m^2\omega_0^2k_1^2k_2^2k^2}(\mathbf{k}\cdot\mathbf{k}_2)\,\delta(\mathbf{k}-\mathbf{k}_1-\mathbf{k}_2)$$

$$\times\int d\mathbf{v}\,\frac{e^{i\mathbf{k}_1\cdot\mathbf{v}\tau}}{\varepsilon(\mathbf{k}_1\cdot\mathbf{v},\mathbf{k}_1)}\,\frac{\partial}{\partial\mathbf{v}}\left(\frac{e^{i\omega(t-\tau)}}{\varepsilon(\omega,\mathbf{k})\,\varepsilon(\omega-(\mathbf{k}_1\cdot\mathbf{v}),\mathbf{k}_2)}\right)_{\omega=\mathbf{k}\cdot\mathbf{v}}\left(\mathbf{k}_1\cdot\frac{\partial f_0}{\partial\mathbf{v}}\right)$$

$$\simeq -8(2\pi)^6\,\mathbf{k}\,\frac{e\varrho_1\varrho_2}{m^2\omega_0^2k_1^2k_2^2k^2}(\mathbf{k}\cdot\mathbf{k}_2)\,(t-\tau)\,\delta(\mathbf{k}-\mathbf{k}_1-\mathbf{k}_2)$$

$$\times\int d\mathbf{v}\,\frac{e^{-i(\mathbf{k}\cdot\mathbf{v})t-i(\mathbf{k}_2\cdot\mathbf{v})\tau}}{\varepsilon((\mathbf{k}_1\cdot\mathbf{v}),\mathbf{k}_1)\,\varepsilon((\mathbf{k}_2\cdot\mathbf{v}),\mathbf{k}_2)\,\varepsilon((\mathbf{k}\cdot\mathbf{v})\mathbf{k})}\left(\mathbf{k}_1\cdot\frac{\partial f_0}{\partial\mathbf{v}}\right). \tag{5.10}$$

The exponent in the integrand may be written as $\exp[-i(\mathbf{k}\cdot\mathbf{v})(t-\tau')]$ with

$$\tau' \equiv \frac{(\mathbf{k}_2\cdot\mathbf{v})}{(\mathbf{k}\cdot\mathbf{v})}\tau. \tag{5.11}$$

This exponent factor is equal to unity as $t = \tau'$ and the field (5.10) reaches its maximum; the velocity integral in (5.10) vanishes for any $t \neq \tau'$ by virtue of the fast oscillations of the exponent.

The vectors $\mathbf{k}_1$, $\mathbf{k}_2$, and $\mathbf{k}$ are related according to (5.8). It is evident that the echoes can occur only if $\tau' > \tau$. The calculations show that this condition holds if $\mathbf{k}_2$ is opposite to $\mathbf{k}_1$ and the absolute value of $\mathbf{k}_2$ is greater than that of $\mathbf{k}_1$. It may be easily verified that the angle between the vectors $\mathbf{k}_1$ and $\mathbf{k}_2$ can differ from π only by a small amount of the order of magnitude $(\tau ks)^{-1}$. No echo can take place if $\mathbf{k}_2$ is parallel to $\mathbf{k}_1$ (since in such a case $\tau' < \tau$).

Echo oscillations in the case of anti-parallel wave vectors of the perturbations

Let the vector $\mathbf{k}_2$ be anti-parallel to $\mathbf{k}_1$ and $k_2 > k_1$. Then $k = k_2 - k_1$ and

$$\tau' = \frac{k_2}{k_2-k_1}\tau. \tag{5.12}$$

Suppose for simplicity $k_1 = \frac{1}{2}k_2$ and $k = k_1$. Integrating (5.10) over the perpendicular to $\mathbf{k}$ velocity components we obtain the following expression for the field of the echo oscillations:

$$\mathbf{E}^{(2)}(\mathbf{r},t) = 32\pi^3\frac{e^3\varrho_1\varrho_2}{m^2\omega_0^2k^3}(t-\tau)\mathbf{k}\,e^{i(\mathbf{k}\cdot\mathbf{r})}\int_{-\infty}^{\infty}dv\,\frac{e^{ikv(t-\tau')}}{\varepsilon(kv,k)\,\varepsilon(-k_1v,k_1)\,\varepsilon(k_2v,k_2)}\,\frac{\partial f_0}{\partial v}, \tag{5.13}$$

where v is the velocity component along $\mathbf{k}$. The physical meaning of the factors $\varepsilon^{-1}(kv, k)$, $\varepsilon^{-1}(-k_1v, k_1)$, and $\varepsilon^{-1}(k_2v, k_2)$ in (5.13) is very simple. They describe the influence of the dielectric plasma properties on the external perturbations and the echo field. If $ak \ll 1$ (a is the Debye length) the factor $\varepsilon^{-1}(kv, k)$ has a sharp maximum as v is close to the phase velocities of the relevant plasma waves.

The integration in (5.13) over the velocity component along $\mathbf{k}$ may be carried out in the complex v-plane, the contour being closed by an infinite semicircle either in the upper or in the lower half-plane. When $t < \tau'$ the contour must be closed in the upper half-plane (the integral along the semicircle vanishes) and we find that only the pole $\varepsilon(-k_1v, k_1) = 0$ inside the contour contributes:

$$\mathbf{E}^{(2)}(\mathbf{r}, t) = 32\pi^4 i \frac{e^3\varrho_1\varrho_2}{m^2\omega_0^2k^3}(t-\tau)\mathbf{k}\, e^{i(\mathbf{k}\cdot\mathbf{r})}$$

$$\times\sum \frac{e^{(\pm i\Omega-\gamma)(\tau'-t)}}{\left.\frac{\partial}{\partial v}\varepsilon(-kv, k)\right|_{v=\pm\frac{\Omega}{k}}\varepsilon(\pm\Omega-i\gamma, k)\,\varepsilon(2(\pm\Omega-i\gamma), 2k)}\left.\frac{\partial f_0}{\partial v}\right|_{v=\pm\frac{\Omega}{k}}. \tag{5.14}$$

If $t > \tau'$, the semicircle must be closed in the lower half-plane v. Then two poles $\varepsilon(kv, k) = 0$ and $\varepsilon(k_2v, k_2) = 0$ contribute and

$$\mathbf{E}^{(2)}(\mathbf{r}, t) = -32\pi^4 i \frac{e^3\varrho_1\varrho_2}{m^2\omega_0^2k^3}(t-\tau)\mathbf{k}\, e^{i(\mathbf{k}\cdot\mathbf{r})}$$

$$\times\left\{\sum \frac{e^{(\pm i\Omega-\gamma)(t-\tau')}}{\left.\frac{\partial}{\partial v}\varepsilon(kv, k)\right|_{v=\pm\frac{\Omega}{k}}\varepsilon(\mp\Omega-i\gamma, k)\,\varepsilon(2(\pm\Omega-i\gamma), 2k)}\left.\frac{\partial f_0}{\partial v}\right|_{v=\pm\frac{\Omega}{k}}\right.$$

$$\left.+\sum \frac{e^{(\pm i\Omega\times\gamma_2)(t-\tau')}}{\left.\frac{\partial}{\partial v}\varepsilon(2kv, 2k)\right|_{v=\pm\frac{\Omega}{2k}}\varepsilon\left(\frac{1}{2}(\pm\Omega-i\gamma_2), k\right)\varepsilon\left(\frac{1}{2}(\mp\Omega-i\gamma_2), k\right)}\left.\frac{\partial f_0}{\partial v}\right|_{v=\pm\frac{\Omega}{2k}}\right\}. \tag{5.15}$$

Notice that the echo signal is time-asymmetric. The echo oscillations grow as $\exp[-\gamma(\tau'-t)]$, while the terms that describe the damping are proportional to $\exp[-\gamma(t-\tau')]$ or $\exp[-\gamma_2(t-\tau')]$, where γ and γ_2 are the Landau damping rates of the waves with wave vectors $\mathbf{k}$ and $\mathbf{k}_2 = 2\mathbf{k}$ respectively. It should be pointed out that the binary collisions cause the damping of the distribution function oscillations and, hence, weaken the amplitude of the plasma echo (Hinton and Oberman, 1968; O'Neil, 1968).

Let us find the shape of the echo signal which is given rise to by three Langmuir oscillations; suppose $ak \ll 1$. Since $\gamma \sim \exp\left(-\frac{1}{2a^2k^2}\right)$ and $k_2 = 2k$, then $\gamma_2 \gg \gamma$ and the contribution of the second term in (5.15) may be neglected. Making use of the explicit expression (2.26) for the dielectric permittivity, we obtain the following echo amplitude:

$$\mathbf{E}^{(2)}(\mathbf{r}, t) = -32\pi^2 i \frac{e\varrho_1\varrho_2}{m\omega_0^2k^2}\Omega\tau \sin\varphi\,\mathbf{k}\, e^{i(\mathbf{k}\cdot\mathbf{r})-\gamma|t-\tau'|}\cos[\Omega(t-\tau')+\varphi], \tag{5.16}$$

where the phase φ is determined by

$$\tan\varphi = \frac{2k}{k-k_1}\frac{\gamma}{\Omega}.$$

In a similar manner we obtain the shape of the echo signal for ion-sound oscillations in a non-isothermal plasma:

$$\mathbf{E}^{(2)}(\mathbf{r}, t) = 8\pi^2 i \frac{\varrho_1\varrho_2\Omega^2\tau}{m\omega_0^2\gamma_s} a^6k^6\mathbf{k}\, e^{i\mathbf{kr}-\gamma_s|t-\tau'|}\cos[kv_s(t-\tau')], \tag{5.17}$$

where γ_s is the ion-sound damping rate. Notice that the ion-sound echo is time-symmetric for any relations between the wave vectors.

The amplitude of the ion-sound echo produced as a result of the superposition of Langmuir oscillations is as follows:

$$\mathbf{E}^{(2)}(\mathbf{r}, t) = -\frac{32\pi^2}{3} i \frac{e\varrho_1\varrho_2}{m\omega_0^2k^2}\Omega\tau\mathbf{k}\, e^{i(\mathbf{k}\cdot\mathbf{r})}$$

$$\times\begin{cases} \sin\varphi\, e^{-\gamma(\tau'-t)}\cos[\Omega(\tau'-t)+\varphi], & t<\tau'; \\ -\frac{3}{2}\frac{\gamma_2}{\Omega}e^{-\frac{1}{2}\gamma_2(t-\tau')}\cos\left[\frac{1}{2}\Omega(t-\tau')\right] & \\ -\frac{3}{4}\frac{k^2v_s^2}{\Omega\gamma_s}a^4k^4\, e^{-\gamma_s(t-\tau')}\cos[kv_s(t-\tau')], & t>\tau'. \end{cases} \tag{5.18}$$

The Langmuir echo resulting from the superposition of a sound oscillation and a Langmuir one has the amplitude

$$\mathbf{E}^{(2)}(\mathbf{r}, t) = -\frac{32\pi^2}{3} i \frac{e\varrho_1\varrho_2}{m\omega_0^2k^2}\Omega\tau\mathbf{k}\, e^{i(\mathbf{k}\cdot\mathbf{r})}$$

$$\times\begin{cases} -\frac{3}{4}\frac{k^2v_s^2}{\Omega\gamma_s}a^4k^4\, e^{-\gamma_s(\tau'-t)}\cos[kv_s(\tau'-t)], & t<\tau'; \\ \sin\varphi\, e^{-\gamma(t-\tau')}\cos[\Omega(t-\tau')+\varphi], & t>\tau'. \end{cases} \tag{5.19}$$

Echo oscillations of the last two types are time-asymmetric.

A variety of echo phenomena may occur in a bounded plasma too (Sitenko, Pavlenko, and Zasenko, 1975a, b, 1976). There may occur additional echoes associated with the reflection of particles and, hence, of the oscillations of the distribution function perturbation, from the boundaries. It goes without saying that the echoes in a bounded plasma are very sensitive to the boundary conditions for the distribution function.

Three-pulse echo oscillations

As an example of a third order non-linear wave interaction under non-resonance conditions we consider the plasma echo resulting from three perturbations, the time segments between which are longer than the characteristic damping times of the relevant oscillations (Sitenko, Nguen van Chong, and Pavlenko, 1970a). We take the perturbations in the plane wave form and assume that the collinearity conditions for the wave vectors of the perturbations are not satisfied. This makes any echoes caused by the interaction of some two perturbations impossible. If the wave vectors of all perturbations are coplanar, there may occur a third-order echo, which has both longitudinal and transverse components. The formation time and the shape of the echo signal depend in an essential way on the time segments between the pulses.

Let the external perturbations be longitudinal with the charge density of the form

$$\varrho_0(\mathbf{r}, t) = \varrho_1 e^{i(\mathbf{k}_1\cdot\mathbf{r})}\delta(\omega_0 t)+\varrho_2 e^{i(\mathbf{k}_2\cdot\mathbf{r})}\delta[\omega_0(t-\tau)]+\varrho_3 e^{i(\mathbf{k}_3\cdot\mathbf{r})}\delta[\omega_0(t-(1+\lambda)\tau)], \quad (5.20)$$

i.e. the pulses are applied to the plasma at $t = 0$, $t = \tau$, and $t = (1+\lambda)\tau$, where $\tau \gg \gamma^{-1}$ and $\lambda \gtrsim 1$. The perturbation (5.20) excites damped oscillations of the electric field $\mathbf{E}_{\mathbf{k}}^{(1)}(t)$, the frequencies of which are equal to the plasma eigenfrequencies, and undamped oscillations of the distribution function $f_{\mathbf{k}}^{(1)}(t)$ at $\omega = (\mathbf{k}\cdot\mathbf{v})$. Since the pulses are longitudinal, there are no transverse oscillations of the electric field in the linear approximation.

If among the wave vectors of the pulses $\mathbf{k}_1$, $\mathbf{k}_2$, and $\mathbf{k}_3$ there is no pair which is antiparallel, the second-order echoes do not occur. Let us see whether it is possible that the echo occurs in the next approximation as a result of the superposition of three pulses. Making use of the basic non-linear equation (2.66) we obtain the following third-order correction to the field amplitude:

$$E_i^{(3)}(\omega, \mathbf{k}) = -\Lambda_{ij}^{-1}(\omega, \mathbf{k}) \sum_{\omega', \omega'', \mathbf{k}', \mathbf{k}''} \varkappa_{jklm}^{(3)}(\omega-\omega', \mathbf{k}-\mathbf{k}'; \omega'-\omega'', \mathbf{k}'-\mathbf{k}''; \omega'', \mathbf{k}'')$$
$$\times E_k^{(1)}(\omega-\omega', \mathbf{k}-\mathbf{k}') E_l^{(1)}(\omega'-\omega'', \mathbf{k}'-\mathbf{k}'') E_m^{(1)}(\omega'', \mathbf{k}''), \quad (5.21)$$

where $\varkappa_{jklm}^{(3)}(\omega-\omega', \mathbf{k}-\mathbf{k}'; \omega'-\omega'', \mathbf{k}'-\mathbf{k}''; \omega'', \mathbf{k}'')$ is the third-order tensor susceptibility (see (2.71) and (2.72)) and $\Lambda_{ij}^{-1}(\omega, \mathbf{k})$ is the inverse of the tensor (2.67):

$$\Lambda_{ij}^{-1}(\omega, \mathbf{k}) = \frac{k_i k_j}{k^2}\varepsilon_l^{-1}(\omega, k) - \left(\delta_{ij} - \frac{k_i k_j}{k^2}\right)(\varepsilon_t(\omega, k)-\eta^2)^{-1}. \quad (5.22)$$

Substituting (2.71), (2.72), and (5.20) into (5.21), we obtain the following third-order echo field:

$$E_i^{(3)}(\mathbf{r}, t) = 32\pi i \frac{e^4\varrho_1\varrho_2\varrho_3}{m^3\omega_0^3 k_1^2 k_2^2 k_3^2 k^2} e^{i(\mathbf{k}\cdot\mathbf{r})}$$
$$\times \int d\mathbf{v} \int_{-\infty}^{\infty} d\omega \int_{-\infty}^{\infty} d\omega' \int_{-\infty}^{\infty} d\omega'' \frac{e^{-i[\omega(t-(1+\lambda)\tau)+\omega'\lambda\tau+\omega''\tau]}}{\varepsilon_l(\omega-\omega', k_1)\,\varepsilon_l(\omega'-\omega'', k_2)\,\varepsilon_l(\omega'', k_3)} \cdot \frac{1}{\omega(\omega'-(\mathbf{k}'\cdot\mathbf{v})+i0)}$$
$$\times \left(\mathbf{k}_3\cdot\frac{\partial}{\partial\mathbf{v}}\right)\left(\frac{\Lambda_{ij}^{-1}(\omega, \mathbf{k})v_j}{\omega-(\mathbf{k}\cdot\mathbf{v})+i0}\right)\left(\mathbf{k}_2\cdot\frac{\partial}{\partial\mathbf{v}}\right)\left(\frac{1}{\omega''-(\mathbf{k}_1\cdot\mathbf{v})+i0}\left(\mathbf{k}_1\cdot\frac{\partial f_0}{\partial\mathbf{v}}\right)\right), \quad (5.23)$$

where $\mathbf{k} = \mathbf{k}_1+\mathbf{k}_2+\mathbf{k}_3$ and $\mathbf{k}' = \mathbf{k}_2+\mathbf{k}_3$. Since the echo may be only the result of three pulses, we have retained only the cross-terms in (5.23). It follows from (5.23) that the wave vector of the resulting signal is equal to the sum of those of the perturbations.

The electric field (5.23) has both longitudinal and transverse components:

$$\mathbf{E}^{(3)} \equiv \frac{\mathbf{k}}{k} E^l + \boldsymbol{\eta} E^t, \tag{5.24}$$

where $\boldsymbol{\eta}$ is a unit vector perpendicular to $\mathbf{k}$. The integrations over ω'', ω', and ω reduce (5.23) to the following form:

$$E^l(\mathbf{r}, t) = -(4\pi)^4 \frac{e^4 \varrho_1 \varrho_2 \varrho_3}{m^3 \omega_0^3 k_1^2 k_2^2 k_3^2 k^2} \mathbf{k}\cdot(\mathbf{k}\cdot\mathbf{k}_3)(\mathbf{k}_1\cdot\mathbf{k}_2)[t-(1+\lambda)\tau]\tau e^{i(\mathbf{k}\cdot\mathbf{r})}$$
$$\times \int d\mathbf{v} \frac{e^{-i[(\mathbf{k}\cdot\mathbf{v})t-(\mathbf{k}_2\cdot\mathbf{v})\tau-(1+\lambda)(\mathbf{k}_3\cdot\mathbf{v})\tau]}}{\varepsilon_l((\mathbf{k}\cdot\mathbf{v}), k)\,\varepsilon_l((\mathbf{k}_1\cdot\mathbf{v}), k_1)\,\varepsilon_l((k_2\cdot\mathbf{v}), k_2)\,\varepsilon_l((\mathbf{k}_3\cdot\mathbf{v}), k_3)} \left(\mathbf{k}_1\cdot\frac{\partial f_0}{\partial \mathbf{v}}\right). \tag{5.25}$$

It is clear that the field oscillations (5.25) reach a maximum at $t = \tau'$, when the factor of v in the exponent vanishes, i.e.

$$\mathbf{k}_1\tau' + \mathbf{k}_2(\tau'-\tau) + \mathbf{k}_3[\tau'-(1+\lambda)\tau] = 0 \tag{5.26}$$

whence we find

$$\tau' = \frac{(\mathbf{k}\cdot\mathbf{k}_2)+(1+\lambda)(\mathbf{k}\cdot\mathbf{k}_3)}{k^2}\tau. \tag{5.27}$$

For a solution of (5.26) to exist, the vectors $\mathbf{k}_1$, $\mathbf{k}_2$, and $\mathbf{k}_3$ must lie in the same plane. Thus the condition of the echo occurrence reduces to the requirement that the wave vectors of the perturbations be coplanar.

The perpendicular to $\mathbf{k}$ projection of the vector equality (5.26) yields

$$k_{2\perp} + (1+\lambda)k_{3\perp} = 0, \tag{5.28}$$

and, besides,

$$k_{1\perp} + k_{2\perp} + k_{3\perp} = 0. \tag{5.29}$$

The relations (5.28) and (5.29) give us the conditions which must be satisfied for the echo to occur. One of the possibilities to realize this is to require

$$k_{1\perp} = \lambda k_{3\perp}, \quad k_{2\perp} = -(1+\lambda)k_{3\perp}. \tag{5.30}$$

As the echoes may arise only after the third perturbation, $\tau' > (1+\lambda)\tau$, and one more requirement is to be satisfied:

$$k_{2\parallel} + (1+\lambda)k_{3\parallel} > (1+\lambda)k. \tag{5.31}$$

If (5.30) and (5.31) are satisfied we can easily carry out the v-integration in (5.25). Restricting ourselves for simplicity to the case when the wave vectors of the perturbations are related as

$$k_{1\parallel} = -k, \quad k_{2\parallel} = k_{3\parallel} = k, \tag{5.32}$$

we obtain the following longitudinal component of the echo electric field:

$$E^l(\mathbf{r}, t) = E_0 e^{i(\mathbf{k}\cdot\mathbf{r})} \begin{cases} |A_0|\, e^{-\gamma(\tau'-t)} \sin(\Omega(t-\tau')+\delta_0), & t < \tau' \\ |A'|\, e^{-\tilde{\gamma}(t-\tau')} \sin(\Omega(t-\tau')+\delta') & \\ + \sum\limits_{i=1,2,3} |A_i''|\, e^{-\gamma(t-\tau')} \sin(\Omega(t-\tau')+\delta_i), & t > \tau', \end{cases} \tag{5.33}$$

where we have put:

$$E_0 = (4\pi)^4 i \frac{e^2 \varrho_1 \varrho_2 \varrho_3 \Omega^5 \tau^2}{m^2 k_1^2 k_2^2 k_3^2 k^2 k_{3\perp} s^4 \omega_0^3} (\mathbf{k}_1 \cdot \mathbf{k}_2)(\mathbf{k}_3 \cdot \mathbf{k}) \exp\left[-\frac{1}{2a^2k^2}\right],$$

$$A_0 = \frac{1}{\varepsilon_l(\Omega + i\tilde{\gamma}, k)\, \varepsilon_l(\Omega + i(\gamma_1+\gamma_2+\tilde{\gamma}), k_3)},$$

$$A' = -\frac{1}{(\lambda+2)\, \varepsilon_l(-\Omega - i\tilde{\gamma}, k)\, \varepsilon_l(\Omega + i(\gamma_2+\gamma_3-\tilde{\gamma}'), k_1)},$$

$$A_1'' = \frac{\lambda \Omega^4}{[(\Omega^2 - \tilde{\omega}_0^2) + 2i\Gamma_1\Omega]\,[(\Omega^2 - \tilde{\omega}_0^2) - 2i\Gamma_2\Omega]},$$

$$A_2'' = \frac{(1+\lambda)\Omega^4}{[(\Omega^2 - \tilde{\omega}_0^2) + 2i\Gamma_1\Omega]\,[(\Omega^2 - \tilde{\omega}_0^2) - 2i\Gamma_3\Omega]},$$

$$A_3'' = \frac{\Omega^4}{[(\Omega^2 - \tilde{\omega}_0^2) + 2i\Gamma_2\Omega]\,[(\Omega^2 - \tilde{\omega}_0^2) - 2i\Gamma_3\Omega]},$$

$$\delta_0 = \arg A_0, \quad \delta' = \arg A', \quad \delta_i = \arg A_i'' \qquad (i = 1, 2, 3),$$

$$\tilde{\gamma} = (1+\lambda)\gamma_1 + \lambda\gamma_2, \quad \tilde{\gamma}' = \frac{1}{\lambda+2}(\gamma_2 + (1+\lambda)\gamma_3),$$

$$\Gamma_1 = \gamma + \tilde{\gamma}, \quad \Gamma_2 = (1+\lambda)\gamma + \gamma_1 - \lambda\gamma_3, \quad \Gamma_2 = (\lambda+2)\gamma - \gamma_2 - (\lambda-1)\gamma_3,$$

$$\tilde{\omega}_0 = \frac{k}{k_{12}}\Omega, \quad k_{12} = -(1+\lambda)k_{1\parallel} - \lambda k_{2\parallel},$$

γ, γ_1, γ_2, and γ_3 are the Landau damping rates of the waves with wave vectors $\mathbf{k}$, $\mathbf{k}_1$, $\mathbf{k}_2$, and $\mathbf{k}_3$, respectively. Notice the resonant nature of the longitudinal echo amplitudes.

In a similar manner we find the transverse component of the echo electric field:

$$E^t(\mathbf{r}, t) = E_0 e^{i(\mathbf{k}\cdot\mathbf{r})} \begin{cases} e^{-\gamma(\tau'-t)}\{|B_0| \sin(\Omega(t-\tau') + \tilde{\delta}_0) \\ \quad + |B_1| \sin((2\lambda+1)\,\Omega(t-\tau') + \tilde{\delta}_1)\}, & t < \tau', \\ e^{-\tilde{\gamma}'(t-\tau')}\{|B_0'| \sin(\Omega(t-\tau') + \tilde{\delta}_0'), \\ \quad + |B_1'| \sin(\Omega(t-\tau') + \tilde{\delta}_1')\}, & t > \tau', \end{cases} \tag{5.34}$$

where

$$B_0 = i \frac{(\gamma_1 - \gamma_2)\Omega}{k k_{3\perp} c^2 \varepsilon_l(\Omega + i(\gamma_1+\gamma_2+\tilde{\gamma}), k_3)},$$

$$B_0' = \frac{\Omega(2\Omega - i\gamma_2 + i\gamma_3)}{\lambda(\lambda+2) k k_{3\perp} c^2 \varepsilon_l(\Omega, k_1)} \exp\left(-\frac{1}{2a^2\varkappa_0^2}\right),$$

$$B_1 = \frac{k(2\lambda+1)\,\Omega(2\Omega - i\gamma_1 - i\gamma_2)}{k_{3\perp}(4(1+\lambda)\lambda\Omega^2 - k^2c^2)\, \varepsilon_l((1-2\lambda)\Omega, k_3)} \exp\left(-\frac{1}{2a^2\varkappa^2}\right),$$

$$B_1' = -i \frac{(\gamma_2 - \gamma_3)\Omega}{(\lambda+2)^2 k k_{3\perp} c^2 \varepsilon_l(\Omega + i(\gamma_2+\gamma_3-\tilde{\gamma}'), k_1)},$$

$$\tilde{\delta}_i = \arg B_i, \quad \tilde{\delta}_i' = \arg B_i' \qquad (i = 0, 1),$$

$$\varkappa_0^{-2} = \frac{4}{(\lambda+2)^2}\left(\frac{1}{k_{3\perp}^2} - \frac{1-\lambda}{k^2}\right), \quad \varkappa^{-2} = 4\left(\frac{1}{k_{3\perp}^2} + \frac{\lambda(1+\lambda)}{k^2}\right).$$

The longitudinal and the transverse echo amplitudes compete in the general case to the same order.

Consider the limiting case $\lambda \gg 1$, when the time segment between the second and the third pulses is much longer than that between the first and the second ones. If $\lambda \gg 1$ the growth rate of the longitudinal echo oscillations $\tilde{\gamma}$ is large as compared to the damping rate γ and the ratio $|A''_{1,2}|/|A_0|$ is of the order of magnitude λ. The growth rate of the transverse echo oscillations is in this case also large in comparison with the damping rate $\tilde{\gamma}'$ and the ratio $|B_0|/|B'_{0,1}|$ is of the order of magnitude λ^2.

In a manner analogous to the analysis of the two-pulse echo we can show that the distortion of the coplanarity of the wave vectors $\mathbf{k}_1$, $\mathbf{k}_2$, and $\mathbf{k}_3$ makes the third-order echo impossible. This means that the angle θ between some of the wave vectors and the plane of the two others must satisfy the inequality

$$\theta \ll (\tau k s)^{-1}. \tag{5.35}$$

Consider the case when $\mathbf{k}_1$, $\mathbf{k}_2$, and $\mathbf{k}_3$ are coplanar but do not satisfy (5.28). It may be easily shown that the echo does not occur if $(k_{2\perp}\tau+(1+\lambda)k_{3\perp}\tau)/k_{3\perp} \sim \tau$. So for the echo to be excited, the following condition must be satisfied:

$$\left| \frac{k_{2\perp}+(1+\lambda)k_{3\perp}}{k_{3\perp}} \gamma\tau \right| \lesssim 1,$$

i.e.

$$\frac{k_{2\perp}+(1+\lambda)k_{3\perp}}{k_{3\perp}} \ll 1. \tag{5.36}$$

Let $k_{2\perp} = -(1+1)k_\perp - \delta k_\perp$, $k_{3\perp} = k_\perp + \delta k_\perp$, and $k_{1\perp} = \lambda k_\perp$. Then (5.36) reduces to the requirement $\delta k_\perp / k_\perp \ll 1$. On the other hand, suppose that

$$\delta k_\perp / k_\perp \gg (\tau k_\perp s)^{-1}. \tag{5.37}$$

It may be shown that if $\delta k_\perp < 0$ and

$$(\gamma_1 \tau)^{-1} \ll |\delta k_\perp / k_\perp| \ll (\lambda \gamma_3 \tau)^{-1}, \tag{5.38}$$

then the echo amplitude sharply grows at $t = \tau_3' \left(\tau_3' = \tau' + \lambda \left| \frac{\delta k_\perp}{k_\perp} \right| \tau \right)$, then decreases as $\exp[-\gamma(t-\tau_3')]$. But, if

$$(\gamma_1 \tau)^{-1} \gg |\delta k_\perp / k_\perp| \gg (\lambda \gamma_3 \tau)^{-1}, \tag{5.39}$$

the transverse echo oscillations grow exponentially until $t = \tau_1' \left(\tau_1' = \tau' - \left| \frac{\delta k_\perp}{k_\perp} \right| \tau \right)$ and then are rapidly damped. The requirements (5.38) and (5.39) may be satisfied by changing the parameter λ. The above example shows that the shape and the nature of the third-order echo may be significantly modulated by means of τ and λ variations and by an appropriate matching of the wave vectors of the perturbations.

5.2. Three-wave Resonant Coupling

Dynamic equations for interacting waves

If the frequencies and the wave vectors of the interacting waves satisfy the resonance conditions, then the amplitudes vary slowly with time. The equations that determine the time development of the amplitudes may be derived from the requirement that the secularities in the set of the successive approximation equations be removed. Consider the resonant coupling of three waves:

$$\left.\begin{aligned} \mathbf{E}_{\mathbf{k}}(t) &= \mathbf{E}_{\mathbf{k}} \cos(\omega_{\mathbf{k}} t + \phi_{\mathbf{k}}), \\ \mathbf{E}_{\mathbf{k}_1}(t) &= \mathbf{E}_{\mathbf{k}_1} \cos(\omega_{\mathbf{k}_1} t + \phi_{\mathbf{k}_1}), \\ \mathbf{E}_{\mathbf{k}_2}(t) &= \mathbf{E}_{\mathbf{k}_2} \cos(\omega_{\mathbf{k}_2} t + \phi_{\mathbf{k}_2}), \end{aligned}\right\} \tag{5.40}$$

the frequencies and the wave vectors of which satisfy the resonance conditions

$$\omega_{\mathbf{k}} = \omega_{\mathbf{k}_1} + \omega_{\mathbf{k}_2}, \quad \mathbf{k} = \mathbf{k}_1 + \mathbf{k}_2. \tag{5.41}$$

Each of the coupled waves is associated with an energy

$$W = \frac{1}{16\pi} \omega_{\mathbf{k}} \frac{\partial \varepsilon(\omega_{\mathbf{k}}, \mathbf{k})}{\partial \omega_{\mathbf{k}}} |\mathbf{E}_{\mathbf{k}}|^2, \tag{5.42}$$

and so on. The energies may be either positive or negative. The character of the wave energy is determined by the sign of the derivative $\dfrac{\partial \varepsilon(\omega_{\mathbf{k}}, \mathbf{k})}{\partial \omega_{\mathbf{k}}}$. (It should be remembered that the frequencies $\omega_{\mathbf{k}}$, $\omega_{\mathbf{k}_1}$, and $\omega_{\mathbf{k}_2}$ were defined positive.) From (2.55) we obtain the following equation for the time-evolution of the amplitude $\mathbf{E}_{\mathbf{k}}$ due to the non-linear resonant interaction of waves:

$$\frac{\partial}{\partial t} E_{\mathbf{k}} e^{-i\phi_{\mathbf{k}}} = \frac{i}{2} \frac{1}{\dfrac{\partial \varepsilon(\omega_{\mathbf{k}}, \mathbf{k})}{\partial \omega_{\mathbf{k}}}} \sum_{\mathbf{k}_1 + \mathbf{k}_2 = \mathbf{k}} \varkappa^{(2)}(\omega_{\mathbf{k}_1}, \mathbf{k}_1; \omega_{\mathbf{k}_2}, \mathbf{k}_2) E_{\mathbf{k}_1} E_{\mathbf{k}_2} e^{-i(\phi_{\mathbf{k}_1} + \phi_{\mathbf{k}_2})}, \tag{5.43}$$

where $\varkappa^{(2)}(\omega_{\mathbf{k}_1}, \mathbf{k}_1; \omega_{\mathbf{k}_2}, \mathbf{k}_2)$ is the second-order plasma electric susceptibility. Notice that the right-hand side of (5.43) takes into account the interaction of waves with different frequencies and wave vectors satisfying the conditions (5.41).

Let the amplitudes $\mathbf{A}_{\mathbf{k}}$ and the sign factors $s_{\mathbf{k}}$ be defined as

$$\mathbf{A}_{\mathbf{k}} \equiv \frac{1}{\sqrt{16\pi}} \sqrt{\left| \frac{\partial \varepsilon(\omega_{\mathbf{k}}, \mathbf{k})}{\partial \omega_{\mathbf{k}}} \right|} \, \mathbf{E}_{\mathbf{k}} e^{-i\phi_{\mathbf{k}}}, \tag{5.44}$$

$$s_{\mathbf{k}} \equiv \operatorname{sgn}\left(\frac{\partial \varepsilon(\omega_{\mathbf{k}}, \mathbf{k})}{\partial \omega_{\mathbf{k}}} \right). \tag{5.45}$$

Substituting (5.44) and (5.45) in (5.42) we have

$$W_{\mathbf{k}} = s_{\mathbf{k}} \omega_{\mathbf{k}} |\mathbf{A}_{\mathbf{k}}|^2. \tag{5.46}$$

It is clear that the sign factor $s_{\mathbf{k}}$ determines the sign of the wave energy. The latter is positive if $s_{\mathbf{k}} > 0$ and negative if $s_{\mathbf{k}} < 0$. The normalization of the amplitude $\mathbf{A_k}$ in (5.44) was chosen to provide $|\mathbf{A_k}|^2$ to be the number of quanta with a given frequency $\omega_{\mathbf{k}}$:

$$N_{\mathbf{k}} \equiv |\mathbf{A_k}|^2. \tag{5.47}$$

Within the context of (5.47) the wave energy (5.46) may be written in the form

$$W_{\mathbf{k}} = s_{\mathbf{k}} \omega_{\mathbf{k}} N_{\mathbf{k}}. \tag{5.48}$$

In a similar manner we define the wave momentum:

$$\mathbf{P_k} = s_{\mathbf{k}} \mathbf{k} N_{\mathbf{k}}. \tag{5.49}$$

The basic equation (5.43), using (5.44) and (5.45), may be reduced to the Schrödinger equation in the interaction representation (Galeev and Karpman, 1963):

$$i \frac{\partial \mathbf{A_k}}{\partial t} = s_{\mathbf{k}} \sum_{\mathbf{k}_1 + \mathbf{k}_2 = \mathbf{k}} V_{\mathbf{k};\mathbf{k}_1,\mathbf{k}_2} \mathbf{A}_{\mathbf{k}_1} \mathbf{A}_{\mathbf{k}_2}, \tag{5.50}$$

where $V_{\mathbf{k};\mathbf{k}_1,\mathbf{k}_2}$ is a matrix element of the interaction:

$$V_{\mathbf{k};\mathbf{k}_1,\mathbf{k}_2} \equiv -\sqrt{4\pi}\, \frac{\varkappa^{(2)}(\omega_{\mathbf{k}_1}, \mathbf{k}_1;\, \omega_{\mathbf{k}_2}, \mathbf{k}_2)}{\sqrt{\varepsilon'_{\mathbf{k}} \varepsilon'_{\mathbf{k}_1} \varepsilon'_{\mathbf{k}_2}}}, \qquad \varepsilon'_{\mathbf{k}} = \frac{\partial \varepsilon(\omega_{\mathbf{k}}, \mathbf{k})}{\partial \omega_{\mathbf{k}}}. \tag{5.51}$$

Equation (5.50) provides a complete description of the dynamics of resonant wave coupling.

It may be easily verified that by virtue of the symmetry properties (2.21) and (2.22) of the second-order longitudinal susceptibility $\varkappa^{(2)}(\omega_{\mathbf{k}_2}\, \mathbf{k}_1;\, \omega_{\mathbf{k}_2}\, \mathbf{k}_2)$ the matrix elements (5.51) satisfy the following conditions:

$$V_{\mathbf{k};\mathbf{k}_1,\mathbf{k}_2} = V_{\mathbf{k};\mathbf{k}_2,\mathbf{k}_1} = V_{\mathbf{k}_1;\mathbf{k},-\mathbf{k}_2}. \tag{5.52}$$

It follows also from (2.20) that

$$V^*_{\mathbf{k};\mathbf{k}_1,\mathbf{k}_2} = -V_{-\mathbf{k};-\mathbf{k}_1,-\mathbf{k}_2}. \tag{5.53}$$

The real part of the non-linear susceptibility $\varkappa^{(2)}(\omega_{\mathbf{k}_1}, \mathbf{k}_1;\, \omega_{\mathbf{k}_2}, \mathbf{k}_2)$ may be neglected in the plasma transparency domain, so that

$$V^*_{\mathbf{k};\mathbf{k}_1,\mathbf{k}_2} = -V_{\mathbf{k};\mathbf{k}_1,\mathbf{k}_2} \tag{5.54}$$

and hence

$$V_{\mathbf{k};\mathbf{k}_1,\mathbf{k}_2} = V_{-\mathbf{k};-\mathbf{k}_1,-\mathbf{k}_2}. \tag{5.55}$$

It should be emphasized that the relations (5.54) and (5.55) are valid only if the linear dissipation due to the resonant wave-particle interaction is negligibly small.

Conservation of energy and momentum of interacting waves

We can easily show that the energy and the total momentum of the coupled wave ensemble are conserved due to the symmetries of the matrix elements of the non-linear interaction. Indeed, (5.46) and (5.50) yield

$$\frac{d}{dt}\sum_{\mathbf{k}} W_{\mathbf{k}} = 2\,\mathrm{Re}(-i) \sum_{\mathbf{k}_1,\mathbf{k}_2,\mathbf{k}} \omega_{\mathbf{k}} V_{\mathbf{k};\mathbf{k}_1,\mathbf{k}_2} \mathbf{A}^*_{\mathbf{k}} \mathbf{A}_{\mathbf{k}_1} \mathbf{A}_{\mathbf{k}_2}. \tag{5.56}$$

It follows from (5.52) that

$$\sum_{\mathbf{k}_1,\mathbf{k}_2,\mathbf{k}} \omega_{\mathbf{k}} V_{\mathbf{k};\mathbf{k}_1,\mathbf{k}_2} \mathbf{A}^*_{\mathbf{k}} \mathbf{A}_{\mathbf{k}_1} \mathbf{A}_{\mathbf{k}_2} = \sum_{\mathbf{k}_1,\mathbf{k}_2,\mathbf{k}} \omega_{\mathbf{k}} V_{\mathbf{k}_1;\mathbf{k},-\mathbf{k}_2} \mathbf{A}^*_{\mathbf{k}} \mathbf{A}_{\mathbf{k}_1} \mathbf{A}_{\mathbf{k}_2}.$$

Let us carry out the interchanges $\mathbf{k} \to -\mathbf{k}_1$ and $\mathbf{k}_1 \to -\mathbf{k}$, $\omega_{\mathbf{k}} \to -\omega_{\mathbf{k}_1}$, and $\omega_{\mathbf{k}_1} \to -\omega_{\mathbf{k}}$, $\phi_{\mathbf{k}} \to -\phi_{\mathbf{k}_1}$, and $\phi_{\mathbf{k}_1} \to -\phi_{\mathbf{k}}$ which do not violate the resonance condition (5.41) or change the sum of the phases in the amplitude product on the right-hand side of (5.56). Then by virtue of (5.55) we have

$$\sum_{\mathbf{k}_1,\mathbf{k}_2,\mathbf{k}} \omega_{\mathbf{k}} V_{\mathbf{k};\mathbf{k}_1,\mathbf{k}_2} \mathbf{A}^*_{\mathbf{k}} \mathbf{A}_{\mathbf{k}_1} \mathbf{A}_{\mathbf{k}_2} = -\sum_{\mathbf{k}_1,\mathbf{k}_2,\mathbf{k}} \omega_{\mathbf{k}_1} V_{\mathbf{k};\mathbf{k}_1;\mathbf{k}_2} \mathbf{A}^*_{\mathbf{k}} \mathbf{A}_{\mathbf{k}_1} \mathbf{A}_{\mathbf{k}_2}.$$

Thus, the right-hand side of (5.56) may be written as

$$\frac{d}{dt}\sum_{\mathbf{k}} W_{\mathbf{k}} = \frac{2}{3}\,\mathrm{Re}(-i) \sum_{\mathbf{k}_1,\mathbf{k}_2,\mathbf{k}} (\omega_{\mathbf{k}} - \omega_{\mathbf{k}_1} - \omega_{\mathbf{k}_2}) V_{\mathbf{k};\mathbf{k}_1,\mathbf{k}_2} \mathbf{A}^*_{\mathbf{k}} \mathbf{A}_{\mathbf{k}_1} \mathbf{A}_{\mathbf{k}_2}$$

and within the context of the resonance condition (5.41) we obtain the following energy conservation law:

$$\frac{d}{dt}\sum_{\mathbf{k}} W_{\mathbf{k}} = 0. \tag{5.57}$$

In a similar manner we derive the relation

$$\frac{d}{dt}\sum_{\mathbf{k}} P_{\mathbf{k}} = \frac{2}{3}\,\mathrm{Re}(-i) \sum_{\mathbf{k}_1,\mathbf{k}_2,\mathbf{k}} (\mathbf{k} - \mathbf{k}_1 - \mathbf{k}_2) V_{\mathbf{k};\mathbf{k}_1,\mathbf{k}_2} \mathbf{A}^*_{\mathbf{k}} \mathbf{A}_{\mathbf{k}_1} \mathbf{A}_{\mathbf{k}_2},$$

and, since $\mathbf{k} = \mathbf{k}_1 + \mathbf{k}_2$, the conservation of the total momentum of interacting waves follows immediately from this:

$$\frac{d}{dt}\sum_{\mathbf{k}} \mathbf{P}_{\mathbf{k}} = 0. \tag{5.58}$$

Notice that the conservation laws (5.57) and (5.58) are valid only in neglect of the linear dissipation (caused by the resonant wave-particle interactions) and are direct consequences of the energy and momentum conservation laws for the elementary processes—the decay of a quantum into two or the fusion of two quanta into a one.

The conservation laws are suitable for the formulation of the stability criteria for a coupled wave ensemble. If all wave energies have the same sign, then the infinite growth of the amplitudes $|\mathbf{A}_{\mathbf{k}}|$ is impossible by virtue of the total energy conservation (5.57). The presence of both negative ($s_{\mathbf{k}'} < 0$) and positive ($s_{\mathbf{k}} > 0$) energy waves drives the system unstable, since the energy may be transferred from the former to the latter and the amplitudes $|\mathbf{A}_{\mathbf{k}'}|$ and $|\mathbf{A}_{\mathbf{k}}|$ can grow infinitely, which does not contradict the total energy being conserved:

$$\sum_{\mathbf{k}} W_{\mathbf{k}} \equiv \sum_{\mathbf{k}} \omega_{\mathbf{k}} |\mathbf{A}_{\mathbf{k}}|^2 - \sum_{\mathbf{k}'} \omega_{\mathbf{k}'} |\mathbf{A}_{\mathbf{k}'}|^2 = \text{const.} \tag{5.59}$$

Resonant three-wave coupling

Let us study in more detail the resonant interaction of three waves with frequencies $\omega_{\mathbf{k}}$, $\omega_{\mathbf{k}_1}$ and $\omega_{\mathbf{k}_2}$ and fixed wave vectors $\mathbf{k}$, $\mathbf{k}_1$, and $\mathbf{k}_2$. The basic dynamic equation (5.50) becomes then simplified

$$i\frac{\partial \mathbf{A}_{\mathbf{k}}}{\partial t} = s_{\mathbf{k}} V_{\mathbf{k};\,\mathbf{k}_1,\,\mathbf{k}_2} \mathbf{A}_{\mathbf{k}_1} \mathbf{A}_{\mathbf{k}_2}; \tag{5.60}$$

two other equations may be easily written within the context of the symmetry properties (5.52):

$$i\frac{\partial \mathbf{A}_{\mathbf{k}_1}}{\partial t} = s_{\mathbf{k}_1} V^*_{\mathbf{k};\,\mathbf{k}_1,\,\mathbf{k}_2} \mathbf{A}_{\mathbf{k}} \mathbf{A}^*_{\mathbf{k}_2}, \tag{5.61}$$

$$i\frac{\partial \mathbf{A}_{\mathbf{k}_2}}{\partial t} = s_{\mathbf{k}_2} V^*_{\mathbf{k};\,\mathbf{k}_1,\,\mathbf{k}_2} \mathbf{A}^*_{\mathbf{k}_1} \mathbf{A}_{\mathbf{k}}. \tag{5.62}$$

The set (5.60)–(5.62) is sufficient to describe the dynamics of three interacting waves with fixed frequencies and wave vectors. Notice that all equations of the set contain the same interaction matrix element $V_{\mathbf{k};\,\mathbf{k}_1,\,\mathbf{k}_2}$. Let the modulus of the latter be denoted by V:

$$|V_{\mathbf{k};\,\mathbf{k}_1,\,\mathbf{k}_2}| \equiv V. \tag{5.63}$$

The immediate consequence of (5.60)–(5.62) is the following conservation laws:

$$|\mathbf{A}_{\mathbf{k}}|^2 + s_{\mathbf{k}} s_{\mathbf{k}_1} |\mathbf{A}_{\mathbf{k}_1}|^2 = P, \tag{5.64}$$

$$|\mathbf{A}_{\mathbf{k}}|^2 + s_{\mathbf{k}} s_{\mathbf{k}_2} |\mathbf{A}_{\mathbf{k}_2}|^2 = Q, \tag{5.65}$$

where P and Q are constants. Those may be related to the total energy of the system:

$$W_k + W_{\mathbf{k}_1} + W_{\mathbf{k}_2} = s_{\mathbf{k}}(\omega_{\mathbf{k}_1} P + \omega_{\mathbf{k}_2} Q). \tag{5.66}$$

The set (5.60)–(5.62) may be transformed to a more suitable form. The time-differentiation of the left-hand and the right-hand sides of (5.60) and the use of (5.61) and (5.62) lead to

$$\frac{\partial^2 \mathbf{A}_{\mathbf{k}}}{\partial t^2} = -V^2 \{ s_{\mathbf{k}} s_{\mathbf{k}_1} |\mathbf{A}_{\mathbf{k}_2}|^2 + s_{\mathbf{k}} s_{\mathbf{k}_2} |\mathbf{A}_{\mathbf{k}_1}|^2 \} \mathbf{A}_{\mathbf{k}}. \tag{5.67}$$

Analogous speculations yield

$$\frac{\partial^2 \mathbf{A}_{\mathbf{k}_1}}{\partial t^2} = -V^2\{s_{\mathbf{k}}s_{\mathbf{k}_1}|\mathbf{A}_{\mathbf{k}_2}|^2 - s_{\mathbf{k}_1}s_{\mathbf{k}_2}|\mathbf{A}_{\mathbf{k}}|^2\}\mathbf{A}_{\mathbf{k}_1}, \tag{5.68}$$

$$\frac{\partial^2 \mathbf{A}_{\mathbf{k}_2}}{\partial t^2} = -V^2\{s_{\mathbf{k}}s_{\mathbf{k}_2}|\mathbf{A}_{\mathbf{k}_1}|^2 - s_{\mathbf{k}_1}s_{\mathbf{k}_2}|\mathbf{A}_{\mathbf{k}}|^2\}\mathbf{A}_{\mathbf{k}_2}. \tag{5.69}$$

The sets of coupled equations (5.60)–(5.62) or (5.67)–(5.69) can be solved exactly (Armstrong, Bloembergen, Ducuing, and Pershan, 1962). Without loss of generality we can restrict ourselves by considering two cases:

$$s_{\mathbf{k}} = s_{\mathbf{k}_1} = s_{\mathbf{k}_2} \tag{5.70}$$

and

$$s_{\mathbf{k}} = -s_{\mathbf{k}_1} = -s_{\mathbf{k}_2}. \tag{5.71}$$

In the first case the non-linear wave coupling drives the so-called decay instability.

Decay instability

If all wave energies have equal signs (positive or negative), i.e. (5.70) is satisfied, then the frequencies match the resonance condition

$$\omega_{\mathbf{k}} = \omega_{\mathbf{k}_1} + \omega_{\mathbf{k}_2}$$

only if $\omega_{\mathbf{k}} > \omega_{\mathbf{k}_1}$ and $\omega_{\mathbf{k}} > \omega_{\mathbf{k}_2}$. Then (5.67)–(5.69) reduce to the following:

$$\left.\begin{aligned}
\frac{\partial^2 \mathbf{A}_{\mathbf{k}}}{\partial t^2} &= -V^2\{|\mathbf{A}_{\mathbf{k}_1}|^2 + |\mathbf{A}_{\mathbf{k}_2}|^2\}\mathbf{A}_{\mathbf{k}},\\
\frac{\partial^2 \mathbf{A}_{\mathbf{k}_1}}{\partial t^2} &= V^2\{|\mathbf{A}_{\mathbf{k}}|^2 - |\mathbf{A}_{\mathbf{k}_2}|^2\}\mathbf{A}_{\mathbf{k}_1},\\
\frac{\partial^2 \mathbf{A}_{\mathbf{k}_2}}{\partial t^2} &= V^2\{|\mathbf{A}_{\mathbf{k}}|^2 - |\mathbf{A}_{\mathbf{k}_1}|^2\}\mathbf{A}_{\mathbf{k}_2}.
\end{aligned}\right\} \tag{5.72}$$

Consider the simplest case. Let the amplitudes of the interacting waves satisfy the following initial conditions at $t = 0$:

$$|\mathbf{A}_{\mathbf{k}}|^2 \gg |\mathbf{A}_{\mathbf{k}_1}|^2 \quad \text{and} \quad |\mathbf{A}_{\mathbf{k}}|^2 \gg |\mathbf{A}_{\mathbf{k}_2}|^2,$$

i.e. we assume that the wave with the frequency $\omega_{\mathbf{k}}$ has a large amplitude $A_{\mathbf{k}}$, while the amplitudes $A_{\mathbf{k}_1}$ and $A_{\mathbf{k}_2}$ of the waves with frequencies $\omega_{\mathbf{k}_1}$ and $\omega_{\mathbf{k}_2}$ are small. It follows from (5.72) that the amplitude $A_{\mathbf{k}}$ varies insignificantly during the initial stage of the time evolution, while the amplitudes $A_{\mathbf{k}_1}$ and $A_{\mathbf{k}_2}$ grow exponentially with time. The growth rate of the latter two waves is governed by the intensity of the wave with the frequency $\omega_{\mathbf{k}}$:

$$|\gamma| \simeq V\sqrt{|\mathbf{A}_{\mathbf{k}}|^2}. \tag{5.73}$$

The inverse of (5.73) is conventionally called the decay time. Thus by virtue of the non-linear coupling the wave with frequency $\omega_{\mathbf{k}}$ decays into two waves with frequencies $\omega_{\mathbf{k}_1}$ and $\omega_{\mathbf{k}_2}$. The first to consider the decay instability were Oraevskii and Sagdeev. They studied the decay of a Langmuir wave in a non-isothermal plasma into a Langmuir wave that propagates in the opposite direction and an ion-sound wave (Oraevskii and Sagdeev, 1963).†

The set of coupled equations (5.72) can be solved exactly. The analysis of the solution shows that the decay process in the three-wave ensemble is time-reversible (Galeev and Sagdeev, 1973). (Reversibility is the direct consequence of the system stability with the relation (5.70) being satisfied.)

Explosive instability

If the wave energies have different signs and satisfy the relations (5.71), the set (5.67)–(5.69) may be written as:

$$\left.\begin{aligned} \frac{\partial^2 \mathbf{A}_{\mathbf{k}}}{\partial t^2} &= V^2\{|\mathbf{A}_{\mathbf{k}_1}|^2+|\mathbf{A}_{\mathbf{k}_2}|^2\}\mathbf{A}_{\mathbf{k}}, \\ \frac{\partial^2 \mathbf{A}_{\mathbf{k}_1}}{\partial t^2} &= V^2\{|\mathbf{A}_{\mathbf{k}}|^2+|\mathbf{A}_{\mathbf{k}_2}|^2\}\mathbf{A}_{\mathbf{k}_1}, \\ \frac{\partial^2 \mathbf{A}_{\mathbf{k}_2}}{\partial t^2} &= V^2\{|\mathbf{A}_{\mathbf{k}}|^2+|\mathbf{A}_{\mathbf{k}_1}|^2\}\mathbf{A}_{\mathbf{k}_2}. \end{aligned}\right\} \tag{5.74}$$

It is clear from this that the amplitudes of the interacting waves may grow infinitely irrespective of the initial conditions.‡

Let the amplitudes be of the form

$$\mathbf{A}_{\mathbf{k}} = |\mathbf{A}_{\mathbf{k}}|\, e^{i\psi_{\mathbf{k}}}, \quad \mathbf{A}_{\mathbf{k}_1} = |\mathbf{A}_{\mathbf{k}_1}|\, e^{i\psi_{\mathbf{k}_1}}, \quad \mathbf{A}_{\mathbf{k}_2} = |\mathbf{A}_{\mathbf{k}_2}|\, e^{i\psi_{\mathbf{k}_2}}, \tag{5.75}$$

where the phases $\psi_{\mathbf{k}}$, $\psi_{\mathbf{k}_1}$, and $\psi_{\mathbf{k}_2}$ are real. It becomes evident from (5.74) that the phase

$$\theta \equiv \psi_{\mathbf{k}}+\psi_{\mathbf{k}_1}+\psi_{\mathbf{k}_2}$$

is governed by the equation

$$\frac{\partial \theta}{\partial t} = \cot\theta \frac{\partial}{\partial t} \ln\{|\mathbf{A}_{\mathbf{k}}|\cdot|\mathbf{A}_{\mathbf{k}_1}|\cdot|\mathbf{A}_{\mathbf{k}_2}|\}. \tag{5.76}$$

This equation has an integral of motion

$$|\mathbf{A}_{\mathbf{k}}|\cdot|\mathbf{A}_{\mathbf{k}_1}|\cdot|\mathbf{A}_{\mathbf{k}_2}| \cos\theta = G, \tag{5.77}$$

where G is a constant.

† Various decay instabilities in plasmas are considered in a paper by Galeev and Sagdeev (1973), which contains a detailed bibliography on the matter.

‡ The explosive instability of coupled positive and negative energy waves was first studied by Dikasov, Rudakov, and Ryutov (1965) on the basis of a kinetic wave equation. The explosive instability of waves with fixed phases was examined by Coppi, Rosenbluth, and Sudan (1969) and Wilhelmsson, Stenflo, and Engelmann (1970).

Substituting (5.75) into the first equation (5.74) and taking the imaginary part we have

$$\frac{\partial}{\partial t}|\mathbf{A}_{\mathbf{k}}| = s_{\mathbf{k}}V|\mathbf{A}_{\mathbf{k}_1}|\cdot|\mathbf{A}_{\mathbf{k}_2}|\sin\theta. \tag{5.78}$$

The quantities $|\mathbf{A}_{\mathbf{k}_1}|$, $|\mathbf{A}_{\mathbf{k}_2}|$ and θ may be expressed in terms of $|A_{\mathbf{k}}|^2$ and the integrals of motion P, Q, and G. The resulting equation

$$\frac{\partial}{\partial t}|\mathbf{A}_{\mathbf{k}}|^2 = 2s_{\mathbf{k}}V\sqrt{|\mathbf{A}_{\mathbf{k}}|^2(|\mathbf{A}_{\mathbf{k}}|^2-P)(|\mathbf{A}_{\mathbf{k}}|^2-Q)-G^2} \tag{5.79}$$

may be integrated:

$$t = \frac{1}{2V}\int\limits_{|\mathbf{A}_{\mathbf{k}}|_0^2}^{|\mathbf{A}_{\mathbf{k}}|^2}\frac{d\beta}{\sqrt{(\beta-\beta_1)(\beta-\beta_2)(\beta-\beta_3)}}, \tag{5.80}$$

where β_1, β_2, and β_3 are the roots of the cubic equation

$$\beta(\beta-P)(\beta-Q)-G^2 = 0 \tag{5.81}$$

and $|\mathbf{A}_{\mathbf{k}}|_0^2$ is the absolute square of the initial amplitude. The solution (5.80) may be rewritten in terms of elliptic integrals (Coppi, Rosenbluth, and Sudan, 1969).

To be definite let us consider the case $s_{\mathbf{k}} = -s_{\mathbf{k}_1} = -s_{\mathbf{k}_2} = 1$ and take such initial conditions that the roots of (5.81) are real and satisfy the condition

$$\beta_1 < \beta_2 < \beta_3.$$

Assuming that

$$N_{\mathbf{k}}(0) \equiv |\mathbf{A}_{\mathbf{k}}|_0^2 \geqslant \beta_3 > Q > P$$

and introducing new variables

$$y(t) = \sqrt{\frac{\beta_3-\beta_1}{N_{\mathbf{k}}(t)-\beta_1}}, \quad \varkappa = \sqrt{\frac{\beta_2-\beta_1}{\beta_3-\beta_1}} \qquad (N_{\mathbf{k}}(t) \equiv |\mathbf{A}_{\mathbf{k}}|^2), \tag{5.82}$$

we can reduce (5.80) to the standard elliptic integral

$$t = -\frac{1}{V\sqrt{\beta_3-\beta_1}}\int\limits_{y(0)}^{y(t)}\frac{dy}{\sqrt{(1-y^2)(1-\varkappa^2y^2)}}. \tag{5.83}$$

Putting

$$y(0) \equiv -\operatorname{sn}\left(V\sqrt{\beta_3-\beta_1}\,t_\infty, \varkappa\right), \tag{5.84}$$

we can present the solution (5.83) in the form

$$N_{\mathbf{k}}(t) = \beta_1 + \frac{\beta_3-\beta_1}{\operatorname{sn}^2\left(V\sqrt{\beta_3-\beta_1}(t_\infty-t), \varkappa\right)}. \tag{5.85}$$

Since $\operatorname{sn}\left(V\sqrt{\beta_3-\beta_1}(t_\infty-t), \varkappa\right)$ vanishes as $t \to t_\infty$, $N_{\mathbf{k}}(t)$ turns to infinity according to a

$(t_\infty - t)^{-2}$ law at $t = t_\infty$. Thus, the instability is of an explosive nature, since the amplitudes grow to infinity during some finite time t_∞. The latter may be found from (5.84). It is a function of the initial wave intensity:

$$t_\infty = -\frac{1}{V\sqrt{\beta_3-\beta_1}}\,\mathrm{sn}^{-1}\left(\frac{\beta_3-\beta_1}{N_{\mathbf{k}}(0)-\beta_1},\ \varkappa\right). \tag{5.86}$$

In particular, if $P = Q = G = 0$, we have

$$N_{\mathbf{k}}(t) = \frac{N_{\mathbf{k}}(0)}{\left(1-\dfrac{t}{t_\infty}\right)^2}, \tag{5.87}$$

where

$$t_\infty \equiv \frac{1}{V\sqrt{N_{\mathbf{k}}(0)}}. \tag{5.88}$$

It is to be kept in mind that these formulas become inadequate for the times about t_∞, when the amplitudes of the interacting waves grow very large.

To describe the saturation of the explosive instability one must take into account the higher-order non-linear wave interactions and the influence of the self-consistent field on the particle distributions. Various mechanisms for the saturation of the explosive instabilities were considered by Stenflo (1970), Hamasaki and Krall (1971), and Oraevskii, Wilhelmsson, Kogan, and Pavlenko (1973).

5.3. Four-wave Resonant Coupling

The resonance condition and dynamic wave equations

In the absence of three-wave resonances, those of four and more waves may play a significant role. Let us consider the four-wave resonant coupling which manifests itself when the matching conditions for the frequencies and the wave vectors of four interacting waves are satisfied. Let $\omega_{\mathbf{k}_1}$, $\omega_{\mathbf{k}_2}$, and $\omega_{\mathbf{k}_3}$ be the frequencies and $\mathbf{k}_1$, $\mathbf{k}_2$, and $\mathbf{k}_3$ be the wave vectors of the three waves, the non-linear coupling of which leads to the resonance with a fourth wave associated with frequency $\omega_{\mathbf{k}}$ and wave vector $\mathbf{k}$. Let the wave functions depend on time and space according to

$$\mathbf{E}(\mathbf{r}, t) = \sum_{\mathbf{k}} \mathbf{E}_{\mathbf{k}}(t)e^{i(\mathbf{k}\cdot\mathbf{r})}, \quad \mathbf{E}_{\mathbf{k}}(t) = \mathbf{E}_{\mathbf{k}}\cos(\omega_{\mathbf{k}}t+\phi_{\mathbf{k}}), \tag{5.89}$$

and so on. Resonance may occur if

$$\omega_{\mathbf{k}} = |\omega_{\mathbf{k}_1} \pm \omega_{\mathbf{k}_2} \pm \omega_{\mathbf{k}_3}|, \tag{5.90}$$

$$\mathbf{k} = \mathbf{k}_1+\mathbf{k}_2+\mathbf{k}_3. \tag{5.91}$$

(The frequencies $\omega_{\mathbf{k}_1}$, $\omega_{\mathbf{k}_2}$, $\omega_{\mathbf{k}_3}$ and $\omega_{\mathbf{k}}$ are assumed to be positive.)

By virtue of the resonant coupling the wave amplitudes are functions of the time t'' (we shall use the notation t instead of t''). This time dependence is governed by the equations which follow immediately from the requirement that the secularities in the equations of the third approximation be removed. The time-dependence of the amplitude E_k due to the non-linear resonant wave coupling is determined, according to (2.61), by the following equation:

$$\frac{\partial E_{\mathbf{k}}}{\partial t} = -\frac{i}{4\varepsilon'_{\mathbf{k}}} \sum_{\mathbf{k}_1+\mathbf{k}_2+\mathbf{k}_3=\mathbf{k}}' A(\omega_{\mathbf{k}_1}, \mathbf{k}_1; \omega_{\mathbf{k}_2}, \mathbf{k}_2; \omega_{\mathbf{k}_3}, \mathbf{k}_3) e^{i(\phi_{\mathbf{k}}-\phi_{\mathbf{k}_1}-\phi_{\mathbf{k}_2}-\phi_{\mathbf{k}_3})} E_{\mathbf{k}_1} E_{\mathbf{k}_2} E_{\mathbf{k}_3}, \quad (5.92)$$

where

$$\varepsilon'_{\mathbf{k}} \equiv \frac{\partial \varepsilon(\omega_{\mathbf{k}}, \mathbf{k})}{\partial \omega_{\mathbf{k}}},$$

$$A(\omega_{\mathbf{k}_1}, \mathbf{k}_1; \omega_{\mathbf{k}_2}, \mathbf{k}_2; \omega_{\mathbf{k}_3}, \mathbf{k}_3) = \tfrac{1}{3}\{\bar{A}(\omega_{\mathbf{k}_1}, \mathbf{k}_1; \omega_{\mathbf{k}_2}, \mathbf{k}_2; \omega_{\mathbf{k}_3}, \mathbf{k}_3) + \bar{A}(\omega_{\mathbf{k}_2}, \mathbf{k}_2; \omega_{\mathbf{k}_1}, \mathbf{k}_1; \omega_{\mathbf{k}_3}, \mathbf{k}_3) + \bar{A}(\omega_{\mathbf{k}_3}, \mathbf{k}_3; \omega_{\mathbf{k}_2}, \mathbf{k}_2; \omega_{\mathbf{k}_1}, \mathbf{k}_1)\},$$

$$\bar{A}(\omega_{\mathbf{k}_1}, \mathbf{k}_1; \omega_{\mathbf{k}_2}, \mathbf{k}_2; \omega_{\mathbf{k}_3}, \mathbf{k}_3) \equiv 2\frac{\varkappa^{(2)}(\omega_{\mathbf{k}_1}, \mathbf{k}_1; \omega_{\mathbf{k}_2}+\omega_{\mathbf{k}_3}, \mathbf{k}_2+\mathbf{k}_3)\, \varkappa^{(2)}(\omega_{\mathbf{k}_2}, \mathbf{k}_2; \omega_{\mathbf{k}_3}, \mathbf{k}_3)}{\varepsilon(\omega_{\mathbf{k}_2}+\omega_{\mathbf{k}_3}, \mathbf{k}_2+\mathbf{k}_3)} - \bar{\varkappa}^{(3)}(\omega_{\mathbf{k}_1}, \mathbf{k}_1; \omega_{\mathbf{k}_2}, \mathbf{k}_2; \omega_{\mathbf{k}_3}, \mathbf{k}_3). \quad (5.93)$$

($\bar{\varkappa}^{(3)}(\omega_{\mathbf{k}_1}, \mathbf{k}_1; \omega_{\mathbf{k}_2}, \mathbf{k}_2; \omega_{\mathbf{k}_3}, \mathbf{k}_3)$ is the third-order non-symmetrized non-linear susceptibility. The ion components of the non-linear susceptibilities may be neglected in (5.93) by virtue of the great ion mass.) Equation (5.92) is valid only within the context of the resonance condition (5.90). It should be remembered that the prime on the sum symbol Σ' in (5.92) implies that all possible realizations of the resonance condition (5.90) are to be taken into account. Analogous equations for the amplitudes $E_{\mathbf{k}_1}$, etc., may be obtained from (5.92) by means of the renotations of the frequencies and the wave vectors. The set thus derived completely describes the dynamics of the interacting waves.

If only the waves of some certain type are present in the plasma (for instance, only weakly damped Langmuir oscillations with frequencies $\omega_{\mathbf{k}} = \sqrt{\Omega^2+k^2s^2}$ may exist in an isothermal plasma), one of the frequencies must enter the right-hand side of (5.90) with the minus sign for the resonance to occur. For example, the resonance is possible if

$$\omega_{\mathbf{k}} = -\omega_{\mathbf{k}_1}+\omega_{\mathbf{k}_2}+\omega_{\mathbf{k}_3}. \quad (5.94)$$

If (5.92) is symmetric relative to $\mathbf{k}_1$, $\mathbf{k}_2$, and $\mathbf{k}_3$, then the resonances $\omega_{\mathbf{k}} = \omega_{\mathbf{k}_1}-\omega_{\mathbf{k}_2}+\omega_{\mathbf{k}_3}$ and $\omega = \omega_{\mathbf{k}_1}+\omega_{\mathbf{k}_2}-\omega_{\mathbf{k}_3}$ may be reduced to (5.94) by renotations.

One of the most important manifestations of the four-wave resonant interaction in a plasma is the eigenfrequency shift. In particular, the resonant coupling of Langmuir waves shifts the Langmuir frequency of the plasma. The non-linear shift of the eigenfrequencies of the Langmuir waves was investigated in a number of papers (see, for example, Sturrock, 1957, 1964; Jackson, 1960; Wilhelmsson, 1961; Gorbunov and Timberbulatov, 1968; Sitenko and Zasenko, 1978b).

Resonant coupling of four waves with fixed phases

Let us consider the resonant coupling of four travelling Langmuir waves with fixed wave vectors $\mathbf{k}_1, \mathbf{k}_2, \mathbf{k}_3$, and $\mathbf{k}$. The amplitudes entering the wave functions in (5.89) are to be taken in the form

$$\mathbf{E}_{\mathbf{k}'}\, e^{-i\phi_{\mathbf{k}}} = (2\pi)^3\, \delta(\mathbf{k}'-\mathbf{k})\mathbf{E}, \tag{5.95}$$

where $\mathbf{E}$ is an arbitrary complex amplitude ($\mathbf{E} \equiv \mathbf{E}_0\, e^{-i\phi_0}$, $\mathbf{E}_0$ is the absolute value of the amplitude, and ϕ_0 is a real phase). Indeed, by virtue of the reality condition $\mathbf{E}_{\mathbf{k}'} = \mathbf{E}^*_{-\mathbf{k}'}$ the space-time Fourier component of the field with such an amplitude may be written as

$$\mathbf{E}_{\mathbf{k}'\omega} = \tfrac{1}{2}(2\pi)^4\, \{\mathbf{E}\delta(\mathbf{k}'-\mathbf{k})\, \delta(\omega-\omega_{\mathbf{k}})+\mathbf{E}^*\delta(\mathbf{k}'+\mathbf{k})\, \delta(\omega+\omega_{\mathbf{k}})\} \tag{5.96}$$

and hence the wave function (5.89) describes a travelling monochromatic plane wave

$$\mathbf{E}(\mathbf{r}, t) = \mathbf{E}_0 \cos(\omega_{\mathbf{k}}t-\mathbf{kr}+\phi_0). \tag{5.97}$$

According to (5.96), the wave vector $\mathbf{k}' = \mathbf{k}$ corresponds in the travelling wave to the frequency $\omega = \omega_{\mathbf{k}}$, so that $\mathbf{k}' = -\mathbf{k}$ is associated with $\omega = -\omega_{\mathbf{k}}$.

Suppose the wave vectors of the four travelling waves under consideration satisfy the condition

$$\mathbf{k} = \mathbf{k}_1+\mathbf{k}_2-\mathbf{k}_3. \tag{5.98}$$

Then the resonant coupling can occur only if

$$\omega_{\mathbf{k}} = \omega_{\mathbf{k}_1}+\omega_{\mathbf{k}_2}-\omega_{\mathbf{k}_3}. \tag{5.99}$$

In neglect of dispersion, i.e. if the frequencies are treated as independent of the wave vectors, all frequencies in (5.99) reduce to the Langmuir frequency Ω and the matching condition is satisfied exactly. Since the Langmuir waves are weakly damped only in the long-wave limiting case, when the thermal corrections to the frequencies are small, the resonance condition holds also if we take into account the thermal corrections to the eigenfrequencies of the interacting waves.

Within the context of (5.98) and (5.99) we can write down the following equation that governs the time-evolution of the amplitude E:

$$\frac{\partial E}{\partial t} = -\frac{i}{4\varepsilon'_{\mathbf{k}}} A(\omega_{\mathbf{k}_1}, \mathbf{k}_1;\, \omega_{\mathbf{k}_2}, \mathbf{k}_2;\, -\omega_{\mathbf{k}_3}, -\mathbf{k}_3)E_1E_2E_3^*. \tag{5.100}$$

We restrict ourselves for simplicity to the case when

$$\mathbf{k}_1 = \mathbf{k}, \quad \mathbf{k}_2 = \mathbf{k}_3 = \mathbf{k}'. \tag{5.101}$$

Then

$$\frac{\partial E}{\partial t} = -\frac{i}{4\varepsilon'_{\mathbf{k}}} A(\omega_{\mathbf{k}}, \mathbf{k};\, \omega_{\mathbf{k}'}, \mathbf{k}';\, -\omega_{\mathbf{k}'}, -\mathbf{k}')\, |E'|^2 E. \tag{5.102}$$

The solution of (5.102) may be written as

$$E(t) = E\, e^{-i\, \Delta\omega_{\mathbf{k}} t}, \tag{5.103}$$

where the frequency shift $\Delta\omega_{\mathbf{k}}$ is of the form

$$\Delta\omega_{\mathbf{k}} = \frac{1}{4\varepsilon'_{\mathbf{k}}} A(\omega_{\mathbf{k}}, \mathbf{k}; \omega_{\mathbf{k}'}, \mathbf{k}'; -\omega_{\mathbf{k}'}, -\mathbf{k}')\, |E'|^2, \tag{5.104}$$

and the explicit expression for $A(\omega_{\mathbf{k}}, \mathbf{k}; \omega_{\mathbf{k}'}, \mathbf{k}'; -\omega_{\mathbf{k}'}, -\mathbf{k}')$ follows immediately from (5.93):

$$\begin{aligned} A(\omega, \mathbf{k}; \omega', \mathbf{k}'; -\omega', -\mathbf{k}') \equiv \frac{2}{3} \Bigg(& \frac{\varkappa^{(2)}(\omega, \mathbf{k}; 0, 0)\, \varkappa^{(2)}(\omega', \mathbf{k}'; -\omega', -\mathbf{k}')}{\varepsilon(0, 0)} \\ & + \frac{\varkappa^{(2)}(\omega', \mathbf{k}'; \omega-\omega', \mathbf{k}-\mathbf{k}')\, \varkappa^{(2)}(-\omega', -\mathbf{k}'; \omega, \mathbf{k})}{\varepsilon(\omega-\omega', \mathbf{k}-\mathbf{k}')} \\ & + \frac{\varkappa^{(2)}(-\omega', \mathbf{k}'; \omega+\omega', \mathbf{k}+\mathbf{k}')\, \varkappa^{(2)}(\omega', \mathbf{k}'; \omega, \mathbf{k})}{\varepsilon(\omega+\omega', \mathbf{k}+\mathbf{k}')} \Bigg) \\ & - \varkappa^{(3)}(\omega, \mathbf{k}; \omega', \mathbf{k}'; -\omega', -\mathbf{k}'). \end{aligned} \tag{5.105}$$

The solution (5.103) is valid if the imaginary part of the frequency shift (5.104) either vanishes or is small. It becomes evident from (5.104) that the non-linear frequency shift of the eigenoscillations is proportional to the modulus square of the amplitude of the coupled waves.

Non-linear frequency shift of eigenmodes in a cold plasma

Let us calculate the non-linear frequency shift for Langmuir oscillations in neglect of the thermal motion of plasma particles. Making use of (2.26), (2.30), and (2.37), we come to the following relation:

$$\begin{aligned} & \bar{A}(\omega_1, \mathbf{k}_1; \omega_2, \mathbf{k}_2; \omega_3, \mathbf{k}_3) \\ & = -\frac{1}{2}\frac{e^2}{m^2} \frac{\Omega^2}{\omega_1\omega_2\omega_3\omega[(\omega_2+\omega_3)^2-\Omega^2]} \frac{1}{k_1 k_2 k_3 k} \\ & \times \Bigg\{ \Bigg((\mathbf{k}_2+\mathbf{k}_3)^2 (\mathbf{k}_2\cdot\mathbf{k}_3) + \frac{\omega_2+\omega_3}{\omega_2} k_2^2(\{\mathbf{k}_2+\mathbf{k}_3\}\cdot\mathbf{k}_3) + \frac{\omega_2+\omega_3}{\omega_3} (\mathbf{k}_2\cdot\{\mathbf{k}_2+\mathbf{k}_3\})\cdot k_3^2 \Bigg) (\mathbf{k}\cdot\mathbf{k}_1) \\ & + \frac{\Omega^2}{(\mathbf{k}_2+\mathbf{k}_3)^2} \Bigg(\frac{k^2}{\omega} (\mathbf{k}_1\cdot\{\mathbf{k}_2+\mathbf{k}_3\}) + \frac{k_1^2}{\omega_1} (\mathbf{k}\cdot\{\mathbf{k}_2+\mathbf{k}_3\}) \Bigg) \\ & \times \Bigg(\frac{k_2^2}{\omega_2} (\{\mathbf{k}_2+\mathbf{k}_3\}\cdot\mathbf{k}_3) + \frac{k_3^2}{\omega_3} (\mathbf{k}_2\cdot\{\mathbf{k}_2+\mathbf{k}_3\}) \Bigg) \\ & + (\omega_2+\omega_3) \Bigg(\frac{k^2}{\omega} (\mathbf{k}_1\cdot\{(\mathbf{k}_2+\mathbf{k}_3\}) + \frac{k_1^2}{\omega_1} (\mathbf{k}\cdot\{\mathbf{k}_2+\mathbf{k}_3\}) \Bigg) (\mathbf{k}_2\cdot\mathbf{k}_3). \end{aligned} \tag{5.106}$$

It follows then from (5.101) that the frequency of a Langmuir wave involved into a resonant coupling with three other Langmuir waves in a cold plasma is shifted by

$$\begin{aligned} \Delta\omega_{\mathbf{k}} = & \frac{1}{36}\frac{e^2}{m^2}\frac{k^2k'^2}{\Omega^3} \\ & \times \frac{(k^2+k'^2)(1+\cos^2\vartheta+2\cos^4\vartheta) - 2kk'(1-5\cos^2\vartheta+4\cos^4\vartheta)\cos\vartheta}{k^4+k'^4+2k^2k'^2(1-2\cos^2\vartheta)} |E'|^2 \end{aligned} \tag{5.107}$$

where ϑ is the angle between $\mathbf{k}$ and $\mathbf{k}'$. The non-linear frequency shift of the Langmuir wave, caused by the resonant interaction of the latter with three other Langmuir waves, depends on the relative orientations of the four wave vectors.

The non-linear shift (5.107) vanishes in the one-dimensional case $\mathbf{k} \| \mathbf{k}'$ and $\vartheta = 0$. This result confirms the conclusion of Akhiezer, Lyubarsky, and Polovin that the frequencies of the one-dimensional non-linear Langmuir waves in the cold collisionless plasma do not depend on the amplitudes (Akhiezer and Lyubarskii, 1951; Akhiezer and Polovin, 1955; Polovin, 1957).†

Effect of electron thermal motion on non-linear frequency shifts

Let us calculate the contribution of the thermal motion to the non-linear shift of the Langmuir frequency. Consider the most interesting one-dimensional case, for which no shift is predicted in neglect of the thermal motion, i.e. let $\mathbf{k} \| \mathbf{k}'$. With the general expressions (2.23), (2.24). and (2.41) substituted for $\varepsilon(\omega, \mathbf{k})$, $\varkappa^{(2)}(\omega_1, \mathbf{k}_1, \omega_2, \mathbf{k}_2)$ and $\bar{\varkappa}^{(3)}(\omega_1, \mathbf{k}_1; \omega_2, \mathbf{k}_2; \omega_3, \mathbf{k}_3)$, and the vectors $\mathbf{k}_1, \mathbf{k}_2$, and $\mathbf{k}_3$ being collinear, the expression for $\bar{A}(\omega_1, \mathbf{k}_1; \omega_2, \mathbf{k}_2; \omega_3, \mathbf{k}_3)$ reduces to the following:

$$
\begin{aligned}
&\bar{A}(\omega_1, k_1; \omega_2, k_2; \omega_3, k_3) \\
&\quad \simeq -\frac{1}{2}\frac{e^2}{m^2}\operatorname{sgn}(k_1k_2k_3)\frac{\tilde{k}^2}{\omega_1\omega_2\omega_3\omega}\Bigg\{\frac{1}{1+\varkappa_e(\tilde{\omega}, \tilde{k})} \\
&\quad \times\left[1+\varkappa_e(\tilde{\omega}, \tilde{k})+\frac{z_1 z}{2\sqrt{\pi}a^2\tilde{k}^2}\int dy\,\frac{1}{(z_1-y)(z-y)}\left(\frac{1}{z_1-y}+\frac{1}{\tilde{z}-y}+\frac{1}{z-y}\right)\frac{e^{-y^2}}{\tilde{z}-y}\right] \\
&\quad \times\left[1+\varkappa_e(\tilde{\omega}, \tilde{k})+\frac{z_2 z_3}{2\sqrt{\pi}a^2\tilde{k}^2}\int dy\,\frac{1}{(z_2-y)(z_3-y)}\left(\frac{1}{z_2-y}+\frac{1}{z_3-y}+\frac{1}{z'-y}\right)\frac{e^{-y^2}}{\tilde{z}-y}\right] \\
&\quad -1-\varkappa_e(\tilde{\omega}, \tilde{k})-\frac{z_1 z}{2\sqrt{\pi}a^2\tilde{k}^2}\int dy\,\frac{1}{(z_1-y)(z-y)}\left(\frac{1}{z_1-y}+\frac{1}{\tilde{z}-y}+\frac{1}{z-y}\right)\frac{e^{-y^2}}{\tilde{z}-y} \\
&\quad -\frac{z_2 z_3}{2\sqrt{\pi}a^2\tilde{k}^2}\int dy\,\frac{1}{(z_2-y)(z_3-y)}\left(\frac{1}{z_2-y}+\frac{1}{z_3-y}+\frac{1}{\tilde{z}-y}\right)\frac{e^{-y^2}}{\tilde{z}-y}+\frac{z_1z_2z_3z}{2\sqrt{\pi}a^2\tilde{k}^2} \\
&\quad \times\int dy\,\frac{1}{(z_1-y)(z_2-y)(z_3-y)(z-y)}\left(\frac{1}{z_1-y}+\frac{1}{z_2-y}+\frac{1}{z_3-y}+\frac{1}{\tilde{z}-y}+\frac{1}{z-y}\right) \\
&\quad \times\frac{e^{-y^2}}{\tilde{z}-y}-\frac{[1+\varkappa_e(\tilde{\omega}, \tilde{k})]\,\varkappa_i(\tilde{\omega}, \tilde{k})}{1+\varkappa_e(\tilde{\omega}, \tilde{k})+\varkappa_i(\tilde{\omega}, \tilde{k})}\Bigg\},
\end{aligned}
\tag{5.108}
$$

† See also Bernstein, Greene, and Kruskal (1957) and Hubbard (1961b). Davidson has shown that the oscillations of the average electric field in a cold plasma occur only at the Langmuir frequencies (Davidson, 1968).

where

$$\varkappa_e(\tilde{\omega}, \tilde{k}) = -\frac{1}{2\sqrt{\pi}a^2\tilde{k}^2}\int dy \frac{e^{-y^2}}{(\tilde{z}-y)^2},$$

$$\varkappa_i(\tilde{\omega}, \tilde{k}) = -\frac{1}{2\sqrt{\pi}a_i^2\tilde{k}^2}\int dy \frac{e^{-y^2}}{(z_i-y)^2},$$

$\tilde{\omega} = \omega_2+\omega_3$, $\tilde{k} = k_2+k_3$ (k_1, k_2, and k_3 are the projections of the vectors $\mathbf{k}_1$, $\mathbf{k}_2$, and $\mathbf{k}_3$ on the $\mathbf{k}$-direction),

$$y = \sqrt{\frac{3}{2}}\frac{v}{s}, \quad z = \sqrt{\frac{3}{2}}\frac{\omega}{ks}, \quad z_q = \sqrt{\frac{3}{2}}\frac{\omega_q}{k_q s}\ (q = 1, 2, 3), \quad \tilde{z} = \sqrt{\frac{3}{2}}\frac{\tilde{\omega}}{\tilde{k}s},$$

$$\text{and} \quad a^2 = T/(4\pi e^2 n_0)$$

(the subscript i labels the relevant quantities for the ions). The last term within the curly brackets in (5.108) is associated with the contribution of the ion thermal motion.

Expand the integrands in (5.108) in power series in the ratio of the particle velocity to the wave-phase velocity and retain the correction terms, proportional to the square of this ratio. Since such a procedure implies the resonant wave-particle interaction to be neglected, this treatment may be regarded as the hydrodynamic approximation (which suggests the quantities z, z_q ($q = 1, 2, 3$) and $\tilde{z}$ to be large as compared to unity). Neglecting the ion contribution, we find the non-symmetrized quantities $\bar{A}(\omega_1, k_1; \omega_2, k_2; \omega_3, k_3)$ to be equal to

$$\bar{A}(\omega_1, k_1; \omega_2, k_2; \omega_3, k_3) = -\frac{1}{2}\frac{e^2}{m^2}\frac{\Omega^2}{\omega_1\omega_2\omega_3\omega(\tilde{\omega}^2-\Omega^2)}$$

$$\times \operatorname{sgn}(k_1k_2k_3)\left\{\Omega^2\left(\frac{k_1}{\omega_1}+\frac{k}{\omega}\right)\left(\frac{k_2}{\omega_2}+\frac{k_3}{\omega_3}\right)+\tilde{\omega}\left(\frac{k_1}{\omega_1}+\frac{k_2}{\omega_2}+\frac{k_3}{\omega_3}+\frac{k}{\omega}\right)\tilde{k}+\tilde{k}^2\right.$$

$$+\left[\left(\Omega^2\left(\frac{k_1}{\omega_1}+\frac{k}{\omega}\right)\left(\frac{k_2}{\omega_2}+\frac{k_3}{\omega_3}\right)+\tilde{\omega}\left(\frac{k_1}{\omega_1}+\frac{k_2}{\omega_2}+\frac{k_3}{\omega_3}+\frac{k}{\omega}\right)\tilde{k}+\tilde{k}^2\right)\right.$$

$$\times\left(\frac{k_1^2}{\omega_1^2}+\frac{k_2^2}{\omega_2^2}+\frac{k_3^2}{\omega_3^2}+\frac{\tilde{k}^2}{\tilde{\omega}^2-\Omega^2}+\frac{k^2}{\omega^2}\right)+\tilde{\omega}\left(\frac{k_1k_2k_3}{\omega_1\omega_2\omega_3}+\frac{k_1k_2k}{\omega_1\omega_2\omega}+\frac{k_1k_3k}{\omega_1\omega_3\omega}+\frac{k_2k_3k}{\omega_2\omega_3\omega}\right)\tilde{k}$$

$$\left.\left.+\left(\frac{k_1k_2}{\omega_1\omega_2}+\frac{k_1k_3}{\omega_1\omega_3}+\frac{k_1k}{\omega_1\omega}+\frac{k_2k_3}{\omega_2\omega_3}+\frac{k_2k}{\omega_2\omega}+\frac{k_3k}{\omega_3\omega}\right)\tilde{k}^2\right]s^2\right\}, \tag{5.109}$$

and therefore

$$\Delta\omega_{\mathbf{k}} = \frac{5}{12}\frac{e^2}{m^2}\frac{k^2k'^2}{\Omega^5}s^2|E'|^2. \tag{5.110}$$

This result is in agreement with that obtained by Wilhelmsson (1961) within the context of the hydrodynamic approach. It is clear from (5.110) that the thermal motion modifies the four-wave resonant interaction which thus causes eigenfrequency shifts even in the one-dimensional case.

Note that the use of the hydrodynamical expression (5.109) for the calculation of the shift (5.110) is only reasonable if the particle thermal velocity is small as compared to the phase velocities both of the waves and of the pulsations. Since those of the Langmuir waves do not satisfy the second requirement, the hydrodynamic approximation is not applicable to the treatment of $\Delta\omega_k$ of the Langmuir waves (Sitenko and Zasenko, 1978b).

Indeed, by virtue of the dispersion law for the Langmuir wave, the dimensionless frequencies $\tilde{z}$ in the expressions for

$$\bar{A}(\omega_{k'}, k'; -\omega_{k'}, -k'; \omega_k, k) \quad \text{and} \quad \bar{A}(\omega_k, k; \omega_{k'}, k'; -\omega_{k'}, -k')$$

$\left(\text{which are equal to } \frac{1}{2}\sqrt{\frac{3}{2}}\frac{(k+k')s}{\Omega} \text{ and } \sqrt{\frac{3}{2}}\frac{k's}{\Omega}, \text{ respectively}\right)$ are small in comparison with unity, hence the expansion in an inverse power series in $\tilde{z}$ cannot be carried out. Let $\tilde{z} \ll 1$ in (5.108) and expand the integrands in an inverse power series in large z and z'. The results in neglect of the ion motion are as follows:

$$\bar{A}(\omega_{k'}, k'; -\omega_{k'}, -k'; \omega_k, k) = \frac{e^2}{m^2}\frac{1}{\Omega^4}\left\{k^2+k'^2+\frac{1}{2}(k^4+5k^3k'-2k^2k'^2+5kk'^3+k'^4)\frac{s^2}{\Omega^2} + i\,\text{sign}(k-k')\frac{3}{4}\sqrt{\frac{3\pi}{2}}k^2k'^2(k+k')\frac{s^3}{\Omega^3}\right\}, \tag{5.111}$$

$$\bar{A}(\omega_{k'}, k'; \omega_{k'}, k'; -\omega_k, -k) = \frac{e^2}{m^2}\frac{kk'}{\Omega^4}\left\{2+(k^2+3kk'+k'^2)\frac{s^2}{\Omega^2}\right\}, \tag{5.112}$$

while

$$\bar{A}(-\omega_{k'}, -k'; \omega_{k'}, k'; \omega_k, k) = -\frac{e^2}{m^2}\frac{1}{\Omega^4}\left\{(k+k')^2+\frac{1}{2}(k^4+7k^3k'+12k^2k'^2+7kk'^3+k'^4)\frac{s^2}{\Omega^2}\right\} \tag{5.113}$$

was calculated within the framework of the hydrodynamic model.

Thus, making use of the general formula (5.104) we find

$$\Delta\omega_k = \frac{1}{6}\frac{e^2}{m^2}\frac{k^2k'^2}{\Omega^5}s^2\left\{1-i\,\text{sign}(k-k')\frac{3}{16}\sqrt{\frac{3\pi}{2}}\frac{(k+k')s}{\Omega}\right\}|E'|^2. \tag{5.114}$$

The real part of (5.114) determines the non-linear eigenfrequency shift, the imaginary part is responsible for the non-linear Landau damping (the non-linear damping rate (5.114) was calculated by Drummond and Pines, 1962). The non-linear eigenfrequency shift (5.114) vanishes for $T = 0$ similarly to the results of the hydrodynamic treatment. However, the numerical values of the shift (5.114) and of the hydrodynamical one are different.

It is to be observed that when the non-symmetrized quantities (5.111), (5.112), and (5.113) are calculated with the use of the general relation (5.108), the terms are compensated, which are $1/a^2k^2$ times greater than the resulting values. The non-linear shift contains the symmetrized sum of the quantities (5.111), (5.112), and (5.113), and again there occurs a compensation of the terms which do not depend on the square of the electron thermal velocity since the mobilities of the electron and of the obscuring cloud are equal in neglect of the ion motion. That is why the wave scattering by the electrons competes with that by the screening space charges. By virtue of this compensation the non-linear shift (5.114) is proportional to the square of the electron thermal velocity.

Influence of the ion motion

The ion motion breaks down the compensation of the scattering effects caused by the electrons and by the screening space charges, since the mobilities are in this case different. Let us examine the influence of the ion motion on the non-linear frequency shift and the non-linear Landau damping. We shall retain now the last term within the curly brackets in (5.108) and let

$$\tilde{z} \equiv \sqrt{\frac{3}{2}}\frac{\tilde{\omega}}{\tilde{k}s} \ll 1 \quad \text{and} \quad \tilde{z}_i \equiv \sqrt{\frac{3}{2}}\frac{\tilde{\omega}}{\tilde{k}s_i} \gg 1.$$

Then the dielectric permittivity is to be written as

$$\varepsilon(\tilde{\omega}, \tilde{k}) = 1+\frac{1}{a^2\tilde{k}^2}(1+i\,\text{sign}\,\tilde{k}\,\sqrt{\pi}\tilde{z})-\frac{\Omega_i^2}{\omega^2}. \tag{5.115}$$

We assume the plasma polarization due to the ion motion to be small as compared to that caused by the electrons, i.e.

$$a^2k^2 > \frac{m}{m_i} \tag{5.116}$$

(remember that $a^2k^2 < 1$ for weakly damped Langmuir waves). Then the ion contributions in (5.111) and (5.112) are as follows:

$$\delta\bar{A}(\omega_{k'}, k'; -\omega_{k'}, -k'; \omega_k, k)$$
$$= -2\frac{e^2}{m}\frac{m}{m_i}\frac{1}{(k+k')^2 s^4}\left[1-i\sqrt{\frac{2\pi}{3}}\,\text{sgn}(k-k')\frac{m}{m_i}\frac{\Omega}{(k+k')s}\right], \tag{5.117}$$

$$\delta\bar{A}(\omega_k, k; \omega_{k'}, k'; -\omega_{k'}, -k') = -\frac{1}{2}\frac{e^2}{m^2}\frac{m}{m_i}\frac{1}{k'^2 s^4}. \tag{5.118}$$

The requirement (5.116) implies the smallness of the quantities (5.117) and (5.118) in comparison with the largest of the terms which were eliminated due to the compensations in the sums (5.111) and (5.112); however, the corrections (5.117) and (5.118) can compete to the same order and even exceed the resulting sums. The non-linear frequency shift is determined by (5.114), i.e. the neglect of the ion thermal motion is reasonable if a more stringent inequality than (5.116) is satisfied:

$$a^2k^2 > \sqrt[3]{\frac{m}{m_i}}. \tag{5.119}$$

Otherwise the ion motion is important and the non-linear frequency shift is determined by the corrections (5.117) and (5.118):

$$\Delta\omega_{\mathbf{k}} = \frac{1}{48}\frac{e^2}{m^2}\frac{m}{m_i}\frac{\Omega}{(k+k')^2 s^4}\left\{5+2\frac{k}{k'}+\frac{k^2}{k'^2} - i4\sqrt{\frac{2\pi}{3}}\,\mathrm{sgn}(k-k')\frac{m}{m_i}\frac{\Omega}{(k+k')s}\right\}|E'|^2. \tag{5.120}$$

The non-linear eigenfrequency shift increases considerably when the scattering by the ions becomes the governing one, i.e. when $\tilde{z}_i < 1$ or

$$a^2k^2 < \frac{m}{m_i}\frac{T_i}{T_e} \tag{5.121}$$

(the inequality (5.121) is contrary to the requirement (5.116) when $T_i = T_e$). The dielectric permittivity of a plasma, for which (5.121) is satisfied, is as follows:

$$\varepsilon(\tilde{\omega}, \tilde{k}) = 1+\frac{1}{a^2\tilde{k}^2}(1+i\,\mathrm{sgn}\,\tilde{k}\sqrt{\pi\tilde{z}})+\frac{1}{a_i^2\tilde{k}^2}(1+i\,\mathrm{sgn}\,\tilde{k}\sqrt{\pi\tilde{z}_i}) \tag{5.122}$$

and

$$\bar{A}(\omega_k, k; -\omega_{k'}, -k'; \omega_k, k) = \frac{3}{2}\frac{e^2}{m^2}\frac{1}{\Omega^2 s^2}\frac{T_e}{T_e+T_i} \times\left\{1+\frac{i}{2}\sqrt{\frac{3\pi}{2}}\,\mathrm{sgn}(k-k')\frac{T_e}{T_e+T_i}\frac{(k+k')s}{\Omega}\left(1+\sqrt{\frac{m_iT_i}{mT_e}}\right)\right\}. \tag{5.123}$$

$$\bar{A}(\omega_k, k; \omega_{k'}, k'; -\omega_{k'}, -k') = \frac{3}{2}\frac{e^2}{m^2}\frac{1}{\Omega^2 s^2}\frac{T_e}{T_e+T_i}. \tag{5.124}$$

The quantity $\bar{A}(-\omega_{k'}, -k'; \omega_{k'}, k'; \omega_k, k)$ is small in comparison with (5.123) and (5.124). Thus we obtain

$$\Delta\omega_{\mathbf{k}} = -\frac{1}{8}\frac{e^2}{m^2}\frac{1}{\Omega s^2}\frac{T_e}{T_e+T_i}\left\{1+\frac{i}{4}\sqrt{\frac{3\pi}{2}}\,\mathrm{sgn}(k-k')\frac{T_e}{T_e+T_i} \times\frac{(k+k')s}{\Omega}\left(1+\sqrt{\frac{m_i}{m}\frac{T_i}{T_e}}\right)\right\}|E'|^2, \quad a^2k^2 < \frac{m}{m_i}\frac{T_i}{T_e}. \tag{5.125}$$

Formulas (5.114), (5.120), and (5.125) concern the coupling of four one-dimensional Langmuir waves, the frequencies and the wave vectors of which match the requirements (5.98) and (5.99).

The non-linear shift (5.125) confirms the results of Gorbunov and Timberbulatov (1967). The non-linear Landau damping rates (5.120) and (5.125) are similar to those calculated by Tsytovich and Shapiro (1965).

Non-linear eigenfrequency shifts of Langmuir waves

We consider now the non-linear shifts of the eigenfrequencies of four Langmuir waves with different wave vectors, involved in the resonant interaction. Let the wave vectors and the frequencies satisfy the following matching conditions:

$$\mathbf{k} = \mathbf{k}_1+\mathbf{k}_2+\mathbf{k}_3 \tag{5.126}$$

and

$$\omega_{\mathbf{k}} = \omega_{\mathbf{k}_1}+\omega_{\mathbf{k}_2}-\omega_{\mathbf{k}_3}. \tag{5.127}$$

Since within the context of (5.126) equation (5.92) is symmetric relative to the permutations of $\mathbf{k}_1$, $\mathbf{k}_2$, and $\mathbf{k}_3$, the resonances $\omega_{\mathbf{k}} = \omega_{\mathbf{k}_1}-\omega_{\mathbf{k}_2}+\omega_{\mathbf{k}_3}$ and $\omega_{\mathbf{k}} = -\omega_{\mathbf{k}_1}+\omega_{\mathbf{k}_2}+\omega_{\mathbf{k}_3}$ may be reduced to (5.127) by means of simple renotations. Keeping this in mind and renoting $\mathbf{k}_3 \to -\mathbf{k}_3$ (we remember that $E_{-\mathbf{k}_3} = E^*_{\mathbf{k}_3}$), we may rewrite (5.92) as

$$\frac{\partial E_{\mathbf{k}}}{\partial t} = -\frac{3i}{4\varepsilon'_{\mathbf{k}}} \sum_{\mathbf{k}_1+\mathbf{k}_2-\mathbf{k}_3=\mathbf{k}} A(\omega_{\mathbf{k}_1}, \mathbf{k}_1; \omega_{\mathbf{k}_2}, \mathbf{k}_2, -\omega_{\mathbf{k}_3}, -\mathbf{k}_3)\, e^{i(\phi_{\mathbf{k}}-\phi_{\mathbf{k}_1}-\phi_{\mathbf{k}_2}+\phi_{\mathbf{k}_3})} \times E_{\mathbf{k}_1}E_{\mathbf{k}_2}E^*_{\mathbf{k}_3}. \tag{5.128}$$

(We have written on the right-hand side an additional factor of 3 that is associated with possible permutations in the resonance condition (5.128) and suppressed the prime at the sum symbol.)

The matching condition for the frequencies (5.127) is always satisfied in cold plasmas. It may be also satisfied in the long-wave limiting case provided the thermal corrections to the frequencies be small. Replace $\mathbf{k}_2$ and $\mathbf{k}_3$ in (5.128) by

$$\mathbf{k}' = \tfrac{1}{2}(\mathbf{k}_2+\mathbf{k}_3), \quad \mathbf{\Delta} = \mathbf{k}_2-\mathbf{k}_3.$$

It is clear that (5.127) may be satisfied also if the values Δ are sufficiently small. Let the phase differences $\phi_{\mathbf{k}}-\phi_{\mathbf{k}_1}$ and $\phi_{\mathbf{k}_2}-\phi_{\mathbf{k}_3}$ be random quantities, then we may take an average over these of the expression that follows the sum symbol in (5.128). Besides, let $\Delta = 0$ in the sum on the right-hand side of (5.128), which then becomes

$$\frac{\partial E_{\mathbf{k}}}{\partial t} = -i\frac{3}{4\varepsilon'_{\mathbf{k}}} \sum_{\mathbf{k}', \mathbf{\Delta}} A(-\omega_{\mathbf{k}'}, -\mathbf{k}'; \omega_{\mathbf{k}'}, \mathbf{k}'; \omega_{\mathbf{k}}, \mathbf{k})\, |E_{\mathbf{k}'}|^2 E_{\mathbf{k}}. \tag{5.129}$$

Taking the solution of the latter equation in the exponent form, we find

$$\Delta\omega_{\mathbf{k}} = \frac{3}{2\varepsilon'_{\mathbf{k}}} V \sum_{\mathbf{k}', \mathbf{\Delta}} A(-\omega_{\mathbf{k}'}, -\mathbf{k}'; \omega_{\mathbf{k}'}, \mathbf{k}'; \omega_{\mathbf{k}}, \mathbf{k}) I_{\mathbf{k}'}, \tag{5.130}$$

where $I_{\mathbf{k}'}$ is the intensity of the oscillations with wave vector $\mathbf{k}'$ that is determined by (4.35). The shift (5.130) may be complex; in such a case the real part should be regarded as the eigenfrequency shift, while the imaginary part is responsible for the non-linear damping. The iteration treatment is evidently applicable only if both the non-linear shift and the damping are small in comparison to the eigenfrequency.

It follows from the resonance condition (5.127) and the dispersion law for the Langmuir waves that in case $\mathbf{\Delta}$ is small the vector $\mathbf{k}'$ can differ from $\mathbf{k}$ only by an amount of the order of magnitude $\mathbf{\Delta}$. So we can substitute $\mathbf{k}$ for $\mathbf{k}'$ under the sum symbol in (5.130), then $V\sum_{\mathbf{k}'} \to 1$ and the summation over $\mathbf{\Delta}$ must be restricted to $\Delta_{\max} \simeq k$. As a result we obtain the following approximate Langmuir frequency shift:

$$\Delta\omega_{\mathbf{k}} \simeq \beta_{\mathbf{k}} I_{\mathbf{k}}, \quad \beta_{\mathbf{k}} \equiv \frac{k^3}{4\pi^2\varepsilon'_{\mathbf{k}}} A(-\omega_{\mathbf{k}}, -\mathbf{k}; \omega_{\mathbf{k}}, \mathbf{k}; \omega_{\mathbf{k}}, \mathbf{k}). \tag{5.131}$$

Thus the non-linear coupling of the Langmuir waves causes an eigenfrequency shift that is proportional to the wave intensity. The proportionality coefficient $\beta_{\mathbf{k}}$ is a function of the non-linear plasma susceptibilities.

Making use of (5.93) we have

$$\beta_{\mathbf{k}} = \frac{k^3}{12\pi^2 \varepsilon_{\mathbf{k}}'} \left\{ 2 \frac{\varkappa^{(2)}(-\omega_{\mathbf{k}}, -\mathbf{k}; 2\omega_{\mathbf{k}}, 2\mathbf{k})\, \varkappa^{(2)}(\omega_{\mathbf{k}}, \mathbf{k}; \omega_{\mathbf{k}}, \mathbf{k})}{\varepsilon(2\omega_{\mathbf{k}}, 2\mathbf{k})} - \bar{\varkappa}^{(3)}(-\omega_{\mathbf{k}}, -\mathbf{k}; \omega_{\mathbf{k}}, \mathbf{k}; \omega_{\mathbf{k}}, \mathbf{k}) \right.$$
$$\left. + 2\left(2 \frac{\varkappa^{(2)}(\omega_{\mathbf{k}}, \mathbf{k}; 0, 0)\, \varkappa^{(2)}(\omega_{\mathbf{k}}, \mathbf{k}; -\omega_{\mathbf{k}}, -\mathbf{k})}{\varepsilon(0, 0)} - \bar{\varkappa}^{(3)}(\omega_{\mathbf{k}}, \mathbf{k}; -\omega_{\mathbf{k}}, -\mathbf{k}; \omega_{\mathbf{k}}, \mathbf{k}) \right) \right\}. \tag{5.132}$$

It may be easily verified within the context of (5.106) that the non-linear shift (5.132) vanishes in a cold plasma. (Note that the coupled waves are practically one-dimensional by virtue of $k' \simeq k$ and $\Delta = 0$.) Hence the non-linear interaction of the Langmuir waves in a cold plasma does not shift the eigenfrequency.

The relation (5.108) is suitable for calculating the thermal contribution to the non-linear eigenfrequency shift. The shift of the Langmuir frequency is equal to

$$\Delta\omega_{\mathbf{k}} \simeq \beta_{\mathbf{k}} I_{\mathbf{k}}, \quad \beta_{\mathbf{k}} = \frac{1}{6\pi^2} \frac{e^2}{m^2} \frac{k^7 s^2}{\Omega^5}. \tag{5.133}$$

A hydrodynamic treatment (Sitenko, 1973) yields the following coefficient $\beta_{\mathbf{k}}$:

$$\beta_{\mathbf{k}} = \frac{5}{12\pi^2} \frac{e^2}{m^2} \frac{k^7 s^2}{\Omega^5}. \tag{5.134}$$

The mean intensity $I_{\mathbf{k}}$ of the eigenoscillations in an equilibrium plasma is governed by the temperature T. Hence, the ratio $\Delta\omega_{\mathbf{k}}/\omega_{\mathbf{k}}$ in an equilibrium plasma is of the order of magnitude $\frac{k^3}{n_0} \frac{k^4 s^4}{\Omega^4}$. However, $I_{\mathbf{k}}$ of a non-equilibrium plasma may considerably differ from the equilibrium thermal level and, therefore, there is no reason to neglect the non-linear frequency shifts.

5.4. Parametric Resonance in a Plasma

Dispersion equation for waves in a pump-wave field

Dispersion plasma properties are modified under the influence of a powerful electromagnetic field and become time-dependent. This effect leads to parametric resonance, i.e. the external field causes an intense growth of the fluctuation fields, which results in the so-called parametric plasma instability. Silin was the first to consider parametric resonance associated with longitudinal perturbations driven by a weak high-frequency electromagnetic field (Silin, 1965). Later on the parametric instability was studied by many authors (see, for example, Nishikawa, 1967; DuBois and Goldman, 1967; Andreev,

Kirii, and Silin, 1970). The theory of parametric effects in plasmas with non-uniform external fields was advanced by Gorbunov (1969), Silin (1973), and Drake, Kaw, Lee, Schmidt, Liu, and Rosenbluth (1974).

The elementary process that drives the parametric instability is, as in the case of the decay instability, the conversion of the incident wave with the frequency ω_0 and the wave vector $\mathbf{k}_0$ into other waves $(\omega_1, \mathbf{k}_1)$ and $(\omega_2, \mathbf{k}_2)$ within the context of the conservation laws

$$\omega_0 = \omega_1 + \omega_2 \quad \text{and} \quad \mathbf{k}_0 = \mathbf{k}_1 + \mathbf{k}_2.$$

However, it should be kept in mind that the frequencies and the wave vectors of the waves, which are generated in a decay process, are governed by the plasma properties and do not depend on the intensity of the incident wave, while those of the waves resulting from the parametric instability are essentially functions of the amplitude of the incident wave. That is why in order to describe the parametric instability we must take into account the third-order effects in the non-linear interaction between the fields.

Suppose a plane monochromatic wave of a large amplitude (the pump wave) propagates in a plasma. Let the amplitude be of the form

$$\mathbf{E}(\mathbf{r}, t) = \mathbf{E}_0 \cos[\omega_0 t - (\mathbf{k}_0 \cdot \mathbf{r})], \tag{5.135}$$

and the frequency ω_0 and the wave vector $\mathbf{k}_0$ of the pump wave satisfy the linear dispersion relation

$$\omega_0^2 = \Omega^2 + k_0^2 c^2. \tag{5.136}$$

We shall consider only non-relativistic effects here (i.e. the amplitude E_0 is assumed to satisfy the inequality $eE_0/mc\omega_0 \ll 1$). Besides, we suppose that the pump-wave field does not violate the particle distributions which are suggested to be stationary and spatially homogeneous.

Under the influence of the electromagnetic field of the pump wave the electrons oscillate with high velocities, while the ions form a neutralizing positive background. If some perturbation (e.g. of the density) with frequency ω and wave-vector $\mathbf{k}$ originates in the plasma, then there appear under the influence of the pump-wave modes with frequencies $\omega - \omega_0$ and $\omega + \omega_0$ and wave vectors $\mathbf{k} - \mathbf{k}_0$ and $\mathbf{k} + \mathbf{k}_0$. The coupling of these with the pump field gives rise to a ponderomotive force that can enhance the initial perturbations.† Thus a plasma may be regarded as a medium with a positive feedback, which can drive the system unstable.

Denote by $\mathbf{E}_{\mathbf{k}\omega}$ the perturbation field in the presence of the pump wave (5.135) in a plasma and consider the low-frequency perturbations $\omega \ll \omega_0$. We start from the general non-linear field equation (2.66) with vanishing external current $\mathbf{j}_0$ and retain only the terms of second order in the external field (5.135) and linear in the perturbation field:

$$T_{ij}(\omega, \mathbf{k}) E_j(\omega, \mathbf{k}) + \tfrac{1}{2}\{\varkappa_{ijk}^{(2)}(\omega - \omega_0, \mathbf{k} - \mathbf{k}_0; \omega_0, \mathbf{k}_0) E_j(\omega - \omega_0, \mathbf{k} - \mathbf{k}_0)$$
$$+ \varkappa_{ijk}^{(2)}(\omega + \omega_0, \mathbf{k} + \mathbf{k}_0, -\omega_0, -\mathbf{k}_0) E_j(\omega + \omega_0, \mathbf{k} + \mathbf{k}_0)\} E_{0k} = 0, \tag{5.137}$$

† This force (which is sometimes called the Miller force) was introduced independently in three papers (Gaponov and Miller, 1958; Sagdeev, 1958; Veksler and Kovrizhnykh, 1959).

where

$$T_{ij}(\omega, \mathbf{k}) \equiv \Lambda_{ij}(\omega, \mathbf{k}) + \tfrac{1}{4}\left(\varkappa^{(3)}_{ijkl}(\omega, \mathbf{k}; \omega_0, \mathbf{k}_0; -\omega_0, -\mathbf{k}_0) + \varkappa^{(3)}_{ijkl}(\omega, \mathbf{k}; -\omega_0, -\mathbf{k}_0; \omega_0, \mathbf{k}_0)\right) E_{0k} E_{0l}. \tag{5.138}$$

We neglected in (5.137) the terms proportional to the square of the external field and those varying in time with twice the frequency of the pump field since they contribute to higher orders in the dispersion equation.

The dispersion equation for the perturbation waves, which is exact up to terms proportional to the square of the pump field intensity, may be written as follows:

$$\begin{aligned}\mathrm{Det}\{T_{ij}(\omega, \mathbf{k}) - \tfrac{1}{4}[\varkappa^{(2)}_{ij'k'}(\omega-\omega_0, \mathbf{k}-\mathbf{k}_0; \omega_0, \mathbf{k}_0)\\ \times T^{-1}_{j'i'}(\omega-\omega_0, \mathbf{k}-\mathbf{k}_0)\,\varkappa^{(2)}_{i'jk}(\omega, \mathbf{k}; -\omega_0, -\mathbf{k}_0) + \varkappa^{(2)}_{ij'k'}(\omega+\omega_0, \mathbf{k}+\mathbf{k}_0; -\omega_0, -\mathbf{k}_0)\\ \times T^{-1}_{j'i'}(\omega+\omega_0, \mathbf{k}+\mathbf{k}_0)\,\varkappa^{(2)}_{i'jk}(\omega, \mathbf{k}; \omega_0, \mathbf{k}_0)]E_{0k}E_{0k'}\} = 0,\end{aligned} \tag{5.139}$$

where T^{-1}_{ij} is the inverse of the tensor (5.138). If the phase velocity of the pump wave is considerably greater than the electron thermal velocity and $\omega_0 \gg ks$, then we may take into account only the electron contributions in the non-linear tensor susceptibilities $\varkappa^{(2)}_{ijk}$ and $\varkappa^{(3)}_{ijkl}$. We have

$$\varkappa^{(2)}_{ijk}(\omega-\omega_0, \mathbf{k}-\mathbf{k}_0; \omega_0, \mathbf{k}_0) = -\varkappa^{(2)}_{kji}(\omega, \mathbf{k}; -\omega_0, -\mathbf{k}_0) \simeq -i\frac{e}{m\omega_0^2}\varkappa^{(e)}(\omega, \mathbf{k})k_i\delta_{jk}, \tag{5.140}$$

$$\begin{aligned}\varkappa^{(3)}_{ijkl}(\omega, \mathbf{k}; \omega_0, \mathbf{k}_0; -\omega_0, -\mathbf{k}_0) + \varkappa^{(3)}_{ijkl}(\omega, \mathbf{k}; -\omega_0, -\mathbf{k}_0; \omega_0, \mathbf{k}_0)\\ \simeq \frac{e^2}{m^2}\frac{k^2}{\omega_0^4}\left(1 - \frac{\omega}{\omega_0}\frac{\mathbf{k}\cdot\mathbf{k}_0}{k^2}\right)\varkappa^{(e)}(\omega, \mathbf{k})\delta_{ik}\delta_{jl}\end{aligned} \tag{5.141}$$

and the tensor (5.138) then written as

$$T_{ij}(\omega\pm\omega_0, \mathbf{k}\pm\mathbf{k}_0) \simeq \Lambda_{ij}(\omega\pm\omega_0, \mathbf{k}\pm\mathbf{k}_0) + \frac{1}{4}\frac{k^2}{\omega_0^4}\left(1 - \frac{\omega}{\omega_0}\frac{\mathbf{k}\cdot\mathbf{k}_0}{k^2}\right)\varkappa^{(e)}(\omega, \mathbf{k})v_{Ei}v_{Ej}, \tag{5.142}$$

where $\mathbf{v}_E \equiv \dfrac{e\mathbf{E}_0}{m\omega_0}$. Within the context of (2.67), (2.75), and (5.140) the dispersion equation (5.139) decouples into two equations: the dispersion equation for the transverse waves, which turns out to be independent of the pump field, and the following dispersion equation for the low-frequency longitudinal perturbations:

$$\varepsilon_l(\omega, \mathbf{k}) - \frac{1}{4}\frac{k^2}{\omega_0^2}|\varkappa^{(e)}(\omega, \mathbf{k})|^2\,[T^{-1}_{ij}(\omega-\omega_0, \mathbf{k}-\mathbf{k}_0) + T^{-1}_{ij}(\omega+\omega_0, \mathbf{k}+\mathbf{k}_0)]v_{Ei}v_{Ej} = 0. \tag{5.143}$$

Retaining the quadratic terms in v_E we obtain

$$\begin{aligned}\frac{1}{\varkappa^{(e)}(\omega, \mathbf{k})} + \frac{1}{1+\varkappa^{(i)}(\omega, \mathbf{k})} + \frac{k^2}{4\omega_0^2}\left\{\frac{1}{(\mathbf{k}-\mathbf{k}_0)^2}\left(\frac{(\{\mathbf{k}-\mathbf{k}_0\}\cdot\mathbf{v}_E)^2}{\varepsilon_l(\omega-\omega_0, \mathbf{k}-\mathbf{k}_0)} + \frac{[\{\mathbf{k}-\mathbf{k}_0\}\cdot\mathbf{v}_E]^2}{\varepsilon_t(\omega-\omega_0, \mathbf{k}-\mathbf{k}_0) - \eta_-^2}\right)\right.\\ \left. + \frac{1}{(\mathbf{k}+\mathbf{k}_0)^2}\left(\frac{(\{\mathbf{k}+\mathbf{k}_0\}\cdot\mathbf{v}_E)^2}{\varepsilon_l(\omega+\omega_0, \mathbf{k}+\mathbf{k}_0)} + \frac{[\{\mathbf{k}+\mathbf{k}_0\}\mathbf{v}_E]^2}{\varepsilon_t(\omega+\omega_0, \mathbf{k}+\mathbf{k}_0) - \eta_+^2}\right)\right\} = 0.\end{aligned} \tag{5.144}$$

This equation governs the wave dispersion in the plasma in the presence of a high-frequency pump wave (ω_0, $\mathbf{k}_0$). It follows from (5.144) that the pump-wave field causes the parametric coupling between the low-frequency perturbation with frequency ω and the wave vector $\mathbf{k}$ and the high-frequency longitudinal and transverse pulsations with frequencies $\omega-\omega_0$ and $\omega+\omega_0$ and wave vectors $\mathbf{k}-\mathbf{k}_0$ and $\mathbf{k}+\mathbf{k}_0$.

If the pump-wave frequency ω_0 is close to the Langmuir frequency Ω, then the terms which correspond to the longitudinal pulsations are dominant within the curly brackets in (5.144), since $\varepsilon_l(\omega\pm\omega_0, \mathbf{k}\pm\mathbf{k}_0) \simeq 0$. In such a case the dispersion equation (5.144) describes (when $ak \ll 1$) the parametric excitation of two longitudinal plasma waves (e.g. a Langmuir and an ion-sound wave) by an incident electromagnetic wave. Such a parametric excitation is accompanied by the anomalous absorption of the pump-wave energy (Silin, 1965).

If the pump-wave frequency ω_0 is such that $\varepsilon_t(\omega\pm\omega_0, \mathbf{k}\pm\mathbf{k}_0) = \eta_\pm^2$, then the terms, which are associated with the pulsations of the transverse field, become the leading ones within the curly brackets in (5.144). The dispersion equation (5.144) describes then the parametric excitation of a longitudinal wave and an electromagnetic wave with the shifted frequency by the incident electromagnetic wave, i.e. the induced combination scattering of the electromagnetic wave.

Induced scattering

Let us consider in detail the induced combination scattering of electromagnetic waves in a plasma (Drake *et al.*, 1974). In neglect of longitudinal pulsations the dispersion equation (5.144) reduces to

$$\frac{1}{\varkappa^{(e)}(\omega, \mathbf{k})}+\frac{1}{1+\varkappa^{(i)}(\omega, \mathbf{k})}+\frac{k^2}{4\omega_0^2}$$
$$\times\left\{\frac{1}{(\mathbf{k}-\mathbf{k}_0)^2}\,\frac{[\{\mathbf{k}-\mathbf{k}_0\}\mathbf{v}_E]^2}{\varepsilon_t(\omega-\omega_0, \mathbf{k}-\mathbf{k}_0)-\eta_-^2}+\frac{1}{(\mathbf{k}+\mathbf{k}_0)^2}\,\frac{[\{\mathbf{k}+\mathbf{k}_0\}\mathbf{v}_E]^2}{\varepsilon_t(\omega+\omega_0, \mathbf{k}+\mathbf{k}_0)-\eta_+^2}\right\} = 0. \quad (5.145)$$

The term with $\omega-\omega_0$ in (5.145) is associated with the induced excitation of the Stokes component, that with $\omega+\omega_0$ is given rise to by the anti-Stokes one.

We shall neglect the spatial dispersion in $\varepsilon_t(\omega, \mathbf{k})$ (while the collisions will be phenomenologically taken into account) and suggest the Stokes component to be only significant in (5.145). It may be verified that the last assumption is reasonable if

$$\omega \ll \frac{(\mathbf{k}\cdot\mathbf{k}_0)}{\omega_0}c^2$$

(which requirement is violated for small values of k and when $\mathbf{k} \perp \mathbf{k}_0$). If $\omega \ll \omega_0$, then the following approximate relation holds:

$$\varepsilon(\omega-\omega_0, \mathbf{k}-\mathbf{k}_0)-\eta_-^2 \simeq \frac{2}{\omega_0}\left(\omega-\frac{(\mathbf{k}\cdot\mathbf{k}_0)}{\omega_0}c^2+\frac{k^2c^2}{2\omega_0}\right).$$

The right-hand side of the above equality is small when $k \simeq 2k_0\cos\vartheta$, where ϑ is the angle between $\mathbf{k}$ and $\mathbf{k}_0$. (Since $\omega-\omega_0 < 0$ for the Stokes component, then $\vartheta = 0$ is associated with the back scattering and $\vartheta = \pi/2$ corresponds to the forward scattering.) Neglecting in (5.145) the non-resonant term (that gives rise to the anti-Stokes scattering) and substituting $2k_0 \cos\vartheta$ for k everywhere except the denominator, we obtain

$$\frac{1}{\varkappa^{(e)}(\omega, \mathbf{k})}+\frac{1}{1+\varkappa^{(i)}(\omega, \mathbf{k})} = \frac{k_0^2 v_E^2}{2\omega_0(\omega-\Delta\omega+i\tilde{\gamma})}\Psi^2(\vartheta, \varphi), \tag{5.146}$$

where $\Psi^2(\vartheta, \varphi) = \sin^2\varphi \cos^2\vartheta$ (φ is the angle between the polarizations of the incident wave and the scattered one), $\Delta\omega \equiv (\mathbf{k}\cdot\mathbf{v}_0)-\frac{c^2k^2}{2\omega_0}$ $\left(\mathbf{v}_0 = \frac{c^2}{\omega_0}\mathbf{k}_0 \text{ is the group velocity of the incident wave}\right)$ and is associated with the electromagnetic wave damping due to the binary collisions.

If the frequency $\omega \gg ks$, then we may neglect the ion component of the susceptibility $\varkappa^{(i)}(\omega, \mathbf{k})$ in (5.146) and substitute the high-frequency limiting value for the electron component $\varkappa^{(e)}(\omega, \mathbf{k})$. Then we have

$$(\omega-\omega_{\mathbf{k}}-i\gamma)(\omega-\Delta\omega+i\tilde{\gamma}) = -\frac{1}{4}\omega_0\Omega\frac{v_E^2}{c^2}\Psi^2, \tag{5.147}$$

where $\omega_{\mathbf{k}} = \sqrt{\Omega^2+k^2s^2}$ and γ is the Landau damping rate of the Langmuir wave. Let $\omega = \omega_{\mathbf{k}}+i\Gamma$ and $\Delta\omega = \omega_{\mathbf{k}}$, then we find the following growth rate Γ:

$$\Gamma = -\frac{1}{2}(\gamma+\tilde{\gamma})\pm\frac{1}{2}\sqrt{(\gamma-\tilde{\gamma})^2+\omega_0\Omega\frac{v_E^2}{c^2}\Psi^2}. \tag{5.148}$$

The condition $\Gamma = 0$ determines the threshold amplitude of the pump wave:

$$\bar{v}_E^2 = \frac{1}{4}\frac{c^2}{\Psi^2}\frac{\gamma}{\Omega}\frac{\tilde{\gamma}}{\omega_0}. \tag{5.149}$$

The growth rate near threshold (for $\gamma \gg \tilde{\gamma}, \Gamma$) is proportional to the square of the pump-wave amplitude:

$$\Gamma = \frac{1}{4}\frac{v_E^2-\bar{v}_E^2}{c^2}\frac{\Omega}{\gamma}\omega_0\Psi^2. \tag{5.150}$$

The maximum growth rate (at $\Omega \gg \Gamma \gg \gamma$) is a linear function of the pump-wave amplitude

$$\Gamma_{\max} = \frac{1}{2}\frac{v_E}{c}\sqrt{\omega_0\Omega}\Psi. \tag{5.151}$$

Notice that the induced combination scattering can occur only if $ak = 2ak_0 \cos\vartheta \ll 1$, which corresponds to the weak damping of the Langmuir wave.

In a highly non-isothermal plasma electromagnetic waves may undergo induced scattering by the ion-sound oscillations. If $s \gg \frac{\omega}{k} \gg s_i$, then we obtain from (5.146)

$$(\omega-kv_s+i\gamma_s)(\omega-\Delta\omega+i\tilde{\gamma}) = \frac{1}{8}\Omega_i^2\frac{v_E^2}{v_s}\frac{k_0}{\omega_0}\frac{\Psi^2}{\cos\vartheta}, \tag{5.152}$$

where γ_s is the ion-sound damping rate. Then the growth rate is equal to

$$\Gamma = -\frac{1}{2}(\gamma_s+\tilde{\gamma})\pm\frac{1}{2}\sqrt{(\gamma_s-\tilde{\gamma})^2+\frac{1}{2}\Omega_i^2\frac{v_E^2}{v_s}\frac{k_0}{\omega_0}\sin^2\varphi\cos\vartheta}. \tag{5.153}$$

The threshold pump-wave amplitude is given by

$$\bar{v}_E^2 = \frac{1}{2}\frac{v_s}{\sin^2\varphi\cos\vartheta}\frac{\omega_0}{k_0}\frac{\gamma_s\tilde{\gamma}}{\Omega_i^2}, \tag{5.154}$$

and we have for the growth rate near threshold (for $\gamma_s \gg \tilde{\gamma}$, Γ):

$$\Gamma = \frac{1}{8}\frac{v_E^2-\bar{v}_E^2}{v_s}\frac{k_0}{\omega_0}\frac{\Omega_i^2}{\gamma_s}\sin^2\varphi\cos\vartheta. \tag{5.155}$$

The maximum growth rate (for $kv_s \gg \Gamma \gg \gamma_s$) is equal to

$$\Gamma_{\max} = \frac{1}{2}\frac{v_E}{v_s}\Omega_i\sqrt{\frac{k_0 v_s}{2\omega_0}}\cos\vartheta\,|\sin\varphi|. \tag{5.156}$$

The threshold amplitude of the electromagnetic wave that undergoes induced scattering both by Langmuir oscillations and by ion sound is a minimum and the growth rate is a maximum for $\vartheta = 0$ and $\varphi = \pi/2$.

Parametric instability

The analysis of the dispersion relation (5.145) shows that a high-frequency electromagnetic wave, besides being induced to scatter in a plasma, may also excite long-wave perturbations which drive the so-called parametric or modulation instability. Suppose $k < 2k_0\cos\vartheta$ and $\omega \gtrsim \Omega_i$. Since ω is in this case of the order of magnitude $(\mathbf{k}\cdot\mathbf{v}_0)$ ($\mathbf{v}_0$ is the group velocity of the pump wave), both the terms with $\omega-\omega_0$ and those with $\omega+\omega_0$ must be retained in (5.145). The above-mentioned perturbations are associated with the ion motion and cause density modulations, the modulation length of which is very large in comparison with k_0^{-1}. If $\omega \ll kc$ and $\varphi=\pi/2$, i.e. $\mathbf{k}\perp\mathbf{E}_0$, then (5.145) may be rewritten as

$$\frac{1}{\varkappa^{(e)}(\omega,\mathbf{k})}+\frac{1}{1+\varkappa^{(i)}(\omega,\mathbf{k})} = -\frac{1}{2}\frac{v_E^2}{c^2}\frac{\delta^2}{\left(\omega-(\mathbf{k}\cdot\mathbf{v}_0)+\frac{i}{2}\tilde{\gamma}\right)^2-\delta_0^2}, \quad \delta \equiv \frac{k^2c^2}{2\omega_0}. \tag{5.157}$$

If $\omega \ll ks_i$, then $\varkappa^{(i)}(\omega,\mathbf{k}) \simeq \dfrac{1}{a_i^2k^2}$ and the solution of (5.157) is of the form

$$\omega = (\mathbf{k}\cdot\mathbf{v}_0)-\frac{i}{2}\tilde{\gamma}\pm\left(1-\frac{1}{2a^2k^2}\frac{1}{1+\dfrac{T_i}{T_e}}\frac{v_E^2}{c^2}\right)\delta. \tag{5.158}$$

This solution is unstable if

$$v_E^2 > \left(1+\frac{T_i}{T_e}\right)\frac{\omega_0}{\delta}\,a^2(\tilde{\gamma}^2+4\delta^2). \tag{5.159}$$

The right-hand side of the inequality (5.159) determines the threshold amplitude which is a minimum at $\tilde{\gamma} \simeq 2\delta$

$$\bar{v}_E^2 = \left(1+\frac{T_i}{T_e}\right)\frac{\tilde{\gamma}}{\omega_0}\,a^2k_0^2c^2. \tag{5.160}$$

If $(\mathbf{k}\cdot\mathbf{v}_0) = 0$, then ω in (5.158) is pure imaginary, i.e. the relevant perturbation grows exponentially.†

If $\Omega_i \gg \omega \gg ks_i$ and $(\mathbf{k}\cdot\mathbf{v}_0) = 0$, then

$$\omega \simeq \sqrt{\frac{1}{2}(k^2v_s^2+\delta^2)\pm\sqrt{(k^2v_s^2-\delta^2)^2+2\frac{v_E^2}{c^2}\Omega_i^2\delta^2}}. \tag{5.161}$$

If $\omega \gg kv_s$ and $|\omega-(\mathbf{k}\cdot\mathbf{v}_0)| \gg \delta$, we have

$$\omega \simeq \frac{1}{2}(\mathbf{k}\cdot\mathbf{v}_0)\pm\sqrt{\frac{1}{4}(\mathbf{k}\cdot\mathbf{v}_0)^2\pm\frac{1}{\sqrt{2}}\frac{v_E}{c}\Omega_i\delta}. \tag{5.162}$$

It is clear that in the above-considered cases the instability may be driven only by the long-wave perturbations and for such angles ϑ that $\cos\vartheta \simeq 0$.

† The plasma instability, driven by the field of an electromagnetic wave, was first discovered by Volkov (1958).

CHAPTER 6

Fluctuations in Plasmas

6.1. Fluctuations in Spatially Homogeneous Stationary Systems

Space-time correlation functions

Various physical quantities which describe the state of a plasma (as well as of any statistical ensemble) may be examined through their deviations from the mean values (fluctuations). The latter are governed by the state of the system.† The mean values of the fluctuations are equal to zero by definition, so the latter are described by means of correlation functions (correlators). These are for spatially distributed quantities defined as the average products of fluctuations in different space points at different times. The averaging procedure is carried out both over the quantum mechanical state of the system and over the statistical distribution of various states. Such correlators are usually called space–time correlation functions. If the medium is spatially homogeneous and only stationary states of the system are involved in the consideration, then the quadratic space–time correlation function depends only on the relative distance and on the absolute value of the time segment between the points at which the fluctuations are examined.

To be definite let the spatially distributed quantity be the density of some particle component of the plasma, e.g. the electron density $n(\mathbf{r}, t)$. The mean density being equal to n_0 (which is constant for stationary states of spatially homogeneous systems), the particle density fluctuation $\delta n(\mathbf{r}, t)$ must be understood as the departure of the density $n(\mathbf{r}, t)$ from the mean value n_0:

$$\delta n(\mathbf{r}, t) \equiv n(\mathbf{r}, t) - n_0. \qquad (6.1$$

Let the quadratic space–time correlation function for the density fluctuations be defined according to

$$\langle \delta n(\mathbf{r}_1, t_1)\, \delta n(\mathbf{r}_2, t_2) \rangle \equiv \langle \delta n^2 \rangle_{\mathbf{r}t}, \qquad (6.2)$$

† The fundamental theory of fluctuations was advanced by Callen and Welton (1951), Leontovich and Rytov (1952), Rytov (1953), and Landau and Lifshitz (1960). Silin (1959) was the first to consider electromagnetic fluctuations in spatially dispersive media. A number of authors discuss the theoretical aspects of plasma fluctuations (see, for example, Kadomtsev, 1957; Salpeter, 1960, 1961; Thompson and Hubbard, 1960; Akhiezer, Akhiezer, and Sitenko, 1961; Rostoker, 1961; Klimontovich and Silin, 1962; Silin 1962; Sitenko, 1966; Sitenko and Gurin, 1966; Sitenko and Radzievskii, 1966; Sitenko and Oraevskii, 1968; Ichimaru and Rosenbluth, 1970). A detailed account of the theory of electromagnetic fluctuations in plasmas is given by Sitenko (1967) and Akhiezer, Akhiezer, Polovin, Sitenko, and Stepanov (1975). In the latter books there are also numerous references to original papers. Equilibrium thermal fluctuations in electrodynamics were investigated by Levin and Rytov (1967).

where $\mathbf{r} = \mathbf{r}_1 - \mathbf{r}_2$ and $t = t_1 - t_2$. The brackets $\langle \ldots \rangle$ on the left-hand side of (6.2) imply statistical averaging.

We shall refer to the space–time Fourier transforms of the correlation function

$$\langle \delta n^2 \rangle_{\mathbf{k}\omega} = \int d\mathbf{r} \int dt\, e^{-i(\mathbf{k}\cdot\mathbf{r})+i\omega t} \langle \delta n^2 \rangle_{\mathbf{r}t} \tag{6.3}$$

as the spectral distribution or the spectral density of the fluctuations. Sometimes we shall call it the correlation function in $(\mathbf{k}, \omega)$-space or the spectral representation of the correlation function. It may be easily verified that the mean product of the fluctuation Fourier components is related to the fluctuation spectral distribution as follows:

$$\langle \delta n_{\mathbf{k}\omega} \delta n^{+}_{\mathbf{k}'\omega'} \rangle = (2\pi)^4\, \delta(\omega - \omega')\, \delta(\mathbf{k} - \mathbf{k}')\, \langle \delta n^2 \rangle_{\mathbf{k}\omega}, \tag{6.4}$$

where the $+$ symbol implies the Hermitian conjugate.

The space correlation function will be defined in terms of the mean product of the fluctuations of some quantity in different space points at the same time:

$$\langle \delta n(\mathbf{r}_1, t)\, \delta n(\mathbf{r}_2, t) \rangle \equiv \langle \delta n^2 \rangle_{\mathbf{r}}. \tag{6.5}$$

There is no difficulty to show that the Fourier transform of the space correlation function coincides with the integral over all frequencies of the spectral distribution of fluctuations:

$$\langle \delta n^2 \rangle_{\mathbf{k}} = \frac{1}{2\pi} \int_{-\infty}^{\infty} d\omega \langle \delta n^2 \rangle_{\mathbf{k}\omega}. \tag{6.6}$$

The mean product of the fluctuations examined in the same space point at different times is referred to as the time correlation (or the autocorrelation) function:

$$\langle \delta n(\mathbf{r}, t_1)\, \delta n(\mathbf{r}, t_2) \rangle \equiv \langle \delta n^2 \rangle_t. \tag{6.7}$$

The Fourier transform of the latter may be obtained by integrating the spectral distribution of fluctuations over the phase space of the wave vectors:

$$\langle \delta n^2 \rangle_{\omega} = \frac{1}{(2\pi)^3} \int d\mathbf{k} \langle \delta n^2 \rangle_{\mathbf{k}\omega}. \tag{6.8}$$

Besides the quadratic correlation function $\langle \delta n^2 \rangle_{\mathbf{r}t}$ we introduce also the higher-order ones, e.g. the cubic (or triple) correlation function, the quaternary correlation function, and so on:

$$\langle \delta n(\mathbf{r}_1, t_1)\, \delta n(\mathbf{r}_2, t_2)\, \delta n(\mathbf{r}_3, t_3) \rangle \equiv \langle \delta n^3 \rangle_{\mathbf{r}_1-\mathbf{r}_2,\, t_1-t_2;\, \mathbf{r}_2-\mathbf{r}_3,\, t_2-t_3}, \tag{6.9}$$

$$\langle \delta n(\mathbf{r}_1, t_1)\, \delta n(\mathbf{r}_2, t_2)\, \delta n(\mathbf{r}_3, t_3)\, \delta n(\mathbf{r}_4, t_4) \rangle \equiv \langle \delta n^4 \rangle_{\mathbf{r}_1-\mathbf{r}_2,\, t_1-t_2;\, \mathbf{r}_2-\mathbf{r}_3,\, t_2-t_3;\, \mathbf{r}_3-\mathbf{r}_4,\, t_3-t_4}. \tag{6.10}$$

By virtue of the spatial homogeneity of the system and the stationarity of the states these functions, similarly to the quadratic correlation function, depend only on the absolute values of the time segments and the relative distances between the points at which the fluctuations are examined.

Various correlation functions for any spatially distributed quantities in a plasma may be introduced similarly. If the quantity under consideration is a vector, then the relevant quadratic correlation function is a second-rank tensor, and the higher-order ones are tensors of corresponding ranks.

Let us derive the space–time correlation functions for the fluctuations of charge and current densities in a plasma. Let the plasma be neutral as a whole and the mean current density be equal to zero. Then the charge and current density fluctuations coincide with the densities

$$\delta\varrho(\mathbf{r}, t) = \varrho(\mathbf{r}, t) \quad \text{and} \quad \delta\mathbf{j}(\mathbf{r}, t) = \mathbf{j}(\mathbf{r}, t).$$

Note the direct relation between the charge density fluctuations and those of the particle densities of various species:

$$\varrho(\mathbf{r}, t) = \sum e\delta n(\mathbf{r}, t). \tag{6.11}$$

The summation in (6.11) extends over the plasma components.

Define the space–time correlation function for the charge density fluctuations as

$$\langle \varrho(\mathbf{r}_1, t_1)\, \varrho(\mathbf{r}_2, t_2)\rangle \equiv \langle \varrho^2\rangle_{\mathbf{r}t}, \tag{6.12}$$

and that for the current density fluctuations as

$$\langle j_i(\mathbf{r}_1, t_1)\, j_j(\mathbf{r}_2, t_2)\rangle \equiv \langle j_i j_j\rangle_{\mathbf{r}t}. \tag{6.13}$$

The spectral distributions of the charge and current density fluctuations are respectively equal to

$$\langle \varrho^2\rangle_{\mathbf{k}\omega} = \int d\mathbf{r} \int dt\, e^{-i(\mathbf{k}\cdot\mathbf{r})+i\omega t} \langle \varrho^2\rangle_{\mathbf{r}t}, \tag{6.14}$$

$$\langle j_i j_j\rangle_{\mathbf{k}\omega} = \int d\mathbf{r} \int dt\, e^{-i(\mathbf{k}\cdot\mathbf{r})+i\omega t} \langle j_i j_j\rangle_{\mathbf{r}t}. \tag{6.15}$$

Correlators for the distribution function fluctuations

The correlation functions that were introduced in the first chapter make an example of the spatial (isochronous) correlators and are directly related to the fluctuations of the distribution functions of plasma particles. Indeed, the one-particle distribution function $f(\mathbf{r}, \mathbf{v}, t)$ is the average of the microscopic density in the phase space $\mathcal{F}(\mathbf{r}, \mathbf{v}, t)$. The latter is defined by means of the relation

$$\mathcal{F}(\mathbf{r}, \mathbf{v}, t) = \sum_{\alpha} \delta(\mathbf{r} - \mathbf{r}_\alpha(t))\, \delta(\mathbf{v} - \mathbf{v}_\alpha(t)) \tag{6.16}$$

and depends on the dynamical variables of all particles. Let the distribution function fluctuation $\delta f(\mathbf{r}, \mathbf{v}, t)$ be defined as the departure of the microscopic density (6.16) from the mean value:

$$\delta f(\mathbf{r}, \mathbf{v}, t) \equiv \mathcal{F}(\mathbf{r}, \mathbf{v}, t) - f(\mathbf{r}, \mathbf{v}, t). \tag{6.17}$$

Since the microscopic density $\mathcal{F}(\mathbf{r}, \mathbf{v}, t)$ is a function of all particle velocities and coordinates, $\delta f(\mathbf{r}, \mathbf{v}, t)$ is a stochastic quantity which is governed by the microscopic state of the system. The mean value of $\delta f(\mathbf{r}, \mathbf{v}, t)$ is evidently equal to zero. It is straightforward to verify that the average product of the distribution function fluctuations in two different space points at the same time may be expressed in terms of the above introduced binary correlation function $g(\mathbf{r}, \mathbf{v}; \mathbf{r}', \mathbf{v}'; t)$ as

$$\langle \delta f(\mathbf{r}, \mathbf{v}, t)\, \delta f(\mathbf{r}', \mathbf{v}', t)\rangle = \delta(\mathbf{r}-\mathbf{r}')\, \delta(\mathbf{v}-\mathbf{v}')\, f(\mathbf{r}, \mathbf{v}, t) + g(\mathbf{r}, \mathbf{v}; \mathbf{r}', \mathbf{v}'; t). \qquad (6.18)$$

In case the plasma distribution is stationary, the right-hand side of (6.18) is time-independent, i.e.

$$\langle \delta f(\mathbf{r}, \mathbf{v}, t)\, \delta f(\mathbf{r}', \mathbf{v}', t)\rangle = \delta(\mathbf{r}-\mathbf{r}')\, \delta(\mathbf{v}-\mathbf{v}')\, f(\mathbf{r}, \mathbf{v}) + g(\mathbf{r}, \mathbf{v}; \mathbf{r}', \mathbf{v}'). \qquad (6.19)$$

(It should be remembered here that the one-particle distribution function of a spatially homogeneous plasma does not depend on $\mathbf{r}$, and the binary correlation function g depends only on the coordinate difference $\mathbf{r}-\mathbf{r}'$.) It follows from the relation (6.19) that the isochronous correlation functions for the fluctuations of any quantities are time-independent if the distributions are stationary.

The time correlation functions play a more important role in the study of electrodynamic plasma properties than the spatial ones. The time correlation functions were defined as the mean products of fluctuations at different times. Among those the space–time correlators for the fluctuations of the particle distribution functions in a plasma are of primary interest:

$$\langle \delta f(\mathbf{r}, \mathbf{v}, t)\, \delta f(\mathbf{r}', \mathbf{v}', t')\rangle \equiv \langle \delta f(\mathbf{v})\, \delta f(\mathbf{v}')\rangle_{\mathbf{r}-\mathbf{r}', t-t'}. \qquad (6.20)$$

The correlation function $\langle \delta f(\mathbf{v})\, \delta f(\mathbf{v}')\rangle_{\mathbf{r}-\mathbf{r}', t-t'}$ for a spatially homogeneous stationary plasma is a function of the distances between the space points and of the time segments. The knowledge of the correlator for the fluctuations of the particle distribution is sufficient to find the correlation functions for any quantities in the plasma.

The binary space–time correlator for the fluctuations of the distribution function may be expressed in terms of the one-particle distribution $f(\mathbf{v}')$ and of the conditional transition probability $W_{\mathbf{r}-\mathbf{r}', t-t'}(\mathbf{v}, \mathbf{v}')$, which determines the probability of an individual particle transition from some state $(\mathbf{r}', \mathbf{v}')$ into another state $(\mathbf{r}, \mathbf{v})$ during the time $t-t'$:

$$\langle \delta f(\mathbf{v})\, \delta f(\mathbf{v}')\rangle_{\mathbf{r}-\mathbf{r}', t-t'} \equiv W_{\mathbf{r}-\mathbf{r}', t-t'}(\mathbf{v}, \mathbf{v}')\, f(\mathbf{v}'). \qquad (6.21)$$

The conditional transition probability $W_{\mathbf{r}-\mathbf{r}', t-t'}(\mathbf{v}, \mathbf{v}')$ is governed by the nature of the particle interactions as well as by the boundary conditions in case the plasma is bounded.† The higher-order space–time correlation functions for pairwise interacting particles may also be expressed in terms of the one-particle distribution function and of the conditional transition probability.

† Fluctuations in a bounded plasma are considered by Sitenko and Yakimenko (1974).

6.2. Fluctuations in a System of Non-interacting Charged Particles

Spectral correlators for the fluctuations of the distribution function

Consider the fluctuations of various quantities in a quasi-neutral system of charged particles under the assumption that we may neglect both Coulomb and electromagnetic interactions between the particles. First of all let us find the quadratic space–time correlation function for the fluctuations of the particle distribution. Inasmuch as non-interacting individual particles move along straight-line trajectories, the microscopic density may be written as

$$\mathcal{F}_0(\mathbf{r}, \mathbf{v}, t) = \sum_{\alpha} \delta(\mathbf{r}-\mathbf{r}_\alpha^0(t))\, \delta(\mathbf{v}-\mathbf{v}_\alpha^0(t)), \tag{6.22}$$

where

$$\mathbf{v}_\alpha^0(t) = \mathbf{v}_\alpha^0, \quad \mathbf{r}_\alpha^0(t) = \mathbf{r}_\alpha^0+\mathbf{v}_\alpha^0(t-t_0) \tag{6.23}$$

($\mathbf{r}_\alpha^0$ and $\mathbf{v}_\alpha^0$ are the radius vector and the velocity of the particle at the initial time t_0.) The distribution function of an ensemble of many non-interacting particles reduces to the product of one-particle functions:

$$D_N^0(\mathbf{r}_1^0, \mathbf{v}_1^0; \mathbf{r}_2^0, \mathbf{v}_2^0; \ldots; \mathbf{r}_N^0, \mathbf{v}_N^0) = \prod_{\alpha=1}^{N} \tilde{f}_0(\mathbf{v}_\alpha^0). \tag{6.24}$$

Since the system is assumed to be spatially homogeneous and stationary, the one-particle distributions $f_0(\mathbf{v})$ are coordinate- and time-independent and functions of the velocity only. If the system is in thermal equilibrium, the Maxwellian function must be taken for $f_0(\mathbf{v})$. The distribution function of a non-equilibrium system $f_0(\mathbf{v})$ is in the general case non-Maxwellian (we postpone the question of how to find $f_0(\mathbf{v})$ till Chapter 9).

The averaging of (6.22) over the distribution (6.24) yields

$$\langle \mathcal{F}_0(\mathbf{r}, \mathbf{v}, t)\rangle = f_0(\mathbf{v}) = N\tilde{f}_0(v). \tag{6.25}$$

Define the fluctuation of the distribution function of non-interacting particles by means of the relation

$$\delta f_0(\mathbf{r}, \mathbf{v}, t) \equiv \mathcal{F}_0(\mathbf{r}, \mathbf{v}, t)-f_0(\mathbf{r}, \mathbf{v}). \tag{6.26}$$

Then it is not difficult to verify that

$$\langle \delta f_0(\mathbf{r}, \mathbf{v}, t)\, \delta f_0(\mathbf{r}', \mathbf{v}', t')\rangle = \left\langle \sum_{\alpha} \delta((\mathbf{r}-\mathbf{r}_\alpha^0(t))\, \delta(\mathbf{v}-\mathbf{v}_\alpha^0)\, \delta(\mathbf{r}'-\mathbf{r}_\alpha^0(t'))\, \delta(\mathbf{v}'-\mathbf{v}_\alpha^0)\right\rangle. \tag{6.27}$$

After the averaging of (6.27) over the distribution (6.24) is carried out, the space–time correlator for the distribution function fluctuations in an ensemble of non-interacting particles becomes of the form

$$\begin{aligned}\langle \delta f_0(\mathbf{r}, \mathbf{v}, t)\, \delta f_0(\mathbf{r}', \mathbf{v}', t)\rangle &\equiv \langle \delta f(\mathbf{v})\, \delta f(\mathbf{v}')\rangle^0_{\mathbf{r}-\mathbf{r}', t-t'} \\ &= \delta(\mathbf{v}-\mathbf{v}')\, \delta[\mathbf{r}-\mathbf{r}'-\mathbf{v}(t-t')]\, f_0(\mathbf{v}).\end{aligned} \tag{6.28}$$

The physical interpretation of the last relation is very simple. The correlations between the distribution function fluctuations in the points, spaced at a distance $\mathbf{r}-\mathbf{r}'$ and examined at times, the segment between which is $t-t'$, are associated with the particle identity and occur inasmuch as a particle with a velocity $\mathbf{v}$ can travel the distance $\mathbf{r}-\mathbf{r}'$ during $t-t'$. The factor of the function $f_0(\mathbf{v})$ on the right-hand side of (6.28) may evidently be regarded as the conditional probability of the particle transition in an unbounded system of particles which do not interact:

$$W^0_{\mathbf{r}-\mathbf{r}', t-t'}(\mathbf{v}, \mathbf{v}') = \delta(\mathbf{v}-\mathbf{v}')\,\delta[\mathbf{r}-\mathbf{r}'-\mathbf{v}(t-t')]. \tag{6.29}$$

Since the motions of different species are independent in neglect of particle interactions, the correlator for the fluctuations of the distribution functions of different species vanishes:

$$\langle \delta f_0(\mathbf{r}, \mathbf{v}, t)\,\delta f_0'(\mathbf{r}', \mathbf{v}', t')\rangle \equiv \langle \delta f(\mathbf{v})\,\delta f'(\mathbf{v}')\rangle^0_{\mathbf{r}-\mathbf{r}', t-t'} = 0. \tag{6.30}$$

The relations (6.28) and (6.30) may be written in a unified form:

$$\langle \delta f(\mathbf{v})\,\delta f'(\mathbf{v}')\rangle^0_{\mathbf{r}-\mathbf{r}', t-t'} = \delta'\delta(\mathbf{v}-\mathbf{v}')\,\delta[\mathbf{r}-\mathbf{r}'-\mathbf{v}(t-t')]\,f_0(\mathbf{v}), \tag{6.31}$$

where $\delta' = \text{I}$ for identical particles and $\delta' = 0$ for different species.

The Fourier transformation of (6.31) yields the following spectral distribution for the fluctuations of the distribution functions in neglect of particle interactions:

$$\langle \delta f(\mathbf{v})\,\delta f'(\mathbf{v}')\rangle^0_{\mathbf{k}\omega} = 2\pi\delta'\delta(\mathbf{v}-\mathbf{v}')\,\delta(\omega-(\mathbf{k}\cdot\mathbf{v}))\,f_0(\mathbf{v}). \tag{6.32}$$

Formula (6.32) is sufficient to find explicitly the spectral distributions for the fluctuations of any quantities in an ensemble of non-interacting charged particles.

Spectral distributions of the fluctuations of particle, charge, and current densities

Integrating (6.32) twice over the velocities, we obtain the spectral distribution of the particle density fluctuations

$$\langle \delta n^2\rangle^0_{\mathbf{k}\omega} = 2\pi \int d\mathbf{v}\,\delta(\omega-(\mathbf{k}\cdot\mathbf{v}))\,f_0(\mathbf{v}). \tag{6.33}$$

There is no correlation between the particle density fluctuations of different species in the absence of particle interactions.

By virtue of (6.33) the frequency dependence of the spectral distribution of particle density fluctuations is governed by the one-particle distribution function. Let us introduce the following notation for the integral of the distribution function over the velocity components, perpendicular to the wave vector $\mathbf{k}$:

$$f_{0||}(v_{||}) \equiv \int d\mathbf{v}_{\perp} f_0(\mathbf{v}_{\perp}, v_{||}). \tag{6.34}$$

This reduces the spectral distribution of the particle density fluctuations to the form

$$\langle \delta n^2\rangle_{\mathbf{k}\omega} = \frac{2\pi}{k} f_{0||}\left(\frac{\omega}{k}\right). \tag{6.35}$$

Since the distribution function was assumed to satisfy the normalization condition, the distribution (6.35) must have a maximum at low frequencies and decrease with increasing frequency. Substituting the Maxwellian distribution for $f_0(\mathbf{v})$, we obtain for an equilibrium system

$$\langle \delta n^2 \rangle^0_{\mathbf{k}\omega} = \sqrt{6\pi}\,\frac{n_0}{ks}\,e^{-\frac{3}{2}\frac{\omega^2}{k^2 s^2}}. \tag{6.36}$$

This distribution has a maximum at zero frequency, the width of which is governed by the

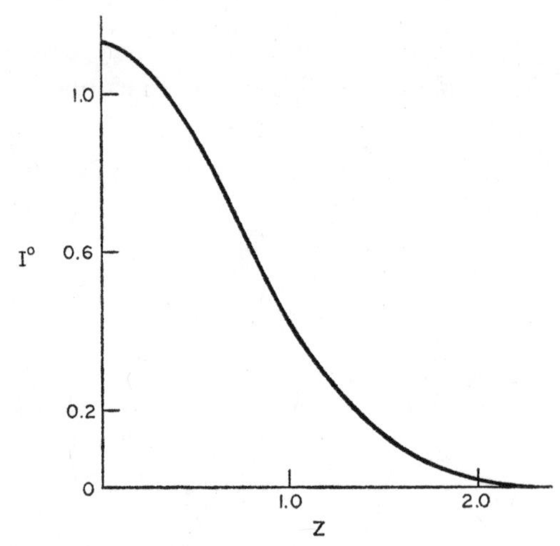

FIG. 4. The spectral distribution of density fluctuations $I^0(z)$ in neglect of particle interactions.

plasma temperature. In Fig. 4 we see the spectral distribution of the density fluctuations $I^0(z) = \langle \delta n^2 \rangle^0_{\mathbf{k}\omega} / 2\pi \langle \delta n^2 \rangle^0_{\mathbf{k}}$ as a function of the dimensionless frequency $z = \sqrt{\frac{3}{2}}\,\frac{\omega}{ks}$.

Making use of (6.33) it is easy to find the spectral distribution of the charge density fluctuations in the absence of particle interactions:

$$\langle \varrho^2 \rangle^0_{\mathbf{k}\omega} = 2\pi \sum e^2 \int d\mathbf{v}\,\delta(\omega - (\mathbf{k}\cdot\mathbf{v}))\,f_0(\mathbf{v}). \tag{6.37}$$

Multiplying (6.32) by the dyadic product of the vectors $\mathbf{v}$ and $\mathbf{v}'$ and integrating over the velocities, we obtain the following spectral distribution of the current density fluctuations in an ensemble of non-interacting particles:

$$\langle j_i j_j \rangle^0_{\mathbf{k}\omega} = 2\pi \sum e^2 \int d\mathbf{v}\; v_i v_j \delta(\omega - \mathbf{k}\cdot\mathbf{v})\,f_0(\mathbf{v}). \tag{6.38}$$

The spectral representations of the correlations between the fluctuations of the distribution function and those of the charge and current densities in neglect of the interaction between particles may be written as:

$$\langle \delta f(\mathbf{v}) \varrho \rangle^0_{\mathbf{k}\omega} = 2\pi e \delta(\omega - (\mathbf{k}\cdot\mathbf{v}))\,f_0(\mathbf{v}), \tag{6.39}$$

$$\langle \delta f(\mathbf{v}) j_i \rangle^0_{\mathbf{k}\omega} = 2\pi e v_i \delta(\omega - (\mathbf{k}\cdot\mathbf{v}))\,f_0(\mathbf{v}). \tag{6.40}$$

Higher-order correlation functions

A similar procedure with the use of (6.26) and (6.24) leads to the following higher-order space–time correlators for the fluctuations of the distribution functions of noninteracting particles:

$$\langle \delta f_0(\mathbf{r}, \mathbf{v}, t)\, \delta f_0(\mathbf{r}', \mathbf{v}', t')\, \delta f(\mathbf{r}'', \mathbf{v}'', t'') \rangle \equiv \langle \delta f(\mathbf{v})\, \delta f(\mathbf{v}')\, \delta f(\mathbf{v}'') \rangle^0_{\mathbf{r}-\mathbf{r}', t-t'; \mathbf{r}'-\mathbf{r}'', t'-t''}$$
$$= \delta(\mathbf{v}-\mathbf{v}')\, \delta(\mathbf{v}'-\mathbf{v}'')\, \delta[\mathbf{r}-\mathbf{r}'-\mathbf{v}(t-t')]\, \delta[\mathbf{r}'-\mathbf{r}''-\mathbf{v}(t'-t'')]\, f_0(\mathbf{v}), \quad (6.41)$$

$$\langle \delta f_0(\mathbf{r}, \mathbf{v}, t)\, \delta f_0(\mathbf{r}', \mathbf{v}', t')\, \delta f_0(\mathbf{r}'', \mathbf{v}'', t'')\, \delta f_0(\mathbf{r}''', \mathbf{v}''', t''') \rangle \equiv$$
$$\equiv \langle \delta f(\mathbf{v})\, \delta f(\mathbf{v}')\, \delta f(\mathbf{v}'')\, \delta f(\mathbf{v}''') \rangle^0_{\mathbf{r}-\mathbf{r}', t-t'; \mathbf{r}'-\mathbf{r}'', t'-t''; \mathbf{r}''-\mathbf{r}''', t''-t'''}$$
$$= \delta(\mathbf{v}-\mathbf{v}')\, \delta(\mathbf{v}'-\mathbf{v}'')\, \delta(\mathbf{v}''-\mathbf{v}''')\, \delta[\mathbf{r}-\mathbf{r}'-\mathbf{v}(t-t')]$$
$$\times\, \delta[\mathbf{r}'-\mathbf{r}''-\mathbf{v}(t'-t'')]\, \delta[\mathbf{r}''-\mathbf{r}'''-\mathbf{v}(t''-t''')]\, f_0(\mathbf{v}). \quad (6.42)$$

As should be expected, the correlation functions of all orders in neglect of particle interactions depend on the one-particle distribution function and the conditional transition probability (6.29).

The space–time Fourier transformations of (6.41) and (6.42) yield the following spectral distributions:

$$\langle \delta f(\mathbf{v})\, \delta f(\mathbf{v}')\, \delta f(\mathbf{v}'') \rangle^0_{\mathbf{k}\omega; \mathbf{k}'\omega'} = (2\pi)^2\, \delta(\mathbf{v}-\mathbf{v}')\, \delta(\mathbf{v}'-\mathbf{v}'')$$
$$\times\, \delta(\omega-(\mathbf{k}\cdot\mathbf{v}))\, \delta(\omega'-(\mathbf{k}'\cdot\mathbf{v}))\, f_0(\mathbf{v}), \quad (6.43)$$

$$\langle \delta f(\mathbf{v})\, \delta f(\mathbf{v}')\, \delta f(\mathbf{v}'')\, \delta f(\mathbf{v}''') \rangle^0_{\mathbf{k}\omega; \mathbf{k}'\omega'; \mathbf{k}''\omega''}$$
$$= \delta(\mathbf{v}-\mathbf{v}')\, \delta(\mathbf{v}'-\mathbf{v}'')\, \delta(\mathbf{v}''-\mathbf{v}''')\, \delta(\omega-(\mathbf{k}\cdot\mathbf{v}))\, \delta(\omega'-(\mathbf{k}'\cdot\mathbf{v}))\, \delta(\omega''-(\mathbf{k}''\cdot\mathbf{v}))\, f_0(\mathbf{v}). \quad (6.44)$$

The velocity integration of (6.43) and (6.44) results in the spectral representations for the higher-order correlation functions for the charge density fluctuations:

$$\langle \varrho^3 \rangle^0_{\mathbf{k}\omega; \mathbf{k}'\omega'} = 4\pi^2 \sum e^3 \int d\mathbf{v}\, \delta(\omega-(\mathbf{k}\cdot\mathbf{v}))\, \delta(\omega'-(\mathbf{k}'\cdot\mathbf{v}))\, f_0(\mathbf{v}), \quad (6.45)$$
$$\langle \varrho^4 \rangle^0_{\mathbf{k}\omega; \mathbf{k}'\omega'; \mathbf{k}''\omega''} = 8\pi^2 \sum e^4 \int d\mathbf{v}\, \delta(\omega-(\mathbf{k}\cdot\mathbf{v}))\, \delta(\omega'-(\mathbf{k}'\cdot\mathbf{v}))\, \delta(\omega''-(\mathbf{k}''\cdot\mathbf{v}))\, f_0(\mathbf{v}). \quad (6.46)$$

The spectral representation of the nth order correlation function for the charge density fluctuations in neglect of particle interactions is of the form:

$$\langle \varrho^n \rangle^0_{\mathbf{k}\omega; \mathbf{k}'\omega'; \ldots\, \mathbf{k}^{(n-1)}\omega^{(n-1)}}$$
$$= (2\pi)^{(n-1)} \sum e^n \int d\mathbf{v}\, \delta(\omega-(\mathbf{k}\cdot\mathbf{v})) \ldots \delta(\omega^{(n-1)}-(\mathbf{k}^{(n-1)}\cdot\mathbf{v}))\, f_0(\mathbf{v}). \quad (6.47)$$

The spectral representations for the correlation functions of current density fluctuations in an ensemble of non-interacting particles may be obtained in a similar manner:

$$\langle j_i j_j j_k \rangle^0_{\mathbf{k}\omega; \mathbf{k}'\omega'} = 4\pi^2 \sum e^3 \int d\mathbf{v} v_i v_j v_k\, \delta(\omega-(\mathbf{k}\cdot\mathbf{v}))\, \delta(\omega'-(\mathbf{k}'\cdot\mathbf{v}))\, f_0(\mathbf{v}), \quad (6.48)$$
$$\langle j_i j_j j_k j_l \rangle^0_{\mathbf{k}\omega; \mathbf{k}'\omega'; \mathbf{k}''\omega''}$$
$$= 8\pi^3 \sum e^4 \int d\mathbf{v} v_i v_j v_k v_l\, \delta(\omega-(\mathbf{k}\cdot\mathbf{v}))\, \delta(\omega'-(\mathbf{k}'\cdot\mathbf{v}))\, \delta(\omega''-(\mathbf{k}''\cdot\mathbf{v}))\, f_0(\mathbf{v}). \quad (6.49)$$

Spectral distributions of fluctuations in the presence of an external magnetic field

In the previous section we derived the correlation functions and the spectral distributions for the fluctuations in an ensemble of charged particles in neglect of particle interactions and in the absence of external fields. The latter being present, the character of the fluctuations in the system may undergo essential changes. For the simplest example consider the fluctuations in an ensemble of non-interacting charged particles exposed to a constant and uniform external magnetic field. (This model describes fluctuations in a magneto-active plasma in neglect of particle interactions.) Individual particles travel under the influence of the magnetic field along helical lines, the trajectory of a labelled particle α being governed by the following equation:

$$\mathbf{r}_\alpha^0(t) = \mathbf{r}_\alpha^0 + \mathbf{R}_\alpha(t - t_0, \mathbf{v}_\alpha^0), \tag{6.50}$$

where

$$\mathbf{R}(t, \mathbf{v}) \equiv \begin{cases} \dfrac{1}{\omega_B}[v_x \sin \omega_B t - v_y(\cos \omega_B t - 1)] \\ \dfrac{1}{\omega_B}[v_x(\cos \omega_B t - 1) + v_y \sin \omega_B t] \\ v_{\|} t \end{cases} \tag{6.51}$$

($\mathbf{r}_\alpha^0$ and $\mathbf{v}_\alpha^0$ are the initial radius vector and the velocity, the z-axis of the coordinate system coincides with the magnetic field.)

The space-time correlator for the fluctuations of the distribution function is similar to (6.27). Averaging (6.27) over the distribution (6.24) and making use of (6.50), we obtain

$$\langle \delta f(\mathbf{v})\, \delta f(\mathbf{v}') \rangle_{\mathbf{r}t}^0 = \delta\left(\mathbf{v} - \frac{\partial}{\partial t}\mathbf{R}(t, \mathbf{v}')\right) \delta(\mathbf{r} - \mathbf{R}(t, \mathbf{v}'))\, f_0(\mathbf{v}'). \tag{6.52}$$

Thus the velocities $\mathbf{v}$ and $\mathbf{v}'$ the phases of which differ by $\omega_B t$ turn out to be correlated in the presence of a magnetic field.

The spectral representation of the distribution function fluctuations is given by the following formula:

$$\langle \delta f(\mathbf{v})\, \delta f(\mathbf{v}') \rangle_{\mathbf{k}\omega}^0 = \int_{-\infty}^{\infty} dt\, e^{-i(\mathbf{k}\cdot\mathbf{R}(t, \mathbf{v}')) + i\omega t} \delta\left(\mathbf{v} - \frac{\partial}{\partial t}\mathbf{R}(t, \mathbf{v})\right) f_0(\mathbf{v}'). \tag{6.53}$$

Substituting (6.51) for $\mathbf{R}(t, \mathbf{v})$ and making use of the expansion (3.4), we can present the spectral distribution (6.53) as a decomposition in the cyclotron frequency harmonics. Integrating the relation thus obtained over the velocities, it is easy to derive explicit expressions for the spectral distributions of the fluctuations of particle, charge and current densities.

Spectral distributions of charge and current density fluctuations in an ensemble of non-interacting charged particles, the distribution function of which is axisymmetric relative to the magnetic field, are as follows:

$$\langle \varrho^2 \rangle^0_{\mathbf{k}\omega} = 2\pi \sum e^2 \int d\mathbf{v} \sum_{n=-\infty}^{\infty} J_n^2\left(\frac{k_\perp v_\perp}{\omega_B}\right) \delta(\omega - k_{||}v_{||} - n\omega_B)\, f_0(v_\perp, v_{||}), \tag{6.54}$$

$$\langle j_i j_j \rangle^0_{\mathbf{k}\omega} = 2\pi \sum e^2 \int d\mathbf{v} \sum_{n=-\infty}^{\infty} \Pi_{ij}^{(n)}(\mathbf{v})\, \delta(\omega - k_{||}v_{||} - n\omega_B)\, f_0(v_\perp, v_{||}). \tag{6.55}$$

The tensor $\Pi_{ij}^{(n)}(\mathbf{v})$ was defined by (3.35). Substituting the Maxwellian distribution for $f_0(\mathbf{v})$, we obtain for an equilibrium system:

$$\langle \varrho^2 \rangle^0_{\mathbf{k}\omega} = \sqrt{6\pi} \sum \frac{e^2 n_0}{|k_{||}|\, s} e^{-\beta} \sum_n I_n(\beta) e^{-z_n^2}, \tag{6.56}$$

$$\langle j_i j_j \rangle^0_{\mathbf{k}\omega} = \sqrt{\frac{2}{3}}\, \pi \sum \frac{e^2 n_0}{|k_{||}|} s e^{-\beta} \sum_n \pi_{ij}(z_n) e^{-z^2}, \tag{6.57}$$

where for $\pi_{ij}(z_n)$ see formula (3.38).

In a similar manner we can find the higher-order correlation functions. For example, we give the spectral representations of the third- and fourth-order correlation functions for the charge density fluctuations:

$$\begin{aligned} \langle \varrho^3 \rangle^3_{\mathbf{k}_1\omega_1;\,\mathbf{k}_2\omega_2} = 4\pi^2 \sum e^3 \int d\mathbf{v} \sum_{n_1, n_1', n_2} & e^{-i(n_1 - n_1')(\varphi_1 - \varphi_2)} J_{n_1}\left(\frac{k_{1\perp} v_\perp}{\omega_B}\right) J_{n_1'}\left(\frac{k_{1\perp} v_\perp}{\omega_B}\right) \\ & \times J_{n_2}\left(\frac{k_{2\perp} v_\perp}{\omega_B}\right) J_{n_1 - n_1' + n_2}\left(\frac{k_{2\perp} v_\perp}{\omega_B}\right) \delta(\omega_1 - k_{1||} v_{||} - n_1 \omega_B) \\ & \times \delta(\omega_1 - k_{2||} v_{||} - n_2 \omega_B)\, f_0(v_\perp, v_{||}), \end{aligned} \tag{6.58}$$

$$\begin{aligned} \langle \varrho^4 \rangle^0_{\mathbf{k}_1\omega_1;\,\mathbf{k}_2\omega_2;\,\mathbf{k}_3\omega_3} = 8\pi^3 \sum e^4 \int d\mathbf{v} \sum_{n_1, n_1', n_2, n_2', n_3} & e^{-i(n_1 - n_1')(\varphi_1 - \varphi_2) - i(n_1 - n_1' + n_2 - n_2')(\varphi_2 - \varphi_3)} \\ & \times J_{n_1}\left(\frac{k_{1\perp} v_\perp}{\omega_B}\right) J_{n_1'}\left(\frac{k_{1\perp} v_\perp}{\omega_B}\right) J_{n_2}\left(\frac{k_{2\perp} v_\perp}{\omega_B}\right) J_{n_2'}\left(\frac{k_{2\perp} v_\perp}{\omega_B}\right) \\ & \times J_{n_3}\left(\frac{k_{3\perp} v_\perp}{\omega_B}\right) J_{n_1 - n_1' + n_2 - n_2' + n_3}\left(\frac{k_{3\perp} v_\perp}{\omega_B}\right) \delta(\omega_1 - k_{1||} v_{||} - n_1 \omega_B) \\ & \times \delta(\omega_2 - k_{2||} v_{||} - n_2 \omega_B)\, \delta(\omega_3 - k_{3||} v_{||} - n_3 \omega_B)\, f_0(v_\perp, v_{||}). \end{aligned} \tag{6.59}$$

6.3. Fluctuations in an Ensemble of Charged Particles Taking into Account Electromagnetic Interaction between Particles (Linear Approximation)

Effect of Coulomb interactions between particles on the fluctuations of the distribution function

Let us consider now the fluctuations of various quantities in a plasma taking into account the self-consistent interaction between particles. It is to be remembered that the microscopic plasma properties are governed by the microscopic distribution function or the microscopic density

$$\mathcal{F}(\mathbf{r}, \mathbf{v}, t) \equiv \sum \delta(\mathbf{r}-\mathbf{r}_\alpha(t))\, \delta(\mathbf{v}-\mathbf{v}_\alpha(t)). \tag{6.60}$$

For simplicity we consider first only the Coulomb interaction between particles. In such a case the evolution of the microscopic plasma density is described by the following set of equations:

$$\left.\begin{aligned} \frac{\partial \mathcal{F}}{\partial t}+\left(\mathbf{v}\cdot\frac{\partial \mathcal{F}}{\partial \mathbf{r}}\right)+\frac{e}{m}\left(\mathbf{E}\cdot\frac{\partial \mathcal{F}}{\partial \mathbf{v}}\right) &= 0,\\ \operatorname{div}\mathbf{E} = 4\pi e\left\{\int d\mathbf{v}\,\mathcal{F}-n_0\right\}, \end{aligned}\right\} \tag{6.61}$$

where $\mathbf{E}$ is the self-consistent microscopic plasma field. Let the one-particle distribution function, which was defined as the average of the microscopic density (6.60) over the Liouville distribution $D_N(\mathbf{r}_1, v_1; \ldots; \mathbf{r}_N, v_N; t)$, be stationary and spatially uniform, i.e.

$$\langle\mathcal{F}(\mathbf{r}, \mathbf{v}, t)\rangle = f_0(\mathbf{v}). \tag{6.62}$$

The fluctuation of the distribution function implies

$$\delta f(\mathbf{r}, \mathbf{v}, t) \equiv \mathcal{F}(\mathbf{r}, \mathbf{v}, t)-f_0(\mathbf{v}). \tag{6.63}$$

The stochastic nature of this quantity is governed by the microscopic nature of the density $\mathcal{F}(\mathbf{r}, \mathbf{v}, t)$.

Besides the exact microscopic density $\mathcal{F}(\mathbf{r}, \mathbf{v}, t)$ we introduce the microscopic density $\mathcal{F}_0(\mathbf{r}, \mathbf{v}, t)$ of the same system in neglect of particle interactions:

$$\mathcal{F}_0(\mathbf{r}, \mathbf{v}, t) \equiv \sum_\alpha \delta(\mathbf{r}-\mathbf{r}_\alpha^0(t-t_0))\, \delta(\mathbf{v}-\mathbf{v}_\alpha^0) \tag{6.64}$$

($\mathbf{r}_\alpha^0$ and $\mathbf{v}_\alpha^0$ are the radius vector and the velocity of the particle at the initial time t_0) and assume the mean value of (6.64) to be similar to (6.62)). The density $\mathcal{F}_0(\mathbf{r}, \mathbf{v}, t)$ satisfies the evident equation

$$\frac{\partial \mathcal{F}_0}{\partial t}+\left(\mathbf{v}\cdot\frac{\partial \mathcal{F}_0}{\partial \mathbf{r}}\right) = 0. \tag{6.65}$$

Let us define the fluctuations of the distribution function for an ensemble of non-interacting particles according to

$$\delta f_0(\mathbf{r}, \mathbf{v}, t) \equiv \mathcal{F}_0(\mathbf{r}, \mathbf{v}, t)-f_0(\mathbf{v}). \tag{6.66}$$

Let the difference between the exact microscopic density $\mathcal{F}(\mathbf{r}, \mathbf{v}, t)$ and the microscopic density in neglect of particle interactions $\mathcal{F}_0(\mathbf{r}, \mathbf{v}, t)$ be denoted by $\tilde{f}(\mathbf{r}, \mathbf{v}, t)$:

$$\tilde{f}(\mathbf{r}, \mathbf{v}, t) \equiv \mathcal{F}(\mathbf{r}, \mathbf{v}, t)-\mathcal{F}_0(\mathbf{r}, \mathbf{v}, t). \tag{6.67}$$

It is clear that this quantity is due to the Coulomb interactions between particles and vanishes if the latter is neglected. Making use of the definitions (6.63) and (6.66), we may easily find from (6.67) that

$$\delta f(\mathbf{r}, \mathbf{v}, t) = \delta f_0(\mathbf{r}, \mathbf{v}, t)+\tilde{f}(\mathbf{r}, \mathbf{v}, t). \tag{6.68}$$

This relation connects the fluctuations of the distribution function with and without taking into account interactions between plasma particles.[†]

Substituting the following expression for the microscopic density $\mathcal{F}(\mathbf{r}, \mathbf{v}, t)$

$$\mathcal{F}(\mathbf{r}, \mathbf{v}, t) = f_0(\mathbf{v}) + \delta f_0(\mathbf{r}, \mathbf{v}, t) + \tilde{f}(\mathbf{r}, \mathbf{v}, t) \tag{6.69}$$

in (6.61) and making use of (6.65), we obtain a set of equations of the form

$$\left.\begin{aligned} &\frac{\partial \tilde{f}}{\partial t} + \left(\mathbf{v} \cdot \frac{\partial \tilde{f}}{\partial \mathbf{r}}\right) + \frac{e}{m}\left(\mathbf{E} \cdot \frac{\partial}{\partial \mathbf{v}}\right)(f_0 + \delta f_0 + \tilde{f}) = 0. \\ &\operatorname{div} \mathbf{E} = 4\pi e \int d\mathbf{v}(\delta f_0 + \tilde{f}). \end{aligned}\right\} \tag{6.70}$$

The space–time Fourier transformation reduces the first equation (6.70) to the following:

$$\begin{aligned} \tilde{f}_{\mathbf{k}\omega} = -i\frac{e}{m}\frac{1}{\omega - (\mathbf{k}\cdot\mathbf{v}) + i0}&\left\{\left(\mathbf{E}_{\mathbf{k}\omega} \cdot \frac{\partial f_0}{\partial \mathbf{v}}\right)\right. \\ &\left. + \sum_{\omega', \mathbf{k}'}\left(\mathbf{E}_{\mathbf{k}'\omega'} \cdot \frac{\partial}{\partial \mathbf{v}}\right)(\delta f^0_{\mathbf{k}-\mathbf{k}', \omega-\omega'} + \tilde{f}_{\mathbf{k}-\mathbf{k}', \omega-\omega'})\right\}. \end{aligned} \tag{6.71}$$

Neglecting the non-linear terms on the right-hand side of (6.71) we have

$$\tilde{f}_{\mathbf{k}\omega} = -i\frac{e}{m}\frac{1}{\omega - (\mathbf{k}\cdot\mathbf{v}) + i0}\left(\mathbf{E}_{\mathbf{k}\omega} \cdot \frac{\partial f_0}{\partial \mathbf{v}}\right). \tag{6.72}$$

Carrying out the space–time Fourier transformation of the second equation (6.70) and eliminating $\tilde{f}_{\mathbf{k}\omega}$ using (6.72), we obtain

$$i\varepsilon(\omega, \mathbf{k})(\mathbf{k}\cdot\mathbf{E}_{\mathbf{k}\omega}) = 4\pi\varrho^0_{\mathbf{k}\omega}, \tag{6.73}$$

where $\varepsilon(\omega, \mathbf{k})$ is the longitudinal plasma dielectric permittivity (2.16) and $\varrho^0_{\mathbf{k}\omega}$ is the Fourier component of the charge density fluctuation in neglect of particle interactions

$$\varrho^0_{\mathbf{k}\omega} = \sum e \int d\mathbf{v}\, \delta f^0_{\mathbf{k}\omega}(\mathbf{v}). \tag{6.74}$$

It should be noted that the charge density fluctuations (6.74) in an ensemble of non-interacting particles are caused by the random motion of the individual plasma particles.

Equation (6.73) directly relates the fluctuating field intensity $\mathbf{E}_{\mathbf{k}\omega}$ and the fluctuations of the motions of individual plasma particles. Inasmuch as the electric field is longitudinal, we have

$$\mathbf{E}_{\mathbf{k}\omega} = -4\pi i\frac{\mathbf{k}}{k^2}\frac{\varrho^0_{\mathbf{k}\omega}}{\varepsilon(\omega, \mathbf{k})}. \tag{6.75}$$

The denominator of the right-hand side of (6.75) contains the dielectric permittivity $\varepsilon(\omega, \mathbf{k})$ by virtue of the screening of the charge density fluctuations due to the plasma polarization.

[†] Cook and Taylor (1973) use a similar approach.

Substituting (6.72) in (6.68) and taking the fluctuating field intensity in the form (6.75) we obtain a relation that connects the fluctuations of the plasma distribution function with the charge density and the distribution function in an ensemble of non-interacting particles:

$$\delta f_{\mathbf{k}\omega}(\mathbf{v}) = \delta f^0_{\mathbf{k}\omega}(\mathbf{v}) - \frac{4\pi e}{mk^2}\frac{1}{\varepsilon(\omega, \mathbf{k})}\frac{1}{\omega - \mathbf{k}\mathbf{v} + i0}\mathbf{k}\frac{\partial f_0(\mathbf{v})}{\partial \mathbf{v}}\varrho^0_{\mathbf{k}\omega}. \tag{6.76}$$

The last relation is suitable to find explicitly the correlator for the fluctuations of the particle distribution function in a plasma taking into account the Coulomb interactions between particles.

Averaging procedure and the correlator for the fluctuations of the distribution function

The quadratic space–time correlator for the fluctuations of the plasma distribution function is determined by the following equality:

$$\langle \delta f(\mathbf{v})\, \delta f'(\mathbf{v}')\rangle_{\mathbf{r}-\mathbf{r}', t-t'} \equiv \int \prod_{\alpha=1}^{N} d\bar{x}_\alpha D_N(\bar{x}_1 \ldots \bar{x}_N; \bar{t})\, \delta f(\mathbf{r}, \mathbf{v}, t)\, \delta f'(\mathbf{r}', \mathbf{v}', t'), \tag{6.77}$$

where $D_N(\bar{x}_1 \ldots \bar{x}_N; \bar{t})$ is the exact Liouville distribution for the ensemble at some time $\bar{t}$, which must in general case satisfy the condition $\bar{t} \leqslant t, t'$. Since the Liouville distribution is conserved along the trajectory of motion of the system as a whole, the averaging in (6.77) at time $\bar{t}$ may be replaced by the averaging over the initial distribution at t_0:

$$\int \prod_{\alpha=1}^{N} d\bar{x}_\alpha D_N(\bar{x}_1 \ldots \bar{x}_N; t) \ldots = \int \prod_{\alpha=1}^{N} dx^0_\alpha D_N(x^0_1 \ldots x^0_N; t^0) \ldots . \tag{6.78}$$

Taking the segments between the initial time t_0 and the times under consideration t and t' to be very long, we may have recourse to the correlation weakening principle. Within the context of (1.67) we may neglect the correlations in the Liouville distribution as $t_0 \to -\infty$,† i.e. to carry out the following replacement:

$$D_N(x^0_1 \ldots x^0_N; t_0) \xrightarrow[t_0 \to \infty]{} D^0_N(x^0_1 \ldots x^0_N) \to \prod_{\alpha=1}^{N} \tilde{f}_0(\mathbf{v}^0_\alpha). \tag{6.79}$$

Here the averaging procedure on the right-hand side of (6.77) is formally the same as that for an ensemble of non-interacting particles. Hence, the averaging process in (6.77) is to be carried out according to:

$$\langle \delta f(\mathbf{v})\, \delta f'(\mathbf{v}')\rangle_{\mathbf{r}-\mathbf{r}', t-t'} \equiv \int \prod_{\alpha=1}^{N} (d\mathbf{v}^0_\alpha \tilde{f}_0(\mathbf{v}^0_\alpha))\, \delta f(\mathbf{r}, \mathbf{v}, t)\, \delta f(\mathbf{r}', \mathbf{v}', t'). \tag{6.80}$$

† Note that when considering the collective plasma properties we should take into account the dependence of the Liouville distribution both on the dynamical variables of N particles and on the oscillation degrees of freedom. However the correlation weakening principle implies that the excitation of the collective degrees of freedom may be neglected at $t_0 \to -\infty$.

The spectral representation of (6.80) is then to be written as

$$\langle \delta f(\mathbf{v})\, \delta f'(\mathbf{v}')\rangle_{\mathbf{k}\omega} \equiv \int \prod_{\alpha=1}^{N} (d\mathbf{v}_\alpha^0 \tilde{f}_0(\mathbf{v}_\alpha^0))\, \delta f_{\mathbf{k}\omega}(\mathbf{v})\, \delta f'^*_{\mathbf{k}\omega}(\mathbf{v}'). \tag{6.81}$$

Thus, within the context of (6.76) we obtain

$$\begin{aligned}\langle \delta f(\mathbf{v})\, \delta f'(\mathbf{v}')\rangle_{\mathbf{k}\omega} = {} & \langle \delta f(\mathbf{v})\, \delta f'(\mathbf{v}')\rangle^0_{\mathbf{k}\omega} \\ & - \frac{4\pi e}{mk^2\varepsilon(\omega, \mathbf{k})} \frac{1}{\omega-(\mathbf{k}\cdot\mathbf{v})+i0}\left(\mathbf{k}\cdot \frac{\partial f_0(\mathbf{v})}{\partial \mathbf{v}}\right)\langle \delta f'(\mathbf{v}')\varrho\rangle^0_{\mathbf{k}\omega} \\ & - \frac{4\pi e'}{m'k^2\varepsilon^*(\omega, \mathbf{k})}\langle \delta f(\mathbf{v})\varrho\rangle^0_{\mathbf{k}\omega} \frac{1}{\omega-(\mathbf{k}\cdot\mathbf{v}')+i0}\left(\mathbf{k}\cdot \frac{\partial f_0'(\mathbf{v}')}{\partial \mathbf{v}'}\right) \\ & + \frac{16\pi e e'}{mm'k^4\,|\varepsilon(\omega, \mathbf{k})|^2} \frac{1}{\omega-(\mathbf{k}\cdot\mathbf{v})+i0}\left(\mathbf{k}\cdot \frac{\partial f_0(\mathbf{v})}{\partial \mathbf{v}}\right) \\ & \times \frac{1}{\omega-(\mathbf{k}\cdot\mathbf{v}')+i0}\left(\mathbf{k}\cdot \frac{\partial f_0'(\mathbf{v}')}{\partial \mathbf{v}'}\right)\langle \varrho^2\rangle^0_{\mathbf{k}\omega},\end{aligned} \tag{6.82}$$

where $\langle \delta f(\mathbf{v})\, \delta f'(\mathbf{v}')\rangle^0_{\mathbf{k}\omega}$, $\langle \delta f(\mathbf{v})\varrho\rangle^0_{\mathbf{k}\omega}$ and $\langle \varrho^2\rangle^0_{\mathbf{k}\omega}$ are the spectral representations of the fluctuations of the distribution functions and the charge densities in an ensemble of non-interacting particles. Substituting (6.32), (6.37), and (6.49) for the above-mentioned spectral representations, we obtain:

$$\begin{aligned}\langle \delta f(\mathbf{v})\, \delta f'(\mathbf{v}')\rangle_{\mathbf{k}\omega} = 2\pi \Bigg\{ & \delta'\delta(\mathbf{v}-\mathbf{v}')\, \delta(\omega-(\mathbf{k}\cdot\mathbf{v}))\, f_0(\mathbf{v}) \\ & - \frac{4\pi e e'}{mk^2\varepsilon(\omega, \mathbf{k})} \frac{1}{\omega-(\mathbf{k}\cdot\mathbf{v})+i0}\left(\mathbf{k}\cdot \frac{\partial f_0(\mathbf{v})}{\partial \mathbf{v}}\right)\delta(\omega-(\mathbf{k}\cdot\mathbf{v}))\, f_0'(\mathbf{v}') \\ & - \frac{4\pi e e'}{m'k^2\varepsilon(\omega, \mathbf{k})}\delta(\omega-(\mathbf{k}\cdot\mathbf{v}))\, f_0(\mathbf{v}) \frac{1}{\omega-(\mathbf{k}\cdot\mathbf{v}')+i0}\left(\mathbf{k}\cdot \frac{\partial f_0'(\mathbf{v}')}{\partial \mathbf{v}'}\right) \\ & + \frac{16\pi^2 e e'}{mm'k^4\,|\varepsilon(\omega, \mathbf{k})|^2} \frac{1}{\omega-(\mathbf{k}\cdot\mathbf{v})+i0}\left(\mathbf{k}\cdot \frac{\partial f_0(\mathbf{v})}{\partial \mathbf{v}}\right)\frac{1}{\omega-(\mathbf{k}\cdot\mathbf{v}')+i0}\cdot \\ & \times \left(\mathbf{k}\cdot \frac{\partial f_0'(\mathbf{v}')}{\partial \mathbf{v}'}\right)\sum e''^2 \int d\mathbf{v}''\, \delta(\omega-(\mathbf{k}\cdot\mathbf{v}''))\, f_0''(\mathbf{v}'')\Bigg\}.\end{aligned} \tag{6.83}$$

The latter formula describes explicitly the spectral correlator for the fluctuations of the particle distribution function in a plasma taking into account the Coulomb interactions between particles. It is to be observed that the last three terms in (6.83) are directly associated with the Coulomb interactions, which is the reason for the fluctuations of the distribution functions both of identical particles and of different species in a plasma to become correlated. It is to be emphasized that relation (6.83) was derived in the linear approximation.

Spectral distributions of particle, charge and current density fluctuations in a plasma

The knowledge of the spectral distribution (6.83) is sufficient to derive explicitly the spectral representations of various correlation functions. For the density fluctuations of identical particles and of different species in a plasma they can be easily found by means of a velocity integration of (6.83).

The spectral distribution for the density fluctuations of some plasma component is as follows:

$$\langle \delta n^2 \rangle_{\mathbf{k}\omega} = \frac{|1+\Sigma' \varkappa'(\omega, \mathbf{k})|^2 \langle \delta n^2 \rangle^0_{\mathbf{k}\omega} + |\varkappa(\omega, \mathbf{k})|^2 \Sigma' \langle \delta n'^2 \rangle^0_{\mathbf{k}\omega}}{|\varepsilon(\omega, \mathbf{k})|^2}. \tag{6.84}$$

Here and below the following notation is used: $\langle \delta n^2 \rangle^0_{\mathbf{k}\omega}$ is the spectral distribution for the fluctuations of some labelled component in neglect of particle interactions:

$$\langle \delta n^2 \rangle^0_{\mathbf{k}\omega} \equiv 2\pi \int d\mathbf{v}\, \delta(\omega - \mathbf{k}\cdot\mathbf{v})\, f_0(\mathbf{v}), \tag{6.85}$$

$\langle \delta n'^2 \rangle^0_{\mathbf{k}\omega}$ is the one for the other species, $\varkappa(\omega, \mathbf{k})$ and $\varkappa'(\omega, \mathbf{k})$ are the partial components of the linear plasma susceptibility (2.16), the symbol Σ' implies the summation over all plasma species except the labelled one.

Analogously, the spectral distribution for the density fluctuations of different plasma components is given by the following expression:

$$\langle \delta n\, \delta n' \rangle_{\mathbf{k}\omega} = \frac{(\varepsilon(\omega, \mathbf{k}) - \varkappa(\omega, \mathbf{k}))\, \varkappa'^*(\omega, \mathbf{k}) \langle \delta n^2 \rangle^0_{\mathbf{k}\omega} + (\varepsilon(\omega, \mathbf{k}) - \varkappa'(\omega, \mathbf{k}))^* \varkappa(\omega, \mathbf{k}) \langle \delta n'^2 \rangle^0_{\mathbf{k}\omega}}{|\varepsilon(\omega, \mathbf{k})|^2}. \tag{6.86}$$

It should be remembered that the density fluctuations of different species are correlated by virtue of the Coulomb interactions between particles.

The spectral distribution of the charge density fluctuations in a plasma is of the form

$$\langle \varrho^2 \rangle_{\mathbf{k}\omega} = \frac{\langle \varrho^2 \rangle^0_{\mathbf{k}\omega}}{|\varepsilon(\omega, \mathbf{k})|^2}, \tag{6.87}$$

where $\langle \varrho^2 \rangle^0_{\mathbf{k}\omega}$ is the spectral distribution of the charge density fluctuations in an ensemble of non-interacting particles which may be expressed in terms of the spectral distributions of the density fluctuations of partial components

$$\langle \varrho^2 \rangle^0_{\mathbf{k}\omega} = \sum e^2 \langle \delta n^2 \rangle^0_{\mathbf{k}\omega}.$$

The factor $\dfrac{1}{|\varepsilon(\omega, \mathbf{k})|^2}$ on the right-hand side of (6.87) describes the charge screening in the plasma due to the self-consistent Coulomb field. The self-consistent interactions give rise to additional maxima in the fluctuation spectrum (6.87) at the frequencies coinciding with the plasma eigenfrequencies.

Making use of the relation (6.75) and carrying out the averaging procedure according to (6.80), we obtain the following spectral distribution of the electric field fluctuations in a plasma:

$$\langle E^2\rangle_{\mathbf{k}\omega} = \frac{16\pi^2}{k^2}\frac{\langle\varrho^2\rangle^0_{\mathbf{k}\omega}}{|\varepsilon(\omega, \mathbf{k})|^2}. \tag{6.88}$$

An observation is to be made that $\langle E^2\rangle_{\mathbf{k}\omega}$ immediately determines the spectral distribution of the fluctuation field energy in the plasma. Similarly to the spectrum (6.87) of the charge density fluctuations, the one, (6.88), of the field fluctuations has a broad maximum in the low-frequency range, which is due to the stochastic motion of the individual charged particles, and a number of sharp maxima that correspond to the plasma eigenfrequencies. These additional sharp maxima in the fluctuation spectrum are associated with the collective fluctuations, i.e. with the random eigenoscillations of the plasma.

The spectral correlation function for the fluctuations of the field and of the electron distribution function is as follows:

$$\langle E\,\delta f(\mathbf{v})\rangle_{\mathbf{k}\omega} = -\frac{4\pi i}{k}\frac{\langle\varrho\,\delta f(\mathbf{v})\rangle^0_{\mathbf{k}\omega}}{\varepsilon(\omega, \mathbf{k})} + i\frac{e}{m}\frac{1}{\omega-(\mathbf{k}\cdot\mathbf{v})+i0}\frac{1}{k}\left(\mathbf{k}\cdot\frac{\partial f_0(\mathbf{v})}{\partial\mathbf{v}}\right)\langle E^2\rangle_{\mathbf{k}\omega}. \tag{6.89}$$

In a plasma with Maxwellian particle distributions (the temperatures of the components may be different in a so-called non-isothermal plasma), the spectral distributions $\langle\delta n^2\rangle^0_{\mathbf{k}\omega}$ and $\langle\varrho^2\rangle^0_{\mathbf{k}\omega}$ may be written in terms of the partial temperatures and the imaginary parts of the partial plasma susceptibilities:

$$\langle\delta n^2_{(e,i)}\rangle^0_{\mathbf{k}\omega} = \frac{k^2}{2\pi e^2\omega}T_{e,i}\,\mathrm{Im}\,\varkappa^{(e,i)}(\omega, \mathbf{k}), \tag{6.90}$$

$$\langle\varrho^2\rangle^0_{\mathbf{k}\omega} = \frac{k^2}{2\pi\omega}\{T_e\,\mathrm{Im}\,\varkappa^{(e)}(\omega, \mathbf{k}) + T_i\,\mathrm{Im}\,\varkappa^{(i)}(\omega, \mathbf{k})\}. \tag{6.91}$$

Thus, the fluctuation spectral distributions (6.84), (6.86), (6.87), and (6.89) depend only on the electron and ion susceptibilities $\varkappa^{(e)}(\omega, \mathbf{k})$ and $\varkappa^{(i)}(\omega, \mathbf{k})$ and on the temperatures T_e and T_i.

The spectral distribution $\langle\varrho^2\rangle^0_{\mathbf{k}\omega}$ for an equilibrium plasma may be expressed in terms of the plasma temperature T and the imaginary part of the dielectric permittivity $\varepsilon(\omega, \mathbf{k})$:

$$\langle\varrho^2\rangle^0_{\mathbf{k}\omega} = \frac{k^2}{2\pi\omega}T\,\mathrm{Im}\,\varepsilon(\omega, \mathbf{k}). \tag{6.92}$$

Hence the knowledge of the dielectric permittivity and the plasma temperature is sufficient to determine completely the spectral distributions of the charge density and electric field fluctuations in an equilibrium plasma $\langle\varrho^2\rangle_{\mathbf{k}\omega}$ and $\langle E^2\rangle_{\mathbf{k}\omega}$:

$$\langle\varrho^2\rangle_{\mathbf{k}\omega} = \frac{k^2}{2\pi}\frac{T}{\omega}\frac{\mathrm{Im}\,\varepsilon(\omega, \mathbf{k})}{|\varepsilon(\omega, \mathbf{k})|^2}, \tag{6.93}$$

$$\langle E^2\rangle_{\mathbf{k}\omega} = 8\pi\frac{T}{\omega}\frac{\mathrm{Im}\,\varepsilon(\omega, \mathbf{k})}{|\varepsilon(\omega, \mathbf{k})|^2}. \tag{6.94}$$

The expressions (6.93) and (6.94) are an extension of the Nyquist fluctuation–dissipation theorem that governs the fluctuation spectra in equilibrium systems (Nyquist, 1928). It should be noted that there is an alternative way to derive (6.94). We have just to compare the correlation function for the field fluctuations with the mean energy that is absorbed by the system by virtue of dissipation.

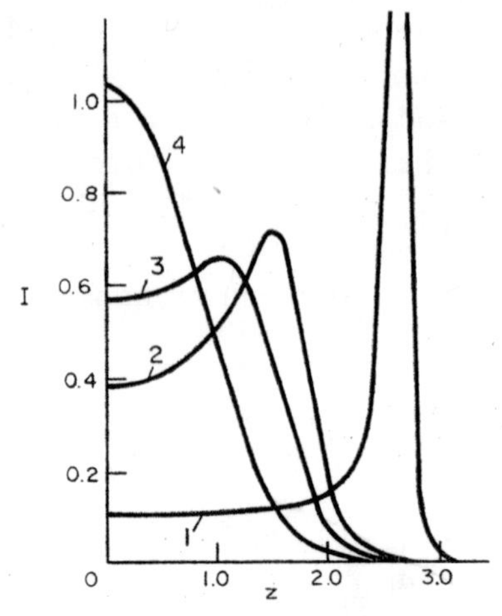

FIG. 5. The spectral distribution of charge density fluctuations in an equilibrium plasma $I(z)$ for $a^2k^2 = 0.1$ (1); 0.5 (2); 1 (3); and 10 (4).

The spectral distributions (6.93) and (6.94) may be integrated over the frequencies within the context of the Kramers–Kronig relations. The result is as follows:

$$\langle \varrho^2 \rangle_{\mathbf{k}} = \frac{k^2}{4\pi} T \left\{ 1 - \frac{1}{\varepsilon(0, \mathbf{k})} \right\}, \tag{6.95}$$

$$\langle E^2 \rangle_{\mathbf{k}} = 4\pi T \left\{ 1 - \frac{1}{\varepsilon(0, \mathbf{k})} \right\}. \tag{6.96}$$

In Fig. 5 we see the spectral distribution of the charge density fluctuations in an equilibrium plasma $I(z) = \langle \varrho^2 \rangle_{\mathbf{k}\omega} / 2\pi \langle \varrho^2 \rangle_{\mathbf{k}}$ as a function of the dimensionless frequency z for various values of the parameter a^2k^2.

The fluctuation spectral distributions (6.84), (6.86), (6.87), and (6.88) were derived for a plasma without external fields. For these formulas to be applicable for the plasma in a constant and uniform external magnetic field, (6.54) and (6.56) are to be substituted for the spectral distributions $\langle \delta n^2 \rangle^0_{\mathbf{k}\omega}$ and $\langle \varrho^2 \rangle^0_{\mathbf{k}\omega}$ and the relevant expressions for the susceptibilities of the magneto-active plasma must be made use of (Sitenko and Kirochkin, 1966).

Effect of electromagnetic interactions between particles on plasma fluctuations

The above consideration of the fluctuations may be extended to take into account the electromagnetic interaction between charged plasma particles. In such a more general case the evolution of the microscopic density $\mathcal{F}(\mathbf{r}, \mathbf{v}, t)$ is described by the following set of equations:

$$\frac{\partial \mathcal{F}}{\partial t}+\left(\mathbf{v}\cdot\frac{\partial \mathcal{F}}{\partial \mathbf{r}}+\frac{e}{m}\left\{\mathbf{E}+\left(\frac{1}{c}[\mathbf{vB}]\right\}\cdot\frac{\partial \mathcal{F}}{\partial \mathbf{v}}\right)=0, \tag{6.97}$$

$$\text{curl curl }\mathbf{E}+\frac{1}{c^2}\frac{\partial^2 \mathbf{E}}{\partial t^2}=-\frac{4\pi}{c}\frac{\partial \mathbf{j}}{\partial t}, \tag{6.98}$$

where

$$\mathbf{j}=\sum e\int d\mathbf{v}\mathbf{v}\mathcal{F}. \tag{6.99}$$

The microscopic density $\mathcal{F}(\mathbf{r}, \mathbf{v}, t)$ may be written in a form, analogous to the one we have analysed when studying the longitudinal case:

$$\mathcal{F}(\mathbf{r}, \mathbf{v}, t)=\mathcal{F}_0(\mathbf{r}, \mathbf{v}, t)+\tilde{f}(\mathbf{r}, \mathbf{v}, t),$$

where $\mathcal{F}_0(\mathbf{r}, \mathbf{v}, t)$ is the microscopic density in neglect of particle interactions and $\tilde{f}(\mathbf{r}, \mathbf{v}, t)$ is governed not by (6.74) but by the following equation that is easy to derive:

$$\tilde{f}_{\mathbf{k}\omega}(\mathbf{v})=-\frac{e}{m}\frac{1}{\omega-(\mathbf{k}\cdot\mathbf{v})+i0}\Bigg\{\left(\left\{\mathbf{E}_{\mathbf{k}\omega}+\frac{1}{\omega}[\mathbf{v}[\mathbf{k}\mathbf{E}_{\mathbf{k}\omega}]]\right\}\cdot\frac{\partial f_0(\mathbf{v})}{\partial \mathbf{v}}\right)$$

$$+\sum_{\omega',\mathbf{k}'}\left(\left\{\mathbf{E}_{\mathbf{k}'\omega'}+\frac{1}{\omega'}[\mathbf{v}[\mathbf{k}'\mathbf{E}_{\mathbf{k}'\omega'}]]\right\}\cdot\frac{\partial}{\partial \mathbf{v}}\right)(\delta f^0_{\mathbf{k}-\mathbf{k}',\,\omega-\omega'}(\mathbf{v})+\tilde{f}_{\mathbf{k}-\mathbf{k}',\,\omega-\omega'}(\mathbf{v}))\Bigg\}. \tag{6.100}$$

Hence we have in the linear approximation

$$\tilde{f}_{\mathbf{k}\omega}(\mathbf{v})=-i\frac{e}{m}\frac{1}{\omega-(\mathbf{k}\cdot\mathbf{v}+i0}\left(\left\{\mathbf{E}_{\mathbf{k}\omega}+\frac{1}{\omega}[\mathbf{v}[\mathbf{k}\mathbf{E}_{\mathbf{k}\omega}]]\right\}\cdot\frac{\partial f_0(\mathbf{v})}{\partial \mathbf{v}}\right). \tag{6.101}$$

Substituting the last expression in (6.98) and (6.99) we obtain the following equation for the intensity of the fluctuating electric field:

$$\Lambda_{ij}(\omega, \mathbf{k})\,E_j(\omega, \mathbf{k})=-\frac{4\pi i}{\omega}j_i^0(\omega, \mathbf{k}), \tag{6.102}$$

where

$$\Lambda_{ij}(\omega, \mathbf{k})=\varepsilon_{ij}(\omega, \mathbf{k})+\left(\frac{k_i k_j}{k^2}-\delta_{ij}\right)\eta^2, \tag{6.103}$$

$\varepsilon_{ij}(\omega, \mathbf{k})$ is the dielectric permittivity tensor and $\mathbf{j}^0(\omega, \mathbf{k})$ is the Fourier component of the fluctuating current associated with the stochastic motion of individual plasma particles:

$$j^0_{\mathbf{k}\omega}=\sum e\int d\mathbf{v}\mathbf{v}\,\delta f^0_{\mathbf{k}\omega}(\mathbf{v}). \tag{6.104}$$

To find the solution of (6.102) we introduce the tensor which is the inverse of (6.103). Then we have

$$E_i(\omega, \mathbf{k}) = -\frac{4\pi i}{\omega} \Lambda_{ij}^{-1}(\omega, \mathbf{k})\, j_j^0(\omega, \mathbf{k}). \tag{6.105}$$

Substituting the last expression in (6.101) and making use of (6.68) we obtain the following formula for the fluctuations of the particle distribution function in a plasma taking into account the electromagnetic particle interaction:

$$\delta f_{\mathbf{k}\omega}(\mathbf{v}) = \delta f^0_{\mathbf{k}\omega}(\mathbf{v}) - \frac{4\pi e}{m\omega} \Lambda_{ij}^{-1}(\omega, \mathbf{k}) \frac{1}{\omega-(\mathbf{k}\cdot\mathbf{v})+i0} \left[\left(1-\frac{(\mathbf{k}\cdot\mathbf{v})}{\omega}\right)\delta_{ik}+\frac{k_k v_i}{\omega}\right] \frac{\partial f_0(\mathbf{v})}{\partial v_k}\, j_j^0(\omega, \mathbf{k}). \tag{6.106}$$

Multiplying the left- and the right-hand sides of (6.106) by the complex conjugates and carrying out the averaging procedure within the context of (6.81), we obtain the general expression for the spectral representation of the distribution function fluctuations:

$$\begin{aligned}
\langle \delta f(\mathbf{v})\, \delta f'(\mathbf{v}')\rangle_{\mathbf{k}\omega} &= \langle \delta f(\mathbf{v})\, \delta f'(\mathbf{v}')\rangle^0_{\mathbf{k}\omega} \\
&- \frac{4\pi e}{m\omega} \Lambda_{ij}^{-1}(\omega, \mathbf{k}) \frac{1}{\omega-(\mathbf{k}\cdot\mathbf{v})+i0} \left[\left(1-\frac{(\mathbf{k}\cdot\mathbf{v})}{\omega}\right)\delta_{ik}+\frac{k_k v_i}{\omega}\right] \frac{\partial f_0(v)}{\partial v_k} \langle \delta f'(\mathbf{v}') j_j\rangle^0_{\mathbf{k}\omega} \\
&- \frac{4\pi e'}{m'\omega} \Lambda_{ij}^{*-1}(\omega, \mathbf{k}) \frac{1}{\omega-(\mathbf{k}\cdot\mathbf{v}')+i0} \left[\left(1-\frac{(\mathbf{k}\cdot\mathbf{v}')}{\omega}\right)\delta_{ik}+\frac{k_k v_i'}{\omega}\right] \frac{\partial f_0'(\mathbf{v}')}{\partial v_k'} \langle \delta f(\mathbf{v}) j_j\rangle^0_{\mathbf{k}\omega} \\
&+ \frac{16\pi^2 ee'}{mm'\omega^2} \Lambda_{ij}^{-1}(\omega, \mathbf{k})\, \Lambda_{i'j'}^{*-1}(\omega, \mathbf{k}) \frac{1}{\omega-(\mathbf{k}\cdot\mathbf{v})+i0} \left[\left(1-\frac{(\mathbf{k}\cdot\mathbf{v})}{\omega}\right)\delta_{ik}+\frac{k_k v_i}{\omega}\right] \frac{\partial f_0(\mathbf{v})}{\partial v_k} \\
&\times \frac{1}{\omega-(\mathbf{k}\cdot\mathbf{v}')+i0} \left[\left(1-\frac{(\mathbf{k}\cdot\mathbf{v}')}{\omega}\right)\delta_{i'k'}+\frac{k_{k'} v_{i'}'}{\omega}\right] \frac{\partial f_0'(\mathbf{v}')}{\partial v_{k'}'} \langle j_j j_{j'}\rangle^0_{\mathbf{k}\omega}.
\end{aligned} \tag{6.107}$$

Substituting (6.32), (6.38), and (6.40) for the spectral distributions $\langle \delta f(\mathbf{v})\, \delta f'(\mathbf{v}')\rangle^0_{\mathbf{k}\omega}$, $\langle j_i j_j\rangle^0_{\mathbf{k}\omega}$, and $\langle \delta f(\mathbf{v}) j_i\rangle^0_{\mathbf{k}\omega}$, respectively, we obtain the following explicit form of the spectral representation for the fluctuations of the particle distribution function:

$$\begin{aligned}
\langle \delta f(\mathbf{v})\, \delta f'(\mathbf{v}')\rangle_{\mathbf{k}\omega} &= 2\pi \Big\{ \delta'\delta(\mathbf{v}-\mathbf{v}')\, \delta(\omega-(\mathbf{k}\cdot\mathbf{v}))\, f_0(\mathbf{v}) \\
&- \frac{4\pi ee'}{m\omega} \frac{1}{\omega-(\mathbf{k}\cdot\mathbf{v})+i0} \left[\left(1-\frac{(\mathbf{k}\cdot\mathbf{v})}{\omega}\right)\delta_{ik}+\frac{k_k v_i}{\omega}\right] \frac{\partial f_0(\mathbf{v})}{\partial v_k} \Lambda_{ij}^{-1}(\omega, \mathbf{k})\, v_j'\delta(\omega-(\mathbf{k}\cdot\mathbf{v}))\, f_0'(\mathbf{v}') \\
&- \frac{4\pi ee'}{m'\omega} \Lambda_{ij}^{*-1}(\omega, \mathbf{k})\, v_j\delta(\omega-(\mathbf{k}\cdot\mathbf{v}))\, f_0(\mathbf{v}) \frac{1}{\omega-(\mathbf{k}\cdot\mathbf{v}')+i0} \left[\left(1-\frac{(\mathbf{k}\cdot\mathbf{v}')}{\omega}\right)\delta_{ik}+\frac{k_k v_i'}{\omega}\right] \frac{\partial f_0'(\mathbf{v}')}{\partial v_k'} \\
&+ \frac{16\pi^2 ee'}{mm'\omega^2} \Lambda_{ij}^{-1}(\omega, \mathbf{k})\, \Lambda_{i'j'}^{*-1}(\omega, \mathbf{k}) \frac{1}{\omega-(\mathbf{k}\cdot\mathbf{v})+i0} \left[\left(1-\frac{(\mathbf{k}\cdot\mathbf{v})}{\omega}\right)\delta_{ik}+\frac{k_k v_i}{\omega}\right] \frac{\partial f_0(\mathbf{v})}{\partial v_k} \\
&\times \frac{1}{\omega-(\mathbf{k}\cdot\mathbf{v}')+i0} \left[\left(1-\frac{(\mathbf{k}\cdot\mathbf{v}')}{\omega}\right)\delta_{i'k'}+\frac{k_{k'} v_{i'}'}{\omega}\right] \frac{\partial f_0'(\mathbf{v}')}{\partial v_{k'}'} \\
&\times \sum e''^2 \int d\mathbf{v}''\, v_j'' v_{j'}''\, \delta(\omega-(\mathbf{k}\cdot\mathbf{v}''))\, f_0''(\mathbf{v}'') \Big\}.
\end{aligned} \tag{6.108}$$

Let the partial current density of some plasma component be denoted by $\mathbf{j}^{\alpha}$ ($\alpha = e, i$) and the notation for the total current be $\mathbf{j}$ as before. Making use of (6.107) (or (6.108)) we may easily write explicit expressions for the spectral distribution of fluctuations of the partial and the total current densities, as well as of the field intensity in a plasma:

$$\langle j_i^{\alpha} j_j^{\beta} \rangle_{\mathbf{k}\omega} = \sum_{\gamma} (\delta_{ii'}\delta_{\alpha\beta} - \varkappa_{ik}^{\alpha}(\omega, \mathbf{k})\, \Lambda_{ki'}^{-1}(\omega, \mathbf{k}))\, (\delta_{jj'}\delta_{\beta\gamma} - \varkappa_{jl}^{\beta}(\omega, \mathbf{k})\, \Lambda_{lj'}^{-1}(\omega, \mathbf{k}))^{*} \langle j_{i'}^{\gamma} j_{j'}^{\gamma} \rangle_{\mathbf{k}\omega}^{0}, \quad (6.109)$$

$$\langle j_i^{\alpha} j_j \rangle_{\mathbf{k}\omega} = \sum_{\beta} (\delta_{ii'}\delta_{\alpha\beta} - \varkappa_{ik}^{\alpha}(\omega, \mathbf{k})\, \Lambda_{ki'}^{-1}(\omega, \mathbf{k}))\, (\delta_{jj'} - \varkappa_{jl}(\omega, \mathbf{k})\, \Lambda_{lj'}^{-1}(\omega, \mathbf{k}))^{*} \langle j_{i'}^{\beta} j_{j'} \rangle_{\mathbf{k}\omega}^{0}, \quad (6.110)$$

$$\langle j_i j_j \rangle_{\mathbf{k}\omega} = (\delta_{ii'} - \varkappa_{ik}(\omega, \mathbf{k})\, \Lambda_{ki'}^{-1}(\omega, \mathbf{k}))\, (\delta_{jj'} - \varkappa_{jl}(\omega, \mathbf{k})\, \Lambda_{lj'}^{-1}(\omega, \mathbf{k}))^{*} \langle j_{i'} j_{j'} \rangle_{\mathbf{k}\omega}^{0}, \quad (6.111)$$

$$\langle E_i E_j \rangle_{\mathbf{k}\omega} = \frac{16\pi^2}{\omega^2} \Lambda_{ik}^{-1}(\omega, \mathbf{k})\, \Lambda_{jl}^{*-1}(\omega, \mathbf{k}) \langle j_k j_l \rangle_{\mathbf{k}\omega}^{0}. \quad (6.112)$$

Correlation functions for the fluctuations in a quasi-equilibrium (non-isothermal) plasma in neglect of particle interactions may be expressed in terms of the partial electric susceptibility and the component temperatures:

$$\langle j_i^{\alpha} j_j^{\alpha} \rangle_{\mathbf{k}\omega}^{0} = -i \frac{\omega T_{\alpha}}{4\pi} \{\varkappa_{ij}^{\alpha}(\omega, \mathbf{k}) - \varkappa_{ji}^{\alpha *}(\omega, \mathbf{k})\}. \quad (6.113)$$

In an equilibrium plasma the spectral distributions (6.111) and (6.112) are functions of the dielectric permittivity tensor $\varepsilon_{ij}(\omega, \mathbf{k})$ and of the temperature T.

Formulas (6.109)–(6.112) provide a complete description of the fluctuations of electrodynamic quantities in a plasma. Note that these results are also adequate for a magnetoactive plasma, in which case appropriate expressions must be substituted for the correlation functions of non-interacting particles and for the partial components of the plasma electric susceptibility.

Electromagnetic fluctuations in an isotropic plasma

Let us derive explicit expressions for the spectral distributions of fluctuations of electrodynamic quantities in an isotropic plasma taking into account the electromagnetic interactions between particles. As in the linear approximation the isotropic plasma field may be decomposed into independent longitudinal and transverse components, the relevant fluctuations are also independent. Denote the longitudinal and the transverse dielectric permittivities by $\varepsilon_l(\omega, \mathbf{k})$, and $\varepsilon_t(\omega, \mathbf{k})$, respectively. Then we can distinguish the corresponding components of the inverse tensor $\Lambda_{ij}^{-1}(\omega, \mathbf{k})$ (as well as of $\Lambda_{ij}(\omega, \mathbf{k})$):

$$\Lambda_{ij}^{-1}(\omega, \mathbf{k}) = \frac{k_i k_j}{k^2} \frac{1}{\varepsilon_l(\omega, \mathbf{k})} + \left(\delta_{ij} - \frac{k_i k_j}{k^2}\right) \frac{1}{\varepsilon_t(\omega, \mathbf{k}) - \eta^2}\,. \quad (6.114)$$

and due to such structure of the tensor it is possible to separate in the correlation function (6.108) the contribution of the longitudinal Coulomb forces from that of the transverse electromagnetic interactions between the particles:

$$\langle \delta f(\mathbf{v})\, \delta f'(\mathbf{v}') \rangle_{\mathbf{k}\omega} = 2\pi\delta' \delta(\mathbf{v} - \mathbf{v}')\, \delta(\omega - (\mathbf{k} \cdot \mathbf{v}))\, f_0(\mathbf{v}) + \langle \delta f(\mathbf{v})\, \delta f'(\mathbf{v}') \rangle_{\mathbf{k}\omega}^{l}$$
$$+ \langle \delta f(\mathbf{v})\, \delta f'(\mathbf{v}) \rangle_{\mathbf{k}\omega}^{t}. \quad (6.115)$$

The sum of the longitudinal term and the correlation function for the non-interacting particles coincides with (6.83) (it is to be kept in mind that $\varepsilon_l(\omega, \mathbf{k}) = \varepsilon(\omega, \mathbf{k})$), the transverse term is as follows:

$$\langle \delta f(\mathbf{v})\, \delta f'(\mathbf{v}')\rangle^t_{\mathbf{k}\omega}$$
$$= \frac{8\pi e e'}{k^2\omega^2}\left\{\frac{1}{m}\,\frac{1}{\varepsilon_t(\omega,\mathbf{k})-\eta^2}\,\frac{\omega}{\omega-(\mathbf{k}\cdot\mathbf{v})+i0}\left([\mathbf{k}[\mathbf{k}\mathbf{v}']]\cdot\frac{\partial f_0(\mathbf{v})}{\partial \mathbf{v}}\right)\delta(\omega-(\mathbf{k}\cdot\mathbf{v}))\, f_0(\mathbf{v})\right.$$
$$+\frac{1}{m'}\,\frac{1}{\varepsilon_t^*(\omega,\mathbf{k})-\eta^2}\,\delta(\omega-(\mathbf{k}\cdot\mathbf{v}))\, f_0(\mathbf{v})\,\frac{\omega}{\omega-(\mathbf{k}\cdot\mathbf{v}')+i0}\left([\mathbf{k}[\mathbf{k}\mathbf{v}]]\cdot\frac{\partial f_0'(\mathbf{v}')}{\partial \mathbf{v}'}\right)$$
$$-\frac{4\pi}{mm'k^2}\,\frac{1}{|\varepsilon_t(\omega,\mathbf{k})-\eta^2|}\,\frac{1}{\omega-(\mathbf{k}\cdot\mathbf{v})+i0}\,\frac{1}{\omega-(\mathbf{k}\cdot\mathbf{v}')+i0}$$
$$\left.\times\sum e''^2\int d\mathbf{v}\left([\mathbf{k}[\mathbf{k}\mathbf{v}'']]\cdot\frac{\partial f_0(\mathbf{v})}{\partial \mathbf{v}}\right)\left([\mathbf{k}[\mathbf{k}\mathbf{v}]]\cdot\frac{\partial f_0'(\mathbf{v}')}{\partial \mathbf{v}'}\right)\delta(\omega-(\mathbf{k}\cdot\mathbf{v}''))\, f_0''(\mathbf{v}'')\right\}. \quad (6.116)$$

Making use of (6.116) we can easily find the spectral distributions of the fluctuations of transverse current density, electric, and magnetic fields:

$$\langle j_t^2\rangle_{\mathbf{k}\omega} = \frac{(1-\eta^2)^2}{|\varepsilon_t(\omega,\mathbf{k})-\eta^2|^2}\langle j_t^2\rangle^0_{\mathbf{k}\omega}, \quad (6.117)$$

$$\langle E_t^2\rangle_{\mathbf{k}\omega} = \frac{16\pi^2}{\omega^2}\,\frac{1}{|\varepsilon_t(\omega,\mathbf{k})-\eta^2|^2}\langle j_t^2\rangle_{\mathbf{k}\omega}, \quad (6.118)$$

$$\langle B^2\rangle_{\mathbf{k}\omega} = \frac{16\pi^2}{\omega^2}\,\frac{\eta^2}{|\varepsilon_t(\omega,\mathbf{k})-\eta^2|^2}\langle j_t^2\rangle_{\mathbf{k}\omega}, \quad (6.119)$$

where

$$\langle j_t^2\rangle^0_{\mathbf{k}\omega} \equiv \sum\frac{2\pi e^2}{k^2}\int d\mathbf{v}[\mathbf{k}\mathbf{v}]^2\,\delta(\omega-(\mathbf{k}\cdot\mathbf{v}))\, f_0(\mathbf{v}). \quad (6.120)$$

Note that the maxima of the transverse fluctuation spectra correspond to the frequencies of the eigenoscillations. These frequencies are determined by the dispersion relation (4.40).

6.4. Collective Fluctuations and the Effective Temperature (Potential Field)

Incoherent and coherent (collective) fluctuations

The spectral distributions of the fluctuations of the field and of other electrodynamic quantities in the plasma transparency domain have sharp delta-like maxima at frequencies ω and wave vectors $\mathbf{k}$ that satisfy the dispersion equation $\mathrm{Re}\,\Lambda(\omega,\mathbf{k}) = 0$. Let us study the spectral distributions near such maxima; consider the potential case first.

It becomes evident from (6.89) that the knowledge of the dielectric permittivity $\varepsilon(\omega, \mathbf{k})$ and of the spectral distribution of the charge density fluctuations in neglect of particle interactions $\langle \varrho^2 \rangle^0_{\mathbf{k}\omega}$ is sufficient to determine the spectral distribution of the electric field fluctuations $\langle E^2 \rangle_{\mathbf{k}\omega}$ in the linear approximation:

$$\langle E^2 \rangle_{\mathbf{k}\omega} = \frac{16\pi^2}{k^2} \frac{\langle \varrho^2 \rangle^0_{\mathbf{k}\omega}}{|\varepsilon(\omega, \mathbf{k})|^2}. \tag{6.121}$$

The spectral distribution $\langle \varrho^2 \rangle^0_{\mathbf{k}\omega}$ has a broad low-frequency maximum, the shape of which is governed by the particle distribution function. Since $\langle E^2 \rangle_{\mathbf{k}\omega}$ is proportional to $\langle \varrho^2 \rangle^0_{\mathbf{k}\omega}$, it also has a broad maximum in the low-frequency range which is due to the stochastic motion of the particles (incoherent fluctuations). The zeros of the denominator, which are associated with the fluctuating oscillations of the electric field at plasma eigenfrequencies, are responsible for additional maxima in the fluctuation spectrum (collective or coherent fluctuations). Note that the frequency ω and the wave vector $\mathbf{k}$ are independent for the incoherent fluctuations, while those for the coherent fluctuations satisfy the usual dispersion relation.

The frequencies of the longitudinal eigenoscillations may be found in the linear approximation from the condition $\text{Re}\, \varepsilon(\omega, \mathbf{k}) = 0$. If $|\text{Im}\, \varepsilon(\omega, \mathbf{k})| \ll |\text{Re}\, \varepsilon(\omega, \mathbf{k})|$ (plasma transparency domain), the eigenoscillations are weakly damped $\gamma_{\mathbf{k}} \ll \omega_{\mathbf{k}}$. By virtue of (6.121) the spectral distribution of the electric field fluctuations in the eigenfrequency band may be written as

$$\langle E^2 \rangle_{\mathbf{k}\omega} = \frac{16\pi^3}{k^2} \frac{\langle \varrho^2 \rangle^0_{\mathbf{k}\omega}}{|\text{Im}\, \varepsilon(\omega, \mathbf{k})|} \delta\{\text{Re}\, \varepsilon(\omega, \mathbf{k})\}. \tag{6.122}$$

Spectral distributions of charge and particle density fluctuations in the plasma transparency domain may be expressed in terms of the spectral distribution of the electric field fluctuations:

$$\langle \varrho^2 \rangle_{\mathbf{k}\omega} = \frac{k^2}{16\pi^2} \langle E^2 \rangle, \tag{6.123}$$

$$\langle \delta n^2 \rangle_{\mathbf{k}\omega} = \frac{k^2}{e^2} (\varkappa^{(e)}(\omega, \mathbf{k}))^2 \langle E^2 \rangle_{\mathbf{k}\omega}, \tag{6.124}$$

where $\varkappa^{(e)}(\omega, \mathbf{k})$ is the partial electric susceptibility of the plasma. Formulas (6.122)–(6.124) hold both for equilibrium and non-equilibrium plasmas.

In an equilibrium plasma the spectral distribution $\langle \varrho^2 \rangle^0_{\mathbf{k}\omega}$ may be written in terms of the temperature T and the imaginary part of the dielectric permittivity $\varepsilon(\omega, \mathbf{k})$ according to (6.92), whence

$$\langle E^2 \rangle_{\mathbf{k}\omega} = 8\pi^2 \frac{T}{|\omega|} \delta\{\text{Re}\, \varepsilon(\omega, \mathbf{k})\}. \tag{6.125}$$

A more suitable form of the spectral distribution (6.125) is

$$\langle E^2 \rangle_{\mathbf{k}\omega} = \pi I_{\mathbf{k}} \{\delta(\omega - \omega_{\mathbf{k}}) + \delta(\omega + \omega_{\mathbf{k}})\}, \tag{6.126}$$

where $I_{\mathbf{k}}$ is the fluctuating intensity which depends on the plasma temperature:

$$I_{\mathbf{k}} = 8\pi \frac{T}{\omega_{\mathbf{k}} \varepsilon'_{\mathbf{k}}}, \quad \varepsilon'_{\mathbf{k}} \equiv \frac{\partial\, \text{Re}\, \varepsilon(\omega_{\mathbf{k}}, \mathbf{k})}{\partial \omega_{\mathbf{k}}}. \tag{6.127}$$

In an equilibrium plasma there may exist weakly damped longitudinal Langmuir oscillations with frequencies $\omega_{\mathbf{k}} = \sqrt{\Omega^2 + k^2 s^2}$. The relation (6.126) describes the spectral distribution for the Langmuir fluctuations of the electric field in a plasma and the intensity of the fluctuating oscillations is $I_{\mathbf{k}} = 4\pi T$. The mean energy of the fluctuating Langmuir oscillations is equal to

$$\frac{1}{8\pi} \langle E^2 \rangle_{\mathbf{k}} = \frac{1}{2} T. \tag{6.128}$$

The relative weight of the collective Langmuir fluctuations in an equilibrium plasma is of the order of magnitude a^2k^2, a denoting the Debye length.

Collective fluctuations in a non-equilibrium plasma and the effective temperature

The relation (6.92) is inadequate for non-equilibrium plasmas. However, the spectral distribution of the electric field fluctuations may be described by a formula similar to (6.126):

$$\langle E^2 \rangle_{\mathbf{k}\omega} = \pi I_{\mathbf{k}}(\omega) \{\delta(\omega - \omega_{\mathbf{k}}) + \delta(\omega + \omega_{\mathbf{k}})\}, \tag{6.129}$$

if we define the fluctuating intensity by

$$I_{\mathbf{k}}(\omega) \equiv 8\pi \frac{\tilde{T}(\omega, \mathbf{k})}{\omega \dfrac{\partial \varepsilon(\omega, \mathbf{k})}{\partial \omega}}, \tag{6.130}$$

We have put here

$$\tilde{T}(\omega, \mathbf{k}) \equiv \frac{2\pi |\omega|}{k^2} \frac{\langle \varrho^2 \rangle^0_{\mathbf{k}\omega}}{|\mathrm{Im}\, \varepsilon(\omega, \mathbf{k}|}. \tag{6.131}$$

In the equilibrium case $\tilde{T}(\omega, \mathbf{k})$ is equal to the plasma temperature T. In the general case $\tilde{T}(\omega, \mathbf{k})$ may be regarded as an effective temperature, which characterizes the mean square of the electric field fluctuation amplitude in the plasma. An observation is to be made that the effective temperature depends on the sign of the frequency, therefore the fluctuating intensities, which correspond to each term in (6.129), are in fact different. The effective temperature satisfies the relation $\tilde{T}(\omega, \mathbf{k}) = \tilde{T}(-\omega, -\mathbf{k})$.

The effective temperature of a non-isothermal plasma is a function of the component temperatures and of the partial plasma susceptibilities:

$$\tilde{T}(\omega, \mathbf{k}) = \frac{\sum_{\alpha} T_{\alpha}\, \mathrm{Im}\, \varkappa^{(\alpha)}(\omega, \mathbf{k})}{\mathrm{Im}\, \varepsilon(\omega, \mathbf{k})}. \tag{6.132}$$

(The summation in (6.132) extends over the different species.) In a non-isothermal plasma there may exist, besides the high-frequency Langmuir fluctuations of the electric field, also low-frequency fluctuations at $\omega_{\mathbf{k}} = kv_s$ (v_s is the non-isothermal sound velocity). The order of magnitude of the mean energy of the sound fluctuations is about a^2k^2 times the Langmuir fluctuation energy.

Critical fluctuations near the instability threshold

The effective temperature of a non-equilibrium plasma may grow very large, especially when the plasma state is close to the threshold of the plasma instability. A plasma may be driven towards a non-equilibrium state as a result of a number of reasons: relative motion of plasma components, anisotropy of the particle velocity distribution, propagation of a fast particle beam, presence of electrical currents, powerful electromagnetic radiation, and so on. In such a non-equilibrium plasma there may be developed various kinetic and hydrodynamic instabilities. The kinetic instability threshold is to be found from the requirement for the imaginary part of the dielectric permittivity to turn to zero ($\mathrm{Im}\,\varepsilon(\omega, \mathbf{k}) = 0$), while the condition for the hydrodynamic instability threshold is the multiplicity of the root of the dispersion equation that corresponds to the eigenfrequency $\left(\mathrm{Re}\,\varepsilon(\omega_{\mathbf{k}}, \mathbf{k}) = 0,\ \frac{\partial}{\partial\omega_{\mathbf{k}}}\mathrm{Re}\,\varepsilon(\omega, \mathbf{k}) = 0\right)$. It follows from (6.130) and (6.131) that the fluctuating intensity steeply grows when approaching the instability domain and in the linear approximation becomes infinite at the threshold (critical fluctuations). The unlimited growth of the fluctuation intensity indicates that the linear approximation is inadequate and non-linear effects must be taken into account for the fluctuations in a non-equilibrium plasma to be described properly.

Note that the higher-order correlation functions may also be very large in the linear approximation when examined near the kinetic instability threshold (Sitenko, 1975). For example consider the cubic correlation function for the electric field fluctuations. By virtue of (6.75) we have

$$\langle E^3\rangle_{\mathbf{k}\omega;\,\mathbf{k}'\omega'} = i\,\frac{64\pi^3}{kk'\,|\mathbf{k}-\mathbf{k}'|}\,\frac{\langle\varrho^3\rangle^0_{\mathbf{k}\omega;\,\mathbf{k}'\omega'}}{\varepsilon(\omega, \mathbf{k})\,\varepsilon^*(\omega', \mathbf{k}')\,\varepsilon^*(\omega-\omega', \mathbf{k}-\mathbf{k}')}, \tag{6.133}$$

where $\langle\varrho^3\rangle^0_{\mathbf{k}\omega;\,\mathbf{k}'\omega'}$ was defined in (6.45). Instead of (6.133) we introduce the correlation function for the fluctuations of the electric field potential $\langle\varphi^3\rangle_{\mathbf{k}\omega;\,\mathbf{k}'\omega'}$ by means of the following relation:

$$\langle E^3\rangle_{\mathbf{k}\omega;\,\mathbf{k}'\omega'} = ikk'\,|\mathbf{k}-\mathbf{k}'|\,\langle\varphi^3\rangle_{\mathbf{k}\omega;\,\mathbf{k}'\omega'}. \tag{6.134}$$

We restrict ourselves for simplicity to the special case $\mathbf{k}' \parallel \mathbf{k}$. Then the cubic correlation function $\langle\varrho^3\rangle_{\mathbf{k}\omega;\,\mathbf{k}'\omega'}$ becomes expressed in terms of the quadratic one $\langle\varrho^2\rangle^0_{\mathbf{k}\omega}$:

$$\langle\varrho^3\rangle^0_{\mathbf{k}\omega;\,\mathbf{k}'\omega'} = 2\pi e\delta\left(\omega' - \frac{k'}{k}\,\omega\right)\langle\varrho^2\rangle^0_{\mathbf{k}\omega}, \tag{6.135}$$

and the correlation function for the fluctuations of the electric field potential may be reduced to the following:

$$\langle\varphi^3\rangle_{\mathbf{k}\omega;\,\mathbf{k}'\omega'} = \frac{128\pi^4 e}{k^2k'^2(\mathbf{k}-\mathbf{k}')^2}\,\frac{1}{\varepsilon^*(\omega-\omega', \mathbf{k}-\mathbf{k}')}\,\delta\left(\omega' - \frac{k'}{k}\,\omega\right)\langle\varrho^2\rangle^0_{\mathbf{k}\omega}. \tag{6.136}$$

Let $\omega' = \omega$, then

$$\langle\varphi^3\rangle_{\mathbf{k}\omega;\,\mathbf{k}'\omega'} \to \frac{8\pi^2 e}{Ck\omega}\,\delta(k-k')\,\langle E^2\rangle_{\mathbf{k}\omega}, \tag{6.137}$$

where $C = \lim\limits_{q\to 0} q^2\varepsilon(0, q)$. If $k' = k$

we have

$$\langle\varphi^3\rangle_{\mathbf{k}\omega;\,\mathbf{k}'\omega'} \to \frac{8\pi^2 e\omega^2}{C'k^4}\,\delta(\omega-\omega')\,\langle E^2\rangle_{\mathbf{k}\omega}, \tag{6.138}$$

where $C' = \lim \Delta\omega^2\varepsilon(\Delta\omega, 0)$. Since the quadratic correlation function for the field $\langle E^2\rangle_{\mathbf{k}\omega}$ tends to infinity as $\operatorname{Im}\varepsilon(\omega, \mathbf{k})\to 0$, the cubic correlation function $\langle\varrho^3\rangle_{\mathbf{k}\omega;\,\mathbf{k}'\omega'}$ anomalously grows in the vicinity of the kinetic instability threshold.

Critical fluctuations in a plasma with a directed particle motion

Formulas (6.129), (6.130), and (6.131) determine the field fluctuations in a plasma with arbitrary non-equilibrium, but stable, particle distributions. The mean directed velocities may be different from zero, therefore these formulas are applicable to the description of a plasma, penetrated by a particle beam, as well as of a plasma, the electrons of which move with respect to the ions. The only requirement is that the beam velocity (or the electron velocity) be less than the critical value that may give rise to the instability.

Let us consider in detail the critical fluctuations in a plasma penetrated by a compensated beam of charged particles. We denote the beam velocity by u and suggest the particles both of the plasma and of the beam to be distributed according to the Maxwell law, the temperatures being T and T' respectively. Then

$$\langle\varrho^2\rangle^0_{\mathbf{k}\omega} = \sqrt{6\pi}\left\{\sum\frac{e^2 n_0}{ks}\,e^{-z^2} + \sum\frac{e'^2 n_0'}{ks'}\,e^{-y^2}\right\}, \tag{6.139}$$

$$\varepsilon(\omega, \mathbf{k}) = 1 + \sum\frac{1}{a^2k^2}\left\{1-\varphi(z)+i\sqrt{\pi}z\,e^{-z^2}\right\}$$
$$+\sum\frac{1}{a'^2k^2}\left\{1-\varphi(y)+i\sqrt{\pi}y\,e^{-y^2}\right\}, \tag{6.140}$$

where $z = \sqrt{\tfrac{3}{2}}\,\omega/ks$ and $y = \sqrt{\tfrac{3}{2}}\,(\omega-\mathbf{k}\cdot\mathbf{u})/ks'$ (the prime labels the quantities, associated with the beam). The summations in (6.139) and (6.140) extend over the various kinds of particles in the plasma and the beam.

The spectral distribution of the electric field fluctuations in the transparency domain of the beam–plasma system is determined by (6.129), (6.130), and (6.131). Neglecting the ion thermal motion and supposing that $(\omega-(\mathbf{k}\cdot\mathbf{v}))^2 \ll k^2 s'^2$, we may present the effective temperature in the form

$$\tilde{T}(\omega, \mathbf{k}) = \frac{a_k}{1-(\mathbf{k}\cdot\mathbf{u})/k\tilde{u}}\,T, \tag{6.141}$$

where

$$a_k = \frac{1+(n_0'/n_0)\,(T/T')^{1/2}\,e^{z_k^2}}{1+(n_0'/n_0)\,(T/T')^{3/2}\,e^{z_k^2}} \tag{6.142}$$

and

$$\tilde{u} = \frac{\omega_k}{k}\left\{1+\frac{n_0}{n_0'}\left(\frac{T'}{T}\right)^{3/2} e^{-z_k^2}\right\}. \tag{6.143}$$

The quantity $\tilde{u}$ may be regarded as the critical velocity of the beam, i.e. the beam–plasma system becomes driven unstable when the beam velocity reaches the value $\tilde{u}$.

Here and below we shall consider low-density beams ($n_0' \ll n_0$). The effect of such a beam on the wave dispersion in the plasma is weak and may be neglected. Nevertheless, the effective temperature of the fluctuating oscillations is very sensitive to the beam. In the high-frequency range we may ignore also the influence of the particle thermal motion on the wave dispersion. The spectral distribution for the Langmuir fluctuations of the fields is of the form

$$\langle E^2\rangle_{\mathbf{k}\omega} = 4\pi^2\tilde{T}(\omega, \mathbf{k})\{\delta(\omega-\Omega)+\delta(\omega-\Omega)\}, \tag{6.144}$$

where the effective temperature is determined by (6.141) at $z_k^2 = \dfrac{1}{2a^2k^2}+\dfrac{3}{2}$. The critical velocity turns out to compete to the same order with the electron thermal velocity for a quiescent plasma:

$$\tilde{u} = \frac{\Omega}{k}\left\{1+\frac{n_0}{n_0'}\left(\frac{T'}{T}\right)^{3/2} e^{-\frac{1}{2a^2k^2}-\frac{3}{2}}\right\}. \tag{6.145}$$

The mean energy of the fluctuating Langmuir oscillations is equal to

$$\frac{1}{8\pi}\langle E^2\rangle_{\mathbf{k}} = \frac{1}{4}\{\tilde{T}(\Omega, \mathbf{k})+\tilde{T}(-\Omega, \mathbf{k})\}. \tag{6.146}$$

This energy may considerably exceed the thermal level if the beam velocity is close to $\tilde{u}$.

In a two-temperature plasma with hot electrons and cold ions there may occur low-frequency fluctuations as well:

$$\langle E^2\rangle_{\mathbf{k}\omega} = 4\pi^2\tilde{T}(\omega, \mathbf{k})\, a^2k^2\{\delta(\omega-kv_s)+\delta(\omega+kv_s)\}, \tag{6.147}$$

where v_s is the non-isothermal sound velocity; the effective temperature $\tilde{T}(\omega, \mathbf{k})$ and the velocity $\tilde{u}$ are respectively equal to

$$\tilde{T}(\omega, \mathbf{k}) = \frac{T}{1-(\mathbf{k}\cdot\mathbf{u})/k\tilde{u}}, \tag{6.148}$$

$$\tilde{u} = \frac{n_0}{n_0'}\left(\frac{T'}{T}\right)^{3/2} v_s. \tag{6.149}$$

By contrast to the high-frequency range, $\tilde{u}$ may be either greater or less than the electron thermal velocity in the low-frequency range. The mean energy of the fluctuating sound oscillations is equal to

$$\frac{1}{8\pi}\langle E^2\rangle_{\mathbf{k}} = \frac{a^2k^2}{4}\{T(kv_s, \mathbf{k})+\tilde{T}(-kv_s, \mathbf{k})\}. \tag{6.150}$$

Within the context of (6.123) and (6.124) the spectral distributions of charge and particle density fluctuations in the high-frequency range are as follows:

$$\langle \varrho^2\rangle_{\mathbf{k}\omega} = \tfrac{1}{4}k^2\tilde{T}(\omega, \mathbf{k})\{\delta(\omega-\Omega)+\delta(\omega+\Omega)\}, \tag{6.151}$$

$$\langle \delta n^2\rangle_{\mathbf{k}\omega} = \frac{k^2}{4e^2}\tilde{T}(\omega, \mathbf{k})\{\delta(\omega-\Omega)+\delta(\omega+\Omega)\}, \tag{6.152}$$

where the effective temperature $\tilde{T}(\omega, \mathbf{k})$ is determined by (6.141). In the low-frequency range we have

$$\langle \varrho^2 \rangle_{\mathbf{k}\omega} = \frac{a^2 k^4}{4} \tilde{T}(\omega, \mathbf{k}) \{\delta(\omega - kv_s) + \delta(\omega + kv_s)\}, \tag{6.153}$$

$$\langle \delta n^2 \rangle_{\mathbf{k}\omega} = \pi n_0 \frac{\tilde{T}(\omega, \mathbf{k})}{T} \{\delta(\omega - kv_s) + \delta(\omega + kv_s)\}. \tag{6.154}$$

Formulas (6.144), (6.147), and (6.151)–(6.154) were derived in the linear approximation and are valid only for the description of fluctuations, the wave vectors $\mathbf{k}$ of which satisfy the requirement $u < \tilde{u}(\mathbf{k})$. Fluctuations with such wave vectors that $u > \tilde{u}(\mathbf{k})$ drive the plasma unstable.

Critical fluctuations in a plasma with an anisotropic velocity distribution

The collective fluctuations in a plasma with an anisotropic velocity distribution in an external magnetic field $\mathbf{B}_0$ make another example of critical fluctuations (Sitenko and Zansenko, 1978a). Denote the temperature that characterizes the distribution of the velocity components in the plane, normal to the magnetic field, by $T_\perp$ and that for the velocity distribution along the field by $T_\parallel$. The dielectric permittivity of such a plasma is in the potential case determined by (3.16). Suppose that the plasma is not too dense (there is hydrodynamic stability provided $\Omega < \omega_B$ (Mikhailovskii, 1974)). In such a plasma there may exist weakly damped oscillations, the wave vector of which is almost perpendicular to the magnetic field. The eigenfrequencies may be found from the condition that the real part of (3.16) be equal to zero:

$$\left.\begin{aligned} \omega_{\mathbf{k}} &= \Omega \sqrt{\frac{a_\perp^2 k_\parallel I_0(\beta_\perp) e^{-\beta_\perp}}{1 + a_\perp^2 k^2 - I_0(\beta_\perp) e^{-\beta_\perp}}}, \\ \gamma_{\mathbf{k}} &= \sqrt{\frac{\pi}{8}} \frac{\omega_{\mathbf{k}}^2 e^{-\beta_\perp}}{(1 + a_\perp^2 k^2 - I_0(\beta_\perp) e^{-\beta_\perp}) |k_\parallel| s_\parallel} \\ &\quad \times \sum_n \left(1 - \frac{n\omega_B}{\omega_k}\left(1 - \frac{T_\parallel}{T_\perp}\right)\right) I_n(\beta_\perp) e^{-\frac{(\omega_k - n\omega_B)^2}{2k_\parallel^2 s_\parallel^2}} \end{aligned}\right\} \tag{6.155}$$

Consider eigenoscillations with frequencies somewhat lower than the electron cyclotron frequency. Assume that $\Omega > \frac{1}{2}\omega_B$ and take the eigenfrequency $\omega_{\mathbf{k}}$ within the band

$$\tfrac{1}{2}\omega_B < \omega_{\mathbf{k}} < \omega_B.$$

The plasma is stable ($\gamma > 0$) if

$$\frac{T_\perp}{T_\parallel} < \frac{\omega_B}{\omega_B - \omega_{\mathbf{k}}}. \tag{6.156}$$

(If the inverse inequality is satisfied, $\gamma_{\mathbf{k}} < 0$ and the plasma is unstable.) The spectral distribution of the electric field fluctuations in the vicinity of the eigenfrequency $\omega_{\mathbf{k}}$ is of the form

$$\langle E^2 \rangle_{\mathbf{k}\omega} = \pi I_{\mathbf{k}}(\omega) \{\delta(\omega - \omega_{\mathbf{k}}) + \delta(\omega + \omega_{\mathbf{k}})\}, \tag{6.157}$$

where

$$I_{\mathbf{k}}(\omega) \equiv \frac{16\pi^2}{k^2} \frac{1}{\left| \dfrac{\partial}{\partial \omega_{\mathbf{k}}} \operatorname{Re} \varepsilon(\omega_{\mathbf{k}}, \mathbf{k}) \right|} \frac{\langle \varrho^2 \rangle^0_{\mathbf{k}\omega}}{|\operatorname{Im}(\omega_{\mathbf{k}}, \mathbf{k})|},$$

$$\langle \varrho^2 \rangle^0_{\mathbf{k}\omega} = \sqrt{\frac{\pi}{2}} \frac{e^2 n_0}{|k_{||}| s_{||}} e^{-\beta_\perp} \sum_n I_n(\beta_\perp) e^{-z_n^2}.$$

Making use of (3.16) for the dielectric permittivity $\varepsilon(\omega, \mathbf{k})$ we obtain the following expression:

$$I_{\mathbf{k}}(\omega) = 16\pi^2 e^2 n_0 a_{||}^2 \frac{a_\perp^2 k^2}{1 + a_\perp^2 k^2 - I_0(\beta_\perp) e^{-\beta_\perp}} \frac{1}{\left| 1 - \left(1 - \dfrac{T_{||}}{T_\perp}\right) \dfrac{\omega_B}{\omega_{\mathbf{k}}} \right|}. \tag{6.158}$$

The threshold of the kinetic instability of the anisotropic plasma is determined by the condition

$$\frac{T_\perp}{T_{||}} = \frac{\omega_B}{\omega_B - \omega_{\mathbf{k}}}.$$

Inasmuch as (6.156) is satisfied, the plasma is stable and then the spectral distribution of the field fluctuations is described by (6.157). The state of the system approaches the kinetic instability threshold as the longitudinal temperature decreases; the fluctuation intensity grows and in the linear approximation turns to infinity at the threshold (critical fluctuations in an anisotropic plasma).

6.5. Collective Electromagnetic Fluctuations

The spectral distribution of fluctuations and the effective temperature

Let us consider collective plasma fluctuations assuming the field to be no longer potential. The general formula for the distribution of the electric field fluctuations is well known:

$$\langle E_i E_j \rangle_{\mathbf{k}\omega} = \frac{16\pi^2}{\omega^2} \Lambda_{ik}^{-1}(\omega, \mathbf{k}) \Lambda_{jl}^{*-1}(\omega, \mathbf{k}) \langle j_k j_l \rangle^0_{\mathbf{k}\omega}. \tag{6.159}$$

In the plasma transparency domain $\operatorname{Im} \Lambda \ll \operatorname{Re} \Lambda$, so we have

$$\Lambda_{ik}^{-1} \Lambda_{jl}^{*-1} \to \pi \frac{\lambda_{ik} \lambda_{jl}^*}{|\operatorname{Im} \Lambda|} \delta(\Lambda), \tag{6.160}$$

hence the spectral distribution of the electric field fluctuations (6.159) has sharp delta-like maxima at plasma eigenfrequencies. It is to be observed that if the frequencies ω and the wave vectors **k** satisfy the dispersion relation, then the following relation occurs:

$$\lambda_{ij} = e_i e_j^* \operatorname{Tr} \lambda, \tag{6.161}$$

where **e** is the polarization vector of the relevant oscillations. Besides,

$$\frac{\operatorname{Tr} \lambda}{\omega} \operatorname{Im} \Lambda > 0, \tag{6.162}$$

Within the context of (6.161) and (6.162) the spectral distribution of the electric field collective fluctuations becomes transformed into

$$\langle E_i E_j \rangle_{\mathbf{k}\omega} = 8\pi^2 e_i e_j^* \tilde{T}(\omega, \mathbf{k}) \frac{|\operatorname{Tr} \lambda(\omega, \mathbf{k})|}{|\omega|} \delta\{\Lambda(\omega, \mathbf{k})\}, \tag{6.163}$$

where the effective temperature $\tilde{T}(\omega, \mathbf{k})$ is equal to

$$\tilde{T}(\omega, \mathbf{k}) = 2\pi \frac{\operatorname{Tr} \lambda(\omega, \mathbf{k})}{\omega \operatorname{Im} \Lambda(\omega, \mathbf{k})} e_i^* e_j \langle j_i j_j \rangle^0_{\mathbf{k}\omega}. \tag{6.164}$$

Upon noting that

$$\operatorname{Im} \Lambda = -\frac{i}{2} (\varepsilon_{ij} - \varepsilon_{ji}^*) e_i^* e_j \operatorname{Tr} \lambda, \tag{6.165}$$

we can reduce the effective temperature (6.164) to the following:

$$\tilde{T}(\omega, \mathbf{k}) = \frac{4\pi i}{\omega} \frac{e_i^* e_j \langle j_i j_j \rangle^0_{\mathbf{k}\omega}}{e_k^* e_l (\varepsilon_{kl} - \varepsilon_{lk}^*)}. \tag{6.166}$$

The effective temperature of a non-isothermal plasma may be written as

$$\tilde{T}(\omega, \mathbf{k}) = \frac{\sum e_i^* e_j (\varkappa_{ij} - \varkappa_{ji}^*) T}{\sum e_k^* e_l (\varkappa_{kl} - \varkappa_{lk}^*)}. \tag{6.167}$$

It is clear that $\tilde{T}(\omega, \mathbf{k})$ coincides with the plasma temperature T in the equilibrium case. In a non-equilibrium plasma $\tilde{T}(\omega, \mathbf{k})$ may grow very large. As the plasma state approaches the kinetic instability threshold, $\operatorname{Im} \Lambda \to 0$ and the effective temperature (6.164) grows infinitely (Sitenko and Kirochkin, 1966).

The spectral distributions of the partial current fluctuations in the plasma transparency domain may be directly related to those of the electric field:

$$\langle j_i^\alpha j_j^\beta \rangle_{\mathbf{k}\omega} = \omega^2 \varkappa_{ik}^{(\alpha)}(\omega, \mathbf{k}) \varkappa_{jl}^{*(\beta)}(\omega, \mathbf{k}) \langle E_k E_l \rangle_{\mathbf{k}\omega}. \tag{6.168}$$

It should be observed that in the plasma transparency domain spectral distributions of any quantities may be expressed in terms of those of the electric field fluctuations.

High-frequency fluctuations in a magneto-active plasma

In order to show how the general formulas of the preceding sections may be applied in practice, we consider the fluctuations in a non-isothermal two-temperature plasma with an external magnetic field $\mathbf{B}_0$. Since the electron contribution in the collective plasma motions is dominant in the high-frequency range, we neglect ion effects. Then the correlation function for the high-frequency fluctuations coincide with those for the equilibrium plasma, the temperature of which is equal to the electron temperature:

$$\langle E_i E_j \rangle_{\mathbf{k}\omega} = 8\pi^2 e_i e_j^* T_e \frac{|\operatorname{Tr} \lambda(\omega, \mathbf{k})|}{|\omega|} \delta\{\Lambda(\omega, \mathbf{k})\}. \tag{6.169}$$

In the high-frequency range the following relation may be made use of:

$$\delta\{\Lambda(\omega, \mathbf{k})\} = \frac{1}{|\operatorname{Tr} \lambda(\omega, \mathbf{k})|} \left\{ \frac{1}{|e|^2 - \dfrac{|(\mathbf{k}\cdot\mathbf{e})|^2}{k^2}} [\delta(\eta^2 - \eta_0^2) + \delta(\eta^2 - \eta_e^2)] + \delta(A) \right\}, \tag{6.170}$$

and the spectral distributions of the field fluctuations associated with ordinary and extraordinary waves reduce to the following form:

$$\langle E_i E_j \rangle_{\mathbf{k}\omega} = 8\pi^2 \frac{e_i e_j^*}{|e|^2 - \dfrac{|(\mathbf{k}\cdot\mathbf{e})|^2}{k^2}} \frac{T_e}{|\omega|} \delta(\eta^2 - \eta_{0,e}^2), \tag{6.171}$$

where the polarization vectors are given by (4.55). The spectral distribution of the Langmuir fluctuations is as follows:

$$\langle E_i E_j \rangle_{\mathbf{k}\omega} = 8\pi^2 \frac{k_i k_j}{k^2} \frac{T_e}{|\omega|} \frac{\eta_L^2}{A_0} \delta(\eta^2 - \eta_L^2); \tag{6.172}$$

in neglect of the Langmuir wave dispersion

$$\langle E_i E_j \rangle_{\mathbf{k}\omega} = 4\pi^2 \frac{k_i k_j}{k^2} T_e \frac{|\omega^2 - \omega_B^2|}{\omega_+^2 - \omega_-^2} \{\delta(\omega - \omega_+) + \delta(\omega + \omega_+) + \delta(\omega - \omega_-) + \delta(\omega + \omega_-)\}. \tag{6.173}$$

To obtain the spectral distribution of the high-frequency fluctuations of the electric field in the absence of the magnetic field one must set $\mathbf{B}_0 = 0$ in (6.171) and (6.172) (or (6.173)).

Low-frequency fluctuations in a magneto-active plasma

The low-frequency fluctuations in a non-isothermal plasma are in an essential way different from the fluctuations in an equilibrium medium. We have pointed out before that in a non-isothermal plasma without external magnetic fields there may occur low-frequency fluctuations associated with the non-isothermal sound oscillations. The weakly damped magneto-hydrodynamic oscillations in a magneto-active non-isothermal plasma cause additional maxima in the low-frequency range of the fluctuation spectrum.

The fluctuations associated with Alfvén and magneto-sonic oscillations in a magneto-active non-isothermal plasma are described by the general formulas (6.163) and (6.167) where appropriate approximate expressions are to be substituted for the susceptibilities. Retaining the leading terms in the large parameter ω_{B_i}/ω we can derive simple formulas for the correlation functions when $\omega \ll \omega_{B_i}$. Thus, the spectral distribution of the electric field fluctuations near the Alfvén frequency may be written as

$$\langle E_i E_j \rangle_{\mathbf{k}\omega} = 4\pi^2 e_i e_j^* T_e \frac{v_A^2}{c^2} \{\delta(\omega - kv_A \cos\vartheta) + \delta(\omega + kv_A \cos\vartheta)\}, \tag{6.174}$$

where for the polarization vector we have

$$\mathbf{e} = \left\{1, \; -i\frac{\omega}{\omega_{B_i}} \cot^2\vartheta, \quad -\frac{v_s^2}{v_A^2}\frac{\omega^2}{\omega_{B_i}^2}\frac{1}{\sin\vartheta\cos\vartheta}\right\}. \tag{6.175}$$

The spectral distributions for the magneto-sonic fluctuations of the electric field are as follows:

$$\langle E_i E_j \rangle_{\mathbf{k}\omega} = 4\pi^2 e_i e_j^* T_e \frac{v_A^2}{c^2} \{\delta(\omega - kv_A) + \delta(\omega + kv_A)\}, \tag{6.176}$$

$$\langle E_i E_j \rangle_{\mathbf{k}\omega} = 4\pi^2 e_i e_j^* T_e a^2 k^2 \{\delta(\omega - kv_s \cos\vartheta) + \delta(\omega + kv_s \cos\vartheta)\}, \tag{6.177}$$

where the polarizations of the fast and slow magneto-sonic waves are, respectively, equal to

$$\mathbf{e} = \left\{-i\frac{\omega}{\omega_{B_i}}\frac{1}{\sin^2\vartheta}, \quad 1, \quad -i\frac{v_s^2}{v_A^2}\frac{\omega}{\omega_{B_i}^2}\sin\vartheta\cos\vartheta\right\}, \tag{6.178}$$

$$\mathbf{e} = \left\{\sin\vartheta, \quad -i\frac{v_s^2}{v_A^2}\frac{\omega_{B_i}}{\omega}\sin\vartheta\cos\vartheta, \quad \cos\vartheta\right\}. \tag{6.179}$$

Note that the spectral distribution of the low-frequency fluctuations is proportional to the square of the ratio of the phase velocity of the relevant wave to the light velocity *in vacuo*.

The spectral distributions for the electric field fluctuations in a magneto-active plasma penetrated by a compensated beam of charged particles, as well as in a magneto-active plasma with electrons that move relative to the ions, were calculated by Sitenko and Kirochkin (1966).

CHAPTER 7

Effect of Non-linear Wave Interactions on Fluctuations in a Plasma

7.1. Equations for the Spectral Correlation Functions (Potential Field)

The non-linear equation for the fluctuation field

As we have shown in the last chapter, the spectral distributions of the plasma fluctuations have a broad maximum in the low-frequency range, which is due to the random motion of the individual charged particles, and a set of sharp maxima at the plasma eigenfrequencies. These latter maxima appear by virtue of the collective fluctuations, i.e. the random eigenoscillations of the plasma. The collective fluctuation level of a plasma in thermal equilibrium is governed by the temperature; it may grow very large in a non-equilibrium plasma, especially when the plasma state approaches the kinetic instability threshold. The fluctuation amplitudes may grow infinite in the linear approximation, which fact indicates that the linear treatment of the non-equilibrium plasma is inadequate and non-linear effects must be taken into account (Kadomtsev, 1965; Galeev and Sagdeev, 1973; Sitenko, 1975). In particular, the latter play the dominant role when a stationary fluctuation spectrum is being established in a turbulent plasma.

When considering fluctuations in a turbulent plasma it is suitable to start from the non-linear field equation that follows immediately from the Maxwell equations and the one for the microscopic density which describes the motions of individual particles. First we examine the simplest case in which the electric field is potential and the plasma is homogeneous and in a stationary state. The set of (6.61) is the most suitable for treating this problem. Separating out the fluctuation contribution in the microscopic density according to (6.60) and solving (6.71) by means of an iteration procedure, we obtain from second equation (6.70) the following non-linear equation for the fluctuation field in the plasma (Sitenko, 1975):

$$
ik\Bigg\{\varepsilon(\omega,\mathbf{k})E_{\mathbf{k}\omega}+\sum_{\substack{\omega_1+\omega_2=\omega\\ \mathbf{k}_1+\mathbf{k}_2=\mathbf{k}}}\varkappa^{(2)}(\omega_1,\mathbf{k}_1;\omega_2,\mathbf{k}_2)E_{\mathbf{k}_1\omega_1}E_{\mathbf{k}_2\omega_2}
+\sum_{\substack{\omega_1+\omega_2+\omega_3=\omega\\ \mathbf{k}_1+\mathbf{k}_2+\mathbf{k}_3=\mathbf{k}}}\varkappa^{(3)}(\omega_1,\mathbf{k}_1;\omega_2,\mathbf{k}_2;\omega_3,\mathbf{k}_3)E_{\mathbf{k}_1\omega_1}E_{\mathbf{k}_2\omega_2}E_{\mathbf{k}_3\omega_3}+\ldots
+\sum_{\omega',\mathbf{k}'}\delta\varepsilon(\omega,\mathbf{k};\omega',\mathbf{k}')E_{\mathbf{k}'\omega'}
+\sum_{\substack{\omega_1+\omega_2=\omega\\ \mathbf{k}_1+\mathbf{k}_2=\mathbf{k}\\ \omega',\mathbf{k}'}}\delta\varkappa^{(2)}(\omega_1,\mathbf{k}_1;\omega_2,\mathbf{k}_2;\omega',\mathbf{k}')E_{\mathbf{k}_1\omega_1}E_{\mathbf{k}'\omega'}+\ldots\Bigg\}=4\pi\varrho^2_{\mathbf{k}\omega}. \tag{7.1}
$$

(We retained in (7.1) only linear, quadratic, and cubic terms in the fluctuation field intensity.) The quantity ϱ^0 on the right-hand side of (7.1) implies the charge density (6.79) associated with the stochastic motion of the individual charged particles. The stochastic particle motion is responsible also for the inhomogeneity of the dielectric plasma properties, as the dielectric permittivity and the non-linear electric susceptibilities are governed by the particle distribution function. The last two terms on the left-hand side of (7.1) are due to this effect. The quantities $\delta\varepsilon(\omega,\mathbf{k};\omega',\mathbf{k}')$ and $\delta\varkappa^{(2)}(\omega_1,\mathbf{k}_1;\omega_2,\mathbf{k}_2;\omega',\mathbf{k}')$ are the variations of the dielectric permittivity and of the second-order electric susceptibility due to the random fluctuations of the distribution function $\delta f_{\mathbf{k}\omega}(\mathbf{v})$ and may be written in the form

$$
\delta\varepsilon(\omega,\mathbf{k};\omega',\mathbf{k}')\equiv\Delta(\omega,\mathbf{k};\omega',\mathbf{k}')\delta f^0_{\mathbf{k}-\mathbf{k}'\omega-\omega'}, \tag{7.2}
$$

$$
\delta\varkappa^{(2)}(\omega_1,\mathbf{k}_1;\omega_2,\mathbf{k}_2;\omega',\mathbf{k}')\equiv\Delta^{(2)}(\omega_1,\mathbf{k}_1;\omega_2,\mathbf{k}_2;\omega',\mathbf{k}')\delta f^0_{\mathbf{k}_2-\mathbf{k}',\,\omega_2-\omega'}, \tag{7.3}
$$

where $\Delta(\omega,\mathbf{k};\omega',\mathbf{k}')$ and $\Delta^{(2)}(\omega_1,\mathbf{k}_1;\omega_2,\mathbf{k}_2;\omega',\mathbf{k}')$ are the following operators:

$$
\Delta(\omega,\mathbf{k};\omega',\mathbf{k}')\delta f^0_{\mathbf{k}-\mathbf{k}',\omega-\omega'}\equiv\sum\frac{\Omega^2}{kk'}\int d\mathbf{v}\,\frac{1}{\omega-(\mathbf{k}\cdot\mathbf{v})+i0}\left(\mathbf{k}'\cdot\frac{\partial}{\partial\mathbf{v}}\right)\delta f^0_{\mathbf{k}-\mathbf{k}',\,\omega-\omega'}(v), \tag{7.4}
$$

$$
\Delta^{(2)}(\omega_1,\mathbf{k}_1;\omega_2,\mathbf{k}_2;\omega',\mathbf{k}')\delta f^0_{\mathbf{k}_2-\mathbf{k}',\,\omega_2-\omega'}\equiv-i\sum\frac{e}{m}\,\frac{\Omega^2}{k_1\,|\mathbf{k}_1+\mathbf{k}_2|\,k'}
\times\int d\mathbf{v}\,\frac{1}{\omega_1+\omega_2-(\{\mathbf{k}_1+\mathbf{k}_2\}\cdot\mathbf{v})+i0}\left(\mathbf{k}_1\cdot\frac{\partial}{\partial\mathbf{v}}\right)\frac{1}{\omega_2-(\mathbf{k}_2\cdot\mathbf{v})+i0}\left(k'\cdot\frac{\partial}{\partial\mathbf{v}}\right)\delta f^0_{\mathbf{k}_2-\mathbf{k}',\,\omega_2-\omega'}(\mathbf{v}). \tag{7.5}
$$

(The variation of the non-linear susceptibility $\varkappa^{(3)}(\omega_1,\mathbf{k}_1;\omega_2,\mathbf{k}_2;\omega_3,\mathbf{k}_3)$ in (7.1) may be neglected at the desired level of description.)

Correlation functions

The non-linear equation (7.1) governs the fluctuation plasma field $E_{\mathbf{k}\omega}$, the mean value of which is equal to zero. In the linear approximation the knowledge of the quadratic correlation function $\langle E^2\rangle_{\mathbf{k}\omega}$ is sufficient for the description of the fluctuation field. In order to study the influence of the non-linear wave interactions we must introduce the third- and fourth-order correlation functions $\langle E^3\rangle_{\mathbf{k}\omega;\mathbf{k}'\omega'}$ and $\langle E^4\rangle_{\mathbf{k}\omega;\mathbf{k}'\omega';\mathbf{k}''\omega''}$. The requested accuracy in the description of the fluctuation field (the cubic terms in the fluctuation field intensity were retained in the non-linear equation) is provided by the sequence of correlation functions $\langle E^2\rangle_{\mathbf{k}\omega}$, $\langle E^3\rangle_{\mathbf{k}\omega;\mathbf{k}'\omega'}$, and $\langle E^4\rangle_{\mathbf{k}\omega;\mathbf{k}'\omega';\mathbf{k}''\omega''}$. We define the correlation functions $\langle E^2\rangle_{\mathbf{k}\omega}$, $\langle E^3\rangle_{\mathbf{k}\omega;\mathbf{k}'\omega'}$, and $\langle E^4\rangle_{\mathbf{k}\omega;\mathbf{k}'\omega';\mathbf{k}''\omega''}$ to be related, respectively, to the average binary, ternary, and quaternary products of the field intensity amplitudes in the following way:

$$\langle E_{\mathbf{k}\omega}E_{\mathbf{k}'\omega'}\rangle \equiv (2\pi)^4\,\delta(\omega+\omega')\,\delta(\mathbf{k}+\mathbf{k}')\langle E^2\rangle_{\mathbf{k}\omega}, \tag{7.6}$$

$$\langle E_{\mathbf{k}\omega}E_{\mathbf{k}'\omega'}E_{\mathbf{k}''\omega''}\rangle \equiv (2\pi)^4\,\delta(\omega+\omega'+\omega'')\,\delta(\mathbf{k}+\mathbf{k}'+\mathbf{k}'')\langle E^3\rangle_{\mathbf{k}\omega;\,\mathbf{k}+\mathbf{k}'\omega+\omega'}, \tag{7.7}$$

$$\begin{aligned}\langle E_{\mathbf{k}\omega}E_{\mathbf{k}'\omega'}E_{\mathbf{k}''\omega''}E_{\mathbf{k}'''\omega'''}\rangle &\equiv (2\pi)^4\,\delta(\omega+\omega'+\omega''+\omega''')\,\delta(\mathbf{k}+\mathbf{k}'+\mathbf{k}''+\mathbf{k}''')\\ &\times\{(2\pi)^4\,\delta(\omega+\omega')\,\delta(\mathbf{k}+\mathbf{k}')\langle E^2\rangle_{\mathbf{k}\omega}\langle E^2\rangle_{\mathbf{k}''\omega''}+(2\pi)^4\,[\delta(\omega+\omega'')\,\delta(\mathbf{k}+\mathbf{k}'')\\ &+\delta(\omega'+\omega'')\,\delta(\mathbf{k}'+\mathbf{k}'')]\,\langle E^2\rangle_{\mathbf{k}\omega}\langle E^2\rangle_{\mathbf{k}'\omega'}+\langle E^4\rangle_{\mathbf{k}\omega;\,\mathbf{k}+\mathbf{k}',\,\omega+\omega';\,\mathbf{k}+\mathbf{k}'+\mathbf{k}'',\,\omega+\omega'+\omega''}\}.\end{aligned} \tag{7.8}$$

To find the correlation functions for the plasma fluctuation field explicitly, we start from the non-linear equation (7.1), where we transfer the terms that contain the fluctuation variations of the dielectric permittivity and of the non-linear susceptibility to the right-hand side. The successive multiplication of the left- and right-hand sides of the equations by themselves and the averaging procedure according to (6.80) yield a set of non-homogeneous integral equations which uniquely determine the succession of the correlation functions for the fluctuation field in the plasma (Sitenko, 1975).

The equation for the quadratic correlation function

We shall restrict ourselves by retaining in the equations thus obtained all terms up to those proportional to the field intensity to the fourth power. The average fourfold products of the field intensity amplitudes will be expressed in terms of the quadratic correlation functions and the irreducible quaternary correlators according to (7.8.). It is to be pointed out that in the assumed approximation it is important to take into account the variations of the dielectric permittivity and of the non-linear susceptibility, caused by the fluctuations of the distribution function, in the equation for the quadratic correlation function; in that for the cubic correlator only the dielectric permittivity variations must be retained; the fluctuation variations both of the dielectric permittivity and of the non-linear susceptibility may be neglected in the equation for the quaternary correlation function. Besides, when calculating the correlation functions (in the right-hand sides of the obtained equations), which describe the correlations between the charge density fluctuations $\varrho^0_{\mathbf{k}\omega}$ and the electric field fluctuations $E_{\mathbf{k}\omega}$, the approximate solution of (7.1) may be substituted for $E_{\mathbf{k}\omega}$:

$$E_{\mathbf{k}\omega} \to -\frac{4\pi i}{k\varepsilon(\omega,\mathbf{k})}\,\varrho^0_{\mathbf{k}\omega}, \tag{7.9}$$

Thus after some rather complicated algebra we obtain the following equation for the quadratic correlation function $\langle E^2\rangle_{\mathbf{k}\omega}$:

$$\begin{aligned}\varepsilon^*(\omega,\mathbf{k})\Big\{\varepsilon(\omega,\mathbf{k})\langle E^2\rangle_{\mathbf{k}\omega}-\frac{2}{\varepsilon^*(\omega,\mathbf{k})}\sum_{\omega',\mathbf{k}'}|\varkappa^{(2)}(\omega-\omega',\mathbf{k}-\mathbf{k}';\omega',\mathbf{k}')|^2\langle E^2\rangle_{\mathbf{k}\omega}\langle E^2\rangle_{\mathbf{k}-\mathbf{k}',\,\omega-\omega'}\\ -\sum_{\omega',\mathbf{k}'}a(\omega,\mathbf{k};\omega',\mathbf{k}')\langle E^2\rangle_{\mathbf{k}'\omega'}\langle E^2\rangle_{\mathbf{k}\omega}\Big\}\\ -\varepsilon(\omega,\mathbf{k})\sum_{\omega',\mathbf{k}'}a^*(\omega,\mathbf{k};\omega',\mathbf{k}')\langle E^2\rangle_{\mathbf{k}'\omega'}\langle E^2\rangle_{\mathbf{k}\omega}\\ =\sum_{\omega',\mathbf{k}'}b(\omega,\mathbf{k};\omega',\mathbf{k}')\langle E^2\rangle_{\mathbf{k}'\omega'}+q_{\mathbf{k}\omega},\end{aligned} \tag{7.10}$$

where the following notation has been used:

$$a(\omega, \mathbf{k}; \omega', \mathbf{k}')$$

$$\equiv 2\left\{2\frac{\varkappa^{(2)}(\omega-\omega', \mathbf{k}-\mathbf{k}'; \omega', \mathbf{k}')\,\varkappa^{(2)}(\omega, \mathbf{k}; -\omega', -\mathbf{k}')}{\varepsilon(\omega-\omega', \mathbf{k}-\mathbf{k}')}-\bar{\varkappa}^{(3)}(\omega', \mathbf{k}'; \omega, \mathbf{k}; -\omega', -\mathbf{k}')\right\}$$

$$+2\frac{\varkappa^{(2)}(\omega, \mathbf{k}; 0, 0)\,\varkappa^{(2)}(\omega', \mathbf{k}'; -\omega', -k')}{\varepsilon(0, 0)}-\varkappa^{(3)}(\omega, \mathbf{k}; \omega', \mathbf{k}'; -\omega', -\mathbf{k}'); \tag{7.11}$$

$$b(\omega, \mathbf{k}; \omega', \mathbf{k}') \equiv \Delta(\omega, \mathbf{k}; \omega', \mathbf{k}')\,\Delta(-\omega, -\mathbf{k}; -\omega', -\mathbf{k}')\,\langle\delta f\,\delta f\rangle^{0}_{\mathbf{k}-\mathbf{k}'\omega-\omega'}$$

$$-16\pi\,\mathrm{Im}\,\frac{\varkappa^{(2)*}(\omega-\omega', \mathbf{k}-\mathbf{k}'; \omega', \mathbf{k}')}{|\mathbf{k}-\mathbf{k}'|\,\varepsilon^{*}(\omega-\omega', \mathbf{k}-\mathbf{k}')}\,\Delta(\omega, \mathbf{k}; \omega', \mathbf{k}')\,\langle\delta f\varrho\rangle^{0}_{\mathbf{k}-\mathbf{k}', \omega-\omega'}; \tag{7.12}$$

$$q_{\mathbf{k}\omega} \equiv q^{(1)}_{\mathbf{k}\omega}+q^{(2)}_{\mathbf{k}\omega}+q^{(3)}_{\mathbf{k}\omega}, \tag{7.13}$$

$$q^{(1)}_{\mathbf{k}\omega} \equiv \frac{16\pi^2}{k^2}\langle\varrho^2\rangle^{0}_{\mathbf{k}\omega}, \tag{7.14}$$

$$q^{(2)}_{\mathbf{k}\omega} \equiv -\frac{32\pi^2}{k}\sum_{\omega', \mathbf{k}'}\frac{1}{k'}\left\{4\pi\,\mathrm{Im}\,\frac{\varkappa^{(2)}(\omega-\omega', \mathbf{k}-\mathbf{k}'; \omega', \mathbf{k}')}{|\mathbf{k}-\mathbf{k}'|\,\varepsilon(\omega-\omega', \mathbf{k}-\mathbf{k}')\,\varepsilon(\omega', \mathbf{k}')}\langle\varrho^3\rangle^{0}_{\mathbf{k}\omega;\mathbf{k}'\omega'}\right.$$

$$\left.+\mathrm{Re}\,\frac{1}{\varepsilon(\omega', \mathbf{k}')}\,\Delta(\omega, \mathbf{k}; \omega', \mathbf{k}')\,\langle\delta f\varrho^2\rangle^{0}_{\mathbf{k}\omega;\mathbf{k}'\omega'}\right\}, \tag{7.15}$$

$$q^{(3)}_{\mathbf{k}\omega} \equiv \frac{32\pi^2}{k^2}P_{\mathbf{k}\omega}\langle\varrho^2\rangle^{0}_{\mathbf{k}\omega}+\frac{8\pi}{k}\,\mathrm{Im}\sum_{\omega', \mathbf{k}'}\Big[\Delta^{(2)}(\omega', \mathbf{k}'; \omega-\omega', \mathbf{k}-\mathbf{k}'; -\omega', -\mathbf{k}')$$

$$-2\frac{\varkappa^{(2)}(\omega-\omega', \mathbf{k}-\mathbf{k}'; \omega', \mathbf{k}')}{\varepsilon(\omega-\omega', \mathbf{k}-\mathbf{k}')}\,\Delta(\omega-\omega', \mathbf{k}-\mathbf{k}'; -\omega', -\mathbf{k}')\Big]\langle E^2\rangle_{\mathbf{k}'\omega'}\langle\delta f\varrho\rangle^{0}_{\mathbf{k}\omega}$$

$$-32\pi^2\,\mathrm{Re}\sum_{\omega', \mathbf{k}'}\frac{1}{kk'\varepsilon(\omega-\omega', \mathbf{k}-\mathbf{k}')\,\varepsilon^{*}(\omega', \mathbf{k}')}\,\Delta(\omega, \mathbf{k}; \omega-\omega', \mathbf{k}-\mathbf{k}')\,\langle\delta f\varrho\rangle^{0}_{\mathbf{k}\omega}$$

$$\times\Delta(-\omega, -\mathbf{k}; -\omega+\omega', -\mathbf{k}+\mathbf{k}')\,\langle\delta f\varrho\rangle^{0}_{\mathbf{k}\omega}$$

$$+16\pi^2\sum_{\omega', \mathbf{k}'}\frac{1}{|\mathbf{k}-\mathbf{k}'|\,k'\varepsilon(\omega', \mathbf{k}')\,\varepsilon^{*}(\omega-\omega', \mathbf{k}-\mathbf{k}')}\,\Delta(\omega, \mathbf{k}; \omega', \mathbf{k}')\,\langle\delta f\varrho\rangle^{0}_{\mathbf{k}-\mathbf{k}', \omega-\omega'}$$

$$\times\Delta(-\omega, -\mathbf{k}; -\omega+\omega', -\mathbf{k}+\mathbf{k}')\,\langle\delta f\varrho\rangle^{0}_{-\mathbf{k}', -\omega'}$$

$$+16\pi^2\sum_{\omega', \mathbf{k}'}\sum_{\omega'', \mathbf{k}''}\frac{\Delta(\omega, \mathbf{k}; \omega', \mathbf{k}')}{k''\varepsilon(\omega', \mathbf{k}')\,\varepsilon(\omega'', \mathbf{k}'')}$$

$$\times\left[\frac{2}{k}\Delta(\omega', \mathbf{k}'; \omega'', \mathbf{k}'')-\frac{1}{k'}\Delta(-\omega, -\mathbf{k}; \omega'', \mathbf{k}'')\right]\langle\delta f^2\varrho^2\rangle_{\mathbf{k}\omega;\mathbf{k}'\omega';\mathbf{k}''\omega''}$$

$$+\frac{128\pi^3}{k}\,\mathrm{Im}\sum_{\omega', \mathbf{k}'}\sum_{\omega'', \mathbf{k}''}\frac{1}{|\mathbf{k}-\mathbf{k}'|\,k''\varepsilon(\omega-\omega', \mathbf{k}-\mathbf{k}')\,\varepsilon(\omega'', \mathbf{k}'')}$$

$$\times\left\{\frac{\varkappa^{(2)}(\omega-\omega', \mathbf{k}-\mathbf{k}'; \omega', \mathbf{k}')}{\varepsilon(\omega', \mathbf{k}')}\left[2\Delta(\omega', \mathbf{k}'; \omega'', \mathbf{k}'')-\frac{k}{k'}\Delta(-\omega, -\mathbf{k}; \omega'', \mathbf{k}'')\right]\right.$$

$$+\frac{\varkappa^{(2)}(\omega-\omega', \mathbf{k}-\mathbf{k}'; \omega'', \mathbf{k}'')\, \Delta(\omega, \mathbf{k}; \omega-\omega'+\omega'', \mathbf{k}-\mathbf{k}'+\mathbf{k}'')}{\varepsilon(\omega-\omega'+\omega'', \mathbf{k}-\mathbf{k}'+\mathbf{k}'')}$$

$$-\Delta^{(2)}(\omega-\omega', \mathbf{k}-\mathbf{k}'; \omega', \mathbf{k}'; \omega'', \mathbf{k}'')\Big\} \langle \delta f \varrho^3\rangle_{\mathbf{k}\omega;\, \mathbf{k}'\omega';\, \mathbf{k}''\omega''}$$

$$+\frac{256\pi^4}{k}\sum_{\omega', \mathbf{k}'}\sum_{\omega'', \mathbf{k}''}\frac{1}{|\mathbf{k}-\mathbf{k}'|\, k''\varepsilon(\omega-\omega', \mathbf{k}-\mathbf{k}')\, \varepsilon^*(\omega'', \mathbf{k}'')}$$

$$\times\Big\{\frac{k}{k'\,|\mathbf{k}-\mathbf{k}''|}\, \frac{\varkappa^{(2)}(\omega-\omega', \mathbf{k}-\mathbf{k}'; \omega', \mathbf{k}')\, \varkappa^{(2)*}(\omega-\omega'', \mathbf{k}-\mathbf{k}''; \omega'', \mathbf{k}'')}{\varepsilon(\omega', \mathbf{k}')\, \varepsilon^*(\omega-\omega', \mathbf{k}-\mathbf{k}')}$$

$$+\frac{2}{|\mathbf{k}'+\mathbf{k}''|\, \varepsilon(\omega'+\omega'', \mathbf{k}'+\mathbf{k}'')}$$

$$\times\Big[2\frac{\varkappa^{(2)}(\omega-\omega', \mathbf{k}-\mathbf{k}'; \omega', \mathbf{k}')\, \varkappa^{(2)}(\omega'+\omega'', \mathbf{k}'+\mathbf{k}''; -\omega'', -\mathbf{k}'')}{\varepsilon(\omega', \mathbf{k}')}$$

$$-\varkappa^{(3)}(\omega-\omega', \mathbf{k}-\mathbf{k}'; \omega'+\omega'', \mathbf{k}'+\mathbf{k}''; -\omega'', -\mathbf{k}'')\Big]\Big\} \langle \varrho^4\rangle_{\mathbf{k}\omega;\, \mathbf{k}'\omega';\, \mathbf{k}''\omega''}, \tag{7.16}$$

$$P_{\mathbf{k}\omega} \equiv 4\pi\, \mathrm{Im}\, \frac{1}{\varepsilon(\omega, \mathbf{k})}\sum_{\omega', \mathbf{k}'}\frac{1}{\varepsilon(\omega', \mathbf{k}')}\Big\{\frac{1}{k'}\Big[2\frac{\varkappa^{(2)}(\omega-\omega', \mathbf{k}-\mathbf{k}'; \omega', \mathbf{k}')}{\varepsilon(\omega-\omega', \mathbf{k}-\mathbf{k}')}$$

$$\times \Delta(\omega-\omega', \mathbf{k}-\mathbf{k}'; \omega', \mathbf{k}')-\Delta^{(2)}(\omega', \mathbf{k}'; \omega-\omega', k-k'; \omega, \mathbf{k})\Big] \langle \delta f \varrho\rangle^0_{-\mathbf{k}', -\omega}$$

$$-2\frac{\varkappa^{(2)}(\omega, \mathbf{k}; \omega'-\omega, \mathbf{k}'-\mathbf{k})}{|\mathbf{k}-\mathbf{k}'|\, \varepsilon(\omega-\omega', \mathbf{k}-\mathbf{k}')}\, \Delta(\omega, \mathbf{k}; \omega', \mathbf{k}')\, \langle \delta f \varrho\rangle^0_{\mathbf{k}-\mathbf{k}', \omega-\omega'}\Big\}$$

$$+\mathrm{Re}\sum_{\omega', \mathbf{k}'}\frac{1}{\varepsilon(\omega, \mathbf{k})\, \varepsilon(\omega', \mathbf{k}')}\, \Delta(\omega, \mathbf{k}; \omega', \mathbf{k}')\, \Delta(\omega', \mathbf{k}'; \omega, \mathbf{k})\, \langle \delta f^2\rangle^0_{\mathbf{k}-\mathbf{k}', \omega-\omega'}. \tag{7.17}$$

(It should be remembered that the quantities $\Delta(\omega, \mathbf{k}; \omega', \mathbf{k}')$ and $\Delta^{(2)}(\omega, \mathbf{k}; \omega', \mathbf{k}'; \omega'', \mathbf{k}'')$ which enter the above expressions for $b(\omega, \mathbf{k}; \omega', \mathbf{k}')$, $q_{\mathbf{k}\omega}$ and $P_{\mathbf{k}\omega}$ are integral operators acting on the correlation functions that follow them, e.g.

$$\Delta(\omega, \mathbf{k}; \omega', \mathbf{k}')\, \langle \delta f \varrho\rangle^0_{\mathbf{k}-\mathbf{k}', \omega-\omega'}$$

$$\equiv \sum \frac{4\pi e^2}{m}\, \frac{1}{kk'}\int d\mathbf{v}\, \frac{1}{\omega-(\mathbf{k}\cdot\mathbf{v})+i0}\left(\mathbf{k}'\cdot\frac{\partial}{\partial \mathbf{v}}\right) \langle \delta f \varrho\rangle^0_{\mathbf{k}-\mathbf{k}', \omega-\omega'}, \tag{7.18}$$

$$\Delta(\omega, \mathbf{k}; \omega', \mathbf{k}')\, \Delta(-\omega, -\mathbf{k}; -\omega', -\mathbf{k}')\, \langle \delta f \varrho\rangle^0_{\mathbf{k}-\mathbf{k}', \omega-\omega'}$$

$$\equiv \sum\sum \frac{4\pi e^2}{m}\, \frac{4\pi e'^2}{m'}\, \frac{1}{k^2k'^2}\int d\mathbf{v}\, \frac{1}{\omega-(\mathbf{k}\cdot\mathbf{v})+i0}\left(\mathbf{k}'\cdot\frac{\partial}{\partial \mathbf{v}}\right)$$

$$\times\int d\mathbf{v}'\, \frac{1}{\omega-(\mathbf{k}\cdot\mathbf{v}')+i0}\left(\mathbf{k}'\cdot\frac{\partial}{\partial \mathbf{v}'}\right) \langle \delta f(\mathbf{v})\, \delta f(\mathbf{v}')\rangle^0_{\mathbf{k}-\mathbf{k}', \omega-\omega'} \tag{7.19}$$

and so on.)

Since the basic non-linear equation (7.1) was derived by means of an expansion in a power series in the electric field, (7.10) holds only if being solved by the iteration process. Having this in mind we may transform (7.10) into an equivalent one, namely, we transfer the last term on the left-hand side to the right-hand side and at the same time substitute the result of the linear approximation for the correlation function $\langle E^2\rangle_{\mathbf{k}\omega}$:

$$\langle E^2\rangle_{\mathbf{k}\omega} \to \frac{16\pi^2}{k^2}\,\frac{\langle\varrho^2\rangle^0_{\mathbf{k}\omega}}{|\varepsilon(\omega, \mathbf{k})|^2}.$$

Thus the equation for the quadratic correlation function becomes of the form

$$\varepsilon(\omega, \mathbf{k})\langle E^2\rangle_{\mathbf{k}\omega} - \frac{2}{\varepsilon^*(\omega, \mathbf{k})}\sum_{\omega', \mathbf{k}'}|\varkappa^{(2)}(\omega-\omega', \mathbf{k}-\mathbf{k}'; \omega', \mathbf{k}')|^2\langle E^2\rangle_{\mathbf{k}'\omega'}\langle E^2\rangle_{\mathbf{k}-\mathbf{k}', \omega-\omega'}$$
$$-\sum_{\omega', \mathbf{k}'} a(\omega, \mathbf{k}; \omega', \mathbf{k}')\langle E^2\rangle_{\mathbf{k}'\omega'}\langle E^2\rangle_{\mathbf{k}\omega}$$
$$= \frac{1}{\varepsilon^*(\omega, \mathbf{k})}\left\{\sum_{\omega', \mathbf{k}'} b(\omega, \mathbf{k}; \omega', \mathbf{k}')\langle E^2\rangle_{\mathbf{k}'\omega'} + q'_{\mathbf{k}\omega}\right\}, \tag{7.20}$$

where

$$q'_{\mathbf{k}\omega} \equiv q_{\mathbf{k}\omega} + \frac{16\pi^2}{k^2\varepsilon^*(\omega, \mathbf{k})}\sum_{\omega', \mathbf{k}'} a^*(\omega, \mathbf{k}; \omega', \mathbf{k}')\langle E^2\rangle_{\mathbf{k}'\omega'}\langle\varrho^2\rangle^0_{\mathbf{k}\omega}. \tag{7.21}$$

Equations for ternary and quaternary correlation functions

The equation for the ternary correlation function $\langle E^3\rangle_{\mathbf{k}\omega; \mathbf{k}'\omega'}$ is as follows:

$$\langle E^3\rangle_{\mathbf{k}\omega; \mathbf{k}'\omega'} + \frac{2}{\varepsilon^*(\omega', \mathbf{k}')}\varkappa^{(2)}(\omega-\omega', \mathbf{k}-\mathbf{k}'; -\omega', -\mathbf{k}')\langle E^2\rangle_{\mathbf{k}-\mathbf{k}', \omega-\omega'}\langle E^2\rangle_{\mathbf{k}\omega}$$
$$+\frac{2}{\varepsilon^*(\omega-\omega', \mathbf{k}-\mathbf{k}')}\varkappa^{(2)}(\omega, \mathbf{k}; -\omega', -\mathbf{k}')\langle E^2\rangle_{\mathbf{k}'\omega'}\langle E^2\rangle_{\mathbf{k}\omega}$$
$$+\frac{2}{\varepsilon(\omega, \mathbf{k})}\varkappa^{(2)}(\omega-\omega', \mathbf{k}-\mathbf{k}'; \omega', \mathbf{k}')\langle E^2\rangle_{\mathbf{k}'\omega'}\langle E^2\rangle_{\mathbf{k}-\mathbf{k}', \omega-\omega'}$$
$$= B(-\omega+\omega', -\mathbf{k}+\mathbf{k}'; -\omega', -\mathbf{k}')\langle E^2\rangle_{\mathbf{k}\omega} + B(\omega, \mathbf{k}; -\omega+\omega', -\mathbf{k}+\mathbf{k}')\langle E^2\rangle_{\mathbf{k}'\omega'}$$
$$+B(-\omega', -\mathbf{k}'; \omega, \mathbf{k})\langle E^2\rangle_{\mathbf{k}-\mathbf{k}', \omega-\omega'} + Q_{\mathbf{k}\omega; \mathbf{k}'\omega'}, \tag{7.22}$$

where we have put

$$B(\omega, \mathbf{k}; \omega', \mathbf{k}') \equiv \frac{4\pi i}{\varepsilon(\omega, \mathbf{k})\,\varepsilon(\omega', \mathbf{k}')}\left\{\frac{1}{k}\Delta(\omega', \mathbf{k}'; \omega+\omega', \mathbf{k}+\mathbf{k}')\langle\delta f\varrho\rangle^0_{-\mathbf{k}, -\omega}\right.$$
$$\left. + \frac{1}{k'}\Delta(\omega, \mathbf{k}; \omega+\omega', \mathbf{k}+\mathbf{k}')\langle\delta f\varrho\rangle^0_{-\mathbf{k}', -\omega'}\right\}, \tag{7.23}$$

$$Q_{\mathbf{k}\omega; \mathbf{k}'\omega'} \equiv \frac{64\pi^3 i}{kk'|\mathbf{k}-\mathbf{k}'|}\,\frac{1}{\varepsilon(\omega, \mathbf{k})\,\varepsilon^*(\omega', \mathbf{k}')\,\varepsilon^*(\omega-\omega'\, \mathbf{k}-\mathbf{k}')}$$
$$\times\left\{\langle\varrho^3\rangle^0_{\mathbf{k}\omega; \mathbf{k}'\omega'} + 4\pi i\sum_{\omega'', \mathbf{k}''}\frac{1}{k''\varepsilon(\omega'', \mathbf{k}'')}\left(\frac{k}{|\mathbf{k}-\mathbf{k}''|}\,\frac{\varkappa^{(2)}(\omega-\omega'', \mathbf{k}-\mathbf{k}''; \omega'', \mathbf{k}'')}{\varepsilon(\omega-\omega'', \mathbf{k}-\mathbf{k}'')}\right.\right.$$
$$+\frac{k'}{|\mathbf{k}'+\mathbf{k}''|}\,\frac{\varkappa^{(2)}(-\omega'-\omega'', -\mathbf{k}'-\mathbf{k}''; \omega'', \mathbf{k}'')}{\varepsilon^*(\omega'+\omega'', \mathbf{k}'+\mathbf{k}'')}$$
$$\left.\left.+\frac{|\mathbf{k}-\mathbf{k}'|}{|\mathbf{k}-\mathbf{k}'+\mathbf{k}''|}\,\frac{\varkappa^{(2)*}(\omega-\omega'+\omega'', \mathbf{k}-\mathbf{k}'+\mathbf{k}''; -\omega'', -\mathbf{k}'')}{\varepsilon(\omega-\omega'+\omega'', \mathbf{k}-\mathbf{k}'+\mathbf{k}'')}\right)\langle\varrho^4\rangle^0_{\mathbf{k}\omega; \mathbf{k}'\omega'; \mathbf{k}''\omega''}\right\}, \tag{7.24}$$

The non-homogeneous parts of (7.20) and (7.22) are written in terms of the correlators for the fluctuations of the distribution function and of the charge density in neglect of the interactions between particles (the explicit expressions are given in Chapter 6, Section 2). The quantities $b(\omega, \mathbf{k}; \omega', \mathbf{k}')$ and $B(\omega, \mathbf{k}; \omega', \mathbf{k}')$ are directly related to the correlation functions $\langle \delta f(\mathbf{v})\, \delta f(\mathbf{v}')\rangle^0_{\mathbf{k}\omega}$ and $\langle \delta f \varrho\rangle^0_{\mathbf{k}\omega}$ (see the definitions (6.32) and (6.39)).

The quaternary correlation function for the field $\langle E^4\rangle_{\mathbf{k}\omega;\mathbf{k}'\omega';\mathbf{k}''\omega''}$ is determined by the quaternary correlation function for the charge density fluctuations associated with the stochastic motion of the individual charged particles in a plasma:

$$\langle E^4\rangle_{\mathbf{k}\omega;\mathbf{k}'\omega';\mathbf{k}''\omega''} = \frac{256\pi^4}{k\,|\mathbf{k}-\mathbf{k}'|\cdot|\mathbf{k}'-\mathbf{k}''|\,k''} \times \frac{\langle \varrho^4\rangle^0_{\mathbf{k}\omega;\mathbf{k}'\omega';\mathbf{k}''\omega''}}{\varepsilon(\omega, \mathbf{k})\,\varepsilon^*(\omega-\omega', \mathbf{k}-\mathbf{k}')\,\varepsilon^*(\omega'-\omega'', \mathbf{k}'-\mathbf{k}'')\,\varepsilon^*(\omega'', \mathbf{k}'')}. \tag{7.25}$$

The set of (7.20), (7.22), and (7.25) provides a complete description of the set of correlation functions for the electric plasma field $\langle E^2\rangle_{\mathbf{k}\omega}$, $\langle E^3\rangle_{\mathbf{k}\omega;\mathbf{k}'\omega'}$, and $\langle E^4\rangle_{\mathbf{k}\omega;\mathbf{k}'\omega';\mathbf{k}''\omega''}$ taking into account the non-linear wave interaction (Sitenko, 1975). The latter shows up significantly in a non-equilibrium plasma, especially near the kinetic instability threshold, as well as in unstable regimes arising in a turbulent plasma.

7.2. The Stationary Fluctuation Spectrum Taking into Account Non-linear Correlations

Iteration process

Let the fluctuation fields be moderate and their spectral distributions be stationary. Then the set of (7.20), (7.22), and (7.25) may be iterated.

Let us present the quadratic correlation function $\langle E^2\rangle_{\mathbf{k}\omega}$ which determines the spectral distribution of the fluctuation electric field energy in the expansion form:

$$\langle E^2\rangle_{\mathbf{k}\omega} = \langle E^2\rangle^{(1)}_{\mathbf{k}\omega} + \langle E^2\rangle^{(2)}_{\mathbf{k}\omega} + \langle E^2\rangle^{(3)}_{\mathbf{k}\omega} + \ldots, \tag{7.26}$$

where $\langle E^2\rangle^{(1)}_{\mathbf{k}\omega}$ corresponds to the linear approximation, $\langle E^2\rangle^{(2)}_{\mathbf{k}\omega}$ is associated with taking into account the quadratic interactions between the fields, $\langle E^2\rangle^{(3)}_{\mathbf{k}\omega}$ is due to the cubic interactions, and so on. Starting from (7.20), we obtain for each term of the expansion (7.26):

$$\langle E^2\rangle^{(1)}_{\mathbf{k}\omega} = \frac{16\pi^2}{k^2}\,\frac{\langle \varrho^2\rangle^0_{\mathbf{k}\omega}}{|\varepsilon(\omega, \mathbf{k})|^2}, \tag{7.27}$$

$$\langle E^2\rangle^{(2)}_{\mathbf{k}\omega} = -\frac{128\pi^3}{k}\,\frac{1}{|\varepsilon(\omega, \mathbf{k})|^2} \times \sum_{\omega', \mathbf{k}'} \frac{1}{k'} \left\{ \mathrm{Im}\, \frac{\varkappa^{(2)}(\omega-\omega', \mathbf{k}-\mathbf{k}'; \omega', \mathbf{k}')}{|\mathbf{k}-\mathbf{k}'|\,\varepsilon(\omega-\omega', \mathbf{k}-\mathbf{k}')\,\varepsilon(\omega', \mathbf{k}')} \langle \varrho^3\rangle^0_{\mathbf{k}\omega;\mathbf{k}'\omega'} + \frac{1}{4\pi}\,\mathrm{Re}\,\frac{1}{\varepsilon(\omega', \mathbf{k}')} \Delta(\omega, \mathbf{k}; \omega', \mathbf{k}')\,\langle \delta f \varrho^2\rangle^0_{\mathbf{k}\omega;\mathbf{k}'\omega'} \right\}, \tag{7.28}$$

$$\langle E^2\rangle^{(3)}_{\mathbf{k}\omega} = \frac{1}{|\varepsilon(\omega,\mathbf{k})|^2}\Big\{2\sum_{\omega',\mathbf{k}'}|\varkappa^{(2)}(\omega-\omega',\mathbf{k}-\mathbf{k}';\omega',\mathbf{k}')|^2\langle E^2\rangle^{(1)}_{\mathbf{k}'\omega'}\langle E^2\rangle^{(2)}_{\mathbf{k}-\mathbf{k}',\omega-\omega'}$$

$$+2\,\mathrm{Re}\,\varepsilon^*(\omega,\mathbf{k})\sum_{\omega',\mathbf{k}'}a(\omega,\mathbf{k};\omega',\mathbf{k}')\langle E^2\rangle^{(1)}_{\mathbf{k}'\omega'}\langle E^2\rangle^{(1)}_{\mathbf{k}\omega}$$

$$+\sum_{\omega',\mathbf{k}'}b(\omega,\mathbf{k};\omega',\mathbf{k}')\langle E^2\rangle^{(1)}_{\mathbf{k}'\omega'}+q^{(3)}_{\mathbf{k}\omega}\Big\}. \tag{7.29}$$

The first term in (7.26), which corresponds to the linear approximation, describes the spectral distribution of the fluctuating electric field energy in neglect of the interactions between the plasma fluctuation fields. The linear spectral distribution (7.27) is similar to the above-derived expression (6.89). According to (7.27) the spectral distribution for the fluctuating electric field in a plasma in the linear approximation $\langle E^2\rangle^{(1)}_{\mathbf{k}\omega}$ is completely determined by the given quadratic correlation function for the charge density fluctuations in an ensemble of non-interacting particles $\langle \varrho^2\rangle^0_{\mathbf{k}\omega}$.

The second term of the expansion (7.26) is due to the quadratic interaction between the fluctuating fields, which is described by the non-linear susceptibility $\varkappa^{(2)}(\omega_1,\mathbf{k}_1;\omega_2,\mathbf{k}_2)$ and by the quantity $\Delta(\omega,\mathbf{k};\omega',\mathbf{k}')$ that is associated with the fluctuation variations of the plasma dielectric permittivity. The spectral distribution of the fluctuation electric field $\langle E^2\rangle^{(2)}_{\mathbf{k}\omega}$ is determined by the ternary correlation function for the charge density fluctuations $\langle \varrho^3\rangle^0_{\mathbf{k}\omega;\mathbf{k}'\omega'}$, calculated without taking into account the particle interactions, and the correlator for the fluctuation variations of the distribution function and the square of the fluctuation charge density in neglect of particle interactions $\langle \delta f\delta^2\rangle^0_{\mathbf{k}\omega;\mathbf{k}'\omega'}$.

Finally, the third term in (7.26) is associated with taking into account the cubic interactions between the fluctuation fields, which is reflected in the third-order susceptibility $\varkappa^{(3)}(\omega_1,\mathbf{k}_1;\omega_2,\mathbf{k}_2;\omega_3,\mathbf{k}_3)$ as well as in the quantity $\Delta^{(2)}(\omega_1,\mathbf{k}_1;\omega_2,\mathbf{k}_2;\omega',\mathbf{k}')$. The latter determines the fluctuation variations of the second order susceptibility $\varkappa^{(2)}(\omega_1,\mathbf{k}_1;\omega_2,\mathbf{k}_2)$. The quadratic and the cubic interactions give rise to the spectral distribution for the fluctuating electric field $\langle E^2\rangle^{(3)}_{\mathbf{k}\omega}$, which depends on the quaternary correlators for the fluctuations of the charge density and the distribution function in a system of non-interacting particles $\langle \varrho^4\rangle^0_{\mathbf{k}\omega;\mathbf{k}'\omega';\mathbf{k}''\omega''}$, $\langle \delta f\varrho^3\rangle^0_{\mathbf{k}\omega;\mathbf{k}'\omega';\mathbf{k}''\omega''}$, and $\langle \delta f^2\varrho^2\rangle^0_{\mathbf{k}\omega;\mathbf{k}'\omega';\mathbf{k}''\omega''}$, as well as on the products of different binary correlation functions for the same ensemble $\langle \varrho^2\rangle^0_{\mathbf{k}\omega}$, $\langle \delta f\varrho\rangle^0_{\mathbf{k}\omega}$, and $\langle \delta f\,\delta f\rangle^0_{\mathbf{k}\omega}$.

If the plasma is in thermal equilibrium, then $\langle E^2\rangle^{(2)}_{\mathbf{k}\omega}$ and $\langle E^2\rangle^{(3)}_{\mathbf{k}\omega}$ make only small corrections to the spectral distribution $\langle E^2\rangle^{(1)}_{\mathbf{k}\omega}$. The relative weight of these contributions is of the order of magnitude $\sqrt{\varepsilon}$ and ε, respectively. Note that the spectral distribution $\langle E^2\rangle^{(3)}_{\mathbf{k}\omega}$ has, besides maxima at the plasma eigenfrequencies $\omega_{\mathbf{k}}$, also additional maxima at the combination frequencies $\omega_{\mathbf{k}-\mathbf{k}'}\pm\omega_{\mathbf{k}'}$.

Induced enhancement of fluctuations

The non-linear wave interactions may cause a considerable enhancement of fluctuations when an external source excites some eigenoscillations in the plasma. Let the spectral distribution of the eigenoscillations, induced by the external source, be denoted by $\langle E^2\rangle^0_{\mathbf{k}\omega}$.

In neglect of damping the spectral distribution $\langle E^2\rangle^0_{\mathbf{k}\omega}$ is of the form of a set of delta-like maxima at the frequencies of the induced oscillations and satisfies the condition

$$\varepsilon(\omega, \mathbf{k})\langle E^2\rangle^{(0)}_{\mathbf{k}\omega} = 0. \tag{7.30}$$

Making use of (7.20) and retaining only the leading terms in the non-linear interactions between the fluctuation field and the field of the induced oscillations (cubic interactions are essential), we obtain the following expression for the spectral distribution of the fluctuation field in the plasma $\langle \tilde{E}^2\rangle_{\mathbf{k}\omega}$:

$$\langle \tilde{E}^2\rangle_{\mathbf{k}\omega} = \left\{1+2\,\mathrm{Re}\,\frac{1}{\varepsilon(\omega, \mathbf{k})}\sum_{\omega', \mathbf{k}'} a(\omega, \mathbf{k}; \omega', \mathbf{k}')\langle E^2\rangle^0_{\mathbf{k}'\omega'}\right\}\langle E^2\rangle^{(1)}_{\mathbf{k}\omega} + \ldots, \tag{7.31}$$

where $\langle E^2\rangle^{(1)}_{\mathbf{k}\omega}$ denotes the spectral distribution of the fluctuating field in the absence of induced oscillations. It is clear that when the intensity of the induced oscillations is large, the fluctuation intensity $\langle \tilde{E}^2\rangle_{\mathbf{k}\omega}$ may be much greater than the intensity of the thermal fluctuations $\langle E^2\rangle^{(1)}_{\mathbf{k}\omega}$, which may occur if

$$\mathrm{Re}\,\frac{1}{\varepsilon(\omega, \mathbf{k})}\sum_{\omega', \mathbf{k}'} a(\omega, \mathbf{k}; \omega', \mathbf{k}')\langle E^2\rangle^{(0)}_{\mathbf{k}'\omega'} \gg 1. \tag{7.32}$$

(We remind ourselves here that formula (7.31) involves the assumption of stationarity of the spectral distribution of fluctuations. This implies that (7.31) holds only for short time segments in comparison with the relaxation period.)

The expansion (7.26) must be treated with care under non-equilibrium conditions, for even the linear approximation leads to divergent results near the plasma instability threshold. We have shown in the previous sections that not only the quadratic correlation function for the field fluctuations turns to infinity (in the linear approximation) near the instability threshold, but also the unlimited growth of the higher-order correlation functions occurs.

7.3. Non-linear Frequency Shifts and Saturation of Fluctuations in a Non-equilibrium Plasma

Non-linear eigenfrequency shift

One of the most important manifestations of non-linear wave interactions in a plasma is that it causes a shift in the eigenfrequency. It is obvious that a proper treatment of the fluctuations in the vicinity of the eigenfrequency is impossible if this effect is ignored; it becomes particularly significant when the fluctuation intensities grow very large.

To consider the plasma eigenoscillations taking into account the non-linear effects, we start from (7.20), the right-hand side of which is set equal to zero.† The eigenfrequencies of the plasma oscillations are determined from the requirement that the real part

† The equation for the frequencies of interacting waves was derived by Pustovalov and Silin (1972a).

of this equation be equal to zero:

$$\mathrm{Re}\left\{\varepsilon(\omega, \mathbf{k})\langle E^2\rangle_{\mathbf{k}\omega} - \frac{2}{\varepsilon^*(\omega, \mathbf{k})}\sum_{\omega', \mathbf{k}'}|\varkappa^{(2)}(\omega-\omega', \mathbf{k}-\mathbf{k}'; \omega', \mathbf{k}')|^2\langle E^2\rangle_{\mathbf{k}'\omega'}\langle E^2\rangle_{\mathbf{k}-\mathbf{k}', \omega-\omega'}\right.$$
$$\left. - \sum_{\omega', \mathbf{k}'} a(\omega, \mathbf{k}; \omega', \mathbf{k}')\langle E^2\rangle_{\mathbf{k}'\omega'}\langle E^2\rangle_{\mathbf{k}\omega}\right\} = 0. \tag{7.33}$$

Let the solutions of the last equation be denoted by $\tilde{\omega}_{\mathbf{k}}$. In the general case the frequencies $\tilde{\omega}_{\mathbf{k}}$ differ from the eigenfrequencies $\omega_{\mathbf{k}}$ of the linear approximation, so that

$$\tilde{\omega}_{\mathbf{k}} = \omega_{\mathbf{k}} + \Delta\omega_{\mathbf{k}}. \tag{7.34}$$

The quantity $\Delta\omega_{\mathbf{k}}$ is the contribution of the non-linear wave interactions to the eigenfrequency $\omega_{\mathbf{k}}$.

Inasmuch as the non-linear shifts are assumed to be small in comparison with the frequencies, we obtain from (7.33) the following approximate expression for $\Delta\omega_{\mathbf{k}}$:

$$\Delta\omega_{\mathbf{k}} = \frac{1}{\varepsilon'_{\mathbf{k}}}\,\mathrm{Re}\sum_{\omega', \mathbf{k}'} a(\omega_{\mathbf{k}}, \mathbf{k}; \omega', \mathbf{k}')\langle E^2\rangle^{(1)}_{\mathbf{k}'\omega'}. \tag{7.35}$$

The spectral distribution in the linear approximation $\langle E^2\rangle^{(1)}_{\mathbf{k}'\omega'}$ is equal to

$$\langle E^2\rangle^{(1)}_{\mathbf{k}'\omega'} = \pi I_{\mathbf{k}'}\{\delta(\omega'-\omega_{\mathbf{k}})+\delta(\omega'+\omega_{\mathbf{k}})\},$$

so that we have for the eigenfrequency shift

$$\Delta\omega_{\mathbf{k}} = \frac{1}{2\varepsilon'_{\mathbf{k}}}\,\mathrm{Re}\sum_{\mathbf{k}'}\{a(\omega_{\mathbf{k}}, \mathbf{k}; \omega', \mathbf{k}')+a(\omega_{\mathbf{k}}, \mathbf{k}; -\omega', -\mathbf{k}')\}I_{\mathbf{k}'}. \tag{7.36}$$

If only certain modes are present in the plasma (e.g. the Langmuir waves), it may be shown within the context of the resonance conditions and the relevant dispersion relations that only $\mathbf{k}' \simeq \mathbf{k}$ contributes to the sum in (7.36). So we can write approximately:

$$\Delta\omega_{\mathbf{k}} \simeq \frac{k^3}{12\pi^2\varepsilon'_{\mathbf{k}}}\,\mathrm{Re}\{a(\omega_{\mathbf{k}}, k; \omega_{\mathbf{k}}, \mathbf{k})+a(\omega_{\mathbf{k}}, \mathbf{k}; -\omega_{\mathbf{k}}, -\mathbf{k})\}I_{\mathbf{k}}.$$

By virtue of (7.11) the last expression reduces to

$$\Delta\omega_{\mathbf{k}} \simeq \frac{k^3}{12\pi^2\varepsilon'_{\mathbf{k}}}$$
$$\times\mathrm{Re}\left\{2\frac{\varkappa^{(2)}(2\omega_{\mathbf{k}}, 2\mathbf{k}; -\omega_{\mathbf{k}}, -\mathbf{k})\,\varkappa^{(2)}(\omega_{\mathbf{k}}, \mathbf{k}; \omega_{\mathbf{k}}, \mathbf{k})}{\varepsilon(2\omega_{\mathbf{k}}, 2\mathbf{k})} - \bar{\varkappa}^{(3)}(-\omega_{\mathbf{k}}, -\mathbf{k}; \omega_{\mathbf{k}}, \mathbf{k}; \omega_{\mathbf{k}}, \mathbf{k})\right.$$
$$\left. +2\left(2\frac{\varkappa^{(2)}(\omega_{\mathbf{k}}, \mathbf{k}; 0, 0)\,\varkappa^{(2)}(\omega_{\mathbf{k}}, \mathbf{k}; -\omega_{\mathbf{k}}, \mathbf{k}-)}{\varepsilon(0, 0)} - \bar{\varkappa}^{(3)}(\omega_{\mathbf{k}}, \mathbf{k}; \omega_{\mathbf{k}}, \mathbf{k}; -\omega_{\mathbf{k}}, -\mathbf{k})\right)\right\}I_{\mathbf{k}}, \tag{7.37}$$

which coincides with the frequency shift we have obtained when considering the resonant four-wave interaction. Note that the correction and the frequency are of equal signs:

$$\Delta\omega = \frac{\omega}{|\omega|}|\Delta\omega|. \tag{7.38}$$

According to (7.37) the eigenfrequency shift is proportional to the wave intensity. We can write then

$$\Delta\omega_{\mathbf{k}} = \beta_{\mathbf{k}} I_{\mathbf{k}}, \tag{7.39}$$

where the coefficient $\beta_{\mathbf{k}}$ is governed by the dispersion of the relevant waves. As the shift grows with increasing oscillation intensity it must necessarily be involved in the consideration when the treatment concerns the fluctuations in a non-equilibrium plasma, in which case the fluctuation intensities may grow infinitely.

Saturation of the non-equilibrium fluctuation level

Let us find the spectral distribution of fluctuations in a stationary non-equilibrium plasma taking into account the non-linear eigenfrequency shifts (Sitenko, 1973). We rewrite the basic equation (7.10) for the quadratic correlation function $\langle E^2\rangle_{\mathbf{k}\omega}$ in the form

$$\Big\{|\varepsilon(\omega, \mathbf{k})|^2 - 2\,\mathrm{Re}\,\varepsilon^*(\omega, \mathbf{k}) \sum_{\omega', \mathbf{k}'} a(\omega, \mathbf{k}; \omega', \mathbf{k}')\langle E^2\rangle_{\mathbf{k}'\omega'}\Big\}\langle E^2\rangle_{\mathbf{k}\omega}$$
$$-2\sum_{\omega', \mathbf{k}'} |\varkappa^{(2)}(\omega-\omega', \mathbf{k}-\mathbf{k}'; \omega', \mathbf{k}')|^2 \langle E^2\rangle_{\mathbf{k}'\omega'}\langle E^2\rangle_{\mathbf{k}-\mathbf{k}', \omega-\omega'}$$
$$= \sum_{\omega', \mathbf{k}'} b(\omega, \mathbf{k}; \omega', \mathbf{k}')\langle E^2\rangle_{\mathbf{k}'\omega'} + q_{\mathbf{k}\omega}. \tag{7.40}$$

In the narrow band around the plasma eigenfrequency and when $|\mathrm{Im}\,\varepsilon(\omega, \mathbf{k})| \ll |\mathrm{Re}\,\varepsilon(\omega, \mathbf{k})|$ the expression within the curly brackets on the left-hand side of (7.40) may be approximately presented as the square of the dielectric permittivity modulus with the shifted frequency of the real part. Suppose the decay conditions are not satisfied and retain only the leading term on the right-hand side of (7.40). Then we may rewrite this equation in the form:

$$\{[\mathrm{Re}\,\varepsilon(\omega-\Delta\omega_{\mathbf{k}}, \mathbf{k})]^2 + [\mathrm{Im}\,\varepsilon(\omega, \mathbf{k})]^2\}\langle E^2\rangle_{\mathbf{k}\omega} = \frac{16\pi^2}{k^2}\langle \varrho^2\rangle^0_{\mathbf{k}\omega} \tag{7.41}$$

and the spectral distribution of the electric field fluctuations near the eigenfrequencies is as follows:

$$\langle E^2\rangle_{\mathbf{k}\omega} = \pi I_{\mathbf{k}}\{\delta(\omega-\tilde{\omega}_{\mathbf{k}}) + \delta(\omega+\tilde{\omega}_{\mathbf{k}})\}, \tag{7.42}$$

where $\tilde{\omega}_{\mathbf{k}}$ is the eigenfrequency taking into account the non-linear wave interactions and

$$I_{\mathbf{k}} = \frac{16\pi^2}{k^2\varepsilon'_{\mathbf{k}}}\,\frac{\langle \varrho^2\rangle^0_{\mathbf{k}\tilde{\omega}_{\mathbf{k}}}}{|\mathrm{Im}\,\varepsilon(\tilde{\omega}_{\mathbf{k}}, \mathbf{k})|}. \tag{7.43}$$

As the eigenfrequency $\tilde{\omega}_{\mathbf{k}}$ calculated taking into account the non-linear wave interactions is a function of $I_{\mathbf{k}}$, the relation (7.43) may serve as an equation for the fluctuation intensity $I_{\mathbf{k}}$ in the region of the plasma instability. A more suitable form of (7.43) is the following:

$$I_{\mathbf{k}}\,|\mathrm{Im}\,\varepsilon(\tilde{\omega}_{\mathbf{k}}(I_{\mathbf{k}}), \mathbf{k})| = I_{\mathbf{k}}^0\,|\mathrm{Im}\,\varepsilon^0(\omega_{\mathbf{k}}, \mathbf{k})|, \qquad (7.44)$$

where $I_{\mathbf{k}}^0$ is the fluctuation intensity far from instability and $\varepsilon^0(\omega, \mathbf{k})$ is the dielectric permittivity of the stable plasma. $I_{\mathbf{k}}^0$ is for an equilibrium plasma governed by the temperature according to (6.127).

The existence of a solution of (7.44) indicates that the wave interactions in an unstable plasma favour a stationary turbulent state to be established; the solution itself determines the stationary level of turbulent fluctuations in the plasma.† It is peculiar for the turbulent state that the spectral distribution of fluctuations is governed by the non-linear wave interactions and weakly depends on the plasma temperature.

Turbulent fluctuations in a beam–plasma system

Expression (7.42) describes the general form of the fluctuation spectral distribution in a non-equilibrium plasma taking into account the non-linear wave interactions. Consider, for example, an electron plasma penetrated by a compensated low-density particle beam (Sitenko, 1973). Denote the beam velocity by u. If the beam velocity is sufficiently low, then we may neglect the influence of the beam on the wave dispersion in the plasma; the thermal particle motion may be also ignored in the high-frequency range. However, both effects contribute to the fluctuation intensity (7.43). Thus, for the Langmuir fluctuations we have

$$I_{\mathbf{k}} = \frac{\tilde{u}(\mathbf{k})+\beta_{\mathbf{k}}' I_{\mathbf{k}}}{|\tilde{u}(\mathbf{k})+\beta_{\mathbf{k}}' I_{\mathbf{k}} - \tilde{u}_{||}|} I_0, \quad \beta_{\mathbf{k}}' \equiv \frac{\beta_{\mathbf{k}}}{k}, \qquad (7.45)$$

where the following notation is used: $I_0 = 4\pi T$ is the equilibrium fluctuation intensity, $\tilde{u}(k) = \dfrac{\omega_{\mathbf{k}}}{k}$, and $u_{||} = \dfrac{(\mathbf{k}\cdot\mathbf{u})}{k}$. The coefficient $\beta_{\mathbf{k}}'$, which determines the non-linear frequency shift, is for the Langmuir oscillations approximately equal to

$$\beta_{\mathbf{k}}' \simeq \frac{1}{4\pi}\frac{e^2}{m^2}\frac{k^6 s^2}{\Omega^5}. \qquad (7.46)$$

In neglect of the non-linear wave interactions ($\beta_{\mathbf{k}}' \to 0$) we obtain from (7.45)

$$I_{\mathbf{k}} = \frac{\tilde{u}(\mathbf{k})}{|\tilde{u}(\mathbf{k})-u_{||}|} I_0. \qquad (7.47)$$

The quantity $\tilde{u}(\mathbf{k})$ may be regarded as a critical velocity. The beam with $u \to u(\mathbf{k})$ causes an infinite growth of the fluctuations in the linear approximation and drives the

† An explanation of the saturation of the collective fluctuation level, involving the assumption of a strongly turbulent plasma state which is characterized by a special ternary correlation function, was proposed by Ichimaru (1970). The stabilization of plasma instabilities by the non-linear shift mechanism was studied by Pustovalov and Silin (1972b), and Bychenkov, Pustovalov, Silin, and Tikhonchuk (1976).

plasma unstable. We shall show now that the non-linear wave interactions lead to the saturation of the fluctuation level for any beam velocities, including those equal to or greater than the critical velocity $\tilde{u}(\mathbf{k})$. Indeed, if $\tilde{u}(\mathbf{k})+\beta'_{\mathbf{k}}I_{\mathbf{k}} > u_{||}$, then, by virtue of (7.45),

$$I_{\mathbf{k}} = \frac{1}{2\beta'_{\mathbf{k}}}\{-(\tilde{u}(\mathbf{k})-\beta'_{\mathbf{k}}I_0-u_{||})\pm\sqrt{(\tilde{u}(\mathbf{k})-\beta'_{\mathbf{k}}I_0-u_{||})^2+4\beta'_{\mathbf{k}}I_0\tilde{u}(\mathbf{k})}\}. \tag{7.48}$$

As $I_{\mathbf{k}}$ is positive definite, we must retain only the plus sign in (7.48). (It may be immediately verified that the inequality $\tilde{u}(\mathbf{k})+\beta'_{\mathbf{k}}I_{\mathbf{k}} > u_{||}$ is always satisfied.)

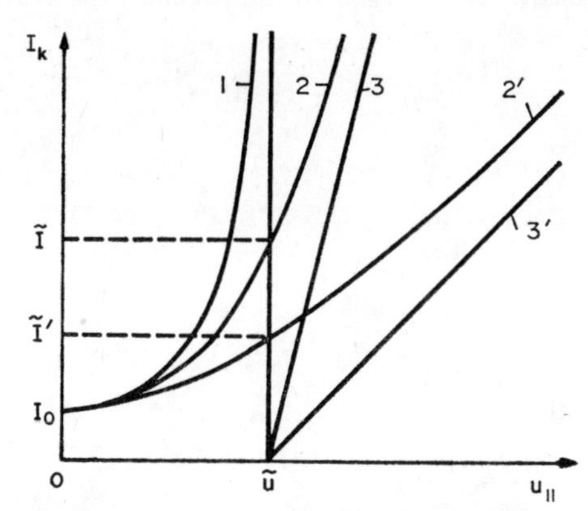

FIG. 6. The intensity of Langmuir fluctuations $I_{\mathbf{k}}$ as a function of the velocity of the ordered particle motion $U_{||}$: 1, (7.47); 2, 2′, (7.48); 3, 3′(7,50); 2, 3, $\beta_{\mathbf{k}}I_0/\tilde{u} = 0.05$; 2′, 3′, $\beta_{\mathbf{k}}I_0/\tilde{u} = 0.2$.

In Fig. 6 we see the Langmuir fluctuation intensity $I_{\mathbf{k}}$ as a function of the beam velocity along the wave propagation $u_{||}$. If the latter is sufficiently small, then the fluctuation intensity is governed by the plasma temperature and is equal to the thermal value I_0 irrespective of the beam temperature, as the beam density was assumed to be small. The fluctuation intensity grows when the beam velocity increases, however, by virtue of the non-linear wave interactions this growth is weaker than was predicted by the linear theory. When the beam velocity reaches the critical value $u = \tilde{u}(\mathbf{k})$ the fluctuation intensity (7.48) becomes modified from the linear result (7.47) and finite:

$$I_{\mathbf{k}} = \frac{1}{2\beta'_{\mathbf{k}}}\{\beta'_{\mathbf{k}}I_0+\sqrt{\beta'^2_{\mathbf{k}}I_0^2+4\beta'_{\mathbf{k}}I_0\tilde{u}(k)}\}. \tag{7.49}$$

The fluctuation level considerably exceeds the thermal one if $4\tilde{u}(\mathbf{k}) \gg \beta'_{\mathbf{k}}I_0$. If the beam velocities are so large that $u_{||}-\tilde{u}(\mathbf{k}) \gg \beta'_{\mathbf{k}}I_0$, then the fluctuation intensity $I_{\mathbf{k}}$ tends to the asymptotic value

$$I_{\mathbf{k}} = \frac{u_{||}-\tilde{u}(\mathbf{k})}{\beta'_{\mathbf{k}}}, \tag{7.50}$$

which does not depend on the thermal level and grows proportionally to the beam velo-ty. Therefore, the fluctuations become characterized by a stationary level even at beam velocities greater than the critical value $\tilde{u}(\mathbf{k})$ after the non-linear wave interactions have been taken into account. The fluctuation level turns out to be temperature-independent when the beam velocity is very large. The values $u_{||} > \tilde{u}(\mathbf{k})$ correspond to turbulent plasma states.

Non-linear stabilization of the hydrodynamic beam–plasma instability

Non-linear shifts of the eigenfrequencies in a plasma with a high level of fluctuation oscillations favour the stabilization of hydrodynamic beam–plasma instabilities also. Let us consider a cold plasma penetrated by a cold electron beam,

$$\varepsilon(\omega, \mathbf{k}) = 1-\frac{\Omega^2}{\omega^2}-\frac{\Omega^2}{[\omega-(\mathbf{k}\cdot\mathbf{u})]^2}\eta^3, \quad \eta^3 \equiv \frac{n_0'}{n_0}.$$

If $(\mathbf{k}\cdot\mathbf{u}) \gg \Omega$, then the dispersion equation $\varepsilon(\omega, \mathbf{k}) = 0$ has four real roots. For example, in the case of a low-density beam ($\eta^3 \ll 1$) two of four roots give rise to Langmuir oscillations, while two others are equal to

$$\omega_{\mathbf{k}} = (\mathbf{k}\cdot\mathbf{u})\pm\eta^{3/2}\frac{\Omega}{\sqrt{1-(\Omega/(\mathbf{k}\cdot\mathbf{u}))^2}}.$$

The hydrodynamic instability threshold is determined by the condition for the multiplicity of the real root of the dispersion equation, i.e.

$$\varepsilon(\omega, \mathbf{k}) = 0 \quad \text{and} \quad \frac{\partial}{\partial\omega}\varepsilon(\omega, \mathbf{k}) = 0.$$

Within the context of the explicit expression for the dielectric permittivity we find the threshold frequency $\omega_k^{\text{thr}} = \Omega\sqrt{1+\eta}$ and the condition for the range of permissible k (the hydrodynamic stability domain):

$$(\mathbf{k}\cdot\mathbf{u}) \geqslant \omega_{\mathbf{k}}^{\text{thr}}(1+\eta).$$

The fluctuation intensity near the instability threshold may be written as

$$I_{\mathbf{k}} = \frac{2I_0}{\omega_{\mathbf{k}}\left|\dfrac{\partial\varepsilon(\omega_{\mathbf{k}}, k)}{\partial\omega_{\mathbf{k}}}\right|} \simeq \frac{1}{3}\frac{\eta\sqrt{1+\eta}\Omega}{\dfrac{(\mathbf{k}\cdot\mathbf{u})}{1+\eta}-\omega_{\mathbf{k}}}I_0. \tag{7.51}$$

The fluctuation oscillation intensity becomes infinite at the threshold. To take the non-linear eigenfrequency shifts into account we substitute $\omega_{\mathbf{k}} \to \tilde{\omega}_{\mathbf{k}} = \omega_{\mathbf{k}}+\beta_{\mathbf{k}}I_{\mathbf{k}}$ in (7.50). We note here that the stabilization is impossible as $\beta_{\mathbf{k}} > 0$ and occurs only if the frequency shift is negative $\beta_{\mathbf{k}} < 0$. For the stationary fluctuation level we find

$$I_{\mathbf{k}} = \frac{1}{2|\beta_{\mathbf{k}}|}\left\{\sqrt{\left(\frac{(\mathbf{k}\cdot\mathbf{u})}{1+\eta}-\omega_{\mathbf{k}}\right)^2+\frac{4}{3}|\beta_{\mathbf{k}}|\Omega\eta\sqrt{1+\eta}I_0}-\left(\frac{(\mathbf{k}\cdot\mathbf{u})}{1+\eta}-\omega_{\mathbf{k}}\right)\right\}$$

and at the instability threshold

$$I_{\mathbf{k}} = \sqrt{\frac{\eta\sqrt{1+\eta}}{3|\beta_{\mathbf{k}}|}\Omega I_0}.$$

The non-linear shift of the Langmuir eigenoscillations is positive if the compensating ions have infinite masses; when computed, taking into account the ion motion, it turns out to be negative if $a^2k^2 < m/M$. Thus, within the context of certain conditions the non-linear frequency shifts can stabilize hydrodynamic instabilities in a beam–plasma system.

Turbulent fluctuations in a plasma with an anisotropic velocity distribution

Let us consider the fluctuations in a magneto-active plasma with an anisotropic velocity distribution (Sitenko and Zasenko, 1978a). According to the linear approximation (6.158) the intensities of the field fluctuations in such a plasma grow infinitely as the transverse temperature $T_\perp$ reaches the critical value $\frac{\omega_B}{\omega_B - \omega_{\mathbf{k}}} T_{||}$ where $\omega_{\mathbf{k}}$ is the eigenfrequency (6.155). Taking into account the non-linear wave interactions we find the eigenfrequency shift

$$\Delta\omega_{\mathbf{k}} = \beta_{\mathbf{k}} I_{\mathbf{k}}, \quad \beta_{\mathbf{k}} \simeq \frac{1}{2\pi^2} \frac{e^2}{m^2} \frac{T_{||}}{m} \frac{k^7}{\Omega^5} \cos\vartheta \quad (\omega_B \gg \Omega).$$

Within the context of (6.158) the equation that determines the stationary level of non-equilibrium fluctuations may be written in the form

$$I_{\mathbf{k}} = \frac{\omega_{\mathbf{k}} + \beta_{\mathbf{k}} I_{\mathbf{k}}}{\left|\omega_{\mathbf{k}} + \beta_{\mathbf{k}} I_{\mathbf{k}} - \omega_B\left(1 - \frac{T_{||}}{T_\perp}\right)\right|} I_0, \tag{7.52}$$

where

$$I_0 \equiv 4\pi T_{||} \frac{a_\perp^2 k^2}{\left|1 + a_\perp^2 k^2 - I_0(\beta_\perp) e^{-\beta_\perp}\right|}.$$

The stationary level of non-equilibrium fluctuations is associated with the positive root of (7.52):

$$I_{\mathbf{k}} = \frac{1}{2\beta_{\mathbf{k}}} \left\{\beta_{\mathbf{k}} I_0 + \omega_B\left(1 - \frac{T_{||}}{T_\perp}\right) - \omega_{\mathbf{k}} + \sqrt{\left(\beta_{\mathbf{k}} I_0 + \omega_B\left(1 - \frac{T_{||}}{T_\perp}\right) - \omega_{\mathbf{k}}\right)^2 + 4\beta_{\mathbf{k}} I_0 \omega_{\mathbf{k}}}\right\}. \tag{7.53}$$

In Fig. 7 we see the fluctuation intensity in a magneto-active plasma with an anisotropic velocity distribution as a function of the temperature ratio $T_\perp/T_{||}$. The dashed line corresponds to the linear approximation; the solid line shows the results which were obtained taking into account the non-linear frequency shift. The instability threshold is associated with $\frac{T_\perp}{T_{||}} = \frac{\omega_B}{\omega_B - \omega_{\mathbf{k}}}$. The large values of this ratio correspond to turbulent plasma states with a fluctuation level (7.53). The fluctuation intensity at the instability threshold is equal to

$$I_{\mathbf{k}} = \frac{1}{2\beta_{\mathbf{k}}} \{\beta_{\mathbf{k}} I_0 + \sqrt{(4\omega_{\mathbf{k}} + \beta_{\mathbf{k}} I_0)\beta_{\mathbf{k}} I_0}\}. \tag{7.54}$$

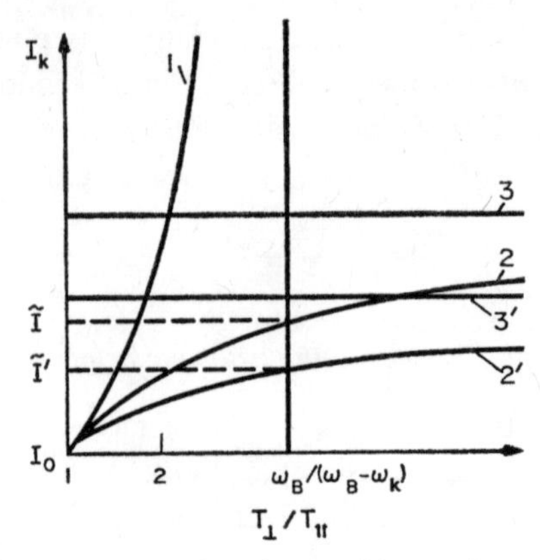

FIG. 7. Fluctuation intensity in a magneto-active plasma with an anisotropic velocity distribution $I_{\mathbf{k}}$ as a function of the temperature ratio $T_\perp/T_{||}$: 1, (7.52) with $\beta_{\mathbf{k}} = 0$; 2, 2′, (7.53); 3, 3′, (7.55); 2, 3, $\beta_{\mathbf{k}} I_0/\omega_B = 0.2$; 2′, 3′, $\beta_{\mathbf{k}} I_0/\omega_B = 0.4$.

With increasing temperature ratio the fluctuation intensity $I_{\mathbf{k}}$ asymptotically tends to the limiting value

$$I_{\mathbf{k}} = \frac{1}{2\beta_{\mathbf{k}}}\{\beta_{\mathbf{k}} I_0 + \omega_B - \omega_{\mathbf{k}} + \sqrt{(\beta_{\mathbf{k}} I_0 + \omega_B - \omega_{\mathbf{k}})^2 + 4\beta_{\mathbf{k}} I_0 \omega_{\mathbf{k}}}\}. \tag{7.55}$$

The limiting value (7.55) depends on the longitudinal temperature $T_{||}$.

7.4. Spectral Correlation Functions for Electromagnetic Fluctuations

The non-linear equation for the fluctuations of the electromagnetic field

Let us consider electromagnetic fluctuations in a stationary plasma without assuming the fluctuation field to be potential (Sitenko and Kocherga, 1977a). Iterating the non-linear equation (6.100) and substituting its solution into (6.98) and (6.99) we obtain the basic non-linear equation for the fluctuation electric field in a plasma:

$$\begin{aligned} \Lambda_{ij}(k) E_j(k) + \sum_{1+2=k} \varkappa^{(2)}_{ijk}(1, 2) E_j(1) E_k(2) + \sum_{1+2+3=k} \varkappa^{(3)}_{ijkl}(1, 2, 3) E_j(1) E_k(2) E_l(3) \\ + \ldots + \sum_{k'} \delta\varepsilon_{ij}(k, k') E_j(k') + \sum_{1+2=k} \sum_{k'} \delta\varkappa^{(2)}_{ijk}(1, 2; k') E_j(1) E_k(k') + \ldots = \\ = -\frac{4\pi i}{\omega} j^0_i(k). \end{aligned} \tag{7.56}$$

Here k denotes the combination of the quantities ω and $\mathbf{k}$, i.e. $k \equiv \omega, \mathbf{k}$; $1 \equiv \omega_1, \mathbf{k}_1$ and so on; the summation over the subscripts that enter twice is implied. We introduce the notation

$$\beta_{ij}(k) \equiv \left(1 - \frac{kv}{\omega}\right)\delta_{ij} + \frac{v_i k_j}{\omega} \tag{7.57}$$

and rewrite the dielectric permittivity tensor and the non-linear tensor susceptibilities as follows:

$$\varepsilon_{ij}(k) = \delta_{ij}+\sum \frac{\Omega^2}{\omega}\int d\mathbf{v}\frac{v_i}{\omega-(\mathbf{k}\cdot\mathbf{v})+i0}\beta_{jj'}(k)\frac{\partial}{\partial v_{j'}}f_0(\mathbf{v}), \tag{7.58}$$

$$\varkappa^{(2)}_{ijk}(1, 2) = \sum \frac{(-i)}{2}\frac{e}{m}\frac{\Omega^2}{\omega_1+\omega_2}\int d\mathbf{v}\frac{v_i}{\omega_1+\omega_2-(\{\mathbf{k}_1+\mathbf{k}_2\}\cdot\mathbf{v})+0i}$$
$$\times\left\{\beta_{jj'}(1)\frac{\partial}{\partial v_j}\frac{1}{\omega_2-(\mathbf{k}_2\cdot\mathbf{v})+i0}\beta_{kk'}(2)\frac{\partial}{\partial v_{k'}}\right.$$
$$\left.+\beta_{kk'}(2)\frac{\partial}{\partial v_{k'}}\frac{1}{\omega_1-(\mathbf{k}_1\cdot\mathbf{v})+i0}\beta_{jj'}(1)\frac{\partial}{\partial v_{j'}}\right\}f_0(\mathbf{v}), \tag{7.59}$$

$$\varkappa^{(3)}_{ijkl}(1, 2, 3) = \tfrac{1}{3}\{\bar{\varkappa}^{(3)}_{ijkl}(1, 2, 3)+\bar{\varkappa}^{(3)}_{ikjl}(2, 1, 3)+\bar{\varkappa}^{(3)}_{ilkj}(3, 2, 1)\}, \tag{7.60}$$

$$\bar{\varkappa}^{(3)}_{ijkl}(1, 2, 3) = \sum \frac{(-i)^2}{2}\left(\frac{e}{m}\right)^2\frac{\Omega^3}{\omega_1+\omega_2+\omega_3}\int d\mathbf{v}\frac{v_i}{\omega_1+\omega_2+\omega_3-(\{\mathbf{k}_1+\mathbf{k}_2+\mathbf{k}_3\}\cdot\mathbf{v})+i0}$$
$$\times\beta_{jj'}(1)\frac{\partial}{\partial v_{j'}}\frac{1}{\omega_2+\omega_3-(\{\mathbf{k}_2+\mathbf{k}_3\}\cdot\mathbf{v})+i0}\left\{\beta_{kk'}(2)\frac{\partial}{\partial v_{k'}}\frac{1}{\omega_3-(\mathbf{k}_3\cdot\mathbf{v})+i0}\right.$$
$$\left.\times\beta_{ll'}(3)\frac{\partial}{\partial v_{l'}}+\beta_{ll'}(3)\frac{\partial}{\partial v_{l'}}\frac{1}{\omega_2-(\mathbf{k}_2\cdot\mathbf{v})+i0}\beta_{kk'}(2)\frac{\partial}{\partial v_{k'}}\right\}f_0(\mathbf{v}). \tag{7.61}$$

The non-linear tensor susceptibilities $\varkappa^{(2)}_{ijk}(1, 2)$ and $\varkappa^{(3)}_{ijkl}(1, 2, 3)$ are symmetric with respect to interchanges of the arguments and of the corresponding subscripts.†

We shall restrict ourselves by retaining in (7.56) the linear, quadratic, and cubic terms in the fluctuation field intensity. Equation (7.56) relates directly the fluctuation field intensity $E(k)$ to the density of the fluctuation current $j^0(k)$ that is due to the stochastic motion of individual plasma particles. Another manifestation of the stochastic particle motion is the inhomogeneity of the dielectric plasma properties. The quantities $\delta\varepsilon_{ij}(k, k')$ and $\delta\varkappa^{(2)}_{ijk}(1, 2, k')$ are the variations of the dielectric permittivity and of the non-linear susceptibilities due to the random fluctuations of the distribution function $\delta f^0_{\mathbf{k}\omega}(\mathbf{v})$. We present these variations in the form:

$$\delta\varepsilon_{ij}(k, k') \equiv \Delta_{ij}(k, k')\,\delta f^0(k-k'), \tag{7.62}$$
$$\delta\varkappa^{(2)}_{ijk}(1, 2; k') \equiv \Delta^{(2)}_{ijk}(1, 2; k')\,\delta f^0(2-k'), \tag{7.63}$$

where the operators $\Delta_{ij}(k, k')$ and $\Delta^{(2)}_{ijk}(1, 2, k')$ are defined as

$$\Delta_{ij}(k, k')\,\delta f^0(k-k') \equiv \sum \frac{\Omega^2}{\omega}\int d\mathbf{v}\frac{v_i}{\omega-(\mathbf{k}\cdot\mathbf{v})+i0}\beta_{jj'}(k')\frac{\partial}{\partial v_{j'}}\delta f^0_{\mathbf{k}-\mathbf{k}',\,\omega-\omega'}(v), \tag{7.64}$$

$$\Delta^{(2)}_{ijk}(1, 2; k')\,\delta f^0(2-k') \equiv \sum(-i)\frac{e}{m}\frac{\Omega^2}{\omega}\int d\mathbf{v}\frac{v_i}{\omega-\mathbf{k}\cdot\mathbf{v}+i0}\beta_{jj'}(1)\frac{\partial}{\partial v_{j'}}$$
$$\times\frac{1}{\omega_2-\mathbf{k}_2\cdot\mathbf{v}+i0}\beta_{kk'}(k')\frac{\partial}{\partial v_{k'}}\delta f^0_{\mathbf{k}_2-\mathbf{k}',\,\omega_2-\omega'}(\mathbf{v}). \tag{7.65}$$

† Equation (7.56) holds for a magneto-active plasma also. For dielectric permittivity tensor and the non-linear tensor susceptibilities of magneto-active plasmas, see Chapter 3.

The variations of the non-linear susceptibility $\varkappa^{(3)}_{ijkl}(1, 2, 3)$ in (7.56) may be neglected in the desired approximation.

It is not difficult to derive a set of equations for the correlation functions of the fluctuation fields starting from the non-linear equation (7.56).

Set of equations for the sequence of correlation functions

To describe the electromagnetic field fluctuations we introduce a set of the correlation functions $\langle E_iE_j\rangle_k$, $\langle E_iE_jE_k\rangle_{k,k'}$ and $\langle E_iE_jE_kE_l\rangle_{k,k',k''}$. We define them to be related to the mean values of the binary, ternary, and quaternary products of the fluctuating field intensity amplitudes according to:

$$\langle E_i(k)\,E_j(k')\rangle \equiv (2\pi)^4\,\delta(k+k')\,\langle E_iE_j\rangle_k, \tag{7.66}$$

$$\langle E_i(k)\,E_j(k')\,E_k(k'')\rangle \equiv (2\pi)^4\,\delta(k+k'+k'')\,\langle E_iE_jE_k\rangle_{k,k+k'}, \tag{7.67}$$

$$\begin{aligned}\langle E_i(k)\,E_j(k')\,E_k(k'')\,E_l(k''')\rangle \equiv (2\pi)^4\,\delta(k+k'+k''+k''')\,\{&(2\pi)^4\,\delta(k+k')\,\langle E_iE_j\rangle_k\,\langle E_kE_l\rangle_{k''}\\ &+(2\pi)^4\,\delta(k+k'')\,\langle E_iE_k\rangle_k\,\langle E_jE_l\rangle_{k'}\\ &+(2\pi)^4\,\delta(k+k''')\,\langle E_iE_l\rangle_k\,\langle E_jE_k\rangle_{k'}\\ &+\langle E_iE_jE_kE_l\rangle_{k,k+k',k+k'+k''}\}.\end{aligned} \tag{7.68}$$

We transfer in (7.56) the terms, due to the fluctuation variations of the dielectric permittivity and of the non-linear susceptibility, into the right-hand side. After having multiplied successively the left- and the right-hand sides of the equation thus obtained by themselves and carried out the averaging procedure according to (6.80), we obtain a set of non-homogeneous integral equations which determine uniquely a sequence of correlation functions for the fluctuating electromagnetic field in a plasma (Sitenko and Kocherga, 1977a).

We restrict ourselves by retaining all terms up to those proportional to the field intensity to the fourth power and carry out averaging in a manner similar to the analysis of the potential case. When calculating the correlation functions in the right-hand sides, which describe the correlations between the current density fluctuation $\mathbf{j}^0(k)$ and the fluctuating electric field $\mathbf{E}(k)$, we may substitute the approximate solution of (7.56) for the field intensity:

$$E_i(k) \to -\frac{4\pi i}{\omega}\,\Lambda^{-1}_{ij}(k)\,j^0_j(k). \tag{7.69}$$

After some straightforward algebra we obtain the following equation for the binary correlation function $\langle E_iE_j\rangle_k$:

$$\begin{aligned}\Lambda^*_{i'j'}(k)\Big\{&\Lambda_{ij}(k)\,\langle E_iE_{j'}\rangle_k - 2\Lambda^{*-1}_{j'i''}(k)\sum \varkappa^{(2)}_{ijk}(k-1, 1)\,\varkappa^{(2)*}_{i''j''k'}(k-1, 1)\,\langle E_kE_{k'}\rangle_1\,\langle E_jE_{j''}\rangle_{k-1}\\ &-\sum_1 a_{ijkk'}(k, 1)\,\langle E_kE_{k'}\rangle_1\,\langle E_jE_{j'}\rangle_k\Big\} - \Lambda_{ij'}(k)\sum_1 a^*_{i'jkk'}(k, 1)\,\langle E_kE_{k'}\rangle^*_1\,\langle E_jE_{j'}\rangle^*_k\\ = &\sum_1 b_{ii'kk'}(k, 1)\,\langle E_kE_{k'}\rangle_1 + q_{ii'}(k),\end{aligned} \tag{7.70}$$

where we have put:

$$a_{ijkk'}(k,1) \equiv 2\{2\varkappa^{(2)}_{ij'k}(k-1,1)\,\Lambda^{-1}_{j'i'}(k-1)\,\varkappa^{(2)}_{i'jk'}(k,-1)-\bar{\varkappa}^{(3)}_{ikjk'}(1,k,-1)\}$$
$$+2\varkappa^{(2)}_{ij'k}(k,0)\,\Lambda^{-1}_{j'i'}(0)\,\varkappa^{(2)}_{i'jk'}(1,-1)-\bar{\varkappa}^{(3)}_{ikjk'}(k,1,-1); \tag{7.71}$$

$$b_{ii'kk'}(k,1) \equiv \Lambda_{ik}(k,1)\,\Lambda^{*}_{i'k'}(k,1)\,\langle\delta f\,\delta f\rangle^{0}_{k-1}$$
$$+\frac{8\pi i}{\omega-\omega_1}\{\Lambda_{ik}(k,1)\,\varkappa^{(2)*}_{i'j'k'}(k-1,1)\,\Lambda^{*-1}_{j'l'}(k-1)\,\langle\delta f j_{l'}\rangle^{0}_{k-1}$$
$$-\varkappa^{(2)}_{ijk}(k-1,1)\,\Lambda^{-1}_{jl}(k-1)\,\Lambda^{*}_{i'k'}(k,1)\,\langle\delta f j_l\rangle^{0*}_{k-1}\}; \tag{7.72}$$

$$q_{ii'}(k) \equiv q^{(1)}_{ii'}(k)+q^{(2)}_{ii'}(k)+q^{(3)}_{ii'}(k), \tag{7.73}$$

$$q^{(1)}_{ii'}(k) \equiv \frac{16\pi^2}{\omega^2}\langle j_i, j_{i'}\rangle^{0}_{k}, \tag{7.74}$$

$$q^{(2)}_{ii'}(k) \equiv \frac{16\pi^2}{\omega^2}\sum_{1}\frac{1}{\omega_1}\Bigg\{\frac{4\pi i}{\omega-\omega_1}\big(\varkappa^{(2)}_{ijk}(k-1,1)\,\Lambda^{-1}_{jl}(k-1)\,\Lambda^{-1}_{km}(1)\,\langle j_m j_l j_{i'}\rangle^{0}_{1,k}$$
$$-\varkappa^{(2)*}_{i'j'k'}(k-1,1)\,\Lambda^{*-1}_{j'l'}(k-1)\,\Lambda^{*-1}_{k'\omega'}(1)\,\langle j_{m'} j_{l'} j_i\rangle^{0*}_{1,k}\big)$$
$$-\Lambda_{ik}(k,1)\,\Lambda^{-1}_{kj}(1)\,\langle\delta f j_j j_{i'}\rangle^{0}_{k-1,k}-\Lambda^{*}_{i'k'}(k,1)\,\Lambda^{*-1}_{k'j'}(1)\,\langle\delta f j_{j'} j_i\rangle^{0*}_{k-1,k}\Bigg\}, \tag{7.75}$$

$$q^{(3)}_{ii'}(k) \equiv \frac{16\pi^2}{\omega^2}\left(P_{ij}(k)\delta_{i'j'}+\delta_{ij}P^{*}_{i'j'}(k)\right)\langle j_j j_{j'}\rangle^{0}_{k}+R_{ii'}(k)+R^{*}_{i'i}(k), \tag{7.76}$$

$$P_{ij}(k) \equiv 4\pi i\sum_{1}\frac{1}{\omega_1}\{\Lambda^{(2)}_{iln}(1,k-1;k)-2\varkappa^{(2)}_{ikl}(k-1,1)\,\Lambda^{-1}_{km}(k-1)\,\Lambda_{mn}(k-1,k)$$
$$-2\Lambda_{ik}(k,k+1)\,\Lambda^{-1}_{km}(k+1)\,\varkappa^{(2)}_{mln}(1,k)\}\,\Lambda^{-1}_{lp}(1)\,\Lambda^{-1}_{nj}(k)\,\langle\delta f j_p\rangle^{0}_{-1}$$
$$+\sum_{1}\Lambda_{ik}(k,1)\,\Lambda^{-1}_{kl}(1)\,\Lambda_{lm}(1,k)\,\Lambda^{-1}_{mj}(k)\,\langle\delta f\,\delta f\rangle^{0}_{k-1}; \tag{7.77}$$

$$R_{ii'}(k) \equiv \frac{4\pi i}{\omega}\sum_{1}\{2\varkappa^{(2)}_{ijk}(k-1,1)\,\Lambda^{-1}_{jl}(k-1)\,\Lambda_{lk'}(k-1,1)-\Lambda^{(2)}_{ikk'}(1,k-1;-1)\,\langle E_k E_{k'}\rangle_1\langle\delta f j_{i'}\rangle^{0}_{k}$$
$$-\frac{16\pi^2}{\omega}\sum_{1}\frac{1}{\omega_1}\Lambda_{ij}(k,k-1)\,\Lambda^{-1}_{jk}(k-1)\,\Lambda_{kl}(k-1,-1)\,\Lambda^{*-1}_{lm}(1)\,\langle\delta f j_m\rangle^{0}_{1}\langle\delta f j_{i'}\rangle^{0}_{k}$$
$$+8\pi^2\sum_{1}\frac{1}{(\omega-\omega_1)\omega_1}\Lambda_{ij}(k,1)\,\Lambda^{-1}_{jk}(1)\,\Lambda^{*}_{i'l}(k,k-1)\,\Lambda^{*}_{lm}(k-1)\,\langle\delta f j_m\rangle^{0}_{k-1}\langle\delta f j_k\rangle^{0}_{1}$$
$$-8\pi^2\sum_{1,2}\frac{1}{\omega_1\omega_2}\Lambda_{ik}(k,1)\,\Lambda_{i'k}(-k,2)\,\Lambda^{-1}_{jl}(1)\,\Lambda^{-1}_{km}(2)\,\langle\delta f\,\delta f j_l j_m\rangle^{0}_{k,1,2}$$
$$+\frac{16\pi^2}{\omega}\sum_{1,2}\frac{1}{\omega_2}\Lambda_{ij}(k,1)\,\Lambda^{-1}_{jk}(1)\,\Lambda_{kl}(1,2)\,\Lambda^{-1}_{lm}(2)\,\langle\delta f\,\delta f j_m j_{i'}\rangle^{0}_{k,1,2}$$
$$+64\pi^3 i\sum_{1,2}\frac{1}{(\omega-\omega_1)\omega_1\omega_2}\varkappa^{(2)}_{ijk}(1,k-1)\,\Lambda_{i'l}(-k,2)$$
$$\times\Lambda^{-1}_{jm}(1)\,\Lambda^{-1}_{kp}(k-1)\,\Lambda^{-1}_{ln}(2)\,\langle\delta f j_m j_n j_p\rangle^{0}_{k,1,2}$$
$$-\frac{64\pi^3 i}{\omega}\sum_{1,2}\frac{1}{(\omega-\omega_1)\omega_1}\{\Lambda_{ik}(k,k-1+2)\,\Lambda^{-1}_{kl}(k-1+2)\,\varkappa^{(2)}_{lmj}(2,k-1)$$

$$+2\varkappa^{(2)}_{ijk}(k-1, 1)\,\Lambda^{-1}_{kl}(1)\,\Delta_{lm}(1, 2)-\Delta^{(2)}_{ijm}(k-1, 1; 2)\}$$
$$\times\Lambda^{-1}_{mn}(2)\,\Lambda^{-1}_{jp}(k-1)\,\langle\delta f j_n j_p j_{i'}\rangle^0_{k,1,2}$$
$$+128\pi^4\sum_{1,2}\frac{1}{(\omega-\omega_1)(\omega+\omega_2)\omega_1\omega_2}\varkappa^{(2)}_{ijk}(1, k-1)\,\varkappa^{(2)}_{i'lm}(2, -k-2)$$
$$\times\Lambda^{-1}_{jn}(1)\,\Lambda^{-1}_{kp}(k-1)\,\Lambda^{-1}_{mr}(2)\,\Lambda^{*-1}_{ns}(k+2)\,\langle j_n j_p j_r j_s\rangle^0_{k,1,2}$$
$$+\frac{256\pi^4}{\omega}\sum_{1,2}\frac{1}{(\omega-\omega_1)(\omega_1-\omega_2)\omega_2}\{2\varkappa^{(2)}_{ijk}(k-1, 1)\,\Lambda^{-1}_{kl}(1)\,\varkappa^{(2)}_{lmn}(2, 1-2)$$
$$-\varkappa^{(3)}_{lmnj}(2, 1-2, k)\}\,\Lambda^{-1}_{mp}(2)\,\Lambda^{-1}_{nr}(1-2)\,\Lambda^{-1}_{js}(k-1)\,\langle j_p j_r j_s j_{i'}\rangle^0_{k,1,2}. \qquad (7.78)$$

It should be remembered that $\Delta_{ij}(k, 1)$ and $\Delta^{(2)}_{ijk}(1, k-1; k)$ that enter the expressions for $b_{ii'kk'}(k, 1)$, $P_{ij}(k)$, and $R_{ii'}(k)$ are integral operators acting on the correlation functions following them. For example:

$$\Delta_{ik}(k, 1)\,\langle\delta f j_l\rangle^0_{k-1}=\sum\frac{\Omega^2}{\omega}\int d\mathbf{v}\,\frac{v_i}{\omega-(\mathbf{k}\cdot\mathbf{v})+i0}\,\beta_{kk'}(1)\,\frac{\partial}{\partial v_{k'}}\langle\delta f(\mathbf{v})j_l\rangle^0_{k-1},$$

$$\Delta_{ik}(k, 1)\,\Delta^*_{i'k'}(k, 1)\,\langle\delta f\,\delta f\rangle^0_{k-1}=\sum\sum\frac{\Omega^2}{\omega}\frac{\Omega'^2}{\omega}\int d\mathbf{v}\,\frac{v_i}{\omega-(\mathbf{k}\cdot\mathbf{v})+i0}\,\beta_{kl}(1)\,\frac{\partial}{\partial v_l}$$
$$\times\int d\mathbf{v}'\,\frac{v'_{i'}}{\omega-(\mathbf{k}\cdot\mathbf{v}')+i0}\,\beta'_{k'l'}(1)\,\frac{\partial}{\partial v'_{l'}}\langle\delta f(\mathbf{v})\,\delta f(\mathbf{v}')\rangle^0_{k-1}.$$

As the basic non-linear equation (7.56) was derived using an expansion in a power series of the electric field intensity, the range of validity of (7.70) is restricted to that of the iteration procedure. So we may reduce the equation for the correlation function $\langle E_iE_j\rangle_k$ to the following equivalent form:

$$\Lambda_{ij}(k)\,\langle E_jE_{j'}\rangle_k-2\Lambda^{*-1}_{j'i'}(k)\sum_1\varkappa^{(2)}_{ijk}(k-1, 1)\,\varkappa^{(2)}_{i'j''k'}(k-1, 1)\,\langle E_kE_{k'}\rangle_1\,\langle E_jE_{j''}\rangle_{k-1}$$
$$-\sum_1 a_{ijkk'}(k, 1)\,\langle E_kE_{k'}\rangle_1\,\langle E_jE_{j'}\rangle_k=\Lambda^{*-1}_{j'i'}(k)\left\{\sum_1 b_{ii'kk'}(k, 1)\,\langle E_kE_{k'}\rangle_1+q'_{ii'}(k)\right\}, \qquad (7.79)$$

where

$$q'_{ii'}(k)\equiv q_{ii'}(k)+\frac{16\pi^2}{\omega^2}\Lambda^{*-1}_{ji''}(k)\sum_1 a^*_{i'jkk'}(k, 1)\,\langle E_kE_{k'}\rangle^*_1\,\langle j_ij_{i''}\rangle^{0*}_k. \qquad (7.80)$$

The equation for the ternary correlation function is as follows:

$$\langle E_iE_{i'}E_{i''}\rangle_{k,k'}+2\Lambda^{-1}_{ij}(k)\,\varkappa^{(2)}_{jkl}(k-k', k')\,\langle E_kE_{i'}\rangle_{k-k'}\,\langle E_lE_{i''}\rangle_{k'}$$
$$+2\Lambda^{*-1}_{i'j}(k-k')\,\varkappa^{(2)}_{jkl}(-k, k')\,\langle E_kE_i\rangle_{-k}\,\langle E_lE_{i''}\rangle_{k'}$$
$$+2\Lambda^{*-1}_{i''j}(k')\,\varkappa^{(2)}_{jkl}(-k, k-k')\,\langle E_kE_i\rangle_{-k}\,\langle E_lE_{i'}\rangle_{k-k'}$$
$$=B_{ii'j}(k, -k+k')\,\langle E_jE_{i''}\rangle_{k'}+B_{ii''j}(-k', k)\,\langle E_jE_{i'}\rangle_{k-k'}$$
$$+B_{i'i''j}(-k+k', -k')\,\langle E_jE_i\rangle_{-k}+\Lambda^{-1}_{ij}(k)\Lambda^{*-1}_{i'j'}(k-k')\Lambda^{*-1}_{i''j''}(k')Q_{jj'j''}(k), \qquad (7.81)$$

where

$$B_{ijk}(k, k') = 4\pi i \Lambda_{il}^{-1}(k)\,\Lambda_{jm}^{-1}(k')\left\{\frac{1}{\omega}\Lambda_{mk}(k', k+k')\,\langle\delta f j_l\rangle_{-k}^{0} + \frac{1}{\omega'}\Lambda_{lk}(k, k+k')\,\langle\delta f j_m\rangle_{-k}^{0}\right\}, \tag{7.82}$$

and

$$\begin{aligned} Q_{ii'i''}(k, k') &= i\frac{(4\pi)^3}{\omega\omega'(\omega-\omega')}\langle j_i j_{i'} j_{i''}\rangle_{k,k'}^{0} \\ &+ i(4\pi)^3\left\{\frac{1}{\omega\omega'}\sum_1 \frac{1}{\omega_1}\Lambda_{i'k}(-k+k', 1)\,\Lambda_{kl}^{-1}(1)\,\langle\delta f j_l j_i j_{i''}\rangle_{k,k',1}^{0}\right. \\ &+ \frac{1}{\omega(\omega-\omega')}\sum_1 \frac{1}{\omega_1}\Lambda_{i''k}(-k', 1)\,\Lambda_{kl}^{-1}(1)\,\langle\delta f j_l j_i j_{i'}\rangle_{k,k',1}^{0} \\ &\left.- \frac{1}{\omega'(\omega-\omega')}\sum_1 \frac{1}{\omega'}\Lambda_{ik}(k, 1)\,\Lambda_{kl}^{-1}(1)\,\langle\delta f j_l j_{i'} j_{i''}\rangle_{k,k',1}^{0}\right\} \\ &- (4\pi)^4\left\{\frac{1}{\omega'(\omega-\omega')}\sum_1 \frac{1}{(\omega-\omega_1)\omega_1}\varkappa_{ikl}^{(2)}(1, k-1)\,\Lambda_{km}^{-1}(1)\,\Lambda_{ln}^{-1}(k-1)\,\langle j_m j_n j_i j_{i''}\rangle_{k,k',1}^{0}\right. \\ &+ \frac{1}{\omega\omega'}\sum_1 \frac{1}{(\omega-\omega'+\omega_1)\omega_1}\varkappa_{i'kl}^{(2)}(1, -k+k'-1)\,\Lambda_{km}^{-1}(1)\,\Lambda_{ln}^{*-1}(k-k'+1)\,\langle j_m j_n j_{i'} j_{i''}\rangle_{k,k',1}^{0} \\ &\left.+ \frac{1}{\omega(\omega-\omega')}\sum_1 \frac{1}{(\omega'+\omega_1)\omega_1}\varkappa_{i''kl}^{(2)}(1, -k'-1)\,\Lambda_{km}^{-1}(1)\,\Lambda_{ln}^{*-1}(k'+1)\,\langle j_m j_n j_i j_{i'}\rangle_{k,k',1}^{0}\right\}. \end{aligned} \tag{7.83}$$

The non-homogeneous parts of (7.79) and (7.81) are expressed in terms of the correlators for the fluctuations of the distribution function and of the charge density in neglect of interactions between particles.

The quaternary correlation function for the field $\langle E_i E_j E_k E_l\rangle_{k,k',k''}$ is directly related to the correlator for the current density associated with the stochastic motion of individual plasma particles:

$$\begin{aligned} \langle E_i E_j E_k E_l\rangle_{k,k',k''} &= -\frac{(4\pi)^4}{\omega(\omega-\omega')(\omega'-\omega'')\omega''}\Lambda_{ii'}^{-1}(k)\,\Lambda_{jj'}^{*-1}(k-k'). \\ &\times \Lambda_{kk'}^{-1}(k'-k'')\,\Lambda_{ll'}^{-1}(k'')\,\langle j_{i'} j_{j'} j_{k'} j_{l'}\rangle_{k,k'k''}^{0}. \end{aligned} \tag{7.84}$$

The set of (7.79), (7.81), and (7.84) completely determines the set of the correlation functions for the fluctuating electric field in the plasma $\langle E_i E_j\rangle_k$, $\langle E_i E_j E_k\rangle_{k,k'}$, and $\langle E_i E_j E_k E_l\rangle_{k,k',k''}$. If the electric field is potential $\mathbf{E} = \frac{\mathbf{k}}{k}E$, then (7.79), (7.81), and (7.84) reduce, respectively, to (7.20), (7.22), and (7.25).

The stationary spectrum of electromagnetic fluctuations

Let us iterate the set of equations for the spectral correlation functions of the electromagnetic field fluctuations similarly to the analysis of the potential case. We expand the pair correlation function $\langle E_i E_j\rangle_k$ which determines the spectral distribution of the flucuating electromagnetic field energy according to

$$\langle E_i E_j \rangle_k = \langle E_i E_j \rangle_k^{(1)} + \langle E_i E_j \rangle_k^{(2)} + \langle E_i E_j \rangle_k^{(3)} + \dots \tag{7.85}$$

Within the context of (7.79) it is not difficult to obtain the following expressions for each term (Sitenko and Kocherga, 1977b):

$$\langle E_i E_j \rangle_k^{(1)} = \frac{16\pi^2}{\omega^2} \Lambda_{ik}^{-1}(k)\, \Lambda_{jl}^{*-1}(k) \langle j_k j_l \rangle_k^0; \tag{7.86}$$

$$\begin{aligned}
\langle E_i E_j \rangle_k^{(2)} = \frac{16\pi^2}{\omega} \Lambda_{ii'}^{-1}(k)\, \Lambda_{jj'}^{*-1}(k) \sum_1 \frac{1}{\omega_1} \\
\times \Bigg\{ \frac{4\pi i}{\omega - \omega_1} \big(\varkappa_{i'kl}^{(2)}(k-1, 1)\, \Lambda_{km}^{-1}(k-1)\, \Lambda_{ln}^{-1}(1) \langle j_{j'} j_m j_n \rangle_{k,1}^0 \\
- \varkappa_{j'kl}^{(2)*}(k-1, 1)\, \Lambda_{km}^{*-1}(k-1)\, \Lambda_{ln}^{*-1}(1) \langle j_{i'} j_m j_n \rangle_{k,1}^0 \big) \\
- \Delta_{i'k}(k, 1)\, \Lambda_{kl}^{-1}(1) \langle \delta f j_{j'} j_l \rangle_{k,1}^0 \\
- \Delta_{j'k}^{*}(k, 1)\, \Lambda_{kl}^{*-1}(1) \langle \delta f j_{j'} j_l \rangle_{k,1}^0 \Bigg\};
\end{aligned} \tag{7.87}$$

$$\begin{aligned}
\langle E_i E_j \rangle_k^{(3)} = \Lambda_{ii'}^{-1}(k)\, \Lambda_{jj'}^{*-1}(k) \\
\times \Bigg\{ 2 \sum_1 \varkappa_{i'kl}^{(2)}(k-1, 1)\, \varkappa_{j'k'l'}^{(1)*}(k-1, 1) \langle E_k E_{k'} \rangle_{k-1}^{(1)} \langle E_l E_{l'} \rangle_1^{(1)} \\
+ \Lambda_{j'k'}^{*}(k) \sum_1 a_{i'kll'}(k, 1) \langle E_l E_{l'} \rangle_1^{(1)} \langle E_k E_{k'} \rangle_k^{(1)} \\
+ \Lambda_{i'k'}(k) \sum_1 a_{j'kll'}^{*}(k, 1) \langle E_l E_{l'} \rangle_1^{(1)*} \langle E_k E_{k'} \rangle_k^{(1)} \\
+ \sum_1 b_{i'j'kk'}(k, 1) \langle E_k E_{k'} \rangle_1^{(1)} + q_{i'j'}^{(3)}(k) \Bigg\}.
\end{aligned} \tag{7.88}$$

The first term of expansion (7.85) is associated with the linear approximation and describes the spectral distribution of the fluctuating electric field energy in neglect of interactions between the fluctuation fields in the plasma. The linear spectral distribution (7.86) is completely determined by the binary correlation function for the current density fluctuations in a system of non-interacting particles $\langle j_i j_j \rangle_k^0$.

The second term of expansion (7.85) is due to the quadratic non-linear interaction between the fluctuating electromagnetic fields. The spectral distribution of the electric field fluctuations (7.87) is determined by the ternary correlation function for the current density fluctuations $\langle j_i j_j j_k \rangle_{k,k'}^0$ and by the correlator between the fluctuation variation of the distribution function and the square of the fluctuation current density $\langle \delta f j_i j_j \rangle_{k,k'}^0$ (both the former and the latter are to be calculated without taking into account the electromagnetic interactions between particles).

Finally, the third term of (7.85) corresponds to the cubic interaction between the fluctuating electromagnetic fields. Quadratic and cubic interactions lead to the spectral distribution (7.88) that depends on the quaternary correlators for the fluctuations of the current density and of the distribution function in the ensemble of non-interacting particles $\langle j_i j_j j_k j_l \rangle^0_{k,k',k''}$, $\langle \delta f j_i j_j j_k \rangle^0_{k,k',k''}$ and $\langle \delta f \, \delta f j_i j_j \rangle^0_{k,k',k''}$, as well as on the products of various pair correlation functions for the same ensemble $\langle j_i j_j \rangle^0_k$, $\langle \delta f j_i \rangle^0_k$, and $\langle \delta f \, \delta f \rangle^0_k$. The spectral distribution $\langle E_i E_j \rangle^{(3)}_k$ has, besides the maxima at the eigenfrequencies $\omega_{\mathbf{k}}$, also additional ones corresponding to the combination frequencies $\omega_{\mathbf{k}-\mathbf{k}'} \pm \omega_{\mathbf{k}'}$.

In a thermally equilibrium plasma the corrections $\langle E_i E_j \rangle^{(2)}_k$ and $\langle E_i E_j \rangle^{(3)}_k$ are small as compared to the spectral distribution $\langle E_i E_j \rangle^{(1)}_k$.

In our analysis of the potential field we have pointed out already that if some eigenmodes are excited in the plasma by external sources, then non-linear wave interactions may lead to a considerable enhancement of fluctuations. Let us denote the spectral distribution of the induced electromagnetic eigenoscillations by $\langle E_i E_j \rangle^{(0)}_k$. This distribution satisfies the condition

$$\Lambda_{ij}(k) \langle E_j E_k \rangle^{(0)}_k = 0.$$

In the vicinity of the eigenfrequencies the spectral distribution of the fluctuation field may be written as

$$\langle E_i E_j \rangle_k = e_i e_j^* \langle E^2 \rangle_k, \tag{7.89}$$

where e_i and e_j are the polarizations of the relevant eigenoscillations. There occurs the following relation between the polarization vectors and the cofactor $\lambda_{ij}(k)$ of the tensor $\Lambda_{ij}(k)$ (the components of the latter are to be taken at the relevant eigenfrequencies):

$$e_i e_j^* = \frac{\lambda_{ij}}{\operatorname{Tr} \lambda}.$$

Make use of (7.79) to retain the terms to the lowest orders in the non-linear coupling between the fluctuating field and that of the induced oscillations (the cubic interaction is important). As a result we obtain the following spectral distribution of the fluctuation field:

$$\langle \tilde{E}^2 \rangle_k = \Big\{ 1 + \Lambda_{ji}^{-1}(k) \sum_1 a_{ijkk'}(k, 1) \langle E_k E_{k'} \rangle^{(0)}_1 + \Lambda_{ij}^{*-1}(k) \sum_1 a^*_{jikk'}(k, 1) \langle E_k E_{k'} \rangle^{(0)*}_1 \Big\} \langle E^2 \rangle^{(1)}_k, \tag{7.90}$$

where $\langle E^2 \rangle^{(1)}_k$ is the spectral distribution of the fluctuating field in the absence of induced oscillations. If the intensity of the induced eigenoscillations $\langle E_i E_j \rangle^0_k$ is sufficiently large, then the fluctuation intensity $\langle \tilde{E}^2 \rangle_k$ may considerably exceed the thermal fluctuation level $\langle E^2 \rangle^{(1)}_k$. It becomes clear that the induced fluctuation enhancement is significant if

$$\left| \Lambda_{ji}^{-1}(k) \sum_1 a_{ijkk'}(k, 1) \langle E_k E_{k'} \rangle^{(0)}_1 + \Lambda_{ij}^{*-1}(k) \sum_1 a^*_{jikk'}(k, 1) \langle E_k E_{k'} \rangle^{(0)*}_1 \right| \gg 1. \tag{7.91}$$

An observation is to be made that the enhancement of the potential fluctuations accompanies the excitation of both potential and non-potential oscillations. In particular, the induced enhancement of the potential fluctuations may be the result of electromagnetic wave pumping.

Non-linear frequency shifts of electromagnetic eigenoscillations

To find the shift of the electromagnetic eigenoscillations due to non-linear interactions we set equal to zero the right-hand side of (7.79):

$$\Lambda_{ij}(k)\langle E_j E_{j'}\rangle - 2\Lambda^{*-1}_{j'i'}(k)\sum_{1}\varkappa^{(2)}_{ijk}(k-1,\,1)\,\varkappa^{(2)}_{i'j''k'}(k-1,1)\langle E_k E_{k'}\rangle_1\langle E_j E_{j''}\rangle_{k-1}$$
$$-\sum_{1} a_{ijkk'}(k,\,1)\langle E_k E_{k'}\rangle\langle E_j E_{j'}\rangle_k = 0. \tag{7.92}$$

The equation for the eigenoscillations in neglect of non-linear effects is of the form

$$\Lambda_{ij}(k)\langle E_j E_k\rangle^{(1)}_k = 0. \tag{7.93}$$

Let us separate out the Hermitian and the anti-Hermitian parts of the tensor $\Lambda_{ij}(k)$

$$\Lambda_{ij}(k) = \Lambda^H_{ij}(k) + \Lambda^A_{ij}(k)$$

and consider the fluctuations in the transparency domain where

$$|\Lambda^A_{ij}(k)| \ll |\Lambda^H_{ij}(k)|. \tag{7.94}$$

The linear eigenfrequencies are determined by the requirement that the determinant, consisting of the elements of the Hermitian part of $\Lambda_{ij}(k)$ vanishes:

$$\mathrm{Det}\,(\Lambda^H_{ij}(k)) = 0, \tag{7.95}$$

which is evidently equivalent to

$$\mathrm{Re}\,\Lambda(k) = 0, \tag{7.96}$$

where $\Lambda(k) = \Lambda_{ij}(k)\,\lambda_{ji}(k)$. The substitution of (7.89) for the spectral distribution $\langle E_i E_j\rangle^{(1)}_k$ in (7.93) and the following contraction yield:

$$\Lambda(k)\langle E^2\rangle^{(1)}_k = 0. \tag{7.97}$$

As $\langle E^2\rangle_k$ is real, (7.96) (which determines the eigenfrequencies in the linear approximation) follows from the requirement that the real part of (7.97) must vanish:

$$\mathrm{Re}\,\Lambda(k)\langle E^2\rangle^{(1)}_k = 0, \tag{7.98}$$

and, within the context of (7.97), the spectral distribution is not equal to zero only if the frequencies coincide with the eigenfrequencies of the electromagnetic plasma oscillations:

$$\langle E^2\rangle_k = B(k)\,\delta\{\mathrm{Re}\,\Lambda(k)\}, \tag{7.99}$$

where the proportionality coefficient B is a function of the effective temperature

$$B(k) = 8\pi^2\frac{|\mathrm{Tr}\,\lambda|}{|\omega|}\,\tilde{T}(k).$$

The eigenfrequencies taking into account the non-linear effects $\tilde{\omega}_{\mathbf{k}}$ may be written as

$$\tilde{\omega}_{\mathbf{k}} = \omega_{\mathbf{k}} + \Delta\omega_{\mathbf{k}}.$$

Assuming that the shifts $\Delta\omega_{\mathbf{k}}$ are small in comparison to the eigenfrequencies we can easily calculate them explicitly. By analogy with (7.98) we shall assume the eigenfrequencies involving the non-linear effects to be the solutions of the equation which follows from the requirement that the real part of the contraction of (7.92) vanishes:

$$\mathrm{Re}\Big\{\Lambda_{ij}(k)\langle E_j E_i\rangle_k - 2\Lambda_{ii'}^{*-1}(k)\sum_1 \varkappa_{ijk}^{(2)}(k-1,\,1)\,\varkappa_{i'j'k'}^{(2)*}(k-1,\,1)\langle E_k E_{k'}\rangle_1\langle E_j E_{j'}\rangle_{k-1}$$
$$-\sum_1 a_{ijkk'}(k,\,1)\langle E_k E_{k'}\rangle_1\langle E_j E_i\rangle_k\Big\} = 0. \tag{7.100}$$

Let the spectral distribution $\langle E_i E_j\rangle_k$ be the following series

$$\langle E_i E_j\rangle_k = e_i e_j^*(\langle E^2\rangle_k^{(1)} + \langle E^2\rangle_k^{(2)} + \ldots)$$

With the last expansion substituted in (7.100) we have

$$\mathrm{Re}\Big\{\Lambda(k) - \sum_1 a_{ijkk'}(k,\,1)\,\lambda_{ij}(k)\langle E_k E_{k'}\rangle_1^{(1)}\Big\}(\langle E^2\rangle_k^{(1)} + \langle E^2\rangle_k^{(2)} + \ldots) = 0. \tag{7.101}$$

If $\Delta\omega_{\mathbf{k}} \ll \omega_{\mathbf{k}}$ then $\Lambda(\tilde{\omega}_{\mathbf{k}}, \mathbf{k})$ may be expanded in a power series of $\Delta\omega_{\mathbf{k}}$:

$$\Lambda(\tilde{\omega}_{\mathbf{k}}, \mathbf{k}) = \Lambda(\omega_{\mathbf{k}}, \mathbf{k}) + \frac{\partial\Lambda(\omega_{\mathbf{k}}, \mathbf{k})}{\partial\omega_{\mathbf{k}}}\Delta\omega_{\mathbf{k}} + \ldots\,. \tag{7.102}$$

Substituting (7.102) into (7.101) and retaining terms of equal orders we obtain the following approximate expression for $\Delta\omega_{\mathbf{k}}$:

$$\Delta\omega_{\mathbf{k}} = \left(\frac{\partial\,\mathrm{Re}\,\Lambda(\omega_{\mathbf{k}}, \mathbf{k})}{\partial\omega_{\mathbf{k}}}\right)^{-1}\sum_{\omega',\mathbf{k}'} a_{ijkk'}(\omega_{\mathbf{k}}, \mathbf{k};\,\omega', \mathbf{k}')\,\lambda_{ji}(\omega_{\mathbf{k}}, \mathbf{k})\langle E_k E_{k'}\rangle_{\mathbf{k}'\omega'}^{(1)}. \tag{7.103}$$

The shift (7.103) was derived in neglect of the second term on the left-hand side of (7.100) which is associated with the resonant three-wave interaction. The non-linear shift (7.103) results from the four-wave resonant interaction. The magnitude of the shift is governed by the wave intensities. It may be shown that when the interaction of certain modes is the only significant one, the dominant contribution in (7.103) comes from the values of $\mathbf{k}'$, close to $\mathbf{k}$.

Saturation of the level of non-equilibrium electromagnetic fluctuations

Let us calculate the spectral distribution of electromagnetic fluctuations in a stationary non-equilibrium plasma taking into account the eigenfrequency shifts caused by the non-linear wave interaction. Suppose the decay conditions are not satisfied. Then the equation for the quadratic correlation function (7.70) may be written as:

$$\Big(\Lambda_{i'j'}(k) - \sum_1 a_{i'j'kk'}(k,\,1)\langle E_k E_{k'}\rangle_1\Big)^*$$
$$\times\Big(\Lambda_{ij}(k) - \sum_1 a_{ijll'}(k,\,1)\langle E_l E_{l'}\rangle_1\Big)\langle E_j E_{j'}\rangle_k = \frac{16\pi^2}{\omega^2}\langle j_i j_{i'}\rangle_k^0. \tag{7.104}$$

(Only the leading term is retained on the right-hand side of (7.104).)

In the vicinity of the plasma eigenfrequencies and upon the assumption that the transparency condition (7.94) is satisfied, the left-hand side of (7.104), within the context of (7.103), may be written as:

$$(\Lambda^H_{i'j'}(\omega-\Delta\omega_{\mathbf{k}},\mathbf{k})+\Lambda^A_{i'j'}(\omega,\mathbf{k}))^*(\Lambda^H_{ij}(\omega-\Delta\omega_{\mathbf{k}},\mathbf{k})+\Lambda^A_{ij}(\omega,\mathbf{k}))\langle E_jE_{j'}\rangle_{\mathbf{k}\omega}$$
$$=\frac{16\pi^2}{\omega^2}\langle j_ij_{i'}\rangle^0_{\mathbf{k}\omega}. \tag{7.105}$$

Thus the effect of the non-linear wave interaction results in shifts of the frequencies that enter the Hermitian part of the tensor $\Lambda_{ij}(k)$. From (7.105) we have

$$\langle E^2\rangle_{\mathbf{k}\omega}=\pi I_{\mathbf{k}}\{\delta(\omega-\tilde{\omega}_{\mathbf{k}})+\delta(\omega+\tilde{\omega}_{\mathbf{k}})\}, \tag{7.106}$$

where $\tilde{\omega}_{\mathbf{k}}$ is the shifted eigenfrequency and

$$I_{\mathbf{k}}=\frac{16\pi^2}{k^2}\frac{\operatorname{Tr}\lambda(\omega_{\mathbf{k}},\mathbf{k})}{\left|\dfrac{\partial\operatorname{Re}\Lambda(\tilde{\omega}_{\mathbf{k}},\mathbf{k})}{\partial\tilde{\omega}_{\mathbf{k}}}\right|}\frac{\langle\varrho^2\rangle^0_{\mathbf{k}\tilde{\omega}_{\mathbf{k}}}}{|\operatorname{Im}\Lambda(\tilde{\omega}_{\mathbf{k}}\,\mathbf{k})|}. \tag{7.107}$$

Thus, the non-linear wave interactions shift the maxima of the spectral distribution of the electric field fluctuations in the plasma. Since the non-linear eigenfrequency $\tilde{\omega}_{\mathbf{k}}$ is a function of $I_{\mathbf{k}}$:

$$\tilde{\omega}_{\mathbf{k}}=\omega_{\mathbf{k}}+\beta_{\mathbf{k}}I_{\mathbf{k}},$$

the relation (7.107) may be regarded as an equation for the fluctuation intensity $I_{\mathbf{k}}$ in the range of plasma instability. Similarly to the potential case, the existence of a solution of (7.107) implies the existence of a stationary turbulent state of the non-equilibrium plasma (which is unstable in the linear approximation). The solution itself determines a stationary level of turbulent electromagnetic fluctuations.

For example, electromagnetic instabilities may occur in a magneto-active plasma with drifting charged particles if the drift velocity is greater than the phase velocities of the relevant electromagnetic waves. The frequencies of the electromagnetic waves that propagate along the magnetic field are equal to

$$\omega_{\mathbf{k}}=\sqrt{\Omega^2+k^2c^2}\pm\omega_B,$$

where + and – are associated with the right- and left-polarized waves respectively. Taking into account the non-linear frequency shifts, we obtain an equation, formally similar to (7.48), for the turbulent fluctuation intensity in the instability range. The level of electromagnetic fluctuations in the instability region may be much greater than the thermal one.

CHAPTER 8

Fluctuation-Dissipation Theorem

8.1. Inversion of the Fluctuation–Dissipation Theorem

Relation between the dielectric permittivity of an equilibrium plasma and the correlation function for fluctuations in a system of non-interacting particles

The fluctuation–dissipation theorem reflects the fundamental connection between the dissipation and the fluctuations of various quantities in thermally equilibrium media. Since the dissipation properties of an electrodynamic system are described by the macroscopic coefficients of the linear relation between the induced charges or currents and the fields, the knowledge of these coefficients suffices to find the spectral distributions of fluctuations of electrodynamic quantities. In particular, the distribution of electromagnetic fluctuations in an equilibrium plasma is completely determined by the dielectric permittivity tensor. (For a general derivation of the fluctuation–dissipation theorem for thermally equilibrium systems, see, for example, Callen and Welton, 1951; Kubo, 1957; Landau and Lifshitz, 1960). And vice versa, the dielectric permittivity of the medium may be found in terms of the known spectrum of electromagnetic fluctuations by means of the inversion of the fluctuation–dissipation theorem. The plasma dielectric permittivity depends only on the pair correlation function for the current density fluctuations computed without taking account of particle interactions (Sitenko, 1967). Such an approach makes it possible to describe completely the electrodynamic properties of equilibrium plasma without utilizing the kinetic equations and thus considerably facilitates the study of a bounded plasma and some other problems (Sitenko and Yakimenko, 1974; Sitenko, 1976).

First of all, let us consider potential fluctuations in a plasma. The linear electrodynamic plasma properties are completely described by the given dielectric permittivity $\varepsilon(\omega, \mathbf{k})$. The fluctuation–dissipation theorem for an equilibrium plasma states that the spectral distributions of the fluctuations of the electric field and the charge density may be written in terms of the imaginary part of the dielectric permittivity $\varepsilon(\omega, \mathbf{k})$ in the following way:

$$\langle E^2 \rangle_{\mathbf{k}\omega} = 8\pi \frac{T}{\omega} \frac{\operatorname{Im} \varepsilon(\omega, \mathbf{k})}{|\varepsilon(\omega, \mathbf{k})|^2}, \tag{8.1}$$

$$\langle \varrho^2 \rangle_{\mathbf{k}\omega} = \frac{k^2}{2\pi} \frac{T}{\omega} \frac{\operatorname{Im} \varepsilon(\omega, \mathbf{k})}{|\varepsilon(\omega, \mathbf{k})|^2}, \tag{8.2}$$

where T is the plasma temperature. The relations (8.1) and (8.2) serve usually to find the spectral distributions of fluctuations from the known plasma dielectric permittivity. They allow in principle to calculate the dielectric permittivity provided the spectral distribution of fluctuations in the plasma is given. In view of the latter purpose the relation (8.2) is to be rewritten as (Kubo, 1957)

$$\operatorname{Im} \frac{1}{\varepsilon^*(\omega, \mathbf{k})} = \frac{2\pi}{k^2} \frac{\omega}{T} \langle \varrho^2 \rangle_{\mathbf{k}\omega}. \tag{8.2'}$$

However, this way of calculating the permittivity is in fact inefficient, for there are no direct methods which would make it possible to find the spectral distribution of fluctuations $\langle \varrho^2 \rangle_{\mathbf{k}\omega}$ taking into account the Coulomb interactions between particles.

We expand the left- and right-hand sides of (8.2) in a power series in e^2 and retain the leading terms only. Since the linear electric susceptibility is proportional to e^2, its imaginary part $\operatorname{Im} \varkappa^{(1)}(\omega, \mathbf{k})$ (which coincides with that of the dielectric permittivity $\operatorname{Im} \varepsilon(\omega, \mathbf{k})$) may be obtained immediately from (8.2):

$$\operatorname{Im} \varkappa^{(1)}(\omega, \mathbf{k}) = \frac{2\pi}{k^2} \frac{\omega}{T} \langle \varrho^2 \rangle^0_{\mathbf{k}\omega}, \tag{8.3}$$

where $\langle \varrho^2 \rangle^0_{\mathbf{k}\omega}$ is the spectral distribution of the charge density fluctuations in neglect of the Coulomb interactions between particles. The last relation may be regarded as the inversion of (8.2). In contrast to (8.2′) the right-hand side of (8.3) involves only the quantity $\langle \varrho^2 \rangle^0_{\mathbf{k}\omega}$, which is easy to calculate.

The electric susceptibility $\varkappa^{(1)}(\omega, \mathbf{k})$ is an analytic function in the complex ω-plane with a cut along the real axis. It is evident that the imaginary part of the function $\varkappa^{(1)}(\omega, \mathbf{k})$ for real values of ω determines the discontinuity of the electric susceptibility at the cut:

$$\operatorname{Im} \varkappa^{(1)}(\omega, \mathbf{k}) = \operatorname{Im}_\omega \varkappa^{(1)}(\omega, \mathbf{k}) \equiv \frac{1}{2i} \{\varkappa^{(1)}(\omega + i0, \mathbf{k}) - \varkappa^{(1)}(\omega - i0, \mathbf{k})\}. \tag{8.4}$$

According to (8.3), the discontinuity of the electric susceptibility at the cut along the real axis in the complex ω-plane is described by the correlation function for the charge density fluctuations $\langle \varrho^2 \rangle^0_{\mathbf{k}\omega}$ computed without taking into account the Coulomb interactions between charged particles:

$$\operatorname{Im}_\omega \varkappa^{(1)}(\omega, \mathbf{k}) = \frac{2\pi}{k^2} \frac{\omega}{T} \langle \varrho^2 \rangle^0_{\mathbf{k}\omega}. \tag{8.5}$$

The last relation is the inversion of the fluctuation–dissipation theorem. The latter together with the Kramers–Kronig dispersion relation

$$\varkappa^{(1)}(\omega, \mathbf{k}) = \frac{1}{\pi} \int_{-\infty}^{\infty} d\omega' \frac{\operatorname{Im}_{\omega'} \varkappa^{(1)}(\omega', \mathbf{k})}{\omega' - \omega - i0} \tag{8.6}$$

yields the linear electric susceptibility of the equilibrium plasma:

$$\varkappa^{(1)}(\omega, \mathbf{k}) = \frac{2}{Tk^2} \int_{-\infty}^{\infty} d\omega' \frac{\omega' \langle \varrho^2 \rangle^0_{\mathbf{k}\omega'}}{\omega' - \omega - i0}. \tag{8.7}$$

Substituting the explicit form for the correlation function $\langle \varrho^2 \rangle^0_{\mathbf{k}\omega}$, one can easily verify that (8.7) is in accordance with the general formula (2.16) if the distribution function in the latter is taken to be Maxwellian.

The arbitrary (not necessarily potential) fluctuations in a plasma may be treated in a similar manner. In this case the inversion of the fluctuation–dissipation theorem makes it possible to express the anti-Hermitian part of the linear electric susceptibility tensor $\varkappa^{(1)}_{ij}(\omega, \mathbf{k})$ (which is equal to that of the plasma dielectric permittivity tensor) in terms of the spectral distribution of current density fluctuations $\langle j_i j_j \rangle^0_{\mathbf{k}\omega}$ calculated without taking into account the electromagnetic interactions between particles:

$$\frac{1}{2i}\{\varkappa^{(1)}_{ij}(\omega, \mathbf{k}) - \varkappa^{(1)*}_{ji}(\omega, \mathbf{k})\} = \frac{2\pi}{T\omega} \langle j_i j_j \rangle^0_{\mathbf{k}\omega}. \tag{8.8}$$

If the plasma is isotropic, then (8.8) becomes reduced to

$$\mathrm{Im}_\omega\, \varkappa^{(1)}_{ij}(\omega, \mathbf{k}) = \frac{2\pi}{T\omega} \langle j_i j_j \rangle^0_{\mathbf{k}\omega}, \tag{8.9}$$

and within the context of the Kramers–Kronig relation we find

$$\varkappa^{(1)}_{ij}(\omega, \mathbf{k}) = \frac{2}{T\omega} \int d\omega' \frac{\langle j_i j_j \rangle^0_{\mathbf{k}\omega'}}{\omega' - \omega - i0}. \tag{8.10}$$

Thus the knowledge of the equilibrium fluctuation spectrum $\langle j_i j_j \rangle^0_{\mathbf{k}\omega}$ is sufficient to find by means of (8.10) the dielectric permittivity tensor which describes in the linear approximation all electrodynamic properties of an equilibrium plasma.

The relations (8.3) and (8.8) are general and may be used to calculate the dielectric permittivity not only of a hot plasma. It is not difficult to find from (8.3) and (8.8) the dielectric permittivities of a degenerate plasma and a superconducting one (then $\langle \varrho^2 \rangle^0_{\mathbf{k}\omega}$ must be replaced by the correlation function for the system of particles with pairing but without Coulomb interactions) of a solid state plasma (in such a case $\langle \varrho^2 \rangle^0_{\mathbf{k}\omega}$ must involve the interaction between the electrons and the lattice), and so on.

Fluctuation–dissipation theorem for a non-equilibrium plasma

The fluctuation–dissipation theorem may be extended to involve into consideration the non-equilibrium (but stationary and stable) plasma states. Indeed, the derivation of the fluctuation–dissipation theorem for equilibrium systems is based on the connection between the correlation function for current density fluctuations and the mean energy which is absorbed by the system by virtue of dissipation. An analogous connection exists also in the absence of thermal equilibrium, therefore we can derive a generalized fluctuation–dissipation theorem that would describe fluctuations in non-equilibrium systems (Sitenko, 1966, 1967).

The generalized fluctuation–dissipation theorem for potential fluctuations in a non-equilibrium plasma may be written in the following form:

$$\langle \varrho^2 \rangle_{\mathbf{k}\omega}^{\hbar\omega} - \langle \varrho^2 \rangle_{\mathbf{k}\omega} = 2\hbar \, \mathrm{Im}_\omega \, \alpha(\omega, \mathbf{k}), \tag{8.11}$$

where $\langle \varrho^2 \rangle_{\mathbf{k}\omega}^{\hbar\omega}$ is the correlation function, the averaging procedure being carried out over an energy distribution shifted by $\hbar\omega$ ($\hbar$ is Planck's constant divided by 2π) and $\alpha(\omega, \mathbf{k})$ is the coefficient of the linear relation between the field potential and the charge density

$$\alpha(\omega, \mathbf{k}) = \frac{k^2}{2\pi} \frac{\varkappa^{(1)}(\omega, \mathbf{k})}{1+\varkappa^{(1)}(\omega, \mathbf{k})}.$$

Formula (8.11) may be simplified in the classical case by expanding the distribution function $f(E-\hbar\omega)$ that enters $\langle \varrho^2 \rangle_{\mathbf{k}\omega}^{\hbar\omega}$ in a power series in the transfer energy $\hbar\omega$. In the limiting case $\hbar \to 0$ we have

$$-\frac{k^2}{m} \frac{\partial}{\partial\omega} \langle \varrho^2 \rangle_{\mathbf{k}\omega} = \mathrm{Im}_\omega \, \alpha(\omega, \mathbf{k}).$$

Making use of the last formula it is not difficult to derive the inversion of the generalized fluctuation–dissipation theorem that connects the linear susceptibility of a non-equilibrium plasma with the correlation function for the charge density fluctuations in neglect of particle interactions:

$$\mathrm{Im}_\omega \, \varkappa^{(1)}(\omega, \mathbf{k}) = -\frac{2\pi}{\omega} \frac{\partial}{\partial\omega} \langle \varrho^2 \rangle_{\mathbf{k}\omega}^{0}. \tag{8.12}$$

It should be noted that another way to derive (8.12) is to compare (2.16) and (6.37).

The knowledge of the discontinuity of the electric susceptibility $\varkappa^{(1)}(\omega, \mathbf{k})$ on the cut along the real axis in the complex ω-plane is sufficient to find the electric susceptibility of a non-equilibrium plasma $\varkappa^{(1)}(\omega, \mathbf{k})$ for any complex ω by means of the dispersion relation (8.6). We have

$$\varkappa^{(1)}(\omega, \mathbf{k}) = -\frac{2}{m} \int_{-\infty}^{\infty} d\omega' \frac{\dfrac{\partial}{\partial\omega'} \langle \varrho^2 \rangle_{\mathbf{k}\omega'}^{0}}{\omega' - \omega - i0}. \tag{8.13}$$

The dielectric permittivity of a non-equilibrium plasma is then as follows:

$$\varepsilon(\omega, \mathbf{k}) = 1 - \frac{2}{m} \int d\omega' \frac{\dfrac{\partial}{\partial\omega'} \langle \varrho^2 \rangle_{\mathbf{k}\omega'}^{0}}{\omega' - \omega - i0}. \tag{8.14}$$

Thus the dielectric permittivities, both of equilibrium and of non-equilibrium plasmas, are completely determined by the quadratic correlation function for the charge density fluctuations calculated without taking into account the Coulomb interaction between particles. It should be kept in mind that to set the dielectric permittivity or the linear electric susceptibility is insufficient for the description of fluctuations in non-equilibrium plasmas (in contrast to the equilibrium case). The spectral distribution of the electric field fluctuations in a non-equilibrium plasma is expressed not only in terms of the dielectric permittivity (as it was in the equilibrium case (8.1)), but involves also the spectral distribution of the charge density fluctuations in neglect of particle interactions:

$$\langle E^2\rangle_{\mathbf{k}\omega} = \frac{16\pi^2}{k^2}\frac{\langle \varrho^2\rangle^0_{\mathbf{k}\omega}}{|\varepsilon(\omega, \mathbf{k})|^2}. \tag{8.15}$$

Hence, the dielectric permittivity describes the electrodynamic properties of a non-equilibrium plasma incompletely. A full description is however, provided, if the spectral distribution of the charge density fluctuations in neglect of particle interactions $\langle\varrho^2\rangle^0_{\mathbf{k}\omega}$ is given, in terms of which both the dielectric permittivity of and the spectral distribution of fluctuations in a non-equilibrium plasma may be expressed directly (Sitenko, 1966).

The generalized fluctuation-dissipation theorem for non-potential fluctuations in a non-equilibrium isotropic plasma may be written in the form

$$\mathrm{Im}_\omega\, \varkappa_{ij}(\omega, \mathbf{k}) = \frac{2\pi}{m}\frac{1}{\omega^2}\left((k_i\delta_{jl}+\delta_{il}k_j)\frac{k_k}{\omega} - \delta_{ik}\delta_{jl}k^2\frac{\partial}{\partial\omega}\right)\langle j_k j_l\rangle^0_{\mathbf{k}\omega}, \tag{8.16}$$

where $\langle j_k j_l\rangle^0_{\mathbf{k}\omega}$ is the spectral distribution of current density fluctuations in a non-equilibrium ensemble of charged particles in neglect of the electromagnetic interactions between the latter. Within the context of (8.16) we obtain the dielectric permittivity tensor of a non-equilibrium plasma in the following form:

$$\varepsilon_{ij}(\omega, \mathbf{k}) = \left(1-\frac{\Omega^2}{\omega^2}\right)\delta_{ij} + \frac{2}{m\omega^2}\int_{-\infty}^{\infty} d\omega' \frac{\left((k_i\delta_{jl}+\delta_{il}k_j)\frac{k_k}{\omega'} - \delta_{ik}\delta_{jl}k^2\frac{\partial}{\partial\omega'}\right)\langle j_k j_l\rangle^0_{\mathbf{k}\omega'}}{\omega'-\omega-i0}. \tag{8.17}$$

For the extension of the above results for a spatially inhomogeneous plasma, see Sitenko and Yakimenko, 1974.

8.2. Non-linear Fluctuation–Dissipation Relations

Relation between non-linear susceptibilities and higher-order correlation functions for plasma fluctuations (potential case)

The fluctuation–dissipation theorem, which was discussed in the preceding section both for equilibrium and non-equilibrium situations, determines the fundamental connection between the linear electric susceptibility and the pair correlation functions for the fluctuations in the system. It is clear that such a connection exists both in linear and non-linear electrodynamic systems.† When investigating non-linear electrodynamic systems one can derive, besides the usual fluctuation–dissipation theorem, also a number of additional relations which connect the non-linear electric susceptibilities with the correlation functions of order higher than two for the fluctuations in the system (Sitenko, 1973*a*, c). In the plasma-like media such additional relations may be written in the form of a generalized fluctuation–dissipation theorem that connects the discontinuities of the non-linear electric susceptibilities $\varkappa^{(n)}(\omega_1, \mathbf{k}_1; \omega_2, \mathbf{k}_2; \ldots \omega_n, \mathbf{k}_n)$ at the cuts in the complex ω-planes ($\omega = \omega_1, \omega_2, \ldots, \omega_n$) with the higher-order spectral correlation functions for the charge density fluctuations in a system of non-interacting particles $\langle \varrho^{n+1} \rangle^0_{\mathbf{k}_1\omega_1; \mathbf{k}_2\omega_2; \ldots; \mathbf{k}_n\omega_n}$:

$$\begin{aligned}\operatorname{Im}_{\omega_1}\{\operatorname{Im}_{\omega_2} \ldots \{\operatorname{Im}_{\omega_n} \varkappa^{(n)}(\omega_1, \mathbf{k}_1; \omega_2, \mathbf{k}_2; \ldots; \omega_n, \mathbf{k}_n)\} \ldots\} \\ = \mathcal{L}^{(n)}(\omega_1, \mathbf{k}_1; \omega_2, \mathbf{k}_2; \ldots; \omega_n, \mathbf{k}_n) \langle \varrho^{n+1} \rangle^0_{\mathbf{k}_1\omega_1; \mathbf{k}_2\omega_2; \ldots; \mathbf{k}_n\omega_n},\end{aligned} \tag{8.18}$$

where $\mathcal{L}^{(n)}(\omega_1, \mathbf{k}_1; \omega_2, \mathbf{k}_2; \ldots; \omega_n, \mathbf{k}_n)$ is an nth order differential operator with respect to the variables $\omega_1, \omega_1, \ldots, \omega_n$, which depends on the parameters $\mathbf{k}_1, \mathbf{k}_2, \ldots, \mathbf{k}_n$. $\operatorname{Im}_{\omega_1}\{\operatorname{Im}_{\omega_2} \ldots \{\operatorname{Im}_{\omega_n} \varkappa^{(n)}(\omega_1, \mathbf{k}_1; \omega_2, \mathbf{k}_2; \ldots; \omega_n, \mathbf{k}_n\} \ldots\}$ is real if n is odd and imaginary if n is even. Note that the conventional fluctuation-dissipation theorem (8.12) is a special case of (8.18) at $n = 1$.

To derive the operator $\mathcal{L}^{(n)}(\omega_1, \mathbf{k}_1; \omega_2, \mathbf{k}_2; \ldots; \omega_n, \mathbf{k}_n)$ explicity we compare the general formula for the nth order non-linear susceptibility (2.19) with the $(n+1)$st order spectral correlation function for the charge density fluctuations in a system of non-interacting particles (6.47):

$$\begin{aligned}&\mathcal{L}^{(n)}(\omega_1, \mathbf{k}_1; \omega_2, \mathbf{k}_2; \ldots; \omega_n, \mathbf{k}_n) \\ &\quad = -\frac{i^{(n-1)}}{n!} \frac{4\pi}{(2m)^n} \frac{1}{k_1 k_2 \ldots k_n k} \mathcal{P}\Big\{(\mathbf{k}_1 \cdot \{\mathbf{k}_1 + \mathbf{k}_2 + \ldots + \mathbf{k}_n\}) \frac{\partial}{\partial \omega_1} \\ &\quad \times \left((\mathbf{k}_1 \cdot \mathbf{k}_2) \frac{\partial}{\partial \omega_1} + (\mathbf{k}_2 \cdot \{\mathbf{k}_2 + \ldots + \mathbf{k}_n\}) \frac{\partial}{\partial \omega_2}\right) \\ &\quad \times \left((\mathbf{k}_1 \cdot \mathbf{k}_3) \frac{\partial}{\partial \omega_1} + (\mathbf{k}_2 \cdot \mathbf{k}_3) \frac{\partial}{\partial \omega_2} + (\mathbf{k}_3 \cdot \{\mathbf{k}_3 + \ldots + \mathbf{k}_n\}) \frac{\partial}{\partial \omega_3}\right) \ldots \times \\ &\quad \times \left((\mathbf{k}_1 \cdot \mathbf{k}_n) \frac{\partial}{\partial \omega_1} + (\mathbf{k}_2 \cdot \mathbf{k}_n) \frac{\partial}{\partial \omega_2} + \ldots + \mathbf{k}_n^2 \frac{\partial}{\partial \omega_n}\right)\Big\},\end{aligned} \tag{8.19}$$

where the symbol $\mathcal{P}$ implies all possible permutations of the pairs $(\omega_i, \mathbf{k}_i)$ and $\mathbf{k} = \mathbf{k}_1 + \mathbf{k}_2 + \ldots + \mathbf{k}_n$.

We give here the lowest orders of $\mathcal{L}^{(n)}$ ($n = 1, 2, 3$):

$$\mathcal{L}^{(1)}(\omega, k) = -\frac{2\pi}{m} \frac{\partial}{\partial \omega},$$

† The validity of the fluctuation–dissipation theorem for non-linear dissipative media was proved by Bernard and Callen (1960) and Bunkin (1962); for extensive discussions of this problem see Levin and Rytov (1967).

$$\mathcal{L}^{(2)}(\omega_1, \mathbf{k}_1; \omega_2, \mathbf{k}_2) = -i\frac{\pi}{2m^2}\frac{1}{k_1k_2\,|\mathbf{k}_1+\mathbf{k}_2|}\left\{(\mathbf{k}_1\cdot\{\mathbf{k}_1+\mathbf{k}_2\})\frac{\partial}{\partial\omega_1}\left((\mathbf{k}_1\cdot\mathbf{k}_2)\frac{\partial}{\partial\omega_1}+k_2^2\frac{\partial}{\partial\omega_2}\right)\right.$$
$$\left.+(\mathbf{k}_2\cdot\{\mathbf{k}_1+\mathbf{k}_2\})\frac{\partial}{\partial\omega_2}\left(k_1^2\frac{\partial}{\partial\omega_1}+(\mathbf{k}_1\cdot\mathbf{k}_2)\frac{\partial}{\partial\omega_2}\right)\right\},$$

$$\mathcal{L}^{(3)}(\omega_1, \mathbf{k}_1; \omega_2, \mathbf{k}_2; \omega_3, \mathbf{k}_3) = \frac{\pi}{12}\frac{1}{m^3}\frac{1}{k_1k_2k_3\,|\mathbf{k}_1+\mathbf{k}_2+\mathbf{k}_3|}\left\{(\mathbf{k}_1\cdot\{\mathbf{k}_1+\mathbf{k}_2+\mathbf{k}_3\})\frac{\partial}{\partial\omega_1}\right.$$
$$\times\left[\left((\mathbf{k}_1\cdot\mathbf{k}_2)\frac{\partial}{\partial\omega_1}+(\mathbf{k}_2\cdot\{\mathbf{k}_2+\mathbf{k}_3\})\frac{\partial}{\partial\omega_2}\right)\right.$$
$$\times\left((\mathbf{k}_1\cdot\mathbf{k}_3)\frac{\partial}{\partial\omega_1}+(\mathbf{k}_2\cdot\mathbf{k}_3)\frac{\partial}{\partial\omega_2}+k_3^2\frac{\partial}{\partial\omega_3}\right)$$
$$+\left\{(\mathbf{k}_1\cdot\mathbf{k}_3)\frac{\partial}{\partial\omega_1}+(\mathbf{k}_3\cdot\{\mathbf{k}_2+\mathbf{k}_3\})\frac{\partial}{\partial\omega_3}\right\}$$
$$\left.\times\left((\mathbf{k}_1\cdot\mathbf{k}_2)\frac{\partial}{\partial\omega_3}+(\mathbf{k}_2\cdot\mathbf{k}_3)\frac{\partial}{\partial\omega_3}+k_2^2\frac{\partial}{\partial\omega_2}\right)\right]$$
$$\left.+(1\rightleftarrows 2)+(1\rightleftarrows 3)\right\}.$$

The dispersion relation, which reflects the causality principle, is for the nth order susceptibility of the form

$$\varkappa^{(n)}(\omega_1, \mathbf{k}_1; \omega_2, \mathbf{k}_2; \ldots; \omega_n, \mathbf{k}_n) = \frac{1}{\pi^n}\int_{-\infty}^{\infty}d\omega_1'\int_{-\infty}^{\infty}d\omega_2'\ldots$$
$$\ldots\int_{-\infty}^{\infty}d\omega_n'\frac{\mathrm{Im}_{\omega_1'}\{\mathrm{Im}_{\omega_2'}\ldots\{\mathrm{Im}_{\omega_n'}\varkappa^{(n)}(\omega_1', \mathbf{k}_1; \omega_2', \mathbf{k}_2; \ldots; \omega_n', \mathbf{k}_n)\}\ldots\}}{(\omega_1'+\omega_2'+\ldots+\omega_n'-\omega_1-\omega_2-\ldots-\omega_n-i0)\,(\omega_2'+\ldots+\omega_n'-\omega_2-\ldots\omega_n-i0)\ldots(\omega_n'-\omega_n-i0)}. \quad (8.20)$$

Making use of this relation we can find the non-linear susceptibility $\varkappa^{(n)}(\omega_1, \mathbf{k}_1; \omega_2, \mathbf{k}_2; \ldots; \omega_n, \mathbf{k}_n)$ for any complex frequencies $\omega_1, \omega_2, \ldots, \omega_n$ from the known susceptibility discontinuities at the cuts in the relevant complex planes. Within the context of (8.18) these discontinuities are determined by the spectral correlation functions for the charge density fluctuations in neglect of the Coulomb interactions between particles.

The formulas (8.18) and (8.20) are general and make it possible to calculate the non-linear susceptibilities for the systems of particles interacting through the Coulomb forces from the given correlation functions for systems without Coulomb interaction. The relation (8.18) may be regarded as the generalized fluctuation–dissipation theorem for a non-linear electrodynamic medium. According to (8.18) and (8.20) the linear and non-linear electrodynamic plasma properties are completely determined by the given set of correlation functions of ascending orders for the charge density fluctuations in neglect of the Coulomb interactions between particles.

Spectral correlation functions

The nth order spectral correlation function for the charge density fluctuations taking into account the Coulomb interaction between particles may be written in the polarization approximation as

$$\langle \varrho^{n+1} \rangle_{\mathbf{k}_1\omega_1;\,\mathbf{k}_2\omega_2;\,\ldots;\,\mathbf{k}_n\omega_n} = \frac{\langle \varrho^{n+1} \rangle^0_{\mathbf{k}_1\omega_1;\,\mathbf{k}_2\omega_2;\,\ldots;\,\mathbf{k}_n\omega_n}}{\varepsilon(\omega_1, \mathbf{k}_1)\,\varepsilon(\omega_2, \mathbf{k}_2)\ldots\varepsilon(\omega_n, \mathbf{k}_n)\,\varepsilon^*(\omega_1+\omega_2+\ldots+\omega_n, \mathbf{k}_1+\mathbf{k}_2+\ldots+\mathbf{k}_n)}. \tag{8.21}$$

Within the context of (8.18) the spectral correlation functions of various orders for the charge density and electric field fluctuations may be expressed in terms of the non-linear susceptibilities. To illustrate this we give here the third-order spectral correlation function:

$$\langle \varrho^3 \rangle_{\mathbf{k}_1\omega_1;\,\mathbf{k}_2\omega_2} = \frac{\{\mathcal{L}^{(2)}(\omega_1, \mathbf{k}_1;\, \omega_2, \mathbf{k}_2)\}^{-1}\,\mathrm{Im}_{\omega_1}\{\mathrm{Im}_{\omega_2}\,\varkappa^{(2)}(\omega_1, \mathbf{k}_1;\, \omega_2, \mathbf{k}_2)\}}{\varepsilon(\omega_1, \mathbf{k}_1)\,\varepsilon(\omega_2, \mathbf{k}_2)\,\varepsilon^*(\omega, \mathbf{k})}. \tag{8.22}$$

If the plasma is in equilibrium, then $\mathcal{L}^{(n)}$ is a multiplication operator. In the $n = 2$ case we have

$$\mathcal{L}^{(2)}(\omega_1, \mathbf{k}_1;\, \omega_2, \mathbf{k}_2) = i\,\frac{\pi}{2T^2}\,\frac{\{\omega_1^2(\mathbf{k}_2\cdot\mathbf{k})+\omega_2^2(\mathbf{k}_1\cdot\mathbf{k})\}\,(\mathbf{k}_1\cdot\mathbf{k}_2)-\omega_1\omega_2(k_1^2\cdot(\mathbf{k}_2\cdot\mathbf{k})+k_2^2\cdot(\mathbf{k}_1\cdot\mathbf{k}))}{k_1k_2k(k_1^2k_2^2-(\mathbf{k}_1\cdot\mathbf{k}_2)^2)} \tag{8.23}$$

and the ternary spectral correlation function for the charge density fluctuations reduces to the following:

$$\langle \varrho^3 \rangle_{\mathbf{k}_1\omega_1;\,\mathbf{k}\omega} = -\frac{2i}{\pi}\,T^2\,\frac{k_1k_2k(k_1^2k_2^2-(\mathbf{k}_1\cdot\mathbf{k}_2)^2)}{(\omega_1^2(\mathbf{k}_2\cdot\mathbf{k})+\omega_2^2(\mathbf{k}_1\cdot\mathbf{k}))\,(\mathbf{k}_1\cdot\mathbf{k}_2)-\omega_1\omega_2(k_1^2(\mathbf{k}_2\cdot\mathbf{k})+k_2^2(\mathbf{k}_1\cdot\mathbf{k}))}$$
$$\times\frac{\mathrm{Im}_{\omega_1}\{\mathrm{Im}_{\omega_2}\,\varkappa^{(2)}(\omega_1, \mathbf{k}_1;\, \omega_2, \mathbf{k}_2)\}}{\varepsilon(\omega_1, \mathbf{k}_1)\,\varepsilon(\omega_2, \mathbf{k}_2)\,\varepsilon^*(\omega, \mathbf{k})}. \tag{8.24}$$

The higher-order spectral correlation functions for the charge density and electric field fluctuations may be written analogously.

Within the context of (8.18) and (8.23) the second-order susceptibility of an equilibrium plasma may be written as

$$\varkappa^{(2)}(\omega_1, \mathbf{k}_1;\, \omega_2, \mathbf{k}_2) = \frac{i}{2\pi T^2}\,\frac{1}{\omega k_1k_2k[k_1^2k_2^2-(\mathbf{k}_1\cdot\mathbf{k}_2)^2]}$$
$$\times\int_{-\infty}^{\infty} d\omega_1' \int_{-\infty}^{\infty} d\omega_2'\,\frac{\omega_1'+\omega_2'}{\omega_1'+\omega_2'-\omega_1-\omega_2-i0}\left\{\frac{\omega_1'}{\omega_1'-\omega_1-i0}\,(\omega_1'(\mathbf{k}_1\cdot\mathbf{k}_2)-\omega_2k_1^2)\,(\mathbf{k}_2\cdot\mathbf{k})\right.$$
$$\left.+\frac{\omega_2'}{\omega_2'-\omega_2-i0}\,(\omega_2'(\mathbf{k}_1\cdot\mathbf{k}_2)-\omega_1k_2^2)\,(\mathbf{k}_1\cdot\mathbf{k})\right\}\langle \varrho^3 \rangle^0_{\mathbf{k}_1\omega_1';\,\mathbf{k}_2\omega_2'}. \tag{8.25}$$

It may be immediately verified making use of this representation that the following relation occurs:

$$\operatorname{Im}\{\omega\varkappa^{(2)}(\omega_1, \mathbf{k}_1; \omega_2, \mathbf{k}_2)-\omega_1\varkappa^{(2)}(\omega, \mathbf{k}; -\omega_2, -\mathbf{k}_2)-\omega_2\varkappa^{(2)}(\omega, \mathbf{k}; -\omega_1, -\mathbf{k}_1)\}$$
$$= \frac{\pi}{T^2}\,\frac{\omega_1\omega_2\omega}{k_1k_2k}\langle\varrho^3\rangle^0_{\mathbf{k}_1\omega_1;\,\mathbf{k}_2\omega_2}. \tag{8.26}$$

The discontinuity of the second-order susceptibility at the cut may easily be calculated explicitly by means of the last relation. Thus we obtain the following formula for the ternary spectral correlation function for the charge density fluctuations in an equilibrium plasma:

$$\langle\varrho^3\rangle_{\mathbf{k}_1\omega_1;\,\mathbf{k}_2\omega_2} = \frac{T^2}{\pi}\,\frac{k_1k_2k}{\varepsilon(\omega_1, \mathbf{k}_1)\,\varepsilon(\omega_2, \mathbf{k}_2)\,\varepsilon^*(\omega, \mathbf{k})}$$
$$\times \operatorname{Im}\left\{\frac{\varkappa^{(2)}(\omega_1, \mathbf{k}_1; \omega_2, \mathbf{k}_2)}{\omega_1\omega_2} - \frac{\varkappa^{(2)}(\omega, \mathbf{k}; -\omega_1, -\mathbf{k}_1)}{\omega_1\omega} - \frac{\varkappa^{(2)}(\omega, \mathbf{k}; -\omega_2, -\mathbf{k}_2)}{\omega_2\omega}\right\}. \tag{8.27}$$

This expression was derived by Golden, Kalman, and Silevich (1972). It follows from (8.27) that the ternary spectral correlation function for the electric field fluctuations is equal to

$$\langle E^2\rangle_{\mathbf{k}_1\omega_1;\,\mathbf{k}_2\omega_2} = -i64\pi^2\,\frac{T^2}{\varepsilon(\omega_1, \mathbf{k}_1)\,\varepsilon(\omega_2, \mathbf{k}_2)\,\varepsilon^*(\omega, \mathbf{k})}$$
$$\times \operatorname{Im}\left\{\frac{\varkappa^{(2)}(\omega_1, \mathbf{k}_1; \omega_2, \mathbf{k}_2)}{\omega_1\omega_2} - \frac{\varkappa^{(2)}(\omega, \mathbf{k}; -\omega_1, -\mathbf{k}_1)}{\omega_1\omega} - \frac{\varkappa^{(2)}(\omega, \mathbf{k}; -\omega_2, -\mathbf{k}_2)}{\omega_2\omega}\right\}. \tag{8.28}$$

It should be pointed out that, being written in the form (8.28), the spectral correlation function for electric field fluctuations is adequate not only for a plasma, but also for any non-linear medium. An analogous formula for the spectral distribution of the field fluctuations in the absence of spatial dispersion was derived by Efremov (1969).

Sum rules

The explicit expressions for the spectral correlation functions for the charge density fluctuations in the absence of Coulomb interactions may be made use of to derive the general integral relation between the linear plasma susceptibility and the non-linear ones. Indeed, according to (6.47) the following relation occurs:

$$\int_{-\infty}^{\infty} d\omega_n\langle\varrho^{n+1}\rangle^0_{\mathbf{k}_1\omega_1;\,\mathbf{k}_1\omega_2;\,\ldots;\,\mathbf{k}_n\omega_n} = 2\pi e\langle\varrho^n\rangle^0_{\mathbf{k}_1\omega_1;\,\mathbf{k}_2\omega_2;\,\ldots;\,\mathbf{k}_{n-1}\omega_{n-1}}. \tag{8.29}$$

After having expressed the spectral correlation functions in terms of the discontinuities of the non-linear susceptibilities on the cuts in the complex planes of the relevant frequencies within the context of (8.18), we obtain the following general formula:

$$\mathscr{L}^{(n-1)}(\omega_1, \mathbf{k}_1; \ldots; \omega_{n-1}, \mathbf{k}_{n-1})\int_{-\infty}^{\infty} d\omega_n\{\mathscr{L}^{(n)}(\omega_1, \mathbf{k}_1; \ldots; \omega_n, \mathbf{k}_n)\}^{-1}$$
$$\times \operatorname{Im}_{\omega_1}\{\operatorname{Im}_{\omega_2}\ldots\{\operatorname{Im}_{\omega_n}\varkappa^{(n)}(\omega_1, \mathbf{k}_1; \ldots; \omega_n, \mathbf{k}_n)\}\ldots\}$$
$$= 2\pi e\operatorname{Im}_{\omega_1}\{\operatorname{Im}_{\omega_2}\ldots\{\operatorname{Im}_{\omega_{n-1}}\varkappa^{(n-1)}(\omega_1, \mathbf{k}_1; \ldots; \omega_{n-1}, \mathbf{k}_{n-1})\}\ldots\}. \tag{8.30}$$

In particular, the dielectric permittivity and the quadratic non-linear susceptibility are related according to

$$\int_{-\infty}^{\infty} d\omega_2 \operatorname{Im}\left\{\frac{\varkappa^{(2)}(\omega_1, \mathbf{k}_1; \omega_2, \mathbf{k}_2)}{\omega_1\omega_2} - \frac{\varkappa^{(2)}(\omega, \mathbf{k}; -\omega_1, -\mathbf{k}_1)}{\omega_1\omega} - \frac{\varkappa^{(2)}(\omega, \mathbf{k}; -\omega_2, -\mathbf{k}_2)}{\omega_2\omega}\right\}$$

$$= \pi \frac{e}{T} \frac{k_1}{k_2 k} \frac{\operatorname{Im} \varepsilon(\omega, \mathbf{k})}{\omega_1}. \tag{8.31}$$

Making use of (6.47) we can derive numerous relations between the non-linear susceptibilities of different orders.

The non-linear electric susceptibilities of a plasma satisfy a number of sum rules in the same way as the dielectric permittivity. Making use of (6.47) we can easily carry out the integration of the spectral correlation function $\langle \varrho^{n+1} \rangle^0_{\mathbf{k}_1\omega_1; \ldots; \mathbf{k}_n\omega_n}$ over all frequencies. After having expressed the spectral correlation function in terms of the non-linear susceptibility $\varkappa^{(n)}(\omega_1, \mathbf{k}_1; \ldots; \omega_n, \mathbf{k}_n)$ with the use of (8.18), we obtain an integral sum rule:

$$\int_{-\infty}^{\infty} d\omega_1 \int_{-\infty}^{\infty} d\omega_2 \ldots \int_{-\infty}^{\infty} d\omega_n \{\mathcal{L}^{(n)}(\omega_1, \mathbf{k}_1; \omega_2, \mathbf{k}_2; \ldots; \omega_n, \mathbf{k}_n)\}^{-1}$$

$$\times \operatorname{Im}_{\omega_1}\{\operatorname{Im}_{\omega_2} \ldots \{\operatorname{Im}_{\omega_n} \varkappa^{(n)}(\omega_1, \mathbf{k}_1; \omega_2, \mathbf{k}_2; \ldots; \omega_n, \mathbf{k}_n)\} \ldots\} = (2\pi)^n e^{n+1} n_0. \tag{8.32}$$

This relation turns out to be suitable for calculations in the case when a plasma is in thermal equilibrium. Then the quantity $\{\mathcal{L}^{(n)}(\omega_1, \mathbf{k}_1; \ldots; \omega_n, \mathbf{k}_n)\}^{-1}$ is a multiplication operator. For example, the dielectric permittivity and the quadratic non-linear susceptibility satisfy the following sum rules:

$$\int_{-\infty}^{\infty} d\omega \frac{\operatorname{Im} \varepsilon(\omega, \mathbf{k})}{\omega} = \frac{4\pi^2 e^2 n_0}{k^2 T}, \tag{8.33}$$

$$\int_{-\infty}^{\infty} d\omega_1 \int_{-\infty}^{\infty} d\omega_2 \operatorname{Im}\left\{\frac{\varkappa^{(2)}(\omega_1, \mathbf{k}_1; \omega_2, \mathbf{k}_2)}{\omega_1\omega_2} - \frac{\varkappa^{(2)}(\omega, \mathbf{k}; -\omega_1, -\mathbf{k}_1)}{\omega_1\omega}\right.$$

$$\left. - \frac{\varkappa^{(2)}(\omega, \mathbf{k}; -\omega_2, -\mathbf{k}_2)}{\omega_2\omega}\right\} = \frac{4\pi^3}{k_1 k_2 k} \frac{e^3 n_0}{T^2}. \tag{8.34}$$

The sum rules of the form

$$\int_{-\infty}^{\infty} d\omega_1 \int_{-\infty}^{\infty} d\omega_2 \ldots \int_{-\infty}^{\infty} d\omega_n\, \omega_1\omega_2 \{\mathcal{L}^{(n)}((\omega_1, \mathbf{k}_1; \omega_2, \mathbf{k}_2; \ldots; \omega_n, \mathbf{k}_n)\}^{-1}$$

$$\times \operatorname{Im}_{\omega_1}\{\operatorname{Im}_{\omega_2} \ldots \{\operatorname{Im}_{\omega_n} \varkappa^{(n)}(\omega_1, \mathbf{k}_1; \omega_2, \mathbf{k}_2; \ldots; \omega_n, \mathbf{k}_n)\} \ldots\}$$

$$= (2\pi)^n e^{(n+1)} n_0 \frac{T}{m} (\mathbf{k}_1 \cdot \mathbf{k}_2) \tag{8.35}$$

hold in an equilibrium plasma. In particular, for the dielectric permittivity and the quadratic non-linear susceptibility, we have:

$$\int_{-\infty}^{\infty} d\omega\, \omega \,\mathrm{Im}\, \varepsilon(\omega, \mathbf{k}) = \frac{4\pi^2 e^2 n_0}{m}, \tag{8.36}$$

$$\int_{-\infty}^{\infty} d\omega_1 \int_{-\infty}^{\infty} d\omega_2 \omega_1 \omega_2 \,\mathrm{Im}\left\{ \frac{\varkappa^{(2)}(\omega_1, \mathbf{k}_1; \omega_2, \mathbf{k}_2)}{\omega_1\omega_2} - \frac{\varkappa^{(2)}(\omega, \mathbf{k}; -\omega_1, -\mathbf{k}_1)}{\omega_1\omega} - \frac{\varkappa^{(2)}(\omega, \mathbf{k}; -\omega_1, -\mathbf{k}_1)}{\omega_2\omega} \right\} = \frac{4\pi^3}{k_1k_2k}\, \frac{e^3 n_0}{mT} (\mathbf{k}_1 \cdot \mathbf{k}_2). \tag{8.37}$$

The sum rules for the third-order non-linear susceptibility may be derived in the same way.

Let us derive the sum rules for the non-linear susceptibilities $\varkappa^{(2)}(1, 2)$ and $\varkappa^{(3)}(1, 2, 3)$ which would be analogous to those for the linear conductivity. The dispersion relation for the quadratic susceptibility $\varkappa^{(2)}(\omega_1, \mathbf{k}_1; \omega_2, \mathbf{k}_2)$ with respect to one of the frequencies, e.g. ω_1, may be written as

$$\varkappa^{(2)}(\omega_1, \mathbf{k}_1; \omega_2, \mathbf{k}_2) = \frac{1}{\pi} \int_{-\infty}^{\infty} d\omega_1' \frac{\mathrm{Im}\, \varkappa^{(2)}(\omega_1', \mathbf{k}_1; \omega_2, \mathbf{k}_2)}{\omega_1' - \omega_1 - i0} \tag{8.38}$$

(the frequency ω_2 is suggested real). Separating the real and the imaginary parts we have

$$\mathrm{Re}\, \varkappa^{(2)}(\omega_1, \mathbf{k}_1; \omega_2, \mathbf{k}_2) = \frac{1}{\pi} \fint_{-\infty}^{\infty} d\omega_1' \frac{\mathrm{Im}\, \varkappa^{(2)}(\omega_1', \mathbf{k}_1; \omega_2, \mathbf{k}_2)}{\omega_1' - \omega_1},$$

$$\mathrm{Im}\, \varkappa^{(2)}(\omega_1, \mathbf{k}_1; \omega_2, \mathbf{k}_2) = -\frac{1}{\pi} \fint_{-\infty}^{\infty} d\omega_1' \frac{\mathrm{Re}\, \varkappa^{(2)}(\omega_1', \mathbf{k}_1; \omega_2, \mathbf{k}_2)}{\omega_1' - \omega_1}. \tag{8.39}$$

The sum rule of interest follows from the second relation (8.39) in the limiting case $\omega_1 \to \infty$ and $\omega_2 \to \infty$. To verify this we expand the integrand on the right-hand side in a power series in the ratio ω_1'/ω_1 and make an observation that the real part of the non-linear susceptibility is an odd function ($\mathrm{Re}\, \varkappa^{(2)}(-\omega_1, \mathbf{k}_1; \omega_2, \mathbf{k}_2) = -\mathrm{Re}\, \varkappa^{(2)}(\omega_1, \mathbf{k}_1; \omega_2, \mathbf{k}_2)$). Then

$$\lim_{\omega_1 \to \infty} \fint_{-\infty}^{\infty} d\omega_1 \frac{\mathrm{Re}\, \varkappa^{(2)}(\omega_1', \mathbf{k}_1; \omega_2 \to \infty, \mathbf{k}_2)}{\omega_1' - \omega_1}$$

$$= -\lim_{\omega_1 \to \infty} \frac{1}{\omega_1^2} \int_{-\omega_1}^{\omega_1} d\omega_1' \omega_1' \,\mathrm{Re}\, \varkappa^{(2)}(\omega_1', \mathbf{k}_1; \omega_2 \to \infty, \mathbf{k}_2).$$

Note that the real part of the non-linear susceptibility that enters the integrand is to be taken at $\omega_2 \to \infty$ and finite ω_1'.

The limiting value of the imaginary part of the non-linear susceptibility $\varkappa^{(2)}(\omega_1, \mathbf{k}_1; \omega_2, \mathbf{k}_2)$ may be found by means of the hydrodynamic formula (2.30):

$$\operatorname{Im} \varkappa^{(2)}(\omega_1 \to \infty, \mathbf{k}_1; \omega_2 \to \infty, \mathbf{k}_2) = -\frac{1}{2} \frac{e}{m} \frac{\Omega^2}{\omega_1^2 \omega_2^2} R(\mathbf{k}_1, \mathbf{k}_2), \tag{8.40}$$

where

$$R(\mathbf{k}_1, \mathbf{k}_2) = \begin{cases} \dfrac{k_1}{k_2 k} (\mathbf{k}_2 \cdot \mathbf{k}), & \dfrac{\omega_2}{\omega_1} \to \infty, \\[2ex] \dfrac{k_2}{k_1 k} (\mathbf{k}_1 \cdot \mathbf{k}), & \dfrac{\omega_1}{\omega_2} \to \infty. \end{cases}$$

Thus we obtain the following sum rules for the second-order non-linear susceptibility:

$$\int_{-\infty}^{\infty} d\omega_1 \omega_1 \operatorname{Re} \varkappa^{(2)}(\omega_1, \mathbf{k}_1; \omega_2 \to \infty, \mathbf{k}_2) = -\frac{\pi}{2} \frac{e}{m} \frac{\Omega^2}{\omega_2^2} \frac{k_1}{k_2 k} (\mathbf{k}_2 \cdot \mathbf{k}),$$

$$\int_{-\infty}^{\infty} d\omega_2 \omega_2 \operatorname{Re} \varkappa^{(2)}(\omega_1 \to \infty, \mathbf{k}_1; \omega_2, \mathbf{k}_2) = -\frac{\pi}{2} \frac{e}{m} \frac{\Omega^2}{\omega_1^2} \frac{k_2}{k_1 k} (\mathbf{k}_1 \cdot \mathbf{k}). \tag{8.41}$$

It may be directly verified that the quadratic non-linear susceptibility (2.16) always satisfies the above sum rules. (The sum rule (8.41) was derived by Golden, Kalman, and Datta, 1975.)

The derivation procedure for the sum rules for the cubic susceptibility is analogous. The result is as follows:

$$\int_{-\infty}^{\infty} d\omega_1 \omega_1 \operatorname{Im} \varkappa^{(3)}(\omega_1, \mathbf{k}_1; \omega_2 \to \infty, \mathbf{k}_1; \omega_3 \to \infty, \mathbf{k}_3) = \frac{\pi}{6} \frac{e^2}{m^2} \frac{\Omega^2}{\omega_2 \omega_3 (\omega_2 + \omega_3)^2} \frac{k_1}{k_2 k_3 k}$$

$$\times \left\{ 2(\mathbf{k}_2 \cdot \mathbf{k}) \cdot (\mathbf{k}_3 \cdot \mathbf{k}) + \frac{\omega_2}{\omega_3} k_3^2 \cdot (\mathbf{k}_2 \cdot \mathbf{k}) + \frac{\omega_3}{\omega_2} k_2^2 \cdot (\mathbf{k}_3 \cdot \mathbf{k}) \right\}, \tag{8.42}$$

and so on.

The non-linear fluctuation–dissipation theorem for electromagnetic fluctuations

The generalized fluctuation–dissipation theorem may be extended to involve considering arbitrary electromagnetic (not necessarily potential) fluctuations (Kocherga, 1975). In the latter case the fluctuation–dissipation theorem for an arbitrary electromagnetic field establishes a relationship between the imaginary parts of the non-linear tensor susceptibilities and the spectral distribution or the current density fluctuations in neglect of electromagnetic particle interactions:

$$\operatorname{Im}_{\omega_1} \left\{ \operatorname{Im}_{\omega_2} \ldots \left\{ \operatorname{Im}_{\omega_n} \varkappa^{(n)}_{ijk \ldots pr} (1, 2, \ldots, n-1, n) \right\} \ldots \right\}$$

$$= \mathcal{L}^{(n)}_{ijk \ldots pr; i'j'k' \ldots p'r'} (1, 2, \ldots, n-1, n) \langle j_{i'} j_{j'} j_{k'} \ldots j_{p'} j_{r'} \rangle^0_{1, 2, \ldots n-1, n}. \tag{8.43}$$

Summation is implied over repeated subscipts. The differential operator is of the following general form:

$$\mathcal{L}^{(n)}_{ijk\ldots pr;i'j'k'\ldots p'r'}(1,2,3,\ldots,n-1,n)$$

$$= \frac{(-i)^{n-1}}{n!}\,\frac{4\pi}{(2m)^n}\,\frac{1}{\omega_1\omega_2\ldots\omega_n\omega}\,\mathcal{P}\left\{\frac{1}{\omega_1}(k_{1i}k_{1i''}\delta_{jj''}+\delta_{ii''}k_{1j}k_{1j''})-\delta_{ii''}\delta_{jj''}(\mathbf{k}_1\cdot\mathbf{k})\frac{\partial}{\partial\omega_1}\right\}$$

$$\times\left\{\frac{1}{\omega_2}[k_{2i''}k_{2i'''}\delta_{j''j'''}\delta_{kk''}+\delta_{i''i'''}k_{2j''}k_{2j'''}\delta_{kk''}+\delta_{i''i'''}\delta_{j''j'''}(\mathbf{k}_2+\ldots+\mathbf{k}_n)_k\,k_{2k''}]\right.$$

$$\left.-\delta_{i''i'''}\delta_{j''j'''}\delta_{kk''}\left[(\mathbf{k}_1\cdot\mathbf{k})\frac{\partial}{\partial\omega_1}+(\mathbf{k}_2\cdot\{\mathbf{k}_2+\mathbf{k}_3+\ldots+\mathbf{k}_n\})\frac{\partial}{\partial\omega_2}\right]\right\}$$

$$\times\left\{\frac{1}{\omega_3}[k_{3i'''}k_{3i''''}\delta_{j'''j''''}\delta_{k''k'''}\delta_{ll''}+\delta_{i'''i''''}k_{3j'''}k_{3j''''}\delta_{k''k'''}\delta_{ll''}\right.$$

$$+\delta_{i'''i''''}\delta_{j'''j''''}k_{3k''}k_{k3'''}\delta_{ll''}+\delta_{i'''i''''}\delta_{j'''j''''}\delta_{k''k'''}(\mathbf{k}_3+\ldots+\mathbf{k}_n)_l\,k_{3l''}]$$

$$\left.-\delta_{i'''i''''}\delta_{j'''j''''}\delta_{k''k'''}\delta_{ll''}\left[(\mathbf{k}_1\cdot\mathbf{k}_3)\frac{\partial}{\partial\omega_1}+(\mathbf{k}_2\cdot\mathbf{k}_3)\frac{\partial}{\partial\omega_2}(\mathbf{k}_3\cdot\{\mathbf{k}_3+\ldots+\mathbf{k}_n\})\frac{\partial}{\partial\omega_3}\right]\right\}\ldots$$

$$\times\left\{\frac{1}{\omega_n}\left[k_{ni^{(n)}}k_{ni'}\delta_{j^{(n)}j'}\delta_{k^{(n-1)}k'}\ldots\delta_{p''p'}\delta_{r'r}+\delta_{i^{(n)}i'}k_{nj^{(n)}}k_{nj'}\delta_{k^{(n)}k'}\ldots\delta_{p''p'}\delta_{r'r}+\ldots\right.\right.$$

$$+\delta_{i^{(n)}i'}\delta_{j^{(n)}j'}\delta_{k^{(n)}k'}\ldots k_{np''}k_{np'}\delta_{rr'}+\delta_{i^{(n)}i'}\delta_{j^{(n)}j'}\delta_{k^{(n-1)}k'}\ldots\delta_{p''p'}k_{nr'}k_{nr}$$

$$\left.\left.-\delta_{i^{(n)}i'}\delta_{j^{(n)}j'}\delta_{k^{(n-1)}k'}\ldots\delta_{p''p'}\delta_{r'r}\left[(\mathbf{k}_1\cdot\mathbf{k}_n)\frac{\partial}{\partial\omega_1}+(\mathbf{k}_2\cdot\mathbf{k}_n)\frac{\partial}{\partial\omega_2}+\ldots+k_n^2\frac{\partial}{\partial\omega_n}\right]\right]\right\},\tag{8.44}$$

where $\mathcal{P}$ implies symmetrization with respect to the arguments $1, 2, \ldots, n$ and the corresponding subscripts $i, j, k, \ldots, p, r$.

The operator (8.44) reduces for $n = 1, 2, 3$, to the following

$$\mathcal{L}^{(1)}_{ij;i'j'}(\omega,\mathbf{k}) = \frac{2\pi}{m}\,\frac{1}{\omega^2}\left\{\frac{1}{\omega}(k_ik_{i'}\delta_{jj'}+\delta_{ii'}k_jk_{j'})-\delta_{ii'}\delta_{jj'}k^2\frac{\partial}{\partial\omega}\right\},\tag{8.45}$$

$$\mathcal{L}^{(2)}_{ijk;i'j'k'}(\omega_1,\mathbf{k}_1;\omega_2,\mathbf{k}_2)$$

$$= -i\,\frac{\pi}{2m^2}\,\frac{1}{\omega_1\omega_2\omega}\,\mathcal{P}\left[\frac{1}{\omega_1}(k_{1i}k_{1i''}\delta_{jj''}+\delta_{ii''}k_jk_{1j''})-\delta_{ii''}\delta_{jj''}(\mathbf{k}_1\cdot\mathbf{k})\frac{\partial}{\partial\omega_1}\right]$$

$$\times\left\{\frac{1}{\omega_1}[k_{2i''}k_{2i'}\delta_{j''j'}\delta_{kk'}+\delta_{i''i'}k_{2j''}k_{2j'}\delta_{kk'}+\delta_{i''i'}\delta_{j''j'}k_{2k}k_{2k'}]\right.$$

$$\left.-\delta_{i''i'}\delta_{j''j'}\delta_{kk'}\left((\mathbf{k}_1\cdot\mathbf{k}_2)\frac{\partial}{\partial\omega_1}+k_2^2\frac{\partial}{\partial\omega_2}\right)\right\},\tag{8.46}$$

$$\mathcal{L}^{(3)}_{ijkl;i'j'k'l'}(\omega_1,\mathbf{k}_1;\omega_2,\mathbf{k}_2;\omega_3,\mathbf{k}_3)$$

$$= -\frac{\pi}{12m^3}\,\frac{1}{\omega_1\omega_2\omega_3\omega}\,\mathcal{P}\left\{\frac{1}{\omega_1}[k_{1i}k_{1i''}\delta_{jj''}+\delta_{ii''}k_jk_{1j''}]-\delta_{ii''}\delta_{jj''}(\mathbf{k}_1\cdot\mathbf{k})\frac{\partial}{\partial\omega_1}\right\}$$

$$\times\left\{\frac{1}{\omega_2}[k_{2i''}k_{2i'''}\delta_{j''j'''}\delta_{kk''}+\delta_{i''i'''}k_{2j''}k_{2j'''}\delta_{kk''}+\delta_{i''i'''}\delta_{j''j'''}\{\mathbf{k}_2+\mathbf{k}_3\}_k\,k_{2k''}]\right.$$

$$
\begin{aligned}
&-\delta_{i''i'''}\delta_{j''j'''}\delta_{kk''}\left[(\mathbf{k}_1\cdot\mathbf{k}_2)\frac{\partial}{\partial\omega_1}+(\mathbf{k}_2\cdot\{\mathbf{k}_2+\mathbf{k}_3\})\frac{\partial}{\partial\omega_2}\right]\Bigg\}\\
&\times\Bigg\{\frac{1}{\omega_3}[k_{3i'''}k_{3i'}\delta_{j'''j'}\delta_{k''k}\delta_{ll'}+\delta_{i'''i'}k_{3j'''}k_{3j'}\delta_{k''k}\delta_{ll'}+\delta_{i'''i'}\delta_{j'''j'}k_{3k''}k_{3k'}\delta_{ll'}\\
&+\delta_{i'''i'}\delta_{j'''j'}\delta_{k''k'}k_{3l}k_{3l'}]-\delta_{i'''i'}\delta_{j'''j'}\delta_{k''k'}\delta_{ll'}\left[(\mathbf{k}_1\cdot\mathbf{k}_3)\frac{\partial}{\partial\omega_1}+(\mathbf{k}_2\cdot\mathbf{k}_3)\frac{\partial}{\partial\omega_2}+k_3^2\frac{\partial}{\partial\omega_3}\right]\Bigg\}.
\end{aligned}
\tag{8.47}
$$

If the plasma is in thermal equilibrium, then $\mathscr{L}^{(n)}_{ij\ldots pr;i'j'\ldots p'r'}(1, 2, \ldots, n)$ becomes a multiplication operator. For example, $\mathscr{L}^{(2)}_{ijk;i'j'k'}(1, 2)$ in an equilibrium plasma

$$
\begin{aligned}
&\mathscr{L}^{(2)}_{ijk;i'j'k'}(\omega_1, \mathbf{k}_1; \omega_2, \mathbf{k}_2)\\
&\quad = i\frac{\pi}{2mT}\frac{1}{\omega_1\omega_2\omega}\frac{1}{k_1^2k_2^2-(\mathbf{k}_1\cdot\mathbf{k}_2)^2}\Big\{[(\mathbf{k}_1\cdot\mathbf{k}_2)((\mathbf{k}_1\cdot\mathbf{k})\cdot k_{2i}-(\mathbf{k}_2\cdot\mathbf{k})\cdot k_{1i})(k_{1i'}-k_{2i'})\delta_{jj'}\delta_{kk'}\\
&\quad +(\mathbf{k}_1\cdot\mathbf{k})((\mathbf{k}_1\cdot\mathbf{k}_2)\cdot k_{2k}-k_2^2k_{1k})(k_{1k'}-k_{2k'})\delta_{ii'}\delta_{jj'}\\
&\quad +(\mathbf{k}_2\cdot\mathbf{k})(k_1^2k_{2j}-(\mathbf{k}_1\cdot\mathbf{k}_2)k_{1j})(k_{1j'}-k_{2j'})\delta_{ii'}\delta_{kk'}]\\
&\quad +\frac{m}{T}[\{(\mathbf{k}_1\cdot\mathbf{k})\omega_2^2+(\mathbf{k}_2\cdot\mathbf{k})\omega_1^2\}\mathbf{k}_1\mathbf{k}_2-\{(\mathbf{k}_1\cdot\mathbf{k})\cdot k_2^2+(\mathbf{k}_2\cdot\mathbf{k})\cdot k_1^2\}\omega_1\omega_2]\delta_{ii'}\delta_{jj'}\delta_{kk'}\Big\}.
\end{aligned}
\tag{8.48}
$$

In contrast to the longitudinal case, $\mathscr{L}^{(2)}_{ijk;i'j'k'}(1, 2)$ contains besides the terms, proportional to T^{-2}, also the ones which are proportional to T^{-1}.

Making use of the causality principle (8.20) we can find the tensor susceptibility $\varkappa^{(n)}_{ij\ldots pr}(1, 2, \ldots, n)$ for any complex frequency from its imaginary part which is determined by the generalized fluctuation–dissipation theorem (8.43). Thus, within the context of (8.20) and (8.43) we can draw any desired information concerning the electromagnetic properties of a non-equilibrium isotropic plasma from the given set of the spectral correlation functions for the current density fluctuations computed without taking into account interactions between charged particles. The extension of the above results for a magneto-active plasma seems to be evident.

CHAPTER 9

Kinetic Equations

9.1. The Kinetic Equation for Particles

Spectral distribution of field fluctuations and the equation for the one-particle function

In the previous chapters we assumed the particle distributions in the plasma to be stationary and spatially homogeneous and studied time-independent and uniform spectral distributions of fluctuation fields. If the plasma is in an equilibrium state, then the particle distributions and the spectral distributions of fluctuating fields are stationary and spatially homogeneous. In a non-equilibrium plasma there can occur slow irreversible processes, the relaxation times of which are considerably longer than the time needed for the particle to cover the mean free path. The distinguishing peculiarities of irreversible phenomena in the plasma are due to the long-range nature of the electromagnetic interactions between charged particles. The dependence of the particle distribution functions on the space coordinates and time are governed by kinetic equations. Electromagnetic correlations between particles are the very reason which gives rise to the collision terms.

We shall restrict ourselves by taking into account the Coulomb interactions between particles and start from equations (6.61) for the microscopic density in configuration space and for the microscopic self-consistent plasma fields. We define the one-particle distribution function $f(\mathbf{r}, \mathbf{v}, t)$ to be the microscopic density averaged over the Liouville distribution; stationarity and spatial homogeneity are no longer required:

$$f(\mathbf{r}, \mathbf{v}, t) \equiv \langle \mathcal{F}(\mathbf{r}, \mathbf{v}, t) \rangle. \tag{9.1}$$

Averaging of the set of microscopic equations (6.61) yields:

$$\frac{\partial f}{\partial t} + \left(\mathbf{v} \cdot \frac{\partial f}{\partial \mathbf{r}}\right) + \frac{e}{m}\left(\overline{\mathbf{E}} \cdot \frac{\partial f}{\partial \mathbf{v}}\right) = -\frac{e}{m}\left\langle\left(\delta\mathbf{E} \cdot \frac{\partial}{\partial \mathbf{v}}\, \delta f\right)\right\rangle,$$
$$\operatorname{div} \overline{\mathbf{E}} = 4\pi e \int d\mathbf{v} f, \tag{9.2}$$

where $\mathbf{E}$ is the averaged self-consistent field, $\delta\mathbf{E}$ is the fluctuating electric field

$$\delta\mathbf{E} = \mathbf{E} - \overline{\mathbf{E}}, \tag{9.3}$$

and δf is the fluctuation of the distribution function

$$\delta f = \mathcal{F} - f. \tag{9.4}$$

The right-hand side of the first equation of (9.2) will be referred to as the collision term.

Here and below we shall assume the averaged self-consistent fields to be weak so that they may be neglected in equations (9.2). Then the equation for the one-particle distribution $f(\mathbf{r}, \mathbf{v}, t)$ reduces to

$$\frac{\partial f}{\partial t} + \left(\mathbf{v} \cdot \frac{\partial f}{\partial \mathbf{r}}\right) = -\frac{e}{m}\left\langle \left(\mathbf{E} \cdot \frac{\partial}{\partial \mathbf{v}}\right) \delta f \right\rangle. \tag{9.5}$$

The right-hand side of the last equation contains the averaged product of the fluctuating electric field $\mathbf{E}$ and the distribution function fluctuation δf both taken at the same instant of time and in the same space point (the averaging and differentiation procedures commutate). The latter may be expressed in terms of the spectral correlation function for the relevant quantities by means of the following relation:

$$\langle \mathbf{E}\, \delta f(\mathbf{v}) \rangle = \mathrm{Re} \sum_{\omega, \mathbf{k}} \langle \mathbf{E}\, \delta f(\mathbf{v}) \rangle_{\mathbf{k}\omega}. \tag{9.6}$$

The spectral correlation function $\langle \mathbf{E}\, \delta f(\mathbf{v}) \rangle_{\mathbf{k}\omega}$ for a stationary and spatially homogeneous particle distribution in a plasma is determined by (6.89), which is evidently suitable for the calculation of the right-hand side of (9.5) (the collision term) if the relaxation time and the parameter of spatial inhomogeneity for $f(\mathbf{r}, \mathbf{v}, t)$, which are governed by (9.5), are large as compared to the characteristic correlation time and length.

The characteristic correlation time for the case of Coulomb interactions between particles is of the order of magnitude of the inverse Langmuir frequency Ω^{-1}, and the correlation length is governed by the Debye length a. Hence, if

$$\tau \gg \Omega^{-1} \quad \text{and} \quad l \gg a,$$

where τ and l are the time and the length of the relaxation for the one-particle distribution function $f(\mathbf{r}, \mathbf{v}, t)$, then (9.5) may be reduced to the kinetic equation

$$\frac{\partial f}{\partial t} + \left(\mathbf{v} \cdot \frac{\partial f}{\partial \mathbf{r}}\right) = J(f), \tag{9.7}$$

where, within the context of (6.89) and (9.6), the collision term $J(f)$ is of the form†

$$J(f) = \frac{4\pi e^2}{m} \sum_{\mathbf{k}} \frac{1}{k^2} \left(\mathbf{k} \cdot \frac{\partial}{\partial \mathbf{v}}\right) \left\{ \mathrm{Im}\, \frac{1}{\varepsilon^*(\mathbf{k}\cdot\mathbf{v}, \mathbf{k})} f(v) + \frac{1}{8\pi m} \langle E^2 \rangle_{\mathbf{k}\omega} \left(\mathbf{k} \cdot \frac{\partial f}{\partial \mathbf{v}}\right) \right\}. \tag{9.8}$$

The spectral distribution of the electric field fluctuations $\langle E^2 \rangle_{\mathbf{k}\omega}$ depends on the distribution function f according to (6.57) and (6.88). The plasma dielectric permittivity $\varepsilon(\omega, \mathbf{k})$ may be also expressed in terms of the one-particle function within the context of (2.15). Hence the collision term $J(f)$ is a functional of the distribution function f, (9.7) is closed and may be regarded as the kinetic equation that completely determines

† The collision term (9.8) was derived by Silin (1962) and Sitenko and Tszyan'Yu-Tai (1963).

the one-particle distribution function $f(\mathbf{r}, \mathbf{v}, t)$. It is to be noted that the space–time dependence of the collision term $J(f)$ is governed by that of the distribution function $f(\mathbf{r}, \mathbf{v}, t)$.

We could have considered the motions of various plasma components, and we would obtain a set of kinetic equations of the form (9.7) for the partial distribution functions. The collision term depends on the one-particle distribution function through the spectral distribution of the electric field fluctuations and the plasma dielectric permittivity. The explicit expressions for the latter are as follows:

$$\langle E^2\rangle_{\mathbf{k}\omega} = \frac{32\pi^2}{k^2}\,\frac{1}{|\varepsilon(\omega, \mathbf{k})|^2}\sum e'^2\int d\mathbf{v}'\,\delta(\omega-(\mathbf{k}\cdot\mathbf{v}'))f'(\mathbf{v}'), \tag{9.9}$$

$$\varepsilon(\omega, \mathbf{k}) = 1+\sum\frac{4\pi e'^2}{m'k^2}\int d\mathbf{v}'\,\frac{1}{\omega-(\mathbf{k}\cdot\mathbf{v}')+i0}\left(\mathbf{k}\cdot\frac{\partial f'(\mathbf{v}')}{\partial\mathbf{v}'}\right). \tag{9.10}$$

(The summation in (9.9) and (9.10) extends over all plasma components.) The set of kinetic equations with the above collision terms provides a complete description of the one-particle distributions in a plasma of particles interacting according to the Coulomb law.

The Balescu–Lénard collision term

Within the context of the relations (9.9) and (9.10) the collision term (9.8) may be expressed directly in terms of the particle distribution functions. Since

$$\operatorname{Im}\varepsilon(\omega, \mathbf{k}) = -\sum\frac{4\pi e'^2}{m'k^2}\int d\mathbf{v}'\,\delta(\omega-(\mathbf{k}\cdot\mathbf{v}'))\left(\mathbf{k}\cdot\frac{\partial f'(\mathbf{v}')}{\partial\mathbf{v}'}\right), \tag{9.11}$$

we have

$$J(f) = \sum\frac{16\pi^3e^2e'^2}{m}\sum_{\mathbf{k}}\frac{1}{k^4}\left(\mathbf{k}\cdot\frac{\partial}{\partial\mathbf{v}}\right)\int d\mathbf{v}'\,\frac{\delta((\mathbf{k}\cdot\mathbf{v})-(\mathbf{k}\cdot\mathbf{v}'))}{|\varepsilon((\mathbf{k}\cdot\mathbf{v}), \mathbf{k})|^2}\cdot$$
$$\times\left\{\frac{1}{m}\left(\mathbf{k}\cdot\frac{\partial f(\mathbf{v})}{\partial\mathbf{v}}\right)f'(\mathbf{v}')-f(\mathbf{v})\frac{1}{m'}\left(\mathbf{k}\cdot\frac{\partial f'(\mathbf{v}')}{\partial\mathbf{v}'}\right)\right\}. \tag{9.12}$$

This form of the collision term for an ensemble of particles with Coulomb interactions was derived by R. Balescu (1960) and A. Lénard (1960), starting from the integral equation for the two-particle correlation function. (The collision term (9.12) is conventionally called the Balescu–Lénard collision term.)

The Balescu–Lénard collision term determines the variations of the one-particle distribution function associated with the Coulomb interactions between charged particles taking into account the dynamical polarization of the plasma.† The latter is shown by the weakening by a factor $\varepsilon((\mathbf{k}\cdot\mathbf{v}), \mathbf{k})$, of the field that originates in the plasma due to the motion of a charged particle with a velocity $\mathbf{v}$ and gives rise to the factor

† The kinetic equations for a magneto-active plasma taking into account the dynamical polarization were considered by Rostoker (1960) and Eleonskii, Zyrianov, and Silin (1962).

$|\varepsilon((\mathbf{k}\cdot\mathbf{v}), \mathbf{k})|^{-2}$ in the integrand in (9.12). It is to be observed that the integral over k in (9.12) diverges at large k as $\ln k$, since large values of k correspond to small distances and the Coulomb interaction cannot be assumed to be weak. Hence, the range of integration over k must be restricted to the upper limit k_{max}, which must be taken to be inversely proportional to the minimum distance of an encounter of two particles. Since the integral diverges logarithmically, the absolute value is only slightly sensitive to the indefinite choice of k_{max}.

In neglect of the dynamical polarization, i.e. if unity is substituted for $\varepsilon((\mathbf{k}\cdot\mathbf{v}), \mathbf{k})$, (9.12) reduces to the Landau collision term. (Then the range of integration over k is also to be limited from below to $k_{min} \simeq 1/a$.) The kinetic equation with the Landau collision term is suitable for calculating electric and thermal conductivities, as well as the viscosity of the plasma, which are governed by the particle collisions. However, it should be kept in mind that in some situations the Balescu–Lénard collision term cannot be reasonably replaced by the simpler Landau one. The dynamical polarization plays a dominant role in a non-equilibrium plasma with a high level of collective fluctuations. (For a more detailed discussion of this problem, see Silin, 1971.)

Consider in brief some basic properties of the collision term (9.12).† It may be immediately verified that (9.12) vanishes if the Maxwellian function

$$f(\mathbf{v}) = n_0\left(\frac{m}{2\pi T}\right)^{3/2} e^{-\frac{m(\mathbf{v}-\mathbf{u})^2}{2T}} \tag{9.13}$$

with equal T and $\mathbf{u}$ for all plasma components is substituted for the one-particle distribution. Hence, the Maxwellian distribution function (9.13) is a solution of the kinetic equation (9.7) with the collision term (9.12). It is not difficult to show within the context of (9.7) that the total entropy of the system grows and reaches its maximum as the distribution functions tend to the equilibrium ones, i.e. when all plasma components relax towards Maxwellian distributions with equal temperatures T and macroscopic velocities $\mathbf{u}$.

Dynamic friction and diffusion coefficients

It is most suitable to write the collision term in the form (9.8), since the kinetic equation (9.7) can be then reduced to the Fokker–Planck equation which allows us to treat the collisions as dynamic friction and diffusion of particles in the velocity space (Sitenko and Tszyan'Yu-tai, 1963). Consider for simplicity the spatially homogeneous case, the Fokker–Planck equation being of the form

$$\frac{\partial f}{\partial t} = -\frac{\partial}{\partial v_i}(D_i f)+\frac{1}{2}\frac{\partial^2}{\partial v_i\,\partial v_j}(D_{ij} f), \tag{9.14}$$

where D_i is the dynamic friction coefficient which is by definition equal to the mean

† The properties of the Balescu–Lénard and other collision terms are considered in detail by Silin (1971) and Klimontovich (1975).

change in the particle velocity per unit time:

$$D_i \equiv \frac{\langle \Delta v_i \rangle}{\Delta t}, \tag{9.15}$$

and D_{ij} is the tensor of the diffusion coefficients in velocity space defined according to the following equality:

$$D_{ij} \equiv \frac{\langle \Delta v_i \, \Delta v_j \rangle}{\Delta t}. \tag{9.16}$$

The derivation of the Fokker–Planck equation is based on the assumption that small velocity variations Δv_i are the most probable ones and that we may neglect powers of Δv_i higher than two. The friction coefficient D_i and the diffusion coefficient D_{ij} may be obtained by comparing the right-hand side of (9.14) to (9.8). Here are the results:

$$D_i \equiv \frac{v_i}{v} D,$$

$$D = \frac{e^2}{2\pi^2 mv} \operatorname{Im} \int d\mathbf{k} \frac{\omega}{k^2 \varepsilon(\omega, \mathbf{k})} + \frac{e^2}{16\pi^3 m^2 v} \int d\mathbf{k}\omega \frac{\partial}{\partial \omega} \langle E^2 \rangle_{\mathbf{k}\omega}, \tag{9.17}$$

$$D_{ij} = \frac{e^2}{8\pi^3 m^2} \int d\mathbf{k} \frac{k_i k_j}{k^2} \langle E^2 \rangle_{\mathbf{k}\omega}, \quad \omega = (\mathbf{k} \cdot \mathbf{v}). \tag{9.18}$$

Thus the knowledge of the spectral distribution of the electric field fluctuations in a plasma is sufficient to determine the dynamic friction and diffusion coefficients. The possibility of reducing the kinetic equation to the Fokker–Planck form owes to the fact that due to the long-range nature of the Coulomb forces in a plasma only the distant collisions are of importance and therefore the angular deflections and velocity changes of the colliding particles are small.

The Fokker–Planck equation with the friction and diffusion coefficients (9.17) and (9.18) can be derived directly from the examination of the motion of a test particle which is involved in the interaction with polarization and fluctuating fields in the plasma. The equation of the test particle motion may be written in the following form:

$$\dot{v}_i(t) = \frac{e}{m} E_i(\mathbf{r}(t), t), \tag{9.19}$$

where $\mathbf{E}(\mathbf{r}(t), t)$ is the electric field in the point where the particle is located at time t, e and m are the charge and the mass of the particle. After the integration procedure has been carried out the equation of motion becomes transformed to

$$\mathbf{r}(t) = \mathbf{r}_0 + \mathbf{v}_0(t - t_0) + \frac{e}{m} \int_{t_0}^{t} dt' \int_{t_0}^{t'} dt'' \mathbf{E}(\mathbf{r}(t''), t''), \tag{9.20}$$

where $\mathbf{r}_0$ and $\mathbf{v}_0$ are the radius-vector and velocity of the particle at the initial time t_0. Taking the time segment to be long as compared to the period of the random field fluctuations in the plasma, but short in comparison with the time during which the particle motion may be modified greatly, we can rewrite (9.19) in the following approximate form:

$$\dot{v}_i(t) = \frac{e}{m} E_i(\mathbf{r}_0(t), t) + \frac{e^2}{m^2} \int_{t_0}^{t} dt' \int_{t_0}^{t'} dt'' E_j(\mathbf{r}_0(t''), t'') \frac{\partial}{\partial r_{0j}} E_i(\mathbf{r}_0(t), t)$$

$$(\mathbf{r}_0(t) \equiv \mathbf{r}_0 + \mathbf{v}_0(t - t_0), \quad t \equiv t_0 + \Delta t). \tag{9.21}$$

Averaging (9.21) we obtain the dynamic friction coefficient (9.17). It is to be noted that the quantity D may be written in terms of the energy losses of the guided particle motion per unit time $\partial\varepsilon/\partial t$

$$D = \frac{1}{mv} \frac{\partial \varepsilon}{\partial t}. \tag{9.22}$$

The first term in the expression (9.17) for D determines the dynamic friction of the test particle by virtue of its interaction with the electric field originating in the plasma as a result of the motion of this very particle. This interaction causes the usual polarization energy losses of the particle, therefore the friction, which is associated with it, may be referred to as the polarization friction.

The second term in (9.17) is due to the additional dynamic friction of the test particle associated with the existence of the space–time correlations between the fluctuating electric fields in the plasma. As we have seen, the presence of such correlations leads to additional energy losses of the guided particle motion. It is to be observed that the dynamic friction of the particle is associated with the fluctuations of the longitudinal electric field only.

If instead of a single particle there is an aggregate of test particles moving through the plasma, which is described by the distribution function $f(\mathbf{v}, t)$, then the evolution of the latter due to polarization and fluctuation losses may be found from the known friction coefficients:

$$\left(\frac{\partial f}{\partial t}\right)_{p+f} = -\frac{\partial}{\partial v_i}(D_i f). \tag{9.23}$$

These time-variations are very slow, since the energy of the interactions between the test particles and the plasma fields was assumed to be small in comparison with the kinetic energy.

However, the quantity $\left(\frac{\partial f}{\partial t}\right)_{p+f}$ determines the variations of the test particle distribution function incompletely, since besides the changes, caused by the two above-mentioned reasons, we must take into account also the influence on the distribution from the diffusion of the test particles in velocity space. The latter is a slow process, which is associated, like the fluctuation energy losses, with the field fluctuations in the plasma. Since the diffusion coefficients are of a purely fluctuation nature, it is sufficient to retain only the linear terms with respect to the field in the equation of the test particle motion

(9.21), to find D_{ij}. It may be easily verified that

$$D_{ij} = \frac{e^2}{m^2} \int_0^{\Delta t} d\xi \langle E_j(r_0(t-\xi), t-\xi)\, E_i(r_0(t), t)\rangle, \tag{9.24}$$

and since Δt was assumed to be much longer than the period of the field fluctuations, it may be assumed to tend to infinity. To yield the requested diffusion tensor (9.18) the expression (9.24) is to be Fourier transformed.

The tensor of the diffusion coefficients for an isotropic plasma may be written in the form

$$D_{ij} = \frac{v_i v_j}{v^2} D_{||} + \left(\delta_{ij} - \frac{v_i v_j}{v^2}\right) D_{\perp}, \tag{9.25}$$

$$D_{||} = \frac{e^2}{8\pi^3 m^2 v} \int d\mathbf{k} \frac{\omega^2}{k^2} \langle E^2 \rangle_{\mathbf{k}\omega}, \tag{9.26}$$

$$D_{\perp} = \frac{e^2}{16\pi^3 m^2} \int d\mathbf{k} \left(1 - \frac{\omega^2}{k^2 v^2}\right) \langle E^2 \rangle_{\mathbf{k}\omega}, \quad \omega = (\mathbf{k}\cdot\mathbf{v}). \tag{9.27}$$

The longitudinal diffusion coefficient $D_{||}$ is associated with the mean variation of the square of the velocity component along the direction of the particle motion; the transverse one $D_{\perp}$ describes the mean variation of the square of the perpendicular component.

The diffusion coefficients D_{ij} determine the evolution of the test particle distribution function due to the diffusion in velocity space:

$$\left(\frac{\partial f}{\partial t}\right)_d = \frac{1}{2} \frac{\partial^2}{\partial v_i \partial v_j} (D_{ij} f). \tag{9.28}$$

The last equation added to (9.23) yields the Fokker–Planck equation, which is in accordance with the Balescu–Lénard kinetic equation.

A kinetic equation of the form (9.14) that takes into account the Coulomb collisions of charged particles was first derived by Landau (1937). The friction and diffusion coefficients for an isotropic plasma in neglect of the ion motion were calculated by Chandrasekhar (1943), Cohen, Spitzer, and Routly (1950), and Spitzer (1956), taking into account only distant encounters with impact parameters less than the plasma Debye length. The results of the latter two papers may be obtained within the framework of the macroscopic approach without introducing an arbitrary cut-off parameter; the friction and diffusion coefficients are to be written in terms of the spectral distribution of the electric field fluctuations in the plasma. This method was proposed by Hubbard (1961a, b), Ichimaru (1962), and Sitenko and Tszyan'Yu-tai (1963). The latter calculated the dynamic friction and diffusion coefficients in the two-temperature plasma.

Quasi-linear approximation

The kinetic equation (9.7) describes the relaxation of the distribution function which is due to taking into account the Coulomb interaction between charged particles in the plasma. The kinetic equation with the collision term (9.8) is suitable also to study the quasi-linear relaxation caused by the reaction of the oscillation field to the smoothed distribution function. The quasi-linear relaxation is the simplest among the non-linear processes in the plasma which cannot be treated without taking into account the effect of the oscillating field on the averaged distribution function, while the non-linear wave–wave interaction is neglected (this approach is conventionally referred to as the quasi-linear approximation). The quasi-linear approximation involves into consideration only the influence of the fluctuation oscillations on the distributions of the resonant particles, the velocities of which are close to the phase velocities of the relevant waves. The resonant particles are involved in the strong interaction with the plasma eigenoscillations, hence they are responsible both for the damping and for the growth of the oscillations. The non-resonant particles do not exchange energy with the waves on average, so the distributions of the non-resonant particles are in fact insensitive to the effect of the oscillations.

As follows from (9.14), the variations of the distribution function are associated with the polarization and fluctuation friction of particles, as well as with the diffusion in the velocity space. In contrast to the polarization friction, the fluctuation friction and diffusion are governed by the fluctuation field that can grow in a non-equilibrium plasma up to the values much greater than the termal level. Therefore, in case the polarization friction is neglected, the kinetic equation for the non-equilibrium (weakly turbulent) plasma may be reduced to the diffusion equation:

$$\frac{\partial f(\mathbf{v}, t)}{\partial t} = \frac{1}{2} \frac{\partial}{\partial v_i} \left\{ D_{ij}(\mathbf{v}, t) \frac{\partial f(\mathbf{v}, t)}{\partial v_j} \right\}. \tag{9.29}$$

(Equation (9.29) was obtained within the context of the relation

$$D_i = \frac{1}{2} \frac{\partial D_{ij}}{\partial v_j},$$

the adequacy of which may be easily verified.) The diffusion coefficient D_{ij} that enters (9.29) is determined by the general formula (9.18).

Assume the spectral distribution of the fluctuation field to be of the form

$$\langle E^2 \rangle_{\mathbf{k}\omega} = \pi I_{\mathbf{k}}(t) \{\delta(\omega - \omega_{\mathbf{k}}) + \delta(\omega + \omega_{\mathbf{k}})\}, \tag{9.30}$$

where the frequency $\omega_{\mathbf{k}}$ of the oscillations with wave vector $\mathbf{k}$ is governed by the linear dispersion relation, and the intensity $I_{\mathbf{k}}(t)$ is assumed to be a slowly varying function of time. The time development of the quantity $I_{\mathbf{k}}(t)$ is governed either by the damping or by the growth of the oscillations due to the non-monotonic dependence of the distribution function on the particle velocities. Note that for the quasi-linear treatment to be adequate the damping or the growth rate $\gamma_{\mathbf{k}}$ are to be sufficiently small, i.e. the following condition must be satisfied:

$$|\gamma_{\mathbf{k}}| \ll \omega_{\mathbf{k}}. \tag{9.31}$$

Besides, the wave packet (9.30) must be sharp enough (in k-space) in order that the number of resonant particles which interact with it be less than the total number of particles.

According to (9.30), the diffusion coefficient (9.18) may be transformed to

$$D_{ij}(\mathbf{v}, t) = \frac{e^2}{4\pi^2 m^2} \int d\mathbf{k} \frac{k_i k_j}{k^2} I_{\mathbf{k}}(t)\, \delta((\mathbf{k}\cdot\mathbf{v}) - \omega_{\mathbf{k}}). \tag{9.32}$$

The presence of the delta-function in (9.32) implies that this expression holds only for resonant particles. The diffusion equation (9.29) with the diffusion tensor (9.32) governs the changes in the distribution function under the influence of the fluctuation oscillations of the plasma field. This equation should be treated together with the one that describes the time development of the fluctuating field intensity

$$\frac{\partial I_{\mathbf{k}}}{\partial t} = 2\gamma_{\mathbf{k}} I_{\mathbf{k}}, \tag{9.33}$$

where the damping or growth rate $\gamma_{\mathbf{k}}$ is to be taken from the linear theory.

The set of (9.29) and (9.33) determines the evolution of the averaged distribution function of the resonant particles, as well as the damping or the enhancement of fluctuations in the quasi-linear approximation.† It may be shown within the context of (9.29) and (9.33) that particle diffusion resulting from the scattering of particles by the fluctuation oscillations leads to the establishment of a stationary state as $t \to \infty$, which is characterized by a fixed distribution of resonant particles and some definite level of fluctuation oscillations.

Multiply both sides of (9.29) by the distribution function f and carry out the integration over the velocities:

$$\frac{1}{2}\frac{\partial}{\partial t}\int d\mathbf{v} f^2 = -\frac{e^2}{8\pi^2 m^2}\int d\mathbf{v} \int d\mathbf{k} \frac{1}{k^2} I_{\mathbf{k}}(t) \left(\mathbf{k}\cdot\frac{\partial f}{\partial \mathbf{v}}\right)^2 \delta((\mathbf{k}\cdot\mathbf{v}) - \omega_{\mathbf{k}}). \tag{9.34}$$

Since the right-hand side of the resulting equation (9.34) is negative, the positive definite quantity $\sigma(t) \equiv \int d\mathbf{v} f^2$ decreases with time and tends to some constant value as $t \to \infty$. Hence $\partial\sigma/\partial t \to 0$ as $t \to \infty$ and, therefore, the right-hand side of (9.34) vanishes. Since the integrand on the right-hand side of (9.34) involves only non-negative quantities, the following condition is to be satisfied as $t \to \infty$:

$$\lim I_{\mathbf{k}}(t) \left(\mathbf{k}\cdot\frac{\partial f}{\partial \mathbf{v}}\right)^2_{(\mathbf{k}\cdot\mathbf{v}) = \omega_{\mathbf{k}}} = 0.$$

Thus, at least one of the two following requirements must be satisfied as $t \to \infty$

$$\lim_{t\to\infty} I_{\mathbf{k}}(t) = 0 \quad \text{or} \quad \lim \left(\mathbf{k}\cdot\frac{\partial f}{\partial \mathbf{v}}\right)_{(\mathbf{k}\cdot\mathbf{v}) = \omega_{\mathbf{k}}} = 0. \tag{9.35}$$

Therefore, either the oscillations of the fluctuation field are damped as $t \to \infty$, or the distribution function forms a plateau (remains constant along the direction of the wave

† A quasi-linear theory of plasma oscillations was worked out by Vedenov, Velikhov, and Sagdeev (1961, 1962), Drummond and Pines (1962), and Romanov and Filippov (1961).

propagation):

$$\lim_{t\to\infty} f(\mathbf{v}, t) = f(\mathbf{v}_\perp).$$

The possibility of a plateau creation was shown for one-dimensional oscillations by Vedenov, Velikhov, and Sagdeev (1961) (see also Pustovalov and Silin, 1972b).

Let us examine the one-dimensional quasi-linear relaxation in more detail. We consider for simplicity Langmuir oscillations. Let the initial (at $t = 0$) spectral density of the fluctuation oscillations and the electron distribution function be of the form

$$I_k(t=0) = \begin{cases} I_0(k) & \text{for} \quad k_1 < k < k_2, \\ 0 & \text{for} \quad k < k_1 \quad \text{or} \quad k > k_2, \end{cases}$$

$$f(v, t=0) = f_0(v), \quad \frac{\Omega}{k_2} < v < \frac{\Omega}{k_1}.$$

Jf $\partial f_0/\partial v < 0$, then the oscillations are damped. The derivative of the distribution function decreases with t until a plateau is formed and energy exchange between the fluctuation oscillations and the particles stops. In Fig. 8 we see the distribution function of the resonant electrons and the spectral density of the fluctuation oscillations at $t = 0$ (solid lines) and at $t \to \infty$ (dashed lines).

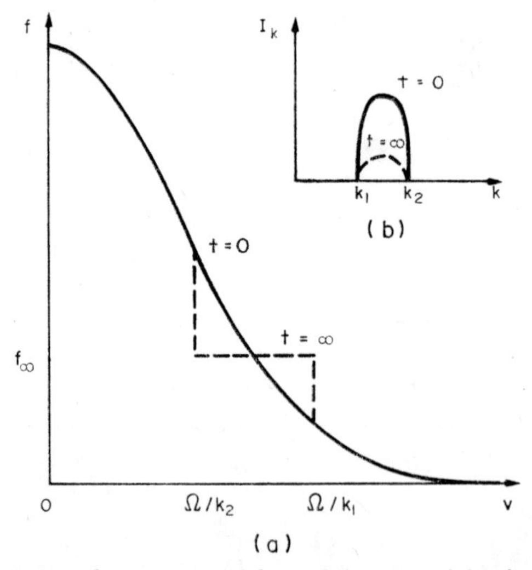

FIG. 8. The distribution function of resonant particles and the spectral density of fluctuation oscillations at the initial time $t = 0$ (solid lines) and as $t \to \infty$ (dashed lines).

It is not difficult to calculate the oscillation energy for a steady state with a plateau. In the one-dimensional case we obtain within the context of (9.29) and (9.33):

$$\frac{\partial}{\partial t}\left\{f(v, t) - \frac{\partial}{\partial v}\left(\frac{\Omega}{8\pi^2 m v^3} I_k(t)\right)\right\} = 0$$

and, hence, the spectral density of the fluctuation oscillations is as follows:

$$I_k(t) = I_k(0) + \frac{8\pi^2 m\Omega^2}{k^3} \int_{\Omega/k_2}^{\Omega/k} dv\{f(v,t) - f_0(v)\}. \tag{9.36}$$

In particular, as $t \to \infty$,

$$I_k(\infty) = I_k(0) + \frac{8\pi^2 m\Omega^2}{k^3} \int_{\Omega/k_2}^{\Omega/k} dv\{f_\infty(v) - f_0(v)\}. \tag{9.37}$$

The height of the plateau f_∞ may be found from the condition that the number of resonant particles is constant

$$f_\infty = \frac{1}{\dfrac{\Omega}{k_1} - \dfrac{\Omega}{k_2}} \int_{\Omega/k_2}^{\Omega/k_1} dv f_0(v). \tag{9.38}$$

As a result of the relaxation the resonant particles gain the following energy:

$$\Delta\varepsilon = \frac{1}{16\pi^2} \int_{k_1}^{k_2} dk\{I_k(0) - I_k(\infty)\}. \tag{9.39}$$

If the initial distribution is unstable, i.e., if there exists such a band of velocities $v' < v < v''$ that $\partial f_0/\partial v > 0$ (e.g. in a plasma penetrated by a low-density particle beam), the oscillations grow with time if their phase velocities belong to the above-mentioned interval and are damped otherwise. The distortion of the distribution function causes a broadening of the range of unstable oscillations with time until there is established a final steady state with a plateau on the distribution function.

9.2. The Kinetic Equation for Waves (Potential Field)

Time evolution of the spectral distribution of the fluctuation field

In the previous sections we considered stationary spectral distributions of the fluctuation plasma fields. The particle distributions were also suggested to be stationary. In an equilibrium plasma all eigenoscillations are damped due to interactions with plasma particles characterized by the thermal distribution of velocities. Therefore there are always a number of particles with velocities fitting the resonance condition $\omega = (\mathbf{k}\cdot\mathbf{v})$ (ω is the frequency, $\mathbf{k}$ is the wave vector, and $\mathbf{v}$ is the particle velocity), which is to be satisfied for the wave-particle interaction to occur. In a non-equilibrium plasma the wave-particle interactions may cause an enhancement of the oscillations. If the intensities of the fluctuation oscillations are very large, the diffusion of particles due to scattering by the oscillations becomes important and leads to a quasi-linear relaxation of the particle distributions. Another process, which is important when the level of the oscillation intensity is high, is the non-linear coupling of the oscillations. The non-linear wave interaction, together with the linear damping or enhancement of oscillations, make effect on the spectral distributions of the fluctuation plasma fields, which are no longer time-independent.

We shall study the time development of the spectral distribution of the fluctuation field by means of a multiple-time scale perturbation analysis. To use this approach we introduce a hierarchy of times which increase successively by a factor α^{-1}, where α is a dimensionless coupling constant. The dependence of some quantity on a slower time scale is determined by the requirement that the secularities to be removed from the equation that governs the time development on a fast time scale. When applied to the field equation (7.1) the procedure reduces to the formal substitution

$$\varepsilon(\omega, \mathbf{k}) \to \varepsilon(\omega, \mathbf{k})+i\frac{\partial}{\partial\omega}\left(\frac{\partial\varepsilon(\omega, \mathbf{k})}{\partial t}\right)+i\frac{\partial\varepsilon(\omega, \mathbf{k})}{\partial\omega}\bigg|_{\omega_{\mathbf{k}}}\left(\frac{\partial}{\partial t}+\left(\frac{d\omega_{\mathbf{k}}}{d\mathbf{k}}\cdot\frac{\partial}{\partial\mathbf{r}}\right)-\frac{i}{2}\frac{d^2\omega_{\mathbf{k}}}{dk^2}\nabla^2+\ldots\right). \tag{9.40}$$

By virtue of spatial dispersion the non-linear interaction makes the field amplitude coordinate-dependent. This independence is significant in the case of strong Langmuir turbulence and may be neglected when other effects are under consideration. Expansion of (7.1) in a power series in the non-linear coupling constant within the context of (9.40) yields immediately the equations of strong turbulence theory.

If the particle distribution is non-stationary and the collision term in the kinetic equation (9.5) does not vanish, we must consider the following non-linear equation instead of (7.1):

$$\begin{aligned} ik\Bigg\{&\varepsilon(\omega, \mathbf{k})E_{\mathbf{k}\omega}+\sum_{\substack{\omega_1+\omega_2=\omega\\ \mathbf{k}_1+\mathbf{k}_2=\mathbf{k}}}\varkappa^{(2)}(\omega_1, \mathbf{k}_1; \omega_2, \mathbf{k}_2)\left(E_{\mathbf{k}_1\omega_1}E_{\mathbf{k}_2\omega_2}-\langle E_{\mathbf{k}_1\omega_1}E_{\mathbf{k}_2\omega_2}\rangle\right)\\ &+\sum_{\substack{\omega_1+\omega_2+\omega_3=\omega\\ \mathbf{k}_1+\mathbf{k}_2+\mathbf{k}_3=\mathbf{k}}}\bar{\varkappa}^{(3)}(\omega_1, \mathbf{k}_1; \omega_2, \mathbf{k}_2; \omega_3, \mathbf{k}_3)\\ &\times\left(E_{\mathbf{k}_1\omega_1}E_{\mathbf{k}_2\omega_2}E_{\mathbf{k}_3\omega_3}-E_{\mathbf{k}_1\omega_1}\langle E_{\mathbf{k}_2\omega_2}E_{\mathbf{k}_3\omega_3}\rangle-\langle E_{\mathbf{k}_1\omega_2}E_{\mathbf{k}_2\omega_2}E_{\mathbf{k}_3\omega_3}\rangle\right)\\ &+\ldots+\sum_{\omega', \mathbf{k}'}\Delta(\omega, \mathbf{k}; \omega', \mathbf{k}')\left(\delta f^0_{\mathbf{k}-\mathbf{k}', \omega-\omega'}E_{\mathbf{k}'\omega'}-\langle\delta f^0_{\mathbf{k}-\mathbf{k}', \omega-\omega'}E_{\mathbf{k}'\omega'}\rangle\right)\\ &+\sum_{\substack{\omega_1+\omega_2=\omega\\ \mathbf{k}_1+\mathbf{k}_2=\mathbf{k}\\ \omega', \mathbf{k}'}}\Delta^{(2)}(\omega_1, \mathbf{k}_1; \omega_2, \mathbf{k}_2; \omega', \mathbf{k}')\left(\delta f^0_{\mathbf{k}_2-\mathbf{k}', \omega_2-\omega'}E_{\mathbf{k}_1\omega_1}E_{\mathbf{k}'\omega'}\right.\\ &\left.-E_{\mathbf{k}_1\omega_1}\langle\delta f^0_{\mathbf{k}_2-\mathbf{k}', \omega-\omega'}E_{\mathbf{k}'\omega'}\rangle\right)-\langle\delta f^0_{\mathbf{k}_2-\mathbf{k}', \omega_2-\omega'}E_{\mathbf{k}_1\omega_1}E_{\mathbf{k}'\omega'}\rangle+\ldots\Bigg\}=4\pi\varrho^0_{\mathbf{k}\omega}. \end{aligned} \tag{9.41}$$

However, it is to be observed that within the context of (9.41) we obtain an equation for the spectral correlation function $\langle E^2\rangle_{\mathbf{k}\omega}$ which exactly coincides with (7.10) or (7.20), though with another coefficient (7.11) (the last two terms are absent).

To take into account the dependence on the slower time scale in the equation for the spectral correlation function $\langle E^2\rangle_{\mathbf{k}\omega}$ the following substitution is to be made:

$$\varepsilon(\omega, \mathbf{k})\langle E^2\rangle_{\mathbf{k}\omega} \to \left\{\varepsilon(\omega, \mathbf{k})+i\frac{\partial}{\partial\omega}\left(\frac{\partial \operatorname{Re}\varepsilon(\omega, \mathbf{k})}{\partial t}\right)+\frac{i}{2}\frac{\partial \operatorname{Re}\varepsilon(\omega, \mathbf{k})}{\partial\omega}\frac{\partial}{\partial t}\right\}\langle E^2\rangle_{\mathbf{k}\omega}. \quad (9.42)$$

Thus, instead of (7.20) we have

$$\left\{\varepsilon(\omega, \mathbf{k})+i\frac{\partial}{\partial\omega}\left(\frac{\partial \operatorname{Re}\varepsilon(\omega, \mathbf{k})}{\partial t}\right)+\frac{i}{2}\frac{\partial \operatorname{Re}\varepsilon(\omega, \mathbf{k})}{\partial\omega}\frac{\partial}{\partial t}\right\}\langle E^2\rangle_{\mathbf{k}\omega}$$
$$-\frac{2}{\varepsilon^*(\omega, \mathbf{k})}\sum_{\omega', \mathbf{k}'}|\varkappa^{(2)}(\omega-\omega', \mathbf{k}-\mathbf{k}'; \omega', \mathbf{k}')|^2\langle E^2\rangle_{\mathbf{k}'\omega'}\langle E^2\rangle_{\mathbf{k}-\mathbf{k}', \omega-\omega'}$$
$$-\sum_{\omega', \mathbf{k}'}a(\omega, \mathbf{k}; \omega', \mathbf{k}')\langle E^2\rangle_{\mathbf{k}'\omega'}\langle E^2\rangle_{\mathbf{k}\omega}=\frac{1}{\varepsilon^*(\omega, \mathbf{k})}\left\{\sum_{\omega', \mathbf{k}'}b(\omega, \mathbf{k}; \omega', \mathbf{k}')\langle E^2\rangle_{\mathbf{k}'\omega'}+q'_{\mathbf{k}\omega}\right\}. \quad (9.43)$$

Equating then the imaginary part of this equation to zero we obtain the following equation that describes the time evolution of the fluctuation field spectral density $\langle E^2\rangle_{\mathbf{k}\omega}$ due to linear dissipation and to the non-linear interactions of the fluctuation fields:

$$\frac{1}{2}\frac{\partial \operatorname{Re}\varepsilon(\omega, \mathbf{k})}{\partial\omega}\frac{\partial}{\partial t}\langle E^2\rangle_{\mathbf{k}\omega}$$
$$=-\left\{\operatorname{Im}\varepsilon(\omega, \mathbf{k})+\frac{\partial}{\partial\omega}\left(\frac{\partial \operatorname{Re}\varepsilon(\omega, \mathbf{k})}{\partial t}\right)\langle E^2\rangle_{\mathbf{k}\omega}\right\}$$
$$+2\operatorname{Im}\frac{1}{\varepsilon^*(\omega, \mathbf{k})}\sum_{\omega', \mathbf{k}'}|\varkappa^{(2)}(\omega-\omega', \mathbf{k}-\mathbf{k}'; \omega', \mathbf{k}')|^2\langle E^2\rangle_{\mathbf{k}'\omega'}\langle E^2\rangle_{\mathbf{k}-\mathbf{k}', \omega-\omega'}$$
$$+\operatorname{Im}\sum_{\omega', \mathbf{k}'}a(\omega, \mathbf{k}; \omega', \mathbf{k}')\langle E^2\rangle_{\mathbf{k}'\omega'}\langle E^2\rangle_{\mathbf{k}\omega}$$
$$+\operatorname{Im}\frac{1}{\varepsilon^*(\omega, \mathbf{k})}\left\{\sum_{\omega', \mathbf{k}'}b(\omega, \mathbf{k}; \omega', \mathbf{k}')\langle E^2\rangle_{\mathbf{k}'\omega'}+q'_{\mathbf{k}\omega}\right\}. \quad (9.44)$$

Equation (9.44) should be treated together with the kinetic equation that governs the time development of the particle distribution function:

$$\frac{\partial f}{\partial t}=\frac{4\pi e^2}{m}\sum_{\mathbf{k}}\frac{1}{k^2}\left(\mathbf{k}\cdot\frac{\partial}{\partial\mathbf{v}}\right)\left\{\operatorname{Im}\frac{1}{\varepsilon^*((\mathbf{k}\cdot\mathbf{v}), \mathbf{k})}f+\frac{1}{8\pi m}\langle E^2\rangle_{\mathbf{k}, (\mathbf{k}\cdot\mathbf{v})}\left(\mathbf{k}\cdot\frac{\partial f}{\partial\mathbf{v}}\right)\right\}. \quad (9.45)$$

This equation takes into account the effect of the fluctuation electric field on the particle distribution function. However, in our considerations below we shall take the distribution functions for fixed.

It may be immediately verified that (9.44) reduces under stationary conditions to (7.10). Indeed, it follows from (9.44) that in the stationary case:

$$0=-\operatorname{Im}\varepsilon(\omega, \mathbf{k})\langle E^2\rangle_{\mathbf{k}\omega}+2\frac{\operatorname{Im}\varepsilon(\omega, \mathbf{k})}{|\varepsilon(\omega, \mathbf{k})|^2}\sum_{\omega', \mathbf{k}'}|\varkappa^{(2)}(\omega-\omega', \mathbf{k}-\mathbf{k}'; \omega', \mathbf{k}')|^2\langle E^2\rangle_{\mathbf{k}'\omega'}$$
$$\times\langle E^2\rangle_{\mathbf{k}-\mathbf{k}', \omega-\omega'}+\operatorname{Im}\sum_{\omega', \mathbf{k}'}a(\omega, \mathbf{k}; \omega', \mathbf{k}')\langle E^2\rangle_{\mathbf{k}'\omega'}\langle E^2\rangle_{\mathbf{k}\omega}$$
$$+\operatorname{Im}\frac{1}{\varepsilon^*(\omega, \mathbf{k})}\left(\sum_{\omega', \mathbf{k}'}b(\omega, \mathbf{k}; \omega', \mathbf{k}')\langle E^2\rangle_{\mathbf{k}'\omega'}+q'_{\mathbf{k}\omega}\right). \quad (9.46)$$

Upon noting that

$$\operatorname{Im} \mathcal{D} + \operatorname{Im} \frac{\varepsilon}{\varepsilon^*} \mathcal{D}^* \simeq \frac{\operatorname{Im} \varepsilon}{|\varepsilon|^2} 2 \operatorname{Re} \varepsilon^* \mathcal{D},$$

$$\mathcal{D} \equiv \sum_{\omega', \mathbf{k}'} a(\omega, \mathbf{k}; \omega', \mathbf{k}') \langle E^2 \rangle_{\mathbf{k}'\omega'} \langle E^2 \rangle_{\mathbf{k}\omega},$$

and that $q'_{\mathbf{k}\omega}$ and $b(\omega, \mathbf{k}; \omega', \mathbf{k}')$ are real, we obtain (7.10) from (9.46) within the context of (7.21).

Linear approximation

Consider first the solution of (9.43) in neglect of the non-linear interactions. Assume the distribution f to be stationary and eliminate the non-linear terms in (9.43). Then

$$\left(\varepsilon(\omega, \mathbf{k}) + \frac{i}{2} \frac{\partial \operatorname{Re} \varepsilon(\omega, \mathbf{k})}{\partial \omega} \frac{\partial}{\partial t} \right) \langle E^2 \rangle^{(1)}_{\mathbf{k}\omega} = \frac{16\pi^2}{k^2} \frac{\langle \varrho^2 \rangle^0_{\mathbf{k}\omega}}{\varepsilon^*(\omega, \mathbf{k})}.$$

From this, equating the real and the imaginary parts of the two sides, we obtain two equations:

$$\operatorname{Re} \varepsilon(\omega, \mathbf{k}) \left\{ \langle E^2 \rangle^{(1)}_{\mathbf{k}\omega} - \frac{16\pi^2}{k^2} \frac{\langle \varrho^2 \rangle^0_{\mathbf{k}\omega}}{|\varepsilon(\omega, \mathbf{k})|^2} \right\} = 0, \tag{9.47}$$

$$\frac{1}{2} \frac{\partial \operatorname{Re} \varepsilon(\omega, \mathbf{k})}{\partial \omega} \frac{\partial}{\partial t} \langle E^2 \rangle^{(1)}_{\mathbf{k}\omega} = -\operatorname{Im} \varepsilon(\omega, \mathbf{k}) \left\{ \langle E^2 \rangle^{(1)}_{\mathbf{k}\omega} - \frac{16\pi^2}{k^2} \frac{\langle \varrho^2 \rangle^0_{\mathbf{k}\omega}}{|\varepsilon(\omega, \mathbf{k})|^2} \right\}. \tag{9.48}$$

The general solution of first equation (9.47) may be presented in the form

$$\langle E^2 \rangle^{(1)}_{\mathbf{k}\omega} = \langle E^2 \rangle^0_{\mathbf{k}\omega} + C_{\mathbf{k}}(t)\, \delta\{\operatorname{Re} \varepsilon(\omega, \mathbf{k})\}, \tag{9.49}$$

where

$$\langle E^2 \rangle^0_{\mathbf{k}\omega} \equiv \frac{16\pi^2}{k^2} \frac{\langle \varrho^2 \rangle^0_{\mathbf{k}\omega}}{|\varepsilon(\omega, \mathbf{k})|^2} \tag{9.50}$$

and $C_{\mathbf{k}}(t)$ is an arbitrary time-dependent function of $\mathbf{k}$. The stationary part of the solution $\langle E^2 \rangle^0_{\mathbf{k}\omega}$ is evidently governed by the non-homogeneous part of (9.47), while the non-stationary part of (9.49) is the solution of the relevant homogeneous equation and depends on the initial conditions. The stationary part of the solution (9.49) involves, by contrast to the non-stationary one, the whole frequency spectrum. Let us choose $C_{\mathbf{k}}(t)$ so that the solution of (9.49) be of the form

$$\langle E^2 \rangle^{(1)}_{\mathbf{k}\omega} = \langle E^2 \rangle^0_{\mathbf{k}\omega} + \pi I_{\mathbf{k}}(t) \{\delta(\omega - \omega_{\mathbf{k}}) + \delta(\omega + \omega_{\mathbf{k}})\}, \tag{9.51}$$

the initial oscillation intensity $I_{\mathbf{k}}(0)$ being governed by the initial conditions.

The time-dependence of the intensity $I_{\mathbf{k}}(t)$ may be easily revealed by means of the second equation (9.48). Indeed, the general solution of (9.48) is a sum of stationary and non-stationary parts:

$$\langle E^2 \rangle^{(1)}_{\mathbf{k}\omega} = \langle E^2 \rangle^0_{\mathbf{k}\omega} + \langle E^2 \rangle^t_{\mathbf{k}\omega}. \tag{9.52}$$

The former is determined by (9.50) as before, and for the latter we have

$$\frac{1}{2}\frac{\partial \operatorname{Re} \varepsilon(\omega, \mathbf{k})}{\partial \omega}\frac{\partial}{\partial t}\langle E^2\rangle^t_{\mathbf{k}\omega} = -\operatorname{Im} \varepsilon(\omega, \mathbf{k})\langle E^2\rangle^t_{\mathbf{k}\omega}.$$

The solution of the last equation yields

$$\langle E^2\rangle^t_{\mathbf{k}\omega} = B_{\mathbf{k}\omega} \exp\left[-\frac{2 \operatorname{Im} \varepsilon(\omega, \mathbf{k})}{\dfrac{\partial \operatorname{Re} \varepsilon(\omega, \mathbf{k})}{\partial \omega}} t\right], \tag{9.53}$$

where the quantity $B_{\mathbf{k}\omega}$ is time-independent. If the frequency ω is close to the plasma eigen-frequency $\omega_{\mathbf{k}}$, then $\operatorname{Im} \varepsilon(\omega, \mathbf{k})\Big/\dfrac{\partial \operatorname{Re} \varepsilon(\omega, \mathbf{k})}{\partial \omega} = \gamma_{\mathbf{k}}$ and $\gamma_{\mathbf{k}} \ll \omega_{\mathbf{k}}$. Thus, the non-stationary part of solution (9.52) is in such a case due to the weakly damped eigenoscillations associated with the initial perturbation. But when ω differs from the eigenfrequencies $\omega_{\mathbf{k}}$. then $\operatorname{Im} \varepsilon(\omega, \mathbf{k})\Big/\dfrac{\partial \operatorname{Re} \varepsilon(\omega, \mathbf{k})}{\partial \omega} \gtrsim \omega$ and the non-stationary addition to solution (9.52) rapidly tends to zero. Therefore the non-stationary solution of (9.53) for times $t \gtrsim \omega_{\mathbf{k}}^{-1}$ may be written as

$$\langle E^2\rangle^t_{\mathbf{k}\omega} = \pi I_{\mathbf{k}}(t)\,\{\delta(\omega-\omega_{\mathbf{k}})+\delta(\omega+\omega_{\mathbf{k}})\}, \tag{9.54}$$

where

$$I_{\mathbf{k}}(t) = I_{\mathbf{k}}(0)e^{-2\gamma_{\mathbf{k}}t} \tag{9.55}$$

($I_{\mathbf{k}}(0)$ is the oscillation intensity at $t = 0$). This solution is in accordance with (9.51).

Thus, the general solution of (9.43) in neglect of the non-linear interactions may be written as follows:

$$\langle E^2\rangle_{\mathbf{k}\omega} = \langle E^2\rangle^0_{\mathbf{k}\omega}+\pi I_{\mathbf{k}}(t)\,\{\delta(\omega-\omega_{\mathbf{k}})+\delta(\omega+\omega_{\mathbf{k}})\}, \tag{9.56}$$

where $\langle E^2\rangle^0_{\mathbf{k}\omega}$ and $I_{\mathbf{k}}(t)$ are determined according to (9.50) and (9.55). The first term of (9.56) is governed by the non-homogeneous part of (9.43) and is due to the stationary level of the electric field fluctuations in the plasma. The second one is the solution of the relevant homogeneous equation and describes the eigenoscillations of the plasma electric field which are governed by the initial conditions. Within the context of (9.55) the intensity of such oscillations decreases exponentially with time. Therefore, the spectral distribution of the electric field oscillations $\langle E^2\rangle_{\mathbf{k}\omega}$ in a stable plasma with a time-independent distribution $f(\mathbf{v})$ tends to a stationary level $\langle E^2\rangle^0_{\mathbf{k}\omega}$ as $t \to \infty$. As we have pointed out before, the stationary spectrum $\langle E^2\rangle^0_{\mathbf{k}\omega}$ consists of incoherent fluctuations and of fluctuation eigen-oscillations. The intensity of the latter in an equilibrium plasma is governed by the temperature (6.127). In a non-equilibrium plasma the stationary fluctuation level is determined by the relevant solution of (7.44) and may considerably exceed the thermal one. If the distribution function is such that $\gamma_{\mathbf{k}} < 0$, then the intensity $I_{\mathbf{k}}(t)$ in (9.56) can reach enormous values. However, the neglect of the non-linear terms in (9.43) is unjustified under such conditions, since the non-linear wave interaction is the dominant effect among those governing the time evolution of $I_{\mathbf{k}}(t)$.

In the general case the stationary fluctuation level $I_{\mathbf{k}}$ depends on the spectrum of the incoherent fluctuations too, in particular, on the non-linear interactions of the fluctuation oscillations with the incoherent fluctuations. However, the latter effect is negligibly small in non-equilibrium plasmas with sufficiently high level of fluctuation oscillations $\left(I_{\mathbf{k}} \gg \int \frac{d\omega}{2\pi} \langle E^2 \rangle^{\text{inc}}_{\mathbf{k}\omega}\right)$. In such a case the time evolution and the stationary fluctuation level are governed by the equation

$$\frac{1}{2} \varepsilon'_{\mathbf{k}} \frac{\partial I_{\mathbf{k}}}{\partial t} = -\text{Im}\ \varepsilon(\omega_{\mathbf{k}}, \mathbf{k}) I_{\mathbf{k}} + \frac{\pi}{\varepsilon'_{\mathbf{k}}} \sum_{\mathbf{k}'} |\varkappa^{(2)}(\omega_{\mathbf{k}} \mp \omega_{\mathbf{k}'}, \mathbf{k}-\mathbf{k}';\ \pm\omega_{\mathbf{k}'}, \mathbf{k}')|^2$$

$$\times \delta(\omega_{\mathbf{k}} \mp \omega_{\mathbf{k}'} \mp \omega_{\mathbf{k}-\mathbf{k}'}) I_{\mathbf{k}'} I_{\mathbf{k}-\mathbf{k}'} + \frac{1}{2}\ \text{Im} \sum_{\mathbf{k}'} a(\omega_{\mathbf{k}}, \mathbf{k};\ \pm\omega_{\mathbf{k}'}, \mathbf{k}') I_{\mathbf{k}'} I_{\mathbf{k}} + \frac{1}{\varepsilon'_{\mathbf{k}}} q_{\mathbf{k}\omega}. \quad (9.57)$$

It is sufficient to substitute $\tilde{\omega}_{\mathbf{k}}$ for $\omega_{\mathbf{k}}$ in (9.57) to take into account the effect of the non-linear frequency shifts.

The non-linear wave interactions in a plasma may result in an appreciable enhancement of the fluctuation level in case some eigenoscillations are excited by external sources. Let the spectral distribution of the eigenoscillations excited by external sources (the pumping field) be denoted by $\langle \tilde{E}^2 \rangle_{\mathbf{k}\omega}$. The effect of the pumping field on the stationary fluctuation level may be described by the following substitution in (9.57):

$$\text{Im}\ \varepsilon(\omega_{\mathbf{k}}, \mathbf{k}) \rightarrow \text{Im}\left\{ \varepsilon(\omega_{\mathbf{k}}, \mathbf{k}) - \sum_{\omega', \mathbf{k}'} a(\omega_{\mathbf{k}}, \mathbf{k}; \omega', \mathbf{k}') \langle \tilde{E}^2 \rangle_{\mathbf{k}'\omega'} \right\}.$$

It is clear that the intensity of the externally excited fluctuation oscillations being sufficiently high, the intensity $I_{\mathbf{k}}$ may be changed; in particular, it may become very large as compared to the thermal level (the induced enhancement of fluctuations). If the origin of the pumping field is radiation, the following quantity is to be substituted in (9.57) for $\varepsilon(\omega, \mathbf{k})$:

$$\varepsilon(\omega, \mathbf{k}) \rightarrow \varepsilon(\omega, \mathbf{k}) + \frac{1}{4} \frac{e^2}{m^2} \varkappa^{(e)}(\omega, \mathbf{k}) (1 + \varkappa^{(i)}(\omega, \mathbf{k})) k^2 \sum_{\omega', \mathbf{k}'} \frac{(k-k')_i \langle E_i E_j \rangle_{\mathbf{k}'\omega'}}{\omega'^4 \varepsilon(\omega-\omega', \mathbf{k}-\mathbf{k}')} \frac{(k-k')_j}{(\mathbf{k}-\mathbf{k}')^2},$$

where $\varkappa^{(e)}$ and $\varkappa^{(i)}$ are the electron and the ion components of the linear plasma susceptibility.

The kinetic equation for longitudinal plasma waves

The kinetic equation for waves may be easily derived from (9.44) if the particle distributions are fixed and there exists a stationary fluctuation level $\langle E^2 \rangle^0_{\mathbf{k}\omega}$. With the following expression substituted for the spectral distribution of the electric field in (9.44),

$$\langle E^2 \rangle_{\mathbf{k}\omega} = \langle E^2 \rangle^0_{\mathbf{k}\omega} + \pi \sum_{\alpha} I^{\alpha}_{\mathbf{k}}(t) \{\delta(\omega - \omega^{\alpha}_{\mathbf{k}}) + \delta(\omega + \omega^{\alpha}_{\mathbf{k}})\},$$

where the summation extends over various longitudinal plasma eigenmodes, (9.44) becomes transformed into an equation for the intensity $I_{\mathbf{k}}^{\alpha}$ of the oscillation mode α:

$$\frac{\varepsilon_{\mathbf{k}}^{\prime\alpha}}{2}\frac{\partial I_{\mathbf{k}}^{\alpha}}{\partial t}=-\operatorname{Im}\varepsilon(\omega_{\mathbf{k}}^{\alpha},\mathbf{k})I_{\mathbf{k}}^{\alpha}$$

$$+\frac{\pi}{\varepsilon_{\mathbf{k}}^{\prime\alpha}}\sum_{\pm\beta}\sum_{\pm\beta'}\sum_{\mathbf{k}'}|\varkappa^{(2)}(\omega_{\mathbf{k}}^{\alpha}\pm\omega_{\mathbf{k}'}^{\beta};\mathbf{k}-\mathbf{k}';\pm\omega_{\mathbf{k}'}^{\beta},\mathbf{k}')|^{2}$$

$$\times\delta(\omega_{\mathbf{k}}^{\alpha}\mp\omega_{\mathbf{k}'}^{\beta}\mp\omega_{\mathbf{k}-\mathbf{k}'}^{\beta'})I_{\mathbf{k}'}^{\beta}I_{\mathbf{k}-\mathbf{k}'}^{\beta'}+\frac{1}{2}\operatorname{Im}\sum_{\pm\beta}\sum_{\mathbf{k}'}a(\omega_{\mathbf{k}'}^{\alpha},\mathbf{k};\pm\omega_{\mathbf{k}'}^{\beta},\mathbf{k}')I_{\mathbf{k}'}^{\beta}I_{\mathbf{k}}^{\alpha}$$

$$+\operatorname{Im}\sum_{\omega',\mathbf{k}'}a(\omega_{\mathbf{k}}^{\alpha},\mathbf{k};\omega',\mathbf{k}')\langle E^{2}\rangle_{\mathbf{k}'\omega'}^{0}I_{\mathbf{k}}^{\alpha}+\frac{1}{2\varepsilon_{\mathbf{k}}^{\prime\alpha}}\sum_{\pm\beta}\sum_{\mathbf{k}'}\Big\{b(\omega_{\mathbf{k}}^{\alpha},\mathbf{k};\pm\omega_{\mathbf{k}'}^{\beta},\mathbf{k}')$$

$$+4|\varkappa^{(2)}(\omega_{\mathbf{k}}^{\alpha}\mp\omega_{\mathbf{k}'}^{\beta};\mathbf{k}-\mathbf{k}';\pm\omega_{\mathbf{k}'}^{\beta},\mathbf{k}')|^{2}\langle E^{2}\rangle_{\mathbf{k}-\mathbf{k}',\,\omega_{\mathbf{k}}^{\alpha}-\omega_{\mathbf{k}'}^{\beta}}^{0}\Big\}I_{\mathbf{k}'}^{\beta}. \tag{9.58}$$

The $\pm$ signs at β and β' imply that the summation over β and β' in (9.58) extends over the frequencies $\omega_{\mathbf{k}'}^{\beta}$ and $\omega_{\mathbf{k}-\mathbf{k}'}^{\beta'}$ of both signs. The quantities $a(\omega,\mathbf{k};\omega',\mathbf{k}')$ and $b(\omega,\mathbf{k};\omega',\mathbf{k}')$ are defined by (7.11) and (7.12) respectively. The notation implies

$$\varepsilon_{\mathbf{k}}^{\prime\alpha}\equiv\frac{\partial\operatorname{Re}\varepsilon(\omega_{\mathbf{k}}^{\alpha},\mathbf{k})}{\partial\omega_{\mathbf{k}}^{\alpha}}.$$

Equation (9.58) describes the wave dynamics in a plasma taking into account wave–wave coupling, as well as the interaction of waves with the fluctuation oscillations. If the fluctuation intensity $\langle E^{2}\rangle_{\mathbf{k}\omega}^{0}$ and the quantity $b(\omega_{\mathbf{k}}^{\alpha},\mathbf{k};\omega_{\mathbf{k}'}^{\beta},\mathbf{k}')$, which is associated with the fluctuations of the particle distributions, are neglected, then (9.58) reduces to the conventional kinetic equation for waves:†

$$\frac{1}{2}\frac{\partial I_{\mathbf{k}}^{\alpha}}{\partial t}=-\gamma_{\mathbf{k}}^{\alpha}I_{\mathbf{k}}^{\alpha}$$

$$+\frac{\pi}{(\varepsilon_{\mathbf{k}}^{\prime\alpha})^{2}}\sum_{\pm\beta}\sum_{\pm\beta'}\sum_{\mathbf{k}'}|\varkappa^{(2)}(\omega_{\mathbf{k}}^{\alpha}\mp\omega_{\mathbf{k}'}^{\beta},\mathbf{k}-\mathbf{k}';\pm\omega_{\mathbf{k}'}^{\beta},\mathbf{k}')|^{2}\delta(\omega_{\mathbf{k}}^{\alpha}\mp\omega_{\mathbf{k}'}^{\beta}\mp\omega_{\mathbf{k}-\mathbf{k}'}^{\beta'})I_{\mathbf{k}'}^{\beta}I_{\mathbf{k}-\mathbf{k}'}^{\beta'}$$

$$+\frac{1}{2\varepsilon_{\mathbf{k}}^{\prime\alpha}}\operatorname{Im}\sum_{\pm\beta}\sum_{\mathbf{k}'}a(\omega_{\mathbf{k}}^{\alpha},\mathbf{k};\pm\omega_{\mathbf{k}'}^{\beta},\mathbf{k}')I_{\mathbf{k}'}^{\beta}I_{\mathbf{k}}^{\alpha}. \tag{9.59}$$

For the quantity $a(\omega_{\mathbf{k}}^{\alpha},\mathbf{k};\omega_{\mathbf{k}'}^{\beta},\mathbf{k}')$, see (7.11).

The kinetic equation (9.59) describes the variations of the spectral density of plasma waves $I_{\mathbf{k}}$ due to the linear dissipation and non-linear interactions. The linear dissipation is associated with linear wave-particle interactions which manifest themselves under the resonance condition $\omega=(\mathbf{k}\cdot\mathbf{v})$. If the particle velocity satisfies the resonance conditions, then the phase of the particle with respect to the wave is constant and the particle

† Kadomtsev and Petviashvili (1963) were the first who derived a kinetic equation for waves, then it was obtained by various approaches, e.g. Drummond and Pines (1964), Silin (1964b), Kadomtsev (1965), Al'tshul and Karpman (1965), Galeev, Karpman, and Sagdeev (1965), Gorbunov, Pustovalov, and Silin (1965), and Kovrizhnych (1966); see also Vedenov (1965), Tsytovich (1970), Davidson (1972), and Galeev and Sagdeev (1973). The kinetic equation for waves taking into account the interaction with the fluctuating plasma field was derived by Sitenko (1975).

is efficiently accelerated by the wave field. The acceleration is accompanied by energy transfer from the field to the plasma particles and leads to a damping of the wave-field amplitude. Since only resonant particles are involved in the linear wave-particle interactions, the damping may be described only by the kinetic theory.

Besides the linear wave-particle interactions there may occur non-linear wave–wave and wave-particle interactions in a plasma. Indeed, inasmuch as the plasma is a non-linear electrodynamic medium, two propagating waves with different frequencies ω_1 and ω_2 and wave vectors $\mathbf{k}_1$ and $\mathbf{k}_2$ excite pulsations with combination frequencies $\omega_1 \pm \omega_2$ and wave vectors $\mathbf{k}_1 \pm \mathbf{k}_2$. The resonance between such a pulsation and a third wave with frequency ω and wave vector $\mathbf{k}$ leads to a decay interaction of waves. The kinetic equation (9.59) involves only the three-wave decay processes, i.e. a confluence of two waves or a decay of one wave into two others. Such processes can occur only for so-called decay spectra $\omega_{\mathbf{k}} = \omega(\mathbf{k})$ which satisfy both the energy and the momentum conservation laws:

$$\omega_{\mathbf{k}} = \omega_{\mathbf{k}_1} + \omega_{\mathbf{k}_2}, \quad \mathbf{k} = \mathbf{k}_1 + \mathbf{k}_2.$$

This coupling is manifested also in the absence of thermal motion and may be described also by a hydrodynamic approach.

If the decay conditions are not satisfied, then the wave-particle scattering processes become of importance. In such a case there may occur a resonance between the pulsations and the particles, the velocities of which satisfy the condition

$$\omega_1 \pm \omega_2 = (\{\mathbf{k}_1 \pm \mathbf{k}_2\} \cdot \mathbf{v}).$$

The resonant particle-pulsation coupling is the reason of the additional wave damping, called non-linear Landau damping. The latter, like the linear damping, can be described only by the kinetic theory. The resonant particle-pulsation interaction is associated with the wave scattering by the particles, i.e. the re-emission of a wave with frequency ω_1 and wave vector $\mathbf{k}_1$ into one with frequency ω_2 and wave vector $\mathbf{k}_2$, while the linear Landau damping corresponds to the absorption of the waves by particles. The non-linear damping is essential when only a few particles are involved into the resonant interaction with a single oscillation, while a great number of particles are in resonance with the pulsations. The effects due to the non-linear wave-particle interactions compete in the general case to the same order of magnitude with the decay processes but become dominant in systems with non-decay spectra.

To reduce the kinetic equation (9.59) to a clearer form we introduce a wave number density $N_{\mathbf{k}}$ instead of the spectral density $I_{\mathbf{k}}$. The spectral energy density $W_{\mathbf{k}}$ is governed by $I_{\mathbf{k}}$:

$$W_{\mathbf{k}} = \frac{1}{16\pi} \omega_{\mathbf{k}} \varepsilon'_{\mathbf{k}} I_{\mathbf{k}}. \tag{9.60}$$

Let the wave number density $N_{\mathbf{k}}$ be defined by means of the following relation:

$$N_{\mathbf{k}} \equiv W_{\mathbf{k}}/s_{\mathbf{k}}\omega_{\mathbf{k}},$$

where $s_{\mathbf{k}} = \operatorname{sgn} \varepsilon'_{\mathbf{k}}$. Hence the following relation for $N_{\mathbf{k}}$ and $I_{\mathbf{k}}$ may be written:

$$N_{\mathbf{k}} = \frac{1}{16\pi} |\varepsilon'_{\mathbf{k}}| I_{\mathbf{k}} \tag{9.61}$$

and therefore (9.59) becomes of the form

$$\frac{\partial N_{\mathbf{k}}^{\alpha}}{\partial t} = -2\gamma_{\mathbf{k}}^{\alpha} N_{\mathbf{k}}^{\alpha} + 32\pi^2 \sum_{\pm\beta} \sum_{\pm\beta'} \sum_{\mathbf{k}'} \frac{1}{|\varepsilon_{\mathbf{k}}'^{\alpha} \varepsilon_{\mathbf{k}'}'^{\beta} \varepsilon_{\mathbf{k}-\mathbf{k}'}'^{\beta'}|} |\varkappa^{(2)}(\omega_{\mathbf{k}}^{\alpha} - \omega_{\mathbf{k}'}^{\beta}, \mathbf{k}-\mathbf{k}'; \omega_{\mathbf{k}'}^{\beta}, \mathbf{k}')^2$$
$$\times \delta(\omega_{\mathbf{k}}^{\alpha} - \omega_{\mathbf{k}'}^{\beta} - \omega_{\mathbf{k}-\mathbf{k}'}^{\beta'}) N_{\mathbf{k}'}^{\beta} N_{\mathbf{k}-\mathbf{k}'}^{\beta'}$$
$$+ 16\pi s_{\mathbf{k}}^{\alpha} \operatorname{Im} \sum_{\pm\beta} \sum_{\mathbf{k}'} \frac{1}{|\varepsilon_{\mathbf{k}}'^{\alpha} \varepsilon_{\mathbf{k}'}'^{\beta}|} a(\omega_{\mathbf{k}}^{\alpha}, \mathbf{k}; \omega_{\mathbf{k}'}^{\beta}, \mathbf{k}') N_{\mathbf{k}'}^{\beta} N_{\mathbf{k}}^{\alpha}. \tag{9.62}$$

Let us separate out the residue and the principal value of the integral in the last term on the right-hand side of (9.62). Within the context of the symmetry properties of the non-linear susceptibility $\varkappa^{(2)}(\omega_1, \mathbf{k}_1; \omega_2, \mathbf{k}_2)$ we obtain the following kinetic equation for waves in a plasma:

$$\frac{\partial N_{\mathbf{k}}^{\alpha}}{\partial t} = -2\gamma_{\mathbf{k}}^{\alpha} N_{\mathbf{k}}^{\alpha} + 32\pi \sum_{\beta} \sum_{\beta'} \sum_{\mathbf{k}'} \frac{|\varkappa^{(2)}(\omega_{\mathbf{k}}^{\alpha} - \omega_{\mathbf{k}'}^{\beta}, \mathbf{k}-\mathbf{k}'; \omega_{\mathbf{k}'}^{\beta}, \mathbf{k}')|^2}{|\varepsilon_{\mathbf{k}}'^{\alpha} \varepsilon_{\mathbf{k}'}'^{\beta} \varepsilon_{\mathbf{k}-\mathbf{k}'}'^{\beta'}|} \delta(\omega_{\mathbf{k}}^{\alpha} - \omega_{\mathbf{k}'}^{\beta} - \omega_{\mathbf{k}-\mathbf{k}'}^{\beta'})$$
$$\times s_{\mathbf{k}}^{\alpha}(s_{\mathbf{k}}^{\alpha} N_{\mathbf{k}'}^{\beta} N_{\mathbf{k}-\mathbf{k}'}^{\beta'} - s_{\mathbf{k}'}^{\beta} N_{\mathbf{k}}^{\alpha} N_{\mathbf{k}-\mathbf{k}'}^{\beta'} - s_{\mathbf{k}-\mathbf{k}'}^{\beta'} N_{\mathbf{k}}^{\alpha} N_{\mathbf{k}'}^{\beta})$$
$$+ 16\pi s_{\mathbf{k}}^{\alpha} \operatorname{Im} \sum_{\beta} \sum_{\mathbf{k}'} \frac{1}{|\varepsilon_{\mathbf{k}}'^{\alpha} \varepsilon_{\mathbf{k}'}'^{\beta}|} \Big\{ 2\varkappa^{(2)}(\omega_{\mathbf{k}}^{\alpha} - \omega_{\mathbf{k}'}^{\beta}, \mathbf{k}-\mathbf{k}'; \omega_{\mathbf{k}'}^{\beta}, \mathbf{k}')$$
$$\times \varkappa^{(2)}(\omega_{\mathbf{k}}^{\alpha}, \mathbf{k}; -\omega_{\mathbf{k}'}^{\beta}, -\mathbf{k}') P \frac{1}{\varepsilon(\omega_{\mathbf{k}}^{\alpha} - \omega_{\mathbf{k}'}^{\beta}, \mathbf{k}-\mathbf{k}')}$$
$$- \bar{\varkappa}^{(3)}(\omega_{\mathbf{k}'}^{\beta}, \mathbf{k}'; \omega_{\mathbf{k}}^{\alpha}, \mathbf{k}; -\omega_{\mathbf{k}'}^{\beta}, \mathbf{k}') \Big\} N_{\mathbf{k}}^{\alpha} N_{\mathbf{k}'}^{\beta}. \tag{9.63}$$

The first term on the right-hand side of (9.63) describes the linear damping or enhancement of oscillations and is associated with resonant wave-particle interactions. The second one is due to three-wave processes: the decay of a wave into two others or the confluence of two waves into one. (The first term within the curly brackets that follow the summation symbol $\sum_{\mathbf{k}'}$ is associated with the spontaneous decay of the incident wave, while the second and the third ones, which correspond to the residues of the last term in (9.62), describe the induced decay processes, since those are proportional to the density of the number of initially decaying waves $N_{\mathbf{k}}^{\alpha}$.) The third term on the right-hand side of (9.63) is due to the induced wave-particle scattering. (The first term within the curly brackets following the summation symbol $\sum_{\mathbf{k}'}$ is related to the decay of the initial oscillation into an oscillation with a close frequency and a pulsation that is in resonance with the particles. The second one is associated with the Compton scattering of the waves by plasma particles.)

The kinetic equation (9.63) may be taken as a basis to describe the turbulent processes in plasmas. However, it is to be observed that (9.63) holds only when the wave intensities are sufficiently large and the plasma fluctuations may be neglected. So the wave scattering and transformation due to interactions with the fluctuating fields in the plasma are not involved in the consideration. In contrast to (9.63), the kinetic equation (9.58) does take into account the interaction of waves with the fluctuating fields, and therefore may be made use of to study the scattering and the transformation of waves by plasma fluctuations.

Non-linear Landau damping

We have already pointed out that by virtue of the non-linearity of the electrodynamic plasma properties the propagation of two waves through the medium is accompanied by pulsations. The resonant coupling of the latter with plasma particles causes non-linear wave damping. Let us consider the non-linear damping of the Langmuir waves, for example (Kadomtsev, 1965).

The non-linear interactions of Langmuir waves results in low-frequency pulsations with phase velocities that may be very small compared to the electron thermal velocity. These pulsations, the absorption of which by particles is the reason of the non-linear damping, interact with the electrons much more intensively than the initial Langmuir waves. If the mean wave number of the wave packet under consideration is considerably less than the inverse Debye length, then the linear Landau damping is exponentially weak and the energy losses of the Langmuir waves are governed mainly by non-linear interactions.

The spectrum of the Langmuir waves in a plasma in equilibrium or close to it does not satisfy the decay conditions, so we have to retain in the kinetic equation (9.63) only the last term that is due to the induced wave-particle scattering. In a strongly non-isothermal plasma the Langmuir spectrum is of the decay form and the coupling between the Langmuir waves and the ion sound drives the decay instability as in the case of waves with fixed phases. Note that the decay time of a Langmuir wave packet, which may be found within the context of (9.63), differs from that of a Langmuir wave with a fixed phase only by a numerical factor. In the absence of decay processes the kinetic equation (9.63) may be written as

$$\frac{\partial N_{\mathbf{k}}}{\partial t} = -2(\gamma_{\mathbf{k}} + \gamma'_{\mathbf{k}}) N_{\mathbf{k}}, \tag{9.64}$$

where $\gamma'_{\mathbf{k}}$ denotes the non-linear Landau damping rate

$$\gamma'_{\mathbf{k}} = -\frac{1}{2\varepsilon'_{\mathbf{k}}} \operatorname{Im} \sum_{\mathbf{k}'} \{a(\omega_{\mathbf{k}}, \mathbf{k}; \omega_{\mathbf{k}'}, \mathbf{k}') + a(\omega_{\mathbf{k}}, \mathbf{k}; -\omega_{\mathbf{k}'}, -\mathbf{k}')\} I_{\mathbf{k}'}. \tag{9.65}$$

(It is of considerable interest to compare the damping rate (9.65) to the non-linear frequency shift of the Langmuir wave (7.36).)

Similarly to the non-linear eigenfrequency shift, the non-linear damping rate (9.65) vanishes if the thermal motion is neglected. Therefore the non-linear damping rate is to be calculated with the use of the general expressions (2.15), (2.16), and (2.18), which take into account thermal effects. Thus we obtain the following approximate expression for the non-linear damping rate of the Langmuir waves:

$$\gamma'_{\mathbf{k}} = \frac{e^2T}{m^3\Omega^5k^2}\sum_{\mathbf{k}'}(\mathbf{k}\cdot\mathbf{k}')^2\,[\mathbf{k},\mathbf{k}']^2\,\mathrm{Im}\,\varepsilon(\omega_{\mathbf{k}}-\omega_{\mathbf{k}'},\mathbf{k}-\mathbf{k}')I_{\mathbf{k}'}. \tag{9.66}$$

An order of magnitude estimate yields

$$\gamma'_{\mathbf{k}} \sim a^3k^3\frac{W}{n_0T}\Omega,$$

where n_0 is the equilibrium electron density, T is the plasma temperature, and W is the total energy density of the Langmuir waves

$$W = \sum_{\mathbf{k}'} W_{\mathbf{k}'}.$$

Within the context of explicit expressions for the linear Langmuir damping rate (4.41) it may be easily verified that non-linear damping becomes dominant when the wave intensities satisfy the condition $I_{\mathbf{k}} > I_{\mathbf{k}\,\min}$, where

$$I_{\mathbf{k}\,\min} \simeq \frac{1}{16\pi}\,\frac{n_0T}{a^6k^6}\,e^{-\frac{1}{2a^2k^2}}.$$

It follows from (9.64) and the explicit formula (9.66) that the number and the energy of the non-linearly damped waves are conserved, while the total momentum of the Langmuir waves decreases by virtue of the non-linear damping. Therefore the damping rate (9.66) determines the damping time of the wave momentum, but not the time needed for the energy to be changed.

In order to calculate the time rate of change of the Langmuir wave energy we have to take into account in (9.65) the higher-order terms in the particle velocity. The result is as follows:

$$\tau_\varepsilon \simeq \frac{1}{a^6k^6}\,\frac{1}{\gamma'_{\mathbf{k}}}. \tag{9.67}$$

Therefore, for the Langmuir energy to be changed a much longer time is needed than the momentum damping time. The reason is that the non-linear damping (in contrast to the linear process) occurs not due to the absorption of the waves by particles, but by virtue of wave-particle scattering, i.e. diffusion in wave-number space. Therefore the "mixing" of waves in the directions of the wave vectors is much more rapid than the decrease of the frequencies. It is to be noted that if the particle distribution function does not decrease with increasing particle energy, non-linear wave-particle interactions may result not in losses but in an increase of the wave energy (Silin, 1964a).

9.3. The Kinetic Equation for Waves (General Case)

Time evolution of the spectral correlation function for the electromagnetic field

Let us derive a kinetic equation for waves in a plasma no longer assuming the fields to be potential. We shall begin with the equation that governs the time evolution of the spectral correlation function for the fluctuations of the electromagnetic field. In a manner analogous to the consideration of the potential field we shall base ourselves on a multiple-time-scale perturbation analysis.

The fluctuating electromagnetic field in the plasma is in the general case governed by the non-linear equation (7.56). If the particle distribution is non-stationary and the collision term in the kinetic equation for the particles does not vanish, then the following non-linear equation is to be made use of instead of (7.56):

$$\begin{aligned} &\Lambda_{ij}(k)\,E_j(k)+\sum_{1+2=k}\varkappa^{(2)}_{ijk}(1,2)\,(E_j(1)\,E_k(2)-\langle E_j(1)\,E_k(2)\rangle)\\ &\quad+\sum_{1+2+3=k}\varkappa^{(3)}_{ijkl}(1,2,3)\,(E_j(1)\,E_k(2)\,E_l(3)-E_j(1)\,\langle E_k(2)\,E_l(3)\rangle\\ &\quad-\langle E_j(1)\,E_k(2)\,E_l(3)\rangle)+\;\dots\;+\sum_{k'}\Delta_{ij}(k,k')\,(\delta f^0(k-k')\,E_j(k')\\ &\quad-\langle\delta f^0(k-k')\,E_j(k')\rangle)+\sum_{\substack{1+2=k\\ k'}}\Delta^{(2)}_{ijk}(1,2;k')\,(\delta f^0(2-k')\,E_j(1)\,E_k(k')\\ &\quad-E_j(1)\,\langle\delta f^0(2-k')\,E_k(k')\rangle-\langle\delta f^0(2-k')\,E_j(1)\,E_k(k')\rangle)+\;\dots\\ &\quad=-\frac{4\pi i}{\omega}\,j^0_i(k). \end{aligned} \tag{9.68}$$

This equation leads to an equation for the spectral correlation function $\langle E_iE_j\rangle_k$ that is the same as (7.70) or (7.79), where the quantity (7.71) does not contain the two last terms.

To involve into consideration the evolution of the field spectral correlation function on a slower time scale the following substitution is to be performed in (7.79):

$$\Lambda_{ij}(k)\,\langle E_jE_i\rangle_k\rightarrow\left\{\Lambda_{ij}e_je_i^*+i\frac{\partial}{\partial\omega}\left(\frac{\partial\,\mathrm{Re}(\varepsilon_{ij}(k)e_je_i^*)}{\partial t}\right)+\frac{i}{2}\,\frac{\partial\,\mathrm{Re}(\Lambda_{ij}(k)e_je_i^*)}{\partial\omega}\,\frac{\partial}{\partial t}\right\}\langle E^2\rangle_k. \tag{9.69}$$

It should be remembered that in the vicinity of the eigenfrequencies the spectral distribution $\langle E_iE_j\rangle_k$ is of the form

$$\langle E_iE_j\rangle_k=e_ie_j^*\langle E^2\rangle_k. \tag{9.70}$$

where e_i and e_j are the polarization vectors of the relevant eigenoscillations.

Equating the imaginary part of the obtained relation to zero we come to the following equation that describes the time development of the spectral correlation function for the electromagnetic field due to the non-linear coupling of the fluctuation fields and the linear dissipation:

$$\frac{1}{2}\frac{\partial\,\mathrm{Re}(\Lambda_{ij}(k)e_je_i^*)}{\partial\omega}\frac{\partial}{\partial t}\langle E^2\rangle_k$$

$$= -\left\{\mathrm{Im}(\varepsilon_{ij}(k)e_je_i^*)+\frac{\partial}{\partial\omega}\frac{\partial\,\mathrm{Re}(\varepsilon_{ij}(k)e_je_i^*)}{\partial t}\right\}\langle E^2\rangle_k$$

$$+2\,\mathrm{Im}\,\Lambda_{ii'}^{*-1}(k)\sum_1 \varkappa_{ijk}^{(2)}(k-1,\,1)\,\varkappa_{i'j'k'}^{(2)*}(k-1,\,1)\langle E_kE_{k'}\rangle_1\langle E_jE_{j'}\rangle_{k-1}$$

$$+\mathrm{Im}\sum_1 a_{ijkk'}(k,\,1)\langle E_kE_{k'}\rangle_1\langle E_jE_{j'}\rangle_k$$

$$+\mathrm{Im}\,\Lambda_{ii'}^{*-1}(k)\left\{\sum_1 b_{ii'kk'}(k,\,1)\langle E_kE_{k'}\rangle_1+q'_{ii'}(k)\right\}. \tag{9.71}$$

This equation should be treated together with the kinetic equations for particles which govern the time evolution of the distribution function. However, we shall assume the particle distributions to be stationary and study the time development of the field spectral distribution only. In the stationary case (9.71) immediately reduces to (7.79). In neglect of the fluctuation sources, i.e. if $b_{ii'kk'}(k, 1)$ and $q'_{ii'}(k)$ vanish, (9.71) reduces to the Gorbunov–Pustovalov–Silin equation (Gorbunov, Pustovalov, and Silin, 1965).

The general solution of (9.71) in neglect of non-linear effects may be written as

$$\langle E_iE_j\rangle_k = \langle E_iE_j\rangle_k^0+\pi\sum_\alpha e_i^\alpha e_j^{\alpha *}I^\alpha(k,\,t)\,\{\delta(\omega-\omega_{\mathbf{k}}^\alpha)+\delta(\omega+\omega_{\mathbf{k}}^\alpha)\}, \tag{9.72}$$

where $\langle E_iE_j\rangle_k^0$ is the stationary distribution of the fluctuating field which is governed by the plasma state, and

$$I^\alpha(\mathbf{k},\,t) = I^\alpha(\mathbf{k},\,0)e^{-2\gamma_{\mathbf{k}}^\alpha t}$$

($I^\alpha(\mathbf{k}, 0)$ is the initial spectral density). The exponential time-dependence of the eigen-oscillation spectral density $I^\alpha(\mathbf{k}, t)$ in (9.72) is a result of the neglect of the non-linear coupling. In the general case the time-dependence of the spectral density of the eigen-oscillations $I^\alpha(\mathbf{k}, t)$ is governed as well by the non-linear wave-particle interactions in the plasma.

The kinetic equation for electromagnetic waves in a plasma

It is not difficult to derive a kinetic equation for arbitrary electromagnetic plasma waves from (9.71) on the assumption that the particle distributions are stationary and there exists a stationary fluctuation level which is described by the spectral correlation function $\langle E_iE_j\rangle_k^0$ and the quantities $b_{ii'kk'}(k, k')$. We substitute (9.72) for the electric field spectral distribution in (9.71) and consider the plasma transparency domain, i.e. assume that

$$\mathrm{Re}\,|\Lambda(k)| \gg |\mathrm{Im}\,\Lambda(k)|.$$

In the transparency domain the following relation holds:

$$\mathrm{Im}\,\frac{1}{\Lambda^*(k)} \rightarrow \pi\sum_\alpha\frac{1}{\Lambda_{\mathbf{k}}'^\alpha}\{\delta(\omega-\omega_{\mathbf{k}}^\alpha)+\delta(\omega-\omega_{\mathbf{k}}^\alpha)\}, \tag{9.73}$$

where

$$\Lambda_{\mathbf{k}}^{\prime\alpha} \equiv \frac{\partial \operatorname{Re} \Lambda(\omega_{\mathbf{k}}^{\alpha}, \mathbf{k})}{\partial \omega_{\mathbf{k}}^{\alpha}}. \tag{9.74}$$

Restricting the consideration to the time evolution of the intensity of some labelled mode $I_{\mathbf{k}}^{\alpha}(t)$ we obtain the following equation (Sitenko and Kocherga, 1977c):

$$\begin{aligned}
\frac{\Lambda_{\mathbf{k}}^{\prime\alpha}}{2}\frac{\partial I_{\mathbf{k}}^{\alpha}}{\partial t} &= -\operatorname{Im}(\varepsilon_{ij}(\omega_{\mathbf{k}}^{\alpha}, \mathbf{k})\,\lambda_{ji}(\omega_{\mathbf{k}}^{\alpha}, \mathbf{k}))I_{\mathbf{k}}^{\alpha} \\
&+\frac{\pi}{\Lambda_{\mathbf{k}}^{\prime\alpha}}\lambda_{i''}^{*}(\omega_{\mathbf{k}}^{\alpha}, \mathbf{k})\sum_{\pm\beta}\sum_{\pm\beta'}\sum_{\mathbf{k}'}\frac{\operatorname{Tr}\lambda_{\mathbf{k}}^{\alpha}}{\operatorname{Tr}\lambda_{\mathbf{k}'}^{\beta}\operatorname{Tr}\lambda_{\mathbf{k}-\mathbf{k}'}^{\beta}}\varkappa_{ijk}^{(2)}(\omega_{\mathbf{k}}^{\alpha}-\omega_{\mathbf{k}'}^{\beta}, \mathbf{k}-\mathbf{k}'; \omega_{\mathbf{k}'}^{\beta}, \mathbf{k}') \\
&\times\varkappa_{i'j'k'}^{(2)*}(\omega_{\mathbf{k}}^{\alpha}-\omega_{\mathbf{k}'}^{\beta}, \mathbf{k}-\mathbf{k}'; \omega_{\mathbf{k}'}^{\beta}, \mathbf{k}')\,\lambda_{jj'}(\omega_{\mathbf{k}-\mathbf{k}'}^{\beta'}, \mathbf{k}-\mathbf{k}')\,\lambda_{kk'}(\omega_{\mathbf{k}'}^{\beta}, \mathbf{k}')\,\delta(\omega_{\mathbf{k}}^{\alpha}-\omega_{\mathbf{k}'}^{\beta}-\omega_{\mathbf{k}-\mathbf{k}'}^{\beta'})I_{\mathbf{k}'}^{\beta}I_{\mathbf{k}-\mathbf{k}'}^{\beta'} \\
&+\frac{1}{2}\operatorname{Im}\sum_{\pm\beta}\sum_{\mathbf{k}'}\frac{1}{\operatorname{Tr}\lambda_{\mathbf{k}'}^{\beta}}a_{ijkk'}(\omega_{\mathbf{k}}^{\alpha}, \mathbf{k}; \omega_{\mathbf{k}'}^{\beta}, \mathbf{k}')\,\lambda_{ji}(\omega_{\mathbf{k}}^{\alpha}, \mathbf{k})\,\lambda_{kk'}(\omega_{\mathbf{k}'}^{\beta}, \mathbf{k}')I_{\mathbf{k}'}^{\beta}I_{\mathbf{k}}^{\alpha} \\
&+\operatorname{Im}\sum_{\omega', \mathbf{k}'}a_{ijkk'}(\omega_{\mathbf{k}}^{\alpha}, \mathbf{k}; \omega', \mathbf{k}')\,\lambda_{ji}(\omega_{\mathbf{k}}^{\alpha}, \mathbf{k})\,\langle E_k E_{k'}\rangle_{\mathbf{k}'\omega'}^{0}\,I_{\mathbf{k}}^{\alpha} \\
&+\frac{1}{2\Lambda_{\mathbf{k}}^{\prime\alpha}}\lambda_{ii'}^{*}(\omega_{\mathbf{k}}^{\alpha}, \mathbf{k})\sum_{\pm\beta}\sum_{\mathbf{k}'}\frac{\operatorname{Tr}\lambda_{\mathbf{k}}^{\alpha}}{\operatorname{Tr}\lambda_{\mathbf{k}'}^{\beta}}\Big\{b_{ii'kk'}(\omega_{\mathbf{k}}^{\alpha}, \mathbf{k}; \omega_{\mathbf{k}'}^{\beta}, \mathbf{k}') \\
&+4\varkappa_{ijk}^{(2)}(\omega_{\mathbf{k}}^{\alpha}-\omega_{\mathbf{k}'}^{\beta}, \mathbf{k}-\mathbf{k}'; \omega_{\mathbf{k}'}^{\beta}, \mathbf{k}')\,\varkappa_{i'j'k'}^{(2)*}(\omega_{\mathbf{k}}^{\alpha}-\omega_{\mathbf{k}'}^{\beta}, \mathbf{k}-\mathbf{k}'; \omega_{\mathbf{k}'}^{\beta}, \mathbf{k}')\,\langle E_j E_{j'}\rangle_{\mathbf{k}-\mathbf{k}', \omega_{\mathbf{k}}^{\alpha}-\omega_{\mathbf{k}'}^{\beta}}^{0}\Big\} \\
&\times\lambda_{kk'}(\omega_{\mathbf{k}'}^{\beta}, \mathbf{k}')I_{\mathbf{k}'}^{\beta}.
\end{aligned} \tag{9.75}$$

When deriving (9.75) use was made of the relations

$$\Lambda(k) \equiv \Lambda_{ij}(k)\,\lambda_{ji}(k) \quad \text{and} \quad e_i^{\alpha}e_j^{\alpha *} = \frac{\lambda_{ij}(\omega_{\mathbf{k}}^{\alpha}, \mathbf{k})}{\operatorname{Tr}\lambda(\omega, \mathbf{k})},$$

and the following notation was introduced:

$$\lambda_{\mathbf{k}}^{\alpha} \equiv \lambda(\omega_{\mathbf{k}}^{\alpha}, \mathbf{k}). \tag{9.76}$$

Equation (9.75) describes the dynamics of the electromagnetic waves in a plasma taking into account the wave interactions both with waves and with fluctuating plasma fields. In neglect of the fluctuation intensity $\langle E_i E_j\rangle_k^0$ and the quantity $b_{ii'kk'}(\omega_{\mathbf{k}}^{\alpha}, \mathbf{k}; \omega_{\mathbf{k}'}^{\beta}, \mathbf{k})$ that is associated with the fluctuations of the particle distributions in the plasma, (9.75) reduces to the commonly used kinetic equation for waves:

$$\begin{aligned}
\frac{1}{2}\frac{\partial I_{\mathbf{k}}^{\alpha}}{\partial t} &= -\gamma_{\mathbf{k}}^{\alpha}I_{\mathbf{k}}^{\alpha} \\
&+\frac{\pi}{(\Lambda_{\mathbf{k}}^{\prime\alpha})^2}\lambda_{ii'}^{*}(\omega_{\mathbf{k}}^{\alpha}, \mathbf{k})\sum_{\pm\beta}\sum_{\pm\beta'}\sum_{\mathbf{k}'}\frac{\operatorname{Tr}\lambda_{\mathbf{k}}^{\alpha}}{\operatorname{Tr}\lambda_{\mathbf{k}'}^{\beta}\operatorname{Tr}\lambda_{\mathbf{k}-\mathbf{k}'}^{\beta'}}\varkappa_{ijk}^{(2)}(\omega_{\mathbf{k}}^{\alpha}-\omega_{\mathbf{k}'}^{\beta}, \mathbf{k}-\mathbf{k}'; \omega_{\mathbf{k}'}^{\beta}, \mathbf{k}') \\
&\times\varkappa_{i'j'k'}^{(2)*}(\omega_{\mathbf{k}}^{\alpha}-\omega_{\mathbf{k}'}^{\beta}, \mathbf{k}-\mathbf{k}'; \omega_{\mathbf{k}'}^{\beta}, \mathbf{k}')\,\lambda_{jj'}(\omega_{\mathbf{k}-\mathbf{k}'}^{\beta'}, \mathbf{k}-\mathbf{k}')\,\lambda_{kk'}(\omega_{\mathbf{k}'}^{\beta}, \mathbf{k}') \\
&\times\delta(\omega_{\mathbf{k}}^{\alpha}-\omega_{\mathbf{k}'}^{\beta}-\omega_{\mathbf{k}-\mathbf{k}'}^{\beta'})I_{\mathbf{k}'}^{\beta}I_{\mathbf{k}-\mathbf{k}'}^{\beta'} \\
&+\frac{1}{2\Lambda_{\mathbf{k}}^{\prime\alpha}}\operatorname{Im}\sum_{\pm\beta}\sum_{\mathbf{k}'}\frac{1}{\operatorname{Tr}\lambda_{\mathbf{k}'}^{\beta}}a_{ijkk'}(\omega_{\mathbf{k}}^{\alpha}, \mathbf{k}; \omega_{\mathbf{k}'}^{\beta}, \mathbf{k}')\,\lambda_{ji}(\omega_{\mathbf{k}}^{\alpha}, \mathbf{k})\,\lambda_{kk'}(\omega_{\mathbf{k}'}^{\beta}, \mathbf{k}')I_{\mathbf{k}'}^{\beta}I_{\mathbf{k}}^{\alpha}.
\end{aligned} \tag{9.77}$$

The kinetic equation (9.77) takes into account the wave intensity variations caused by the linear dissipation, the decay interaction of the waves, and the induced wave-particle scattering.

Within the context of (4.25) the wave energy density $W_{\mathbf{k}}^{\alpha}$ is related to $I_{\mathbf{k}}^{\alpha}$ according to

$$W_{\mathbf{k}}^{\alpha} = \frac{1}{16\pi} \frac{\omega_{\mathbf{k}}^{\alpha}}{\operatorname{Tr} \lambda_{\mathbf{k}}^{\alpha}} \Lambda_{\mathbf{k}}^{\prime\alpha} I_{\mathbf{k}}^{\alpha}. \tag{9.78}$$

With the wave number density $N_{\mathbf{k}}^{\alpha}$ introduced by means of the relation

$$N_{\mathbf{k}}^{\alpha} \equiv W_{\mathbf{k}}^{\alpha} / s_{\mathbf{k}}^{\alpha} \omega_{\mathbf{k}}^{\alpha}, \tag{9.79}$$

where $s_{\mathbf{k}}^{\alpha} = \operatorname{sgn} \Lambda_{\mathbf{k}}^{\prime\alpha}$, the kinetic equation (9.77) becomes transformed into the following:

$$\begin{aligned} \frac{\partial N_{\mathbf{k}}^{\alpha}}{\partial t} &= -2\gamma_{\mathbf{k}}^{\alpha} N_{\mathbf{k}}^{\alpha} \\ &\quad + 32\pi^2 \lambda_{ii'}^{*}(\omega_{\mathbf{k}}^{\alpha}, \mathbf{k}) \sum_{\pm\beta} \sum_{\pm\beta'} \sum_{\mathbf{k}'} \frac{1}{\left| \Lambda_{\mathbf{k}}^{\prime\alpha} \Lambda_{\mathbf{k}'}^{\prime\beta} \Lambda_{\mathbf{k}-\mathbf{k}'}^{\prime\beta'} \right|} \varkappa_{ijk}^{(2)}(\omega_{\mathbf{k}}^{\alpha} - \omega_{\mathbf{k}'}^{\beta}, \mathbf{k}-\mathbf{k}'; \omega_{\mathbf{k}'}^{\beta}, \mathbf{k}') \\ &\quad \times \varkappa_{i'j'k'}^{(2)*}(\omega_{\mathbf{k}}^{\alpha} - \omega_{\mathbf{k}'}^{\beta}, \mathbf{k}-\mathbf{k}'; \omega_{\mathbf{k}'}^{\beta}, \mathbf{k}') \, \lambda_{jj'}(\omega_{\mathbf{k}-\mathbf{k}'}^{\beta'}, \mathbf{k}-\mathbf{k}') \, \lambda_{kk'}(\omega_{\mathbf{k}}^{\alpha}, \mathbf{k}') \\ &\quad \times \delta(\omega_{\mathbf{k}}^{\alpha} - \omega_{\mathbf{k}'}^{\beta} - \omega_{\mathbf{k}-\mathbf{k}'}^{\beta'}) N_{\alpha'}^{\beta} N_{\mathbf{k}-\mathbf{k}'}^{\beta'} + 16\pi s_{\mathbf{k}}^{\alpha} \operatorname{Im} \sum_{\pm\beta} \sum_{\mathbf{k}'} \frac{1}{\left| \Lambda_{\mathbf{k}}^{\prime\alpha} \Lambda_{\mathbf{k}'}^{\prime\beta} \right|} a_{ijkk'}(\omega_{\mathbf{k}}^{\alpha}, \mathbf{k}; \omega_{\mathbf{k}'}^{\beta}, \mathbf{k}') \\ &\quad \times \lambda_{ji}(\omega_{\mathbf{k}}^{\alpha}, \mathbf{k}) \, \lambda_{kk'}(\omega_{\mathbf{k}'}^{\beta}, \mathbf{k}') N_{\mathbf{k}'}^{\beta} N_{\mathbf{k}}^{\alpha}. \end{aligned} \tag{9.80}$$

In contrast to the kinetic equation (9.77), (9.75) may be regarded as a unified kinetic equation. The latter, besides the non-linear effects which are described by the usual kinetic equation, takes into account also the interaction of waves with the fluctuating field and the fluctuating variations of the particle distribution function. These effects give rise to the last two terms on the right-hand side of (9.75). The one but last term in (9.75) is associated with the non-linear wave damping due to the coupling with the fluctuating plasma fields. The last term on the right-hand side of (9.75) describes the scattering and the transformation of waves by the fluctuations of the field and of the particle distributions. These processes may become dominant in a non-equilibrium plasma with a high fluctuation level.

Parametric instability and turbulence

Parametric plasma instabilities develop under the influence of electromagnetic radiation (pumping field). The non-linear stabilization of these results in turbulent states. Indeed, the parametric resonance in a plasma exposed to radiation enhances the fluctuation oscillations. As soon as the radiation intensity exceeds a threshold value, the fluctuations grow infinite in the linear approximation and the plasma becomes unstable (Silin, 1965). The non-linear effects lead to a saturation of the fluctuation level and a stabilization of the instability. The turbulent plasma state is characterized by a stationary level of fluctuation oscillations which depends on the nature of the latter and is determined by the non-linear stabilization mechanism. It depends on the frequency, which of the following processes of pumping-wave transformation occur: $t \to l+a$ (aperiodic instability), $t \to l+s$ (in a non-isothermal plasma $T_e \gg T_i$), $t \to l+i$ (in a plasma with close ion and electron temperatures), $t \to l+l'$, $t \to t'+s$, $t \to t'+l$, and so on.

The main non-linear mechanisms that drive a parametrically unstable plasma towards a stationary turbulent state are the induced wave scattering by plasma particles and the non-linear eigenfrequency shift of plasma waves. The former process stabilizes the instability, for it leads to a transfer of the induced oscillations from the range of wave vectors where these are excited to the damping range. The stabilizing effect of the non-linear eigenfrequency shift is due to its influence on the instability growth rate. The non-linear eigenfrequency shift causes an increase in the wave numbers of the waves corresponding to a maximum growth rate and hence by virtue of a large growth in dissipation, causes stabilization of the instability.

The non-linear equation (9.57) is suitable for the study of the time evolution of parametric turbulence. In particular, one can find the stationary level of turbulent fluctuations, the effective collision frequency, and the time needed for the relaxation towards a stationary turbulent state.

(a) Aperiodic instability (Pustovalov and Silin, 1972b). Let the field of the electromagnetic pumping wave be denoted by

$$\mathbf{E} = \mathbf{E}_0 \cos((\mathbf{k}_0 \cdot \mathbf{r}) - \omega_0 t). \tag{9.81}$$

If the pumping-wave frequency ω_0 is less than that of the Langmuir oscillations $\omega_{\mathbf{k}}$, there may occur an aperiodic instability associated with the excitation of Langmuir oscillations and an aperiodic perturbation that exponentially grows during the linear stage. The increment is a maximum for the perturbations which propagate along the electric field vector of the pumping wave. The minimum threshold pumping field in the case of dissipation through the electron-ion Coulomb collisions is equal to

$$(r_E^2)_{\text{thresh}} = 4\frac{\nu_{ei}}{\Omega}(a^2 + a_i^2).$$

Near the threshold the growth rate is close to its maximum and is described by the following expression:

$$-\gamma_{\mathbf{k}} = \frac{\nu_{ei}}{2}\left[p^2\cos^2\vartheta - 1 - 6\frac{\Omega}{\nu_{ei}}\frac{p^2 + 2(\omega_0 - \Omega)/\nu_{ei}}{p^2}a^2(k - k_m)^2\right],$$

where p is the ratio of the pumping field to the threshold value. The maximum growth rate corresponds to the wave number

$$k_m = \frac{p^2\nu_{ei} + 2(\omega_0 - \Omega)}{3a^2\Omega}.$$

Since the growth rate strongly depends on the pumping-wave frequency mismatch, the parametric instability may be stabilized by means of a small non-linear shift of the Langmuir oscillation frequency. The correction to the frequency near threshold is determined by the Langmuir wave field ($k_a < s_i/s_e$):

$$\delta\omega = -\frac{E^2}{32\pi n_0 T_e}\left(1+\frac{a_i^2}{a^2}\right)\Omega, \quad E^2 = \sum_{\mathbf{k}} I_{\mathbf{k}}.$$

Taking into account the influence of the growing perturbation on the plasma oscillation frequency we find the instability stabilized. The stationary value of the turbulent field of Langmuir oscillations is given by

$$E^2 = \sqrt{2(E_0^2 - E_{\text{thresh}}^2)}E_0. \tag{9.82}$$

The evolution equation for the spectral distribution of turbulent fluctuations may be solved in the general non-stationary case. The characteristic time of the relaxation towards the stationary state is determined by the time of the linear parametric growth of the oscillations times the logarithm of the ratio of the total turbulent stationary spectrum to the spontaneous one. The time needed for the turbulent spectrum to reach a stationary level becomes longer as the pumping field decreases to its threshold value.

(b) Ion-acoustic parametric instability (Bychenkov, Pustovalov, Silin, and Tikhonchuk, 1976). If the pumping-wave frequency ω_0 is somewhat greater than the Langmuir frequency Ω, then, provided $T_e \gg T_i$, there develops a parametric instability in the plasma associated with the pumping-wave decay in a high-frequency plasma wave and a low-frequency ion-acoustic one. The growth rate calculated taking into account the non-linear shift of the Langmuir wave frequency $\delta\omega_k$ is given by the following expression:

$$\gamma_{\mathbf{k}} = \tilde{\gamma}_{\mathbf{k}}^s - \frac{1}{4}\frac{(\mathbf{k}\cdot\mathbf{r})^2}{k^2a^2}\frac{\omega_s^2\omega_0\tilde{\gamma}_{\mathbf{k}}(\delta\omega_{\mathbf{k}} - \delta\omega_{\mathbf{k}})}{[(\delta\omega_{\mathbf{k}} - \Delta\omega_{\mathbf{k}})^2 - \omega_s^2] + 4\omega_s^2\tilde{\gamma}_{\mathbf{k}}^2}, \tag{9.83}$$

where $\delta\omega_{\mathbf{k}}$ is the difference between ω_0 and the Langmuir wave frequency $\omega_{\mathbf{k}}$. Within the context of the decay conditions $\delta\omega_{\mathbf{k}}$ coincides with the frequency of the ion-acoustic wave: $\delta\omega_{\mathbf{k}} = kv_s$. According to (9.83), the effective frequency mismatch $\delta\omega - \Delta\omega\,(\Delta\omega < 0)$ increases with the fluctuation level. Thus, the maximum of the growth rate shifts from the decay wave number towards greater wave numbers, while the maximum value of the growth rate decreases due to an increase of the high-frequency damping rate $\tilde{\gamma}_k$, which leads to a saturation of the instability. The condition for the turbulent plasma fluctuations to be stationary implies $\gamma_{k_m} = 0$, where k_m corresponds to the maximum growth rate (the value of k_m is determined by the modified decay condition $\delta\omega_{\mathbf{k}} - \Delta\omega_{\mathbf{k}} = kv_s$). The Langmuir fluctuation level is given by the approximate expression

$$\frac{1}{8\pi}E_c^2 = 4k_0^3a^3\sqrt{\frac{\Omega_i^2}{\Omega^2} + 6\frac{\omega_0 - \Omega}{\Omega}}\ln[p^2(1 + a^2k_0^2\ln p^2)]n_0T. \tag{9.84}$$

The level of the ion-acoustic fluctuations is related to the latter as

$$E_s^2 = p^2\frac{\tilde{\gamma}(k_0)}{\tilde{\gamma}_s(k_0)}\frac{\omega_s(k_0)}{\omega_0}E_c^2. \tag{9.85}$$

The fluctuation saturation level due to the transfer across the spectrum is much higher. Development of the parametric instability and saturation of the turbulent level are accompanied by an anomalous absorption of the electromagnetic radiation by the plasma, which is usually characterized by an effective collision frequency.

(c) Non-linear transformation of radiation into plasma waves (Pustovalov, Silin, and Tikhonchuk, 1974). If the pumping-field frequency is close to twice the Langmuir frequency, the electromagnetic radiation is transformed into plasma waves. Assume that

$$\omega_0 = 2\Omega(1+\delta), \quad |\delta| \ll 1.$$

Upon the assumption that the wavelengths of the excited waves are appreciably less than that of the pumping wave ($k \gg k_0$), we find the following frequency and damping rate of the plasma waves:

$$\omega_{\mathbf{k}} = \Omega(1+\Delta+\tfrac{3}{2}a^2(\mathbf{k}\cdot\mathbf{k}_0)), \quad \gamma_{\mathbf{k}} = \tilde{\gamma}_{\mathbf{k}} - \Omega\sqrt{(\mathbf{k}\cdot\mathbf{r}_E)^2(\mathbf{k}\cdot\mathbf{k}_0)^2 k^{-4} - (\delta - \tfrac{3}{2}a^2k^2)^2}. \tag{9.86}$$

The pumping field being sufficiently powerful, the plasma is parametrically unstable ($\gamma_{\mathbf{k}} < 0$). To find the threshold value of the pumping field we require the damping coefficient to be equal to zero $\gamma_{\mathbf{k}} = 0$. The minimum value of the threshold field E_m corresponds to $k = k_m = \sqrt{\dfrac{2\delta}{3a^2}}$. In neglect of non-linear interactions the level of the plasma fluctuation oscillations becomes infinite as the pumping-field reaches the threshold value. The dominant non-linear effect is the induced scattering of plasma waves by ions (the third term on the right-hand side of (9.57)). Then

$$a(\omega_{\mathbf{k}}, \mathbf{k}; \omega_{\mathbf{k}'}, \mathbf{k}') = \frac{\sqrt{2\pi}\Omega}{2n_0T_e} \frac{(\mathbf{kk}')^2}{(kk')^2} \frac{a^2a_i^2}{(a^2+a_i^2)^2} \frac{\omega-\omega'}{|\mathbf{k}-\mathbf{k}'|\, s_i} e^{-\frac{3}{2}\frac{(\omega-\omega')^2}{(\mathbf{k}-\mathbf{k})^2 s_i^2}}.$$

The pumping of the plasma wave energy over the spectrum from the region of excitation to the absorption domain may result in the stationary level of fluctuation oscillations. If the pumping field is moderate ($p \ll (\Omega/\tilde{\gamma})\,\delta$, where p is the ratio of the pumping field to the minimum threshold value), the parametric instability growth rate steeply depends on the wave vector mismatch $k^2 - k_m^2$. Hence, the spectral transfer of the plasma waves occurs mainly with respect to the direction of $\mathbf{k}$ without changing its absolute value. Solution of the two-dimensional integral equation for the effective (integrated over $\mathbf{k}$) field of turbulent fluctuations yields:

$$\begin{aligned} \frac{1}{8\pi}E^2 &= \frac{81}{\pi p\sqrt{2}}(p-1)^2 \frac{\Omega\tilde{\gamma}}{\Omega_i^2}\frac{s^2}{c^2}\left(1+\frac{a^2}{a_i^2}\right)n_0T_e, && p-1 \gg \frac{16}{9}\frac{c^2s_i^2}{s^4}, \\ &\simeq \frac{64\sqrt{\pi}}{9}\frac{\Omega\tilde{\gamma}}{\Omega_i^2}\frac{cs_i^2}{s^4}\left(1+\frac{a^2}{a_i^2}\right)\sqrt{p(p-1)}\,n_0T_e, && p-1 \ll \frac{16}{9}\frac{c^2s^2}{s^4} \end{aligned} \tag{9.87}$$

Saturation of the fluctuation level for parametric excitation may be caused as well by the non-linear eigenfrequency shift. Since the correction to the plasma oscillation frequency taking into account the ion motion is negative

$$\Delta\omega = -\frac{\Omega}{4n_0T}\frac{a^2}{a^2+a_i^2}\sum_{\mathbf{k}'}\frac{(\mathbf{k}\cdot\mathbf{k}')^2}{(k\cdot k')^2}I_{\mathbf{k}'},$$

the effect of the shift is manifested through the decrease of the mismatch δ in the growth-rate. The effective field of the turbulent fluctuations is given by the formula

$$\frac{1}{8\pi}E^2 = 8(p-1)\left(1+\frac{a_i^2}{a^2}\right)\left(\frac{3}{4}\frac{s^2}{c^2}+\frac{1}{3}\frac{c^2}{s^2}\frac{\nu_{ei}^2}{\Omega^2}\right)n_0T_e, \qquad |\delta| < \frac{3}{4}\frac{s^2}{c^2}. \tag{9.88}$$

The fluctuation oscillations can grow considerably higher than the thermal level only provided the instability threshold is greatly exceeded. It is to be noted that, the mismatch being great $\delta > \frac{3}{4}\frac{s^2}{c^2}$, the transfer over the spectrum is a more effective stabilizing mechanism. A peculiar feature of the fluctuation levels is the weak dependence on the spontaneous emission of plasma waves by the particles. The stationary fluctuation level (9.88) is associated with a very large coefficient of radiation transformation into plasma waves:

$$K = \frac{1}{6}(p-1)\left(1+\frac{a_i^2}{a^2}\right); \tag{9.89}$$

thus external radiation may be the reason of a high level of fluctuation oscillations. The intensities of the latter compete to the same order with the radiation intensity.

CHAPTER 10

Scattering and Radiation of Waves

10.1. Scattering and Transformation of Waves in a Plasma

Derivation of scattering and transformation cross-sections from the kinetic equation for waves

Let us consider the scattering of electromagnetic waves propagating in a plasma. This process occurs due to the fluctuations of the density and other quantities that perturb the dielectric plasma properties. Various modes, existing in the plasma, are mutually independent in the linear approximation. Under realistic conditions each of these modes may interact with other waves or particles and be scattered or transformed into some other wave by virtue of the non-linearity of the material equations. A process is called scattering if the radiated mode is of the same kind as the initial one and is called a transformation if the resulting wave is of a different kind. The fluctuation spectrum consists of the main maximum, which is associated with the stochastic motion of the individual particles, and a set of sharp maxima corresponding to the frequencies of the eigenoscillations. Therefore, besides incoherent scattering of electromagnetic waves by individual plasma particles, which causes a small frequency shift, there may occur also combination scattering, when the difference between the frequencies of the scattered wave and the incident one is equal to a plasma eigenfrequency. The intensities of combination scattering and transformation are governed by the fluctuation level. These intensities may grow enormously in non-equilibrium plasma near a kinetic instability threshold. A study of wave scattering and transformation is one of the most efficient methods of plasma diagnostics (the detection of the parameters that govern the plasma state),

Wave scattering and transformation in a plasma were studied by a number of authors. Combination scattering and the transformation of waves were first predicted by Akhiezer et al. (1958). A detailed theory of wave scattering and transformation in a magneto-active plasma was extended by Sitenko and Kirochkin (1966).†

When treating wave scattering one usually introduces a current, which induces the scattered waves, on the basis of either kinetic or hydrodynamic equations. The intensity and the distribution of the scattered waves are then described in terms of this inducing

† An account of the theory of wave scattering and transformation in a plasma is given by Bekefi, (1966), Sitenko (1967), Akhiezer, Akhiezer, Polovin, Sitenko, and Stepanov (1975), and Sheffield (1975). Various possibilities of how to use these processes for plasma diagnostics are discussed by Heald and Wharton (1965) and Golant (1968).

current. The formalism of non-linear wave interactions, which was worked out in the previous sections, is most suitable for the study of wave scattering and transformation. In particular, the unified kinetic equation for waves of Chapter 9 allows us to describe these processes completely.†

If the intensities of the incident waves are moderate, then we may neglect in the kinetic equation (9.75) the induced scattering and transformation and consider only spontaneous processes which are due to the interaction of the incident wave with the fluctuation field and the fluctuation variations of the particle distributions. This implies that only the last two non-linear terms must be retained on the right-hand side of (9.75):

$$\frac{\partial I_{\mathbf{k}}^{\alpha}}{\partial t} = -2\gamma_{\mathbf{k}}^{\alpha} I_{\mathbf{k}}^{\alpha} + \frac{2}{\Lambda_{\mathbf{k}}'^{\alpha}} \operatorname{Im} \sum_{\omega', \mathbf{k}'} a_{ijkk'}(\omega_{\mathbf{k}}^{\alpha}, \mathbf{k}; \omega', \mathbf{k}')\, \lambda_{ji}(\omega_{\mathbf{k}}^{\alpha}, \mathbf{k}) \langle E_k E_{k'} \rangle_{\mathbf{k}'\omega'}^{0} I_{\mathbf{k}}^{\alpha}$$
$$+ \frac{1}{(\Lambda_{\mathbf{k}}'^{\alpha})^2} \lambda_{ii'}^{*}(\omega_{\mathbf{k}}^{\alpha}, \mathbf{k}) \sum_{\pm\alpha'} \sum_{\mathbf{k}'} \frac{\operatorname{Tr} \lambda_{\mathbf{k}}^{\alpha}}{\operatorname{Tr} \lambda_{\mathbf{k}'}^{\alpha'}} \Big\{ b_{ii'kk'}(\omega_{\mathbf{k}}^{\alpha}, \mathbf{k}; \omega_{\mathbf{k}'}^{\alpha'}, \mathbf{k}')$$
$$+ 4\varkappa_{ijk}^{(2)}(\omega_{\mathbf{k}}^{\alpha} - \omega_{\mathbf{k}'}^{\alpha'}, \mathbf{k} - \mathbf{k}'; \omega_{\mathbf{k}'}^{\alpha'}, \mathbf{k}')\, \varkappa_{i'j'k'}^{(2)*}(\omega_{\mathbf{k}}^{\alpha} - \omega_{\mathbf{k}'}^{\alpha'}, \mathbf{k} - \mathbf{k}'; \omega_{\mathbf{k}'}^{\alpha'}, \mathbf{k}')$$
$$\times \langle E_j E_{j'} \rangle_{\mathbf{k}-\mathbf{k}', \omega_{\mathbf{k}}^{\alpha} - \omega_{\mathbf{k}'}^{\alpha'}}^{0} \Big\} \lambda_{kk'}(\omega_{\mathbf{k}'}^{\alpha'}, \mathbf{k}') I_{\mathbf{k}'}^{\alpha'}. \tag{10.1}$$

Let the initial condition for (10.1) be of the form

$$I_{\mathbf{k}'}^{\alpha'}(0) = (2\pi)^3\, \delta(\mathbf{k}' - \mathbf{k}_0) \delta_{\alpha'\alpha_0} I^{\alpha_0}, \tag{10.2}$$

where I^{α_0}, ω_0 and $\mathbf{k}_0$ are the intensity, the frequency, and the wave vector of the incident wave α_0 respectively.

Upon noting that the non-linear coupling is weak we may substitute the initial values for the wave intensities into the right-hand side of (10.1). Besides, we neglect in (10.1) the linear and non-linear damping of the scattered wave α. Then (10.1) becomes:

$$\frac{\partial I_{\mathbf{k}}^{\alpha}}{\partial t} = \frac{1}{(\Lambda_{\mathbf{k}}'^{\alpha})^2} \lambda_{ii'}^{*}(\omega, \mathbf{k}) \frac{\operatorname{Tr} \lambda_{\mathbf{k}}^{\alpha}}{\operatorname{Tr} \lambda_{\mathbf{k}_0}^{\alpha_0}} \{ b_{ii'kk'}(\omega, \mathbf{k}; \omega_0, \mathbf{k}_0)$$
$$+ 4\varkappa_{ijk}^{(2)}(\omega - \omega_0, \mathbf{k} - \mathbf{k}_0; \omega_0, \mathbf{k}_0)\, \varkappa_{i'j'k'}^{(2)*}(\omega - \omega_0, \mathbf{k} - \mathbf{k}_0; \omega_0, \mathbf{k}_0) \langle E_j E_{j'} \rangle_{\mathbf{k}-\mathbf{k}_0, \omega - \omega_0}^{0} \}, \tag{10.3}$$

where ω and $\mathbf{k}$ are the frequency and wave vector of the scattered wave α. Equation (10.3) immediately determines the energy gained by the wave α owing to the interaction of the incident wave α_0 with the fluctuation field and the fluctuation variations of the particle distribution function in the plasma. Indeed, the energy density of the scattered wave $W_{\mathbf{k}}^{\alpha}$ is related to $I_{\mathbf{k}}^{\alpha}$ by

$$W_{\mathbf{k}}^{\alpha} = \frac{1}{16\pi} \frac{\omega_{\mathbf{k}}^{\alpha}}{\operatorname{Tr} \lambda_{\mathbf{k}}^{\alpha}} \Lambda_{\mathbf{k}}'^{\alpha} I_{\mathbf{k}}^{\alpha}, \tag{10.4}$$

whence

$$\frac{\partial W_{\mathbf{k}}^{\alpha}}{\partial t} = \frac{1}{16\pi} \frac{\omega_{\mathbf{k}}^{\alpha}}{\operatorname{Tr} \lambda_{\mathbf{k}}^{\alpha}} \Lambda_{\mathbf{k}}'^{\alpha} \frac{\partial I_{\mathbf{k}}^{\alpha}}{\partial t}. \tag{10.5}$$

† Note that an attempt was made to describe scattering within the context of non-linear wave coupling theory (Pustovalov and Silin, 1972a), which reduced in fact to introducing in the kinetic equation an additional current inducing scattered waves.

Scattering and transformation processes are conventionally described by a cross-section, the latter being defined as the ratio of the increase of the energy density of the scattered waves with fixed wave vectors to the energy flux density of the incident waves. The latter is equal to the product of the energy density of the incident waves

$$W_{\mathbf{k}_0}^{\alpha_0} = \frac{1}{16\pi} \frac{\omega_{\mathbf{k}_0}^{\alpha_0}}{\operatorname{Tr} \lambda_{\mathbf{k}_0}^{\alpha_0}} \Lambda_{\mathbf{k}_0}^{\prime\alpha_0} I^{\alpha_0}. \tag{10.6}$$

and their group velocity

$$v_g^{\alpha_0}(\omega_{\mathbf{k}_0}^{\alpha_0}) = \frac{\partial \omega_{\mathbf{k}_0}^{\alpha_0}}{\partial \mathbf{k}_0}. \tag{10.7}$$

Thus the differential cross-section for wave scattering or transformation is equal to

$$d\Sigma_{\alpha_0 \to \alpha} \equiv \frac{\dfrac{\partial W_{\mathbf{k}}^{\alpha}}{\partial t}}{v_g^{\alpha_0}(\omega_0) W_{\mathbf{k}_0}^{\alpha_0}} \frac{d\mathbf{k}}{(2\pi)^3}, \tag{10.8}$$

where $d\mathbf{k}/(2\pi)^3$ is an elementary volume in the wave vector space of the scattered waves. Making use of the relations (10.3), (10.5), and (10.6) we come to the following general formula for the cross-section of wave scattering or transformation (Sitenko and Kocherga 1977c):

$$\begin{aligned} d\Sigma_{\alpha_0 \to \alpha} = {} & \frac{\omega}{\omega_0} \frac{\lambda_{ii'}^{*}(\omega, \mathbf{k})\, \lambda_{kk'}(\omega_0, \mathbf{k}_0)}{\Lambda_{\mathbf{k}_0}^{\prime\alpha_0} \Lambda_{\mathbf{k}}^{\prime\alpha} v_g^{\alpha_0}(\omega_0)} \{b_{ii'kk'}(\omega, \mathbf{k}; \omega_0, \mathbf{k}_0) \\ & + 4\varkappa_{ijk}^{(2)}(\omega-\omega_0, \mathbf{k}-\mathbf{k}_0; \omega_0, \mathbf{k}_0)\, \varkappa_{i'j'k'}^{(2)*}(\omega-\omega_0, \mathbf{k}-\mathbf{k}_0; \omega_0, \mathbf{k}_0) \\ & \times \langle E_j E_{j'} \rangle_{\mathbf{k}-\mathbf{k}_0, \omega-\omega_0}^{0} \} \frac{d\mathbf{k}}{(2\pi)^3}. \end{aligned} \tag{10.9}$$

When calculating the cross-sections for specific processes the explicit expressions (7.59) and (7.72) must be substituted for the non-linear tensor susceptibility $\varkappa_{ijk}^{(2)}(\omega-\omega_0, \mathbf{k}-\mathbf{k}_0; \omega_0, \mathbf{k}_0)$ and the tensor quantity $b_{ii'kk'}(\omega, \mathbf{k}; \omega_0, \mathbf{k}_0)$. If the phase velocities are large in comparison with the electron thermal velocities, then the expansion in power series in the ratio of the latter to the former may be carried out in (7.59) and (7.72). The cross-sections will be governed mainly by $b_{ii'kk'}(\omega, \mathbf{k}; \omega_0, \mathbf{k}_0)$, which, in turn, may be expressed in terms of the spectral distribution of the electron density fluctuations.

The expression (10.9) determines the scattering and transformation cross-sections in the general case and may be used to study processes of interest in plasmas both with and without external fields. It reduces to a much simpler formula when an isotropic plasma without external fields is under consideration.

Scattering and transformation cross-sections in an isotropic plasma

We shall derive now general expressions for the cross-sections of specific scattering and transformation processes in the field—free isotropic plasma. The tensor quantities $\lambda_{ij}(\omega, \mathbf{k})$ of such a plasma are of the form

$$\lambda_{ij}(\omega, \mathbf{k}) = (\varepsilon_t - \eta^2) \left[\varepsilon_l \left(\delta_{ij} - \frac{k_i k_j}{k^2} \right) + (\varepsilon_t - \eta^2) \frac{k_i k_j}{k^2} \right], \tag{10.10}$$

where $\varepsilon_l(\omega, \mathbf{k})$ and $\varepsilon_t(\omega, \mathbf{k})$ are the longitudinal and the transverse plasma dielectric permittivities, $\eta^2 = (k^2c^2)/\omega^2$ is the refractive index. The quantity $\Lambda(\omega, \mathbf{k})$ which describes the wave dispersion is equal to

$$\Lambda(\omega, \mathbf{k}) = \varepsilon_l(\varepsilon_t - \eta^2)^2. \tag{10.11}$$

The following product of the tensors $\lambda^*_{ii'}(\omega, \mathbf{k})$ and $\lambda_{kk'}(\omega_0, \mathbf{k}_0)$ enters the cross-section (10.9):

$$\begin{aligned}\lambda^*_{ii'}(\omega, \mathbf{k})\, \lambda_{kk'}(\omega_0, \mathbf{k}_0) \equiv{} & (\varepsilon_t(\omega_0, \mathbf{k}_0) - \eta_0^2)\,(\varepsilon_t(\omega, \mathbf{k}) - \eta^2) \\ & \times \Big\{ \varepsilon_l(\omega_0, \mathbf{k}_0)\, \varepsilon_l^*(\omega, \mathbf{k}) \left(\delta_{ii'} - \frac{k_i k_{i'}}{k^2}\right)\left(\delta_{kk'} - \frac{k_{0k}k_{0k'}}{k_0^2}\right) \\ & + \varepsilon_l(\omega_0, \mathbf{k}_0)\,(\varepsilon_t(\omega, \mathbf{k}) - \eta^2)^* \frac{k_i k_{i'}}{k^2}\left(\delta_{kk'} - \frac{k_{0k}k_{0k'}}{k_0^2}\right) \\ & + (\varepsilon_t(\omega_0, \mathbf{k}_0) - \eta_0^2)\, \varepsilon_l^*(\omega, \mathbf{k}) \left(\delta_{ii'} - \frac{k_i k_{i'}}{k^2}\right) \frac{k_{0k}k_{0k'}}{k_0^2} \\ & + (\varepsilon_t(\omega_0, \mathbf{k}_0) - \eta^2)\,(\varepsilon_t(\omega, \mathbf{k}) - \eta^2)^* \frac{k_i k_{i'}}{k^2} \frac{k_{0k}k_{0k'}}{k_0^2} \Big\}. \end{aligned} \tag{10.12}$$

It is evident that the expression within the curly brackets in (10.12) is in fact a projection operator, the terms of which are associated with the different specific scattering and transformation processes that may occur in an isotropic plasma.

The first term in the curly brackets of (10.12) separates out the scattering of transverse waves ($t_0 \to t$), the second one is associated with the transformation of a transverse wave into a longitudinal one ($t_0 \to l$), the third term is due to the transformation of a longitudinal wave into a transverse wave ($l_0 \to t$) and the fourth one corresponds to the scattering of longitudinal waves ($l_0 \to l$). Upon noting that

$$\Lambda'^{\alpha_0}_{\mathbf{k}_0} = (\varepsilon_t(\omega_0, k_0) - \eta_0^2)\left[(\varepsilon_t(\omega_0, \mathbf{k}_0) - \eta_0^2)\frac{\partial \varepsilon_l(\omega_0, \mathbf{k}_0)}{\partial \omega_0} + 2\varepsilon_l(\omega_0, \mathbf{k}_0)\frac{\partial}{\partial \omega_0}(\varepsilon_t(\omega_0, \mathbf{k}_0) - \eta_0^2)\right],$$

$$\Lambda'^{\alpha}_{k} = (\varepsilon_t(\omega, \mathbf{k}) - \eta^2)\left[(\varepsilon_t(\omega, \mathbf{k}) - \eta^2)\frac{\partial \varepsilon_l(\omega, \mathbf{k})}{\partial \omega} + 2\varepsilon_l(\omega, \mathbf{k})\frac{\partial}{\partial \omega}(\varepsilon_t(\omega, \mathbf{k}) - \eta^2)\right],$$

we come to the following cross-sections for these specific processes:

$$\begin{aligned} d\Sigma_{t_0 \to t} ={} & \frac{1}{4}\frac{\omega}{\omega_0} \frac{1}{\dfrac{\partial}{\partial \omega_0}(\varepsilon_t^0 - \eta_0^2)\dfrac{\partial}{\partial \omega}(\varepsilon_t - \eta^2) v_g^0} \\ & \times \left(\delta_{ii'} - \frac{k_i k_{i'}}{k^2}\right)\left(\delta_{kk'} - \frac{k_{0k}k_{0k'}}{k_0^2}\right) B_{i'ikk'}(\omega, \mathbf{k}; \omega_0, \mathbf{k}_0)\frac{d\mathbf{k}}{(2\pi)^3}. \end{aligned} \tag{10.13}$$

$$d\Sigma_{t_0 \to l} = \frac{1}{2}\frac{\omega}{\omega_0} \frac{1}{\dfrac{\partial}{\partial \omega_0}(\varepsilon_t^0 - \eta_0^2)\dfrac{\partial \varepsilon_l}{\partial \omega} v_g^0} \frac{k_i k_{i'}}{k^2}\left(\delta_{kk'} - \frac{k_{0k}k_{0k'}}{k_0^2}\right) B_{ii'kk'}(\omega, \mathbf{k}; \omega_0, \mathbf{k}_0)\frac{d\mathbf{k}}{(2\pi)^3}, \tag{10.14}$$

$$d\Sigma_{l\to t_0} = \frac{1}{2}\frac{\omega}{\omega_0}\frac{1}{\dfrac{\partial\varepsilon_l^0}{\partial\omega_0}\dfrac{\partial}{\partial\omega}(\varepsilon_t-\eta^2)v_g^0}\left(\delta_{ii'}-\frac{k_ik_{i'}}{k^2}\right)\frac{k_{0k}k_{0k'}}{k_0^2}B_{ii'kk'}(\omega, \mathbf{k};\, \omega_0, \mathbf{k}_0)\frac{d\mathbf{k}}{(2\pi)^3}. \tag{10.15}$$

$$d\Sigma_{l\to l_0} = \frac{\omega}{\omega_0}\frac{1}{\dfrac{\partial\varepsilon_l^0}{\partial\omega_0}\dfrac{\partial\varepsilon_l}{\partial\omega}v_g^0}\frac{k_ik_{i'}}{k^2}\frac{k_{0k}k_{0k'}}{k_0^2}B_{ii'kk'}(\omega, \mathbf{k};\, \omega_0, \mathbf{k}_0)\frac{d\mathbf{k}}{(2\pi)^3}. \tag{10.16}$$

where the following notation is used:

$$\begin{aligned}B_{ii'kk'}(\omega, \mathbf{k};\, \omega_0, \mathbf{k}_0) \equiv\; & b_{ii'kk'}(\omega, \mathbf{k};\, \omega_0, \mathbf{k}_0)\\ & +4\varkappa_{ijk}^{(2)}(\omega-\omega_0, \mathbf{k}-\mathbf{k}_0;\, \omega_0, \mathbf{k}_0)\,\varkappa_{i'j'k'}^{(2)*}(\omega-\omega_0, \mathbf{k}-\mathbf{k}_0;\, \omega_0, \mathbf{k}_0)\\ & \times\langle E_jE_{j'}\rangle^0_{\mathbf{k}-\mathbf{k}_0,\, \omega-\omega_0}.\end{aligned} \tag{10.17}$$

It should be remembered that the quantity $b_{ii'kk'}(\omega, \mathbf{k};\, \omega_0, \mathbf{k}_0)$ is the following functional of the correlation functions $\langle\delta f\, \delta f\rangle^0_{\mathbf{k}\omega}$ and $\langle\delta fj\rangle^0_{\mathbf{k}\omega}$:

$$\begin{aligned}b_{ii'kk'}(\omega, \mathbf{k};\, \omega_0, \mathbf{k}_0) =\; & \Lambda_{ik}(\omega, \mathbf{k};\, \omega_0, \mathbf{k}_0)\,\Lambda^*_{i'k'}(\omega, \mathbf{k};\, \omega_0, \mathbf{k}_0)\,\langle\delta f\, \delta f\rangle^0_{\mathbf{k}-\mathbf{k}_0,\, \omega-\omega_0}\\ & +\frac{8\pi i}{\omega-\omega_0}\{\Lambda_{ik}(\omega, \mathbf{k};\, \omega_0, \mathbf{k}_0)\,\varkappa^{(2)*}_{i'j'k'}(\omega-\omega_0, \mathbf{k}-\mathbf{k}_0;\, \omega_0, \mathbf{k}_0)\\ & \times\Lambda^{*-1}_{j'l'}(\omega-\omega_0, \mathbf{k}-\mathbf{k}_0)\,\langle\delta f j_{l'}\rangle^0_{\mathbf{k}-\mathbf{k}_0,\, \omega-\omega_0}\\ & -\varkappa^{(2)}_{ijk}(\omega-\omega_0\, \mathbf{k}-\mathbf{k}_0;\, \omega_0, \mathbf{k}_0)\,\Lambda^{-1}_{jl}(\omega-\omega_0, \mathbf{k}-\mathbf{k}_0)\\ & \times\Lambda^*_{i'k'}(\omega, \mathbf{k};\, \omega_0, \mathbf{k}_0)\,\langle\delta f j_l\rangle^0_{\mathbf{k}-\mathbf{k}_0,\, \omega-\omega_0}\}.\end{aligned} \tag{10.18}$$

The formulae (10.13)–(10.16) are exact and describe all processes of wave scattering and transformation by fluctuations that can occur in an isotropic plasma.

10.2. Scattering and Transformation of Transverse Waves

Scattering of transverse electromagnetic waves in a plasma

Let us apply the results of the previous section to study in detail some specific processes. First of all, we shall consider the scattering of high-frequency electromagnetic waves which is the most interesting and important process from the point of view of applications.

High-frequency electromagnetic waves may be treated in neglect of the transverse dielectric permittivity dependence on the wave vectors. Thus

$$\varepsilon_t(\omega, \mathbf{k}) = \varepsilon(\omega),$$

where

$$\varepsilon(\omega) \equiv 1-\frac{\Omega^2}{\omega^2}. \tag{10.19}$$

The relation between the frequency and the wave vector of high-frequency electromagnetic waves is as follows:

$$\omega^2 = \Omega^2 + k^2c^2, \tag{10.20}$$

and may be rewritten in the form

$$k = \frac{\Omega}{c}\sqrt{\varepsilon(\omega)}. \tag{10.21}$$

The group velocity of an electromagnetic wave is equal to

$$v_g = c\sqrt{\varepsilon(\omega)}. \tag{10.22}$$

Since $\omega > \Omega$ for propagating waves, the group velocity is less than that the speed of light.

Upon noting that $\frac{\partial}{\partial\omega}(\varepsilon_t - \eta^2) = \frac{2}{\omega}$ and $d\mathbf{k} = \frac{\sqrt{\varepsilon(\omega)}}{c^3}\omega^2\, d\omega\, do$ where do is an element of solid angle we may rewrite the cross-section (10.13) as

$$d\Sigma_{t_0\to t} = \frac{1}{64\pi^3}\frac{\omega^4}{c^4}\sqrt{\frac{\varepsilon(\omega)}{\varepsilon(\omega_0)}}\left(\delta_{ii'} - \frac{k_ik_{i'}}{k^2}\right)\left(\delta_{kk'} - \frac{k_{0k}k_{0k'}}{k_0^2}\right)\{b_{ii'kk'}(\omega, \mathbf{k}; \omega_0, \mathbf{k}_0)$$
$$+ 4\varkappa_{ijk}^{(2)}(\Delta\omega, \mathbf{q}; \omega_0, \mathbf{k}_0)\,\varkappa_{i'j'k'}^{(2)*}(\Delta\omega, \mathbf{q}; \omega_0, \mathbf{k}_0)\langle E_jE_{j'}\rangle^0_{\mathbf{q}\,\Delta\omega}\}\, d\omega\, do, \tag{10.23}$$

where $\Delta\omega \equiv \omega - \omega_0$ and $\mathbf{q} \equiv \mathbf{k} - \mathbf{k}_0$ are the changes in frequency and wave vector due to the scattering. As we do not discuss wave polarization (the incident waves are also suggested unpolarized), an additional factor of 2 was included in (10 23), which is associated with the two probable polarizations of the scattered waves.

The scattering of high-frequency electromagnetic waves is governed mainly by the electrons, so we shall take into account only the electron components of Δ_{ik} and $\varkappa_{ijk}^{(2)}$ in (10.18) and (10.23), which implies we suppress quantities of the order of m/m_i. However, it is to be noted that the fluctuation field $\mathbf{E}$ in (10.23) is governed by the fluctuations both of the electron current and of the ion one.

Since the phase velocities of the high-frequency electromagnetic waves are much greater than the electron thermal velocity, we may expand the integrands in $b_{ii'kk'}(\omega, \mathbf{k}; \omega_0, \mathbf{k}_0)$ and $\varkappa_{ijk}^{(2)}(\Delta\omega, \mathbf{q}; \omega_0, \mathbf{k}_0)$ in power series in the ratio of the latter to the former. The result for that part of $b_{ii'kk'}(\omega, \mathbf{k}; \omega_0, \mathbf{k}_0)$ that does not concern the self-consistent interaction between the charged particles is as follows:

$$b^{(1)}_{ii'kk'}(\omega, \mathbf{k}; \omega_0, \mathbf{k}_0) \equiv \Delta_{ik}(\omega, \mathbf{k}; \omega_0, \mathbf{k}_0)\,\Delta^*_{i'k'}(\omega, \mathbf{k}; \omega_0, \mathbf{k}_0)\langle\delta f\,\delta f\rangle^0_{\mathbf{q}\,\Delta\omega}$$
$$= 2\pi\frac{\Omega^4}{\omega^4}\int d\mathbf{v}\,\delta(\Delta\omega - (\mathbf{q}\cdot\mathbf{v}))\left(\delta_{ik} + \frac{k_{0i}v_k - (\mathbf{k}_0\cdot\mathbf{v})\delta_{ik}}{\omega_0} + \frac{k_kv_i + (\mathbf{k}\cdot\mathbf{v})\delta_{ik}}{\omega}\right)$$
$$\times\left(\delta_{i'k'} + \frac{k_{0i'}v_{k'} - (\mathbf{k}_0\cdot\mathbf{v})\delta_{i'k'}}{\omega_0} + \frac{k_{k'}v_{i'} + (\mathbf{k}\cdot\mathbf{v})\delta_{i'k'}}{\omega}\right). \tag{10.24}$$

If our considerations are restricted to scattering by longitudinal fluctuations, only the relevant velocity components ($\mathbf{v}\,||\,\mathbf{q}$) are to be retained in the integrand in (10.24). Then

(10.24) immediately becomes expressed in terms of the spectral correlation function for the electron density fluctuations:

$$b^{(1)}_{ii'kk'}(\omega, \mathbf{k}; \omega_0, \mathbf{k}_0) = \frac{\Omega^4}{\omega^4}\left(\delta_{ik}+\frac{\Delta\omega}{\omega_0}\,\frac{k_{0i}q_k-(\mathbf{k}_0\cdot\mathbf{q})\delta_{ik}}{q^2}+\frac{\Delta\omega}{\omega}\,\frac{k_kq_i+(\mathbf{k}\cdot\mathbf{q})\delta_{ik}}{q^2}\right)$$
$$\times\left(\delta_{i'k'}+\frac{\Delta\omega}{\omega_0}\,\frac{k_{0i'}q_{k'}-(\mathbf{k}_0\cdot\mathbf{q})\delta_{i'k'}}{q^2}+\frac{\Delta\omega}{\omega}\,\frac{k_{k'}q_{i'}+(\mathbf{k}\cdot\mathbf{q})\delta_{i'k'}}{q^2}\right)\langle\delta n_e^2\rangle^0_{\mathbf{q}\,\Delta\omega}. \tag{10.25}$$

Analogous expansions may be carried out to calculate the rest of $b_{ii'kk'}(\omega, \mathbf{k}; \omega_0, \mathbf{k}_0)$ (i.e. the part associated with self-consistent interactions between plasma particles) and the non-linear susceptibilities $\varkappa^{(2)}_{ijk}(\Delta\omega, \mathbf{q}; \omega_0, \mathbf{k}_0)$ which enter the last term within the curly brackets in (10.23). Taking into account only the longitudinal fluctuation field we express it in terms of the charge density fluctuations. Some straightforward algebra yields:

$$b^{(2)}_{ii'kk'}(\omega, \mathbf{k}; \omega_0, \mathbf{k}_0)$$
$$=\frac{\Omega^4}{\omega_0^2\omega^2}\left\{2\left(\delta_{ik}-\frac{\Delta\omega^2}{\omega}\,\frac{k_iq_k+(\mathbf{k}\cdot\mathbf{q})\delta_{ik}}{q^2}\right)\left(\delta_{i'k'}-\frac{\Delta\omega^2}{\omega^2}\,\frac{k_{i'}q_{k'}+(\mathbf{k}\cdot\mathbf{q})\delta_{i'k'}}{q^2}\right)(1-\mathrm{Re}\,\varepsilon(\Delta\omega, \mathbf{q}))\right.$$
$$+\frac{\Omega^2}{\omega^2}\left[\left(\delta_{ik}-\frac{\Delta\omega^2}{\omega^2}\,\frac{k_iq_k+(\mathbf{k}\cdot\mathbf{q})\delta_{ik}}{q^2}\right)\frac{k_{k'}q_{i'}+(\mathbf{k}\cdot\mathbf{q})\delta_{i'k'}}{q^2}\,\varepsilon^*(\Delta\omega, \mathbf{q})\right.$$
$$\left.\left.+\frac{k_kq_i+(\mathbf{k}\cdot\mathbf{q})\delta_{ik}}{q^2}\left(\delta_{i'k'}-\frac{\Delta\omega^2}{\omega^2}\,\frac{k_{i'}q_{k'}+(\mathbf{k}\cdot\mathbf{q})\delta_{i'k'}}{q^2}\right)\varepsilon(\Delta\omega, \mathbf{q})\right]\right\}\frac{\langle\delta n_e^2\rangle_{\mathbf{q}\,\Delta\omega}}{|\varepsilon(\Delta\omega, \mathbf{q})|^2}, \tag{10.26}$$

$$4\varkappa^{(2)}_{ijk}(\Delta\omega, \mathbf{q}; \omega_0, \mathbf{k}_0)\,\varkappa^{(2)*}_{i'j'k'}(\Delta\omega, \mathbf{q}; \omega_0, \mathbf{k}_0)\,\langle E_jE_{j'}\rangle^0_{\mathbf{q}\,\Delta\omega}$$
$$=\frac{\Omega^4}{\omega_0^2\omega^2}\left(\frac{\Omega^2}{\omega^2}\,\frac{k_iq_k+(\mathbf{k}\cdot\mathbf{q})\delta_{ik}}{q^2}+\left[\delta_{ik}-\frac{\Delta\omega^2}{\omega^2}\,\frac{k_iq_k+(\mathbf{k}\cdot\mathbf{q})\delta_{ik}}{q^2}\right](\varepsilon(\Delta\omega, \mathbf{q})-1)\right)$$
$$\times\left(\frac{\Omega^2}{\omega^2}\,\frac{k_{i'}q_{k'}+(\mathbf{k}\cdot\mathbf{q})\delta_{i'q'}}{q^2}+\left[\delta_{i'k'}-\frac{\Delta\omega^2}{\omega^2}\,\frac{k_{i'}q_{k'}+(\mathbf{k}\cdot\mathbf{q})\delta_{i'k'}}{q^2}\right](\varepsilon^*(\Delta\omega, \mathbf{q})-1)\right)$$
$$\times\frac{\langle(\delta n_e-\delta n_i)\rangle^0_{\mathbf{q}\,\Delta\omega}}{|\varepsilon(\Delta\omega, \mathbf{q})|^2}. \tag{10.27}$$

Within the context of (10.25)–(10.27) the expression in the curly brackets in the cross-section (10.23) becomes transformed into the following:

$$\{\ldots\}=\frac{\Omega^4}{\omega_0^2\omega^2}\left\langle\left(\delta_{ik}\delta n_e-\frac{\Delta\omega^2}{\omega^2}\,\frac{k_kq_i+(\mathbf{k}\cdot\mathbf{q})\delta_{ik}}{q^2}\,\delta n_e+\frac{\Omega^2}{\omega^2}\,\frac{k_kq_i+(\mathbf{k}\cdot\mathbf{q})\delta_{ik}}{q^2}(\delta n_e-\delta n_i)\right)\right.$$
$$\left.\times\left(\delta_{ik}\delta n_e-\frac{\Delta\omega^2}{\omega^2}\,\frac{k_{k'}q_{i'}+(\mathbf{k}\cdot\mathbf{q})\delta_{i'k'}}{q^2}\,\delta n_e+\frac{\Omega^2}{\omega^2}\,\frac{k_{k'}q_{i'}+(\mathbf{k}\cdot\mathbf{q})\delta_{i'k'}}{q^2}(\delta n_e-\delta n_i)\right)^*\right\rangle_{\mathbf{q}\,\Delta\omega}. \tag{10.28}$$

Let us show that the last two terms of each of the factors within the averaging brackets in (10.28) may be neglected. First of all it is to be remembered that longitudinal fluctuations in a plasma occur mainly with frequencies $\Delta\omega$, which are either much lower than the Langmuir frequency Ω or close to Ω.

We begin with fluctuations with $\Delta\omega \ll \Omega$. Their spectral density satisfies the condition

$$\langle(\delta n_e - \delta n_i)^2\rangle_{\mathbf{q}\,\Delta\omega} \simeq a^4 q^4 \langle \delta n_e^2\rangle_{\mathbf{q}\,\Delta\omega}.$$

Hence, if $\Delta\omega \ll \Omega$, then the ratio of the third term to the first one in each brace in (10.28) is of the order of magnitude k^2s^2/ω^2, which is very small compared with unity (it is assumed that $\omega/k \gg s$). The second terms in the braces may be also neglected when $\Delta\omega \ll \Omega$, since they contain $(\Delta\omega^2/\omega^2)\,(\omega > \Omega)$. Suppose now the fluctuations to be of the Langmuir type, i.e. $\Delta\omega \simeq \Omega$. Then $\delta n_i \ll \delta n_e$ and the second term in each brace cancels out with the third one. Thus the expression within the curly braces in (10.28) may be approximately written as

$$\{\ldots\} \simeq \frac{\Omega^4}{\omega^2\omega_0^2}\,\delta_{ik}\delta_{i'k'}\langle\delta n_e^2\rangle_{\mathbf{q}\,\Delta\omega}.$$

which yields for the cross-section (10.23) (Akhiezer et al., 1961):

$$d\Sigma_{t_0\to t} = \frac{1}{4\pi}\left(\frac{e^2}{mc^2}\right)^2 \frac{\omega^2}{\omega_0^2}\sqrt{\frac{\varepsilon(\omega)}{\varepsilon(\omega_0)}}\,(1+\cos^2\vartheta)\,\langle\delta n_e^2\rangle_{\mathbf{q}\,\Delta\omega}\,d\omega\,do. \tag{10.29}$$

where ϑ is the angle between the vectors $\mathbf{k}$ and $\mathbf{k}_0$.

Formula (10.29) determines the differential cross-section of the scattering of unpolarized electromagnetic waves in an elementary volume do and a frequency band $d\omega$. This formula is adequate for arbitrary changes in frequency; the only assumption is that the frequencies ω_0 and ω exceed Ω. If $\Delta\omega \ll \omega$, then the factor $\dfrac{\omega^2}{\omega_0^2}\sqrt{\dfrac{\varepsilon(\omega)}{\varepsilon(\omega_0)}}$ becomes equal to unity and the cross-section (10.29) reduces to the well-known expression which describes the cross-section for scattering by density fluctuations with small changes in frequency. It is to be emphasized that (10.29) is valid only provided the phase velocities of the incident and the scattered waves be very large as compared to the mean electron velocity.

Following from (10.29), the spectral distribution of the scattered waves is governed by that of the electron density fluctuations. However, an observation is to be made that the spectral distribution of the electron density fluctuations $\langle\delta n_e^2\rangle_{\mathbf{q}\,\Delta\omega}$ is sensitive to the ion motion by virtue of the self-consistent electron-ion interaction.

The spectrum of the scattered radiation in an equilibrium plasma contains a Doppler-broadened main line $\Delta\omega \lesssim qs_i$ (s_i is the ion thermal velocity) and sharp maxima at $\omega \simeq \pm\Omega$ (upon the assumption $aq \ll 1$). If the frequencies of the incident waves are high, $\omega_0 \gg \Omega$, the factor $\dfrac{\omega^2}{\omega_0^2}\sqrt{\dfrac{\varepsilon(\omega)}{\varepsilon(\omega_0)}}$ in (10.27) may be taken equal to unity and the integration by parts may be carried out within the context of the Kramers–Kronig relations. The resulting scattering cross-section for transverse electromagnetic waves in an equilibrium plasma into an element of solid angle is the following:

$$d\Sigma_{t_0\to t} = \frac{1}{2}\,n_0\left(\frac{e}{mc^2}\right)^2 \frac{1+a^2q^2}{2+a^2q^2}\,(1+\cos^2\vartheta)\,do, \tag{10.30}$$

where $q = 2\frac{\omega}{c}\sin\frac{\vartheta}{2}$, ϑ is the scattering angle, and a is the Debye length. The angle-integration of (10.30) yields the total scattering cross-section

$$\Sigma_{t_0\to t} = \left\{1+\frac{3\ln(1+2a^2k_0^2)}{8a^2k_0^2}+\frac{3}{4a^2k_0^2}\left[(1-a^2k_0^2)\frac{\arctan(\sqrt{2}ak_0)}{\sqrt{2}ak_0}-1\right]\right\}n_0\sigma_0, \quad (10.31)$$

where $\sigma_0 = \frac{8\pi}{3}\left(\frac{e^2}{mc^2}\right)^2$ is the Thomson cross-section for the scattering of electromagnetic waves by a free electron. The limiting cases of very short and very long wavelengths for the total cross-section are:

$$\Sigma_{t_0\to t} = n_0\sigma_0, \quad ak_0 \gg 1; \qquad \Sigma_{t_0\to t} = \tfrac{1}{2}n_0\sigma_0, \quad ak_0 \ll 1. \quad (10.32)$$

When deriving the cross-section (10.29) we took into consideration only the longitudinal fluctuations in the plasma. However, there may occur such situations when the electromagnetic wave scattering by the vorticity current fluctuations is of considerable importance (Gorbunov and Silin, 1966). The correction to the cross-section (10.29) taking this effect into account may be easily calculated by means of general formula (10.23).

Indeed, the transverse velocity components in the integrand of (10.24) give rise to the following addition to the quantity (10.25):

$$b^{(1)\perp}_{ii'kk'}(\omega, \mathbf{k}; \omega_0, \mathbf{k}_0) = \frac{1}{2}\frac{\Omega^4}{\omega_0^2\omega^4}\left[k_{0i}k_{0i'}\left(\delta_{kk'}-\frac{q_kq_{k'}}{q^2}\right)+k_{0i}k_{k'}\left(\delta_{i'k}-\frac{q_{i'}q_k}{q^2}\right)\right.$$
$$\left.+k_{0i'}k_k\left(\delta_{ik'}-\frac{q_iq_{k'}}{q^2}\right)+k_kk_{k'}\left(\delta_{ii'}-\frac{q_iq_{i'}}{q^2}\right)\right]\langle v_{e\perp}^2\rangle_{\mathbf{q}\,\Delta\omega}, \quad (10.33)$$

where $\langle v_{e\perp}^2\rangle^0_{\mathbf{q}\,\Delta\omega}$ is the spectral correlation function for the fluctuations of the electron vortex velocity

$$\langle v_{e\perp}^2\rangle^0_{\mathbf{q}\,\Delta\omega} = 2\pi\int d\mathbf{v}v_\perp^2\,\delta((\mathbf{q}\cdot\mathbf{v})-\Delta\omega)\,f_0(\mathbf{v}). \quad (10.34)$$

(For simplicity we considered only the $\Delta\omega \ll \omega_0$ limiting case in (10.33).)

Within the context of (10.33) we obtain the following differential cross-section of the scattering of electromagnetic in a plasma:

$$d\Sigma_{t_0\to t} = \frac{1}{4}\left(\frac{e^2}{mc^2}\right)^2\left\{(1+\cos^2\vartheta)\langle\delta n_e^2\rangle_{\mathbf{q}\,\Delta\omega}+\frac{1}{\omega_0^2}\left[\left(1-\frac{1}{2}\cos^2\vartheta\right)q^2-\frac{(k^2-k_0^2)^2}{2q^2}\cos^2\vartheta\right.\right.$$
$$\left.\left.+\frac{k_0k\cos\vartheta}{q^2}\left(\frac{(\mathbf{k}_0\cdot\mathbf{q})^2}{k_0^2}+\frac{(\mathbf{k}\cdot\mathbf{q})^2}{k^2}\right)\right]\langle v_{e\perp}^2\rangle^0_{\mathbf{q}\,\Delta\omega}\right\}d\omega\,do. \quad (10.35)$$

In contrast to (10.29), (10.35) takes into account the scattering of electromagnetic waves both by longitudinal and by transverse plasma fluctuations. It is evident from (10.35) that the scattering of electromagnetic waves by transverse fluctuations is determined by the spectral distribution of fluctuations of the electron vortex velocity. Since the latter are practically independent of the self-consistent interaction in the low-frequency range, the formula for the cross-section (which is written for the wave scattering with small frequency shifts) contains the correlation function for the electron vortex velocity fluctuations in neglect of interactions between the particles. Gorbunov and Silin (1966) have discussed some situations when the scattering of electromagnetic waves by fluctuations of the electron vortex velocity is of great importance. It should be observed that the relative contribution of the last effect to the total cross-section is of the order of $(k_0^2 s^2)/\omega_0^2$.

Spectral distribution of the scattered radiation

Let us consider in some more detail the spectral distribution of the scattered waves in the plasma.† The spectral distribution of the density fluctuations $\langle \delta n_e^2 \rangle_{\mathbf{q}\,\Delta\omega}$ is in the case of short wavelengths $aq \gg 1$ a Gaussian function of the frequency $\Delta\omega$, therefore the spectral distribution of scattered waves is also Gaussian:

$$d\Sigma = \sqrt{\frac{3}{8\pi}}\, n_0 \left(\frac{e^2}{mc^2}\right)^2 \frac{1}{qs}(1+\cos^2\vartheta)e^{-\frac{3}{2}\frac{\Delta\omega^2}{q^2h^2}}\, d\omega\, do, \quad a^2q^2 \gg 1. \tag{10.36}$$

This formula is valid provided $\Delta\omega \neq \Omega_i$. According to (10.36) the Doppler broadening of the line is governed by the electron thermal velocity. The total cross-section (10.32) is equal to the sum of the cross-sections for scattering by individual electrons.The Coulomb electron-ion interaction is insignificant and the scattering is incoherent.‡

The limiting case of long wavelength $a^2k_0^2 \ll 1$ is of more interest, as collective plasma properties come into play. Let us consider the spectral distributions of the scattered radiation in various frequency bands assuming that $a^2k_0^2 \ll 1$. The possibility that the electron and the ion temperatures T_e and T_i differ will also be taken into account.

If the change in frequency in the scattering process is small, $\Delta\omega \ll qs_i$, where s_i is the ion thermal velocity, the differential cross-section is of the form

$$d\Sigma = \frac{1}{\sqrt{8\pi}}\, n_0 \left(\frac{e^2}{mc^2}\right)^2 \frac{\sqrt{m}T_e^{3/2} + \sqrt{m_i}T_i^{3/2}}{q(T_e+T_i)}(1+\cos^2\vartheta)\, d\omega\, do. \tag{10.37}$$

In the case of a strongly non-isothermal plasma the scattering cross-section for small changes in frequency is $\frac{1}{4}\sqrt{\frac{m_i}{m}}$ times less than the relevant cross-section for an isothermal plasma with temperature T_e.

If $\Delta\omega \lesssim qs_i$, then we have the following cross-section for an isothermal plasma:

$$d\Sigma = \sqrt{\frac{3}{8\pi}}\, n_0 \left(\frac{e^2}{mc^2}\right)^2 \frac{1}{qs_i}\, \frac{e^{-z^2}}{[2-\varphi(z)]^2+\pi z^2 e^{-2z^2}}(1+\cos^2\vartheta)\, d\omega\, do, \tag{10.38}$$

where

$$z = \sqrt{\frac{3}{2}}\, \frac{\Delta\omega}{qs_i}.$$

† This problem was treated for an equilibrium plasma by Dougherty and Farley (1960) and Akhiezer, Akhiezer, and Sitenko (1961). The latter authors considered also the case of a two-temperature plasma.

‡ The effect of the electron collisions on the plasma fluctuations and incoherent electromagnetic wave scattering was studied by Sitenko and Gurin (1966) and Hagfors and Brockelman (1971).

The differential cross-section steeply decreases when $\Delta\omega \simeq qs_i$; therefore the quantity $\Delta\omega \simeq qs_i$ characterizes the width of the spectral distribution of the scattered radiation in an isothermal plasma. Note that this value is governed by the ion thermal velocity notwithstanding that the scattering is caused by the electrons.

If $qs_i < \Delta\omega \ll qs_e$, the scattering cross-section is very small in an isothermal plasma. In a non-isothermal plasma the scattering cross-section has additional maxima when the frequency shift $\Delta\omega$ coincides with the frequency of the low-frequency plasma oscillations with wave vector $\mathbf{q}$.

In particular, if $T_e \gg T_i$ and $a^2q^2 \ll 1$, the maxima correspond to frequency shifts $\Delta\omega = \pm qv_s$, where v_s is the non-isothermal sound velocity. The differential cross-section near the maxima is of the form

$$d\Sigma = \frac{1}{4} n_0 \left(\frac{e^2}{mc^2}\right)^2 (1+\cos^2\vartheta)\{\delta(\Delta\omega - qv_s) + \delta(\Delta\omega + qv_s)\}\, d\omega\, do. \tag{10.39}$$

Integrating (10.39) over angles and frequencies we find the total scattering cross-section in a non-isothermal plasma:

$$\Sigma \simeq n_0\sigma_0, \quad a^2k_0^2 \ll 1. \tag{10.40}$$

Note that the total cross-section (10.40) is twice the scattering cross-section in an isothermal plasma (10.32).

In the case of large changes in frequency ($\Delta\omega > qs_e$) the scattering cross-section has sharp maxima at $\Delta\omega \simeq \pm\Omega$; these are associated with the scattering of electromagnetic waves by longitudinal electron oscillations. The differential scattering cross-section is for this frequency band irrespective of the proportion between the electron and ion temperatures given by the formula

$$d\Sigma = \frac{1}{4} n_0 \left(\frac{e^2}{mc^2}\right)^2 \frac{\omega^2}{\omega_0^2} \sqrt{\frac{\varepsilon(\omega)}{\varepsilon(\omega_0)}}\, a^2q^2(1+\cos^2\vartheta)\{\delta(\Delta\omega - \Omega) + \delta(\Delta\omega + \Omega)\}\, d\omega\, do. \tag{10.41}$$

The total cross-section of the scattering of electromagnetic waves by Langmuir oscillations is for $\omega_0 \gg \Omega$ of the form

$$\Sigma = 2a^2k_0^2n_0\sigma_0, \quad a^2k_0^2 \ll 1. \tag{10.42}$$

Thus the collective effects are manifested in the scattering mostly when $a^2k^2 \ll 1$. The spectra of the scattered electromagnetic waves are in an essential way different in isothermal and non-isothermal plasmas.

The spectrum of scattered radiation in an isothermal plasma has a central maximum caused by the incoherent scattering by the electron density fluctuations, the Doppler width of which is governed by the ion velocities, and satellites associated with the scattering by the electron longitudinal oscillations. The relative weight of the satellites (with respect to the central maximum) is about $2a^2k_0^2$.

There is no central maximum in the spectrum for a non-isothermal plasma. There are two maxima, symmetric relative to $\Delta\omega = 0$, which are associated with the scattering by the ion acoustic oscillations, and satellites due to the scattering by the Langmuir oscillations. The weight of the satellites relative to the sound maxima is about $2a^2k_0^2$.

The spectral distributions of the scattered electromagnetic waves for intermediate situations are shown in Fig. 9. The curves correspond to ratios of the temperatures T_e and T_i equal to 1, 2, 5, and 10. Note that the experimental study of the spectral distribution of scattered electromagnetic waves can draw important information concerning the parameters which govern the plasma state. In particular, it enables us to reach conclusions not only on the electron density and temperature, but also on how strongly non-isothermal the plasma is.

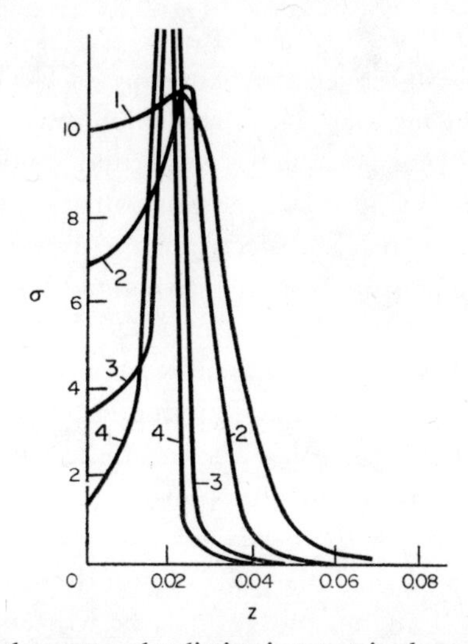

FIG. 9. Spectral distribution of the scattered radiation in a non-isothermal plasma when the temperature ratio $t = T_e/T_i$ is equal to 1 (1), 2 (2), 5 (3) and 10 (4); $a^2q^2 = 0.1$; $\sigma \equiv \frac{1}{\Sigma_0}\frac{d\Sigma}{dz\,dO}$,

$$\Sigma_0 = \frac{n_0}{2\sqrt{\pi}}\left(\frac{e^2}{mc^2}\right)^2 \frac{\omega^2}{\omega_0^2}\sqrt{\frac{\varepsilon(\omega)}{\varepsilon(\omega_0)}}\,(1+\cos^2\vartheta).$$

The cross-sections of the scattering of electromagnetic waves by collective fluctuations may grow enormously in a non-equilibrium plasma if the state of the latter approaches the kinetic instability threshold. The absolute value of the cross-section in a turbulent plasma is determined by the stationary level of non-equilibrium fluctuations. In particular, the cross-section of the scattering of electromagnetic waves by ion-sound fluctuations in a turbulent plasma is equal to

$$d\Sigma = \frac{1}{16\pi}n_0\left(\frac{e^2}{mc^2}\right)^2\frac{I_q}{T_e}(1+\cos^2\vartheta)\{\delta(\Delta\omega-qv_s)+\delta(\Delta\omega+qv_s)\}\,d\omega\,do, \qquad (10.43)$$

where I_q is the intensity of the non-equilibrium fluctuations. To obtain the cross-section of the Raman scattering of electromagnetic waves by Langmuir fluctuation oscillations in a turbulent plasma we have to multiply the right-hand side of the cross-section (10.41) by $I_q/4\pi T_e$.

The enormous growth of the cross-section near the instability threshold was predicted by Ichimaru, Pines, and Rostoker (1962). The anomalous behaviour of the cross-sections of scattering and transformation of electromagnetic waves in a magneto-active plasma was investigated by Sitenko and Kirochkin (1966), and Sitenko and Radzievskii (1966). A detailed account of the theory of scattering of electromagnetic waves in the ionosphere is given by Yeh and Liu (1972).

Transformation of transverse electromagnetic waves into longitudinal Langmuir and ion-sound waves

In the prevous section we discussed the scattering of electromagnetic waves, propagating in a plasma, due to the interaction of the incident waves with plasma fluctuations. Besides that, a transverse electromagnetic wave may undergo transformation into a longitudinal one. The cross-sections of the transformation processes are described by the general formula (10.14). Transverse electromagnetic waves which propagate in an isotropic plasma may be transformed either into longitudinal Langmuir waves or into longitudinal ion-sound ones.

To describe the propagation of a longitudinal wave we must take into account spatial dispersion, i.e. the **k**-dependence of the longitudinal plasma dielectric permittivity ε_l (in contrast to the case of the transverse waves). Otherwise the group velocity of the longitudinal wave vanishes. In an isotropic plasma there can propagate longitudinal high-frequency Langmuir and low-frequency ion-acoustic waves (the latter are weakly damped in a strongly non-isothermal plasma). Within the context of the approximation (4.41) the dispersion law for Langmuir waves may be written as

$$k^2 = \frac{\omega^2}{s^2}\,\varepsilon(\omega), \tag{10.44}$$

where $\varepsilon(\omega) = 1-(\Omega^2/\omega^2)$ and s is the electron thermal velocity. The dispersion law for ion-acoustic waves (in a strongly non-isothermal plasma) is as follows:

$$k = \frac{\omega}{v_s}, \tag{10.45}$$

where $v_s = \sqrt{\dfrac{T_e}{m_i}}$ is the non-isothermal sound velocity.

The differential cross-section, or the coefficient, of the transformation of a transverse wave into a longitudinal one is determined by the following general formula:

$$d\Sigma_{t_0\to l} = \frac{1}{32\pi^3}\,\frac{\omega}{c\sqrt{\varepsilon(\omega_0)}\,\dfrac{\partial\varepsilon_l}{\partial\omega}}\,\frac{k_i k_{i'}}{k^2}\left(\delta_{kk'}-\frac{k_{0k}k_{0k'}}{k_0^2}\right)\{b_{ii'kk'}(\omega, \mathbf{k};\, \omega_0, \mathbf{k}_0)$$
$$+4\varkappa^{(2)}_{ijk}(\Delta\omega, \mathbf{q};\, \omega_0, \mathbf{k}_0)\,\varkappa^{(2)*}_{i'j'k'}(\Delta\omega, \mathbf{q};\, \omega_0, \mathbf{k}_0)\langle E_j E_{j'}\rangle^0_{\mathbf{q}\,\Delta\omega}\}\,d\mathbf{k}, \tag{10.46}$$

where the quantity $b_{ii'kk'}(\omega, \mathbf{k};\, \omega_0, \mathbf{k}_0)$ is given by (10.18). If the electromagnetic waves are transformed into high-frequency Langmuir waves, the following expression must be substituted for $d\mathbf{k}$ in (10.46):

$$d\mathbf{k} = \frac{\sqrt{\varepsilon(\omega)}}{s^3}\,\omega^2\,d\omega\,do. \tag{10.47}$$

If the resulting waves are low-frequency ion-sound, then

$$d\mathbf{k} = \frac{\omega^2}{v_s^3}\, d\omega\, do. \tag{10.48}$$

Let us consider the transformation of electromagnetic waves into Langmuir waves. As in the case of scattering, the dominant contribution in the quantities $b_{ii'kk'}$ and $\varkappa_{ijk}^{(2)}$ is given by the electron component and the longitudinal fluctuations are the most important. Since in the plasma transparency domain the phase velocity of a Langmuir wave is much greater than the electron thermal velocity, the integrands in $b_{ii'kk'}$ and $\varkappa_{ijk}^{(2)}$ may be expanded in power series in the ratio of the electron velocity to the wave phase velocity. The result for the expression within the curly brackets in (10.46) is as follows:

$$\{\ldots\} = \frac{\Omega^2}{\omega_0^2\omega^2}\left\langle \left| \left(1+\frac{\Delta\omega}{\omega}\frac{k^2}{q^2}\right)\delta n_e - 2\frac{\Delta\omega^2}{\omega^2}\frac{(\mathbf{k}\cdot\mathbf{q})}{q^2}\delta n_e \right.\right.$$
$$\left.\left. +2\frac{\Omega^2}{\omega^2}\frac{(\mathbf{k}\cdot\mathbf{q})}{q^2}(\delta n_e - \delta n_i)\right|^2 \right\rangle_{\mathbf{q}\,\Delta\omega} \delta_{ik}\delta_{i'k'}. \tag{10.49}$$

Estimates of the different terms within the modulus symbol in (10.49) may be made in a manner similar to the analysis of the scattering. It becomes evident that only the first term is to be retained in (10.49). Thus we obtain the following differential coefficient for the transformation of a transverse electromagnetic wave into a longitudinal Langmuir one (Sitenko, 1967):

$$d\Sigma_{t_0\to l} = \frac{1}{4\pi}\left(\frac{e^2}{mc^2}\right)^2 \frac{c^3}{s^3}\frac{\omega^2}{\omega_0^2}\left(1+\frac{\Delta\omega}{\omega}\frac{k^2}{q^2}\right)^2 \sin^2\vartheta \langle \delta n_e^2\rangle_{\mathbf{q}\,\Delta\omega}\, d\omega\, do. \tag{10.50}$$

The last expression is valid only provided the frequencies ω are close to the Langmuir frequency Ω, since within this band the damping of the Langmuir waves is negligibly weak.

The ratio of the transformation coefficient (10.50) to the scattering cross-section (10.29) is equal to

$$\frac{d\Sigma_{t_0\to l}}{d\Sigma_{t_0\to t}} = \frac{\sin^2\vartheta}{1+\cos^2\vartheta}\left(1+\frac{\Delta\omega}{\omega}\frac{k^2}{q^2}\right)\frac{c^3}{s^3}. \tag{10.51}$$

Therefore, within a narrow frequency band around the Langmuir frequency absorption of electromagnetic waves by virtue of a transformation into longitudinal waves dominates over the scattering.

If the frequencies of the incident electromagnetic waves are high, $\omega_0 \simeq \Omega$, the transformation process occurs mainly due to low-frequency fluctuations. Thus the transformation of an electromagnetic wave into a longitudinal one in an isothermal plasma is governed by the low-frequency incoherent fluctuations of the electron density only. In a non-isothermal plasma the process under consideration is determined by the electron density fluctuations associated with the low-frequency ion acoustic waves:

$$d\Sigma_{t_0\to l} = \frac{1}{16\pi} n_0\left(\frac{e^2}{mc^2}\right)^2 \frac{c^3}{s^3}\sqrt{\frac{\varepsilon(\omega)}{\varepsilon(\omega_0)}}\frac{I_q}{T_e}\sin^2\vartheta\{\delta(\Delta\omega - qv_s)+\delta(\Delta\omega+qv_s)\}\, d\omega\, do, \tag{10.52}$$

where I_q is the intensity of the ion-sound fluctuation oscillations. In a non-equilibrium plasma (e.g. when the electrons move relative to the ions) the transformation coefficient may grow strongly, as the level of the non-equilibrium collective fluctuations is very high (Zheleznyakov, 1970).

When calculating the coefficient $d\Sigma_{t_0\to l}$ of the transformation of an electromagnetic wave into an ion-acoustic wave we take advantage of the ion-sound phase velocity being small in comparison with the electron thermal velocity (in a highly non-isothermal plasma). Once the electromagnetic waves are assumed to be high-frequency waves, the transformation of the latter into ion-sound may occur only provided the frequency of the incident wave is close to the Langmuir frequency. The transformation is then governed by the interaction of the incident electromagnetic waves with Langmuir fluctuations. The coefficient of the transformation of transverse electromagnetic waves into longitudinal ion-acoustic waves is given by the formula (Sitenko and Kocherga, 1977c):

$$d\Sigma_{t_0\to l} = \frac{3}{32} n_0 \left(\frac{e^2}{mc^2}\right)^2 \frac{c^2}{s^2} \frac{\omega}{\omega_0} \frac{k}{k_0} \frac{\sin^2\vartheta}{a^2q^2} \frac{I_q}{T_e} \{\delta(\Delta\omega-\Omega)+\delta(\Delta\omega+\Omega)\}\, d\omega\, do, \quad (10.53)$$

where I_q is the Langmuir fluctuation intensity. In a two-temperature plasma I_q depends on the electron temperature; in a non-equilibrium plasma I_q can considerably exceed the thermal level.

10.3. Scattering and Transformation of Longitudinal Waves

Scattering of longitudinal Langmuir waves

Let us turn now to the examination of scattering and transformation of longitudinal waves in a plasma. We consider first of all the most interesting process—the scattering of longitudinal Langmuir waves by longitudinal plasma fluctuations. Within the context of the dispersion laws for the incident and the scattered Langmuir waves,

$$k_0^2 = \frac{\omega_0^2}{s^2}\varepsilon(\omega), \quad k^2 = \frac{\omega^2}{s^2}\varepsilon(\omega),$$

and since the group velocity of the incident waves is

$$v_g^0 = s\sqrt{\varepsilon(\omega_0)}, \quad (10.54)$$

we have the following general formula for the scattering cross-section:

$$d\Sigma_{l_0\to l} = \frac{1}{32\pi^3}\frac{\omega^4}{s^4}\sqrt{\frac{\varepsilon(\omega)}{\varepsilon(\omega_0)}}\frac{k_ik_{i'}}{k^2}\frac{k_{0k}k_{0k'}}{k_0^2}\{b_{ii'kk'}(\omega,\mathbf{k};\omega_0,\mathbf{k}_0)$$
$$+4\varkappa_{ijk}^{(2)}(\Delta\omega,\mathbf{q};\omega_0,\mathbf{k}_0)\varkappa_{i'j'k'}^{(2)*}(\Delta\omega,\mathbf{q};\omega_0,\mathbf{k}_0)\langle E_jE_{j'}\rangle_{\mathbf{q}\,\Delta\omega}^0\}\, d\omega\, do. \quad (10.55)$$

As the phase velocities of the Langmuir waves are much greater than the electron thermal velocity, the expression within the curly brackets in (10.55) may be calculated in the same way as it was done when we considered the scattering of electromagnetic waves. Taking into account only scattering by longitudinal fluctuations, we reduce the cross-section (10.55) to the form (Sitenko, 1967):

$$d\Sigma_{l_0\to l} = \frac{1}{2\pi}\left(\frac{e^2}{mc^2}\right)^2 \frac{c^4}{s^4}\frac{\omega^2}{\omega_0^2}\sqrt{\frac{\varepsilon(\omega)}{\varepsilon(\omega_0)}}\left(\cos\vartheta + \frac{\Delta\omega}{\omega_0}\frac{k_0k - k_0^2\cos\vartheta}{q^2} + \frac{\Delta\omega}{\omega}\frac{k^2\cos\vartheta - k_0k}{q^2}\right)^2$$
$$\times\langle\delta n_e^2\rangle_{\mathbf{q}\,\Delta\omega}\, d\omega\, do, \tag{10.56}$$

i.e. the cross-section is in this case determined only by the spectral distribution of the electron density fluctuations.

Since the difference between the frequencies of the incident and the scattered waves is very small, Langmuir wave scattering in an isothermal plasma is governed by incoherent fluctuations. In a highly non-isothermal plasma Langmuir waves may be scattered also by ion-sound fluctuations. The cross-section of the latter process is as follows:

$$d\Sigma_{l_0\to l} = \frac{1}{8\pi} n_0\left(\frac{e^2}{mc^2}\right)^2 \frac{c^4}{s^4}\frac{\omega^2}{\omega_0^2}\sqrt{\frac{\varepsilon(\omega)}{\varepsilon(\omega_0)}}\frac{I_q}{T_e}$$
$$\times\left(\cos\vartheta + \frac{\Delta\omega}{\omega}\frac{k_0k - k_0^2\cos\vartheta}{q^2} + \frac{\Delta\omega}{\omega}\frac{k^2\cos\vartheta - k_0k}{q^2}\right)^2$$
$$\times\{\delta(\Delta\omega - qv_s) + \delta(\Delta\omega + qv_s)\}\, d\omega\, do. \tag{10.57}$$

where I_q is the intensity of the ion-sound fluctuations. The scattering cross-section for Langmuir waves in a two-temperature plasma is, as is that for the electromagnetic waves, governed by the electron temperature; the cross-section (10.57) may grow very large as I_q increases. The anomalous growth of the Langmuir wave scattering cross-section in the presence of relative electron-ion motion in a plasma was investigated by Akhiezer and Bolotin (1963).

The cross-section of the scattering of Langmuir waves taking into account the interaction with the fluctuations of the electron vortex velocity is given by the following formula:

$$d\Sigma_{l_0\to l} = \frac{1}{2\pi}\left(\frac{e^2}{mc^2}\right)^2 \frac{c^4}{s^4}\cos^2\vartheta$$
$$\times\left\{\langle\delta n_e^2\rangle_{\mathbf{q}\,\Delta\omega} + \frac{2}{\omega_0^2}\left(\mathbf{k} - \frac{kq}{q^2}\mathbf{q}\right)^2\langle v_{e\perp}^2\rangle_{\mathbf{q}\,\Delta\omega}\right\} d\omega\, do. \tag{10.58}$$

This expression concerns the simplest case $\Delta\omega \ll \omega_0$.

Transformation of longitudinal Langmuir waves into longitudinal ion-sound

In a strongly non-isothermal plasma there may exist, besides Langmuir waves, also longitudinal ion-acoustic waves, and therefore there may occur mutual transformation of different longitudinal modes. In particular, a longitudinal Langmuir wave propagating in a strongly non-isothermal plasma may be involved in interactions with Langmuir fluctuation oscillations and thus be transformed into a longitudinal ion-acoustic wave.

An approximate cross-section of the latter process may be calculated on the basis of the general formula (10.16) upon noting that the phase velocity of the ion sound is very small compared with the electron thermal velocity. The differential cross-section, or the coefficient, of the transformation of a longitudinal Langmuir wave into a longitudinal ion-acoustic one is as follows (Sitenko and Kocherga, 1977c):

$$d\Sigma_{l_0\to l} = \frac{3}{16} n_0 \left(\frac{e^2}{mc^2}\right)^2 \frac{c^4}{s^4} \frac{\omega}{\omega_0} \frac{k}{k_0} \frac{1}{a^2q^2} \frac{I_q}{T_e} \{\delta(\Delta\omega - \Omega) + \delta(\Delta\omega + \Omega)\}\, d\omega\, do. \quad (10.59)$$

It is evident that the transformation coefficient (10.59) may grow enormously under non-equilibrium conditions.

Transformation of longitudinal Langmuir waves into transverse electromagnetic waves

The general formula (10.15) in the case of the transformation of a Langmuir wave into an electromagnetic wave reduces within the context of the relevant dispersion laws to the form

$$d\Sigma_{l_0\to t} = \frac{1}{32\pi^3} \frac{\omega^4}{c^3 s} \sqrt{\frac{\varepsilon(\omega)}{\varepsilon(\omega_0)}} \left(\delta_{ii'} - \frac{k_i k_{i'}}{k^2}\right) \frac{k_{0k} k_{0k'}}{k_0^2} \{b_{ii'kk'}(\omega, \mathbf{k}; \omega_0, \mathbf{k}_0)$$
$$+ 4\varkappa^{(2)}_{ijk}(\Delta\omega, \mathbf{q}; \omega_0, \mathbf{k}_0)\, \varkappa^{(2)*}_{i'j'k'}(\Delta\omega, \mathbf{q}; \omega_0, \mathbf{k}_0) \langle E_j E_{j'} \rangle^0_{\mathbf{q}\,\Delta\omega}\}\, d\omega\, do. \quad (10.60)$$

Since the phase velocities of the incident and the scattered waves are much greater than the electron thermal velocity, the calculations may be carried out in a manner analogous to the analysis in the previous sections. The resulting differential transformation coefficient becomes (Sitenko, 1967):

$$d\Sigma_{l_0\to t} = \frac{1}{2\pi} \left(\frac{e^2}{mc^2}\right)^2 \frac{c}{s} \frac{\omega^2}{\omega_0^2} \sqrt{\frac{\varepsilon(\omega)}{\varepsilon(\omega_0)}} \left(1 - \frac{\Delta\omega}{\omega_0} \frac{k_0^2}{q^2}\right)^2 \sin^2\vartheta \langle \delta n_e^2 \rangle_{\mathbf{q}\,\Delta\omega}\, d\omega\, do. \quad (10.61)$$

The ratio of the coefficient of the transformation of a Langmuir wave into an electromagnetic wave (10.61) to the scattering cross-section (10.56) for the case of small changes in frequency is equal to $(s^3/c^3)\tan^2\vartheta$.

Since the spectral distribution of the electron density fluctuations $\langle \delta n_e^2 \rangle_{\mathbf{q}\,\Delta\omega}$ reaches its maximum values at $\Delta\omega = 0$ and $\Delta\omega = \Omega$, there are radiated primarily electromagnetic waves with frequencies Ω and 2Ω. In the first case the transformation is governed by the incoherent fluctuations associated with the stochastic motion of the individual plasma particles; in the second case the transformation occurs due to the interaction of the incident Langmuir waves with the longitudinal Langmuir fluctuations in a plasma. The transformation coefficient for the second process may be written in the form

$$d\Sigma_{l_0\to t} = \frac{1}{8\pi} n_0 \left(\frac{e^2}{mc^2}\right)^2 \frac{c}{s} \frac{\omega^2}{\omega_0^2} \sqrt{\frac{\varepsilon(\omega)}{\varepsilon(\omega_0)}}\, a^2q^2 \left(1 - \frac{\Delta\omega}{\omega_0} \frac{k_0^2}{q^2}\right)^2 \sin^2\vartheta\, \frac{I_{\mathbf{q}}}{T}$$
$$\times \{\delta(\Delta\omega - \Omega) + \delta(\Delta\omega + \Omega)\}\, d\omega\, do. \quad (10.62)$$

The coefficient of the transformation of a Langmuir wave into an electromagnetic one may grow very large in a non-equilibrium plasma.

A detailed study of the scattering and the transformation of waves by fluctuations in a magneto-active plasma was carried out by Sitenko and Kirochkin (1966).

10.4. Radiation from a Plasma

Plasma bremsstrahlung

The unified kinetic equation is not only suitable for the description of the scattering and the transformation of waves by the fluctuations of the plasma density and the velocity, but makes it also possible to study the scattering of charged particles by the fluctuations of the plasma electric field. The latter process is accompanied by radiation which is nothing but plasma bremsstrahlung, since the electric field fluctuations are the immediate consequence of the microscopic motion of the individual charged particles.†

In order to describe plasma bremsstrahlung we start from (9.71), which determines the time-evolution of the electromagnetic field energy. Let the particle distribution be stationary and the damping rate of the wave, as well as the non-linear interaction of the radiated waves, be negligibly weak. Thus we retain the second and the last terms on the right-hand side of (9.71) and the latter gets reduced to

$$\frac{1}{2}\frac{\partial\,\mathrm{Re}(\Lambda_{ij}(k)e_j e_i^*)}{\partial\omega}\frac{\partial}{\partial t}\langle E^2\rangle_k = \mathrm{Im}\,\Lambda_{ii'}^*(k)\Big\{\sum_1 (b_{ii'kk'}(k,1)$$
$$+2\varkappa_{ijk}^{(2)}(k-1,1)\,\varkappa_{i'j'k'}^{(2)}(k-1,1)\langle E_j E_{j'}\rangle_{k-1})\langle E_k E_{k'}\rangle_1 + q'_{ii'}(k)\Big\}. \tag{10.63}$$

The relation (10.63) is the energy balance equation. Once we neglect the dissipation associated with the linear wave damping, the increase of the field energy is governed by the radiation sources. We consider here the emission of transverse electromagnetic waves in an isotropic plasma. The increase of the energy density of the transverse electromagnetic field is by virtue of (10.63) determined by the following equation:

$$\frac{\partial W_{\mathbf{k}}}{\partial t} = \frac{1}{8\pi}\frac{\omega_{\mathbf{k}}\,\mathrm{Tr}\,\lambda_{\mathbf{k}}}{\Lambda'_{\mathbf{k}}}e_i^* e_{i'}\Big\{\sum_1 (b_{ii'kk'}(k,1)$$
$$+2\varkappa_{ijk}^{(2)}(k-1,1)\,\varkappa_{i'j'k'}^{(2)}(k-1,1)\langle E_j E_{j'}\rangle_{k-1})\langle E_k E_{k'}\rangle_1 + q'_{ii'}(k)\Big\}. \tag{10.64}$$

In the non-relativistic limiting case we may restrict the consideration by taking into account on the right-hand side of (10.64) only the longitudinal fluctuations. Besides, only the electron component contributes in (10.64) by virtue of the great difference between the masses of the electrons and the ions.

† Plasma bremsstrahlung was studied by Dawson and Oberman (1962, 1963), Birmingham, Dawson, and Oberman, (1965), Dawson (1968), Dupree (1963), Tidman and Dupree (1965), Akopian and Tsytovich (1975).

Let us consider the radiation intensity with respect to a unit frequency band and a unit solid angle $dP/(d\omega\, do)$ which is connected to $\partial W_{\mathbf{k}}/\partial t$ by means of the relation

$$\frac{dP}{d\omega\, do} \equiv \frac{1}{(2\pi)^3}\frac{\omega^2}{c^2}\sqrt{\varepsilon(\omega)}\frac{\partial W_{\mathbf{k}}}{\partial t}. \tag{10.65}$$

As in previous considerations, we carry out the expansion of the integrands in the integral representations of $b_{ii'kk'}$ and $q'_{ii'}$ in power series in the ratio of the electron velocity to the wave phase velocity. Then $b_{ii'kk'}$ reduces to the spectral correlation function for the electron density fluctuations, and the source densities $q'_{ii'}$ become expressed in terms of the spectral functions which describe the correlations between the particle and the charge density fluctuations. As a result we obtain the following spectral distribution of the radiation:

$$\frac{dP}{d\omega\, do} = \frac{1}{\pi}\frac{e^4}{m^2c^3}\sqrt{\varepsilon(\omega)}\sum_{\mathbf{k}'}\frac{[\mathbf{k}\mathbf{k}']^2}{k^2k'^2}$$
$$\times\int d\omega'\left\{\frac{1}{k'^2}\langle\delta n_e^2\rangle_{\mathbf{k}-\mathbf{k}',\,\omega-\omega'}\langle\varrho^2\rangle_{\mathbf{k}'\omega'}-\frac{1}{(\mathbf{k}-\mathbf{k}')^2}\,\mathrm{Re}\langle\delta n_e\varrho\rangle_{\mathbf{k}-\mathbf{k}',\,\omega-\omega'}\langle\varrho\,\delta n_e\rangle_{\mathbf{k}'\omega'}\right\}, \tag{10.66}$$

where $\langle\delta n_e^2\rangle_{\mathbf{k}\omega}$, $\langle\delta n_e\varrho\rangle_{\mathbf{k}\omega}$ and $\langle\varrho^2\rangle_{\mathbf{k}\omega}$ are the spectral correlation functions describing the self-consistent interaction of charged particles in the plasma.

The charge density fluctuations ϱ of a two-component plasma of electrons with charge e and ions with charge $-Ze$ are connected with the electron and ion density fluctuations δn_e and δn_i by means of the relation

$$\varrho = e(\delta n_e - Z\,\delta n_i). \tag{10.67}$$

Hence, the right-hand side of (10.66) may be expressed in terms of the correlation functions for the electron and ion-density fluctuations.

The wave vector $\mathbf{k}$ in (10.66) is associated with the radiated wave, $\mathbf{k}'$ corresponds to the longitudinal plasma fluctuations. Assuming that $k \ll k'$ (dipole approximation), we may expand the right-hand side of (10.66) in power series in k/k'. This yields the following spectral distribution of the bremsstrahlung:

$$\frac{dP}{d\omega\, do} = \frac{1}{\pi}\frac{Z^2e^6}{m^2c^3}\sqrt{\varepsilon(\omega)}\sum_{\mathbf{k}'}\frac{[\mathbf{k}\mathbf{k}']^2}{k^2k'^4}$$
$$\times\int d\omega\{\langle\delta n_e^2\rangle_{-\mathbf{k}',\,\omega-\omega'}\langle\delta n_i^2\rangle_{\mathbf{k}',\,\omega'}-\langle\delta n_e\,\delta n_i\rangle_{-\mathbf{k}',\,\omega-\omega'}\langle\delta n_i\,\delta n_e\rangle_{\mathbf{k}'\omega'}\}. \tag{10.68}$$

Since the dipole moment of the system remains unchanged after the electron–electron scattering, (10.68) does not contain any terms proportional to the product of $\langle\delta n_e^2\rangle_{\mathbf{k}-\mathbf{k}',\,\omega-\omega'}$ and $\langle\delta n_e^2\rangle_{\mathbf{k}'\omega'}$.

According to (10.68), the spectral distribution of the bremsstrahlung intensity depends on the correlation between the ions. It was shown by Dawson and Oberman (1963) that the effect of the ion-thermal correlations on the spectral distribution of the radiated waves is weak. However, the non-thermal ion correlations, e.g. large amplitude ion-acoustic waves, can considerably modify the radiation.

In neglect of the ion motion the spectral distribution (10.68) becomes simplified. In the limiting case $m_i \to \infty$ the spectral correlation functions reduce to

$$\left.\begin{aligned} \langle \delta n_e^2 \rangle_{\mathbf{k}\omega} &= 2\pi n_0 \left\{ \sqrt{\frac{3}{2\pi}} \frac{1}{ks} \frac{e^{-\frac{3}{2}\frac{\omega^2}{k^2 s^2}}}{|\varepsilon(\omega, \mathbf{k})|^2} + \frac{Z}{1+a^2k^2} \delta(\omega) \right\}, \\ \langle \delta n_i^2 \rangle_{\mathbf{k}\omega} &= 2\pi \frac{n_0}{Z} \delta(\omega), \end{aligned}\right\} \tag{10.69}$$

$$\langle \delta n_e \, \delta n_i \rangle_{\mathbf{k}\omega} = \langle \delta n_i \, \delta n_e \rangle = \frac{n_0}{1+a^2k^2} \delta(\omega),$$

and then

$$\frac{dP}{d\omega \, do} = 2\sqrt{6\pi} \frac{Ze^6 n_0^2}{m^2 c^3 s} \sum_{\mathbf{k}'} \frac{[\mathbf{k}\mathbf{k}']^2}{k^2 k'^5} \frac{e^{-\frac{3}{2}\frac{\omega^2}{k^2 s^2}}}{|\varepsilon(\omega, \mathbf{k}')|^2}. \tag{10.70}$$

After passing from summation over $\mathbf{k}'$ to integration and carrying out the integration procedure with respect to the angles, we have

$$\frac{dP}{d\omega \, do} = \frac{2}{\pi} \sqrt{\frac{2}{3\pi}} \frac{Ze^6 n_0^2}{m^2 c^3 s} \int_0^{k_m} dk' \frac{e^{-\frac{3}{2}\frac{\omega^2}{k^2 s^2}}}{k' |\varepsilon(\omega, \mathbf{k}')|^2}. \tag{10.71}$$

To suppress the logarithmic divergence we have introduced a cut-off in the integration through the upper limit k_m. In the classical case k_m is determined by the inverse distance of the closest encounter of two charged particles

$$k_m = \frac{2}{\gamma} \frac{ms^2}{Ze^2},$$

where γ is the Euler constant ($\gamma = 1{,}78 \ldots$). In the quantum case $\left(s \gtrsim \frac{Ze^2}{\hbar} \right)$ it should be taken to be

$$k_m = \frac{2ms}{\hbar}.$$

It is clear that the frequency ω of the radiated wave must be higher than the Langmuir frequency Ω. On the other hand, the integrand in (10.71) does not vanish only if $k's > \omega$. Therefore, the range of the effective values of k' satisfies the condition $ak' > 1$. Since $\varepsilon(\omega, k') \simeq 1$ as $ak' > 1$, the effect of the collective plasma properties on the bremsstrahlung may be neglected. Upon noting that

$$\int_\alpha^\infty dy \frac{e^{-y}}{y} = -\ln(\gamma\alpha) + \alpha - \frac{\alpha^2}{4} + \ldots,$$

we obtain

$$\frac{dP}{d\omega \, do} = \frac{2}{\pi} \sqrt{\frac{2}{3\pi}} \frac{Ze^6 n_0^2}{m^2 c^3 s} \ln\left(\sqrt{\frac{2}{3\gamma}} \frac{k_m s}{\omega} \right). \tag{10.72}$$

Hence the spectral distribution of the bremsstrahlung is characterized by only a weak logarithmic dependence on the frequency.

It is to be pointed out that besides the transverse waves associated with the bremsstrahlung, there may be radiated also longitudinal waves. The intensity of the longitudinal radiation within a narrow band around the Langmuir frequency is $(c/s)^3$ times greater than that of the bremsstrahlung.

The numerical integration of (10.71) reveals in the spectral distribution of the bremsstrahlung a weak maximum at $\omega \simeq \Omega$ associated with a possibility of Langmuir wave emission. The enhancement of the bremsstrahlung in the frequency band $\omega \simeq \Omega$ may be great in a non-equilibrium plasma (Tidman and Dupree, 1965). The spectral distribution of the bremsstrahlung in a non-equilibrium plasma was studied by Hirshfield, (1973), taking into account the ion motion. The bremsstrahlung in an anisotropic plasma was treated by Canuto and Chiu (1970) and Ichimaru and Starr (1970).

Spontaneous emission by a non-equilibrium plasma

Besides the bremsstrahlung caused by the scattering of particles by the field fluctuations, an additional emission by a non-equilibrium plasma may occur, associated with the conversion of two longitudinal fluctuation waves into a transverse electromagnetic one. This spontaneous emission is a peculiar effect for a non-equilibrium plasma, since the whole of the radiation from the equilibrium plasma reduces to Rayleigh emission. The intensity of the spontaneous emission from a non-equilibrium plasma turns out to be anomalously large if the plasma state approaches the instability threshold.†

The intensity of the spontaneous emission may be found starting from the general equation (9.71). Retaining only the second term on the right-hand side of (9.71) and assuming the field fluctuations to be longitudinal, we come to the following formula that determines the intensity of the radiated electromagnetic waves:

$$\frac{dP}{d\omega\, do} = \frac{1}{128\pi^5}\frac{\omega^4}{c^3}\sqrt{\varepsilon(\omega)}\sum_{\mathbf{k}'+\mathbf{k}''=\mathbf{k}}\frac{1}{k'^2k''^2}\int d\omega'$$
$$\times\left|e_i\varkappa^{(2)}_{ijk}(\omega',\mathbf{k}';\omega-\omega',\mathbf{k}'')k'_jk''_k\right|^2\langle E^2\rangle_{\mathbf{k}'\omega'}\langle E^2\rangle_{\mathbf{k}'',\omega-\omega'}. \tag{10.73}$$

In an isotropic strongly non-isothermal plasma there may occur longitudinal Langmuir and ion-acoustic oscillations. It is evident that when the fluctuation Langmuir waves are scattered by Langmuir waves the frequency of the radiation is close to twice the Langmuir frequency; if the fluctuation Langmuir waves are scattered by the fluctuation ion sound, the electromagnetic waves are radiated at nearly Langmuir frequencies.

Let us consider a two-temperature plasma, the electrons of which move relative to the ions. Let us find the intensity of the electromagnetic radiation resulting from the scattering of the fluctuation Langmuir waves by the low-frequency ion-acoustic fluctuations. Substituting the approximate expression

$$e_i\varkappa^{(2)}_{ijk}(\omega',\mathbf{k}';\omega'',\mathbf{k}'')k'_jk''_k \simeq i\frac{e}{T}\frac{|[\mathbf{k}\mathbf{k}']|}{k},\quad \omega'\gg\omega''.$$

† The spontaneous emission by a non-equilibrium plasma was studied by Akhiezer, Daneliya, and Tsyntsadze (1964); that by a semi-bounded plasma was treated by Yakimenko (1976).

for the non-linear susceptibility in the general formulae for the spectral distribution of fluctuations (6.144) and (6.147), we obtain

$$\frac{dP}{d\omega\, do} = \frac{1}{2}\frac{e^4 n_0 T}{m^2 c^3}\frac{\omega^2}{\Omega}\sqrt{\varepsilon(\omega)}\sum_{\mathbf{k}'}\frac{[\mathbf{k}\mathbf{k}']^2}{k^2 k'^2}$$

$$\times q^2 v_s^2 \int d\omega' \frac{1}{|\Delta\omega - (\mathbf{q}\cdot\mathbf{u})|}\,\delta(\Delta\omega^2 - q^2 v_s^2)\,\delta\{(\omega' - \mathbf{k}'\mathbf{u})^2 - \Omega^2\}, \qquad (10.74)$$

where $\Delta\omega = \omega - \omega'$ and $\mathbf{q} = \mathbf{k} - \mathbf{k}'$. It becomes evident from (10.74) that the intensity of the spontaneous emission grows strongly as $(\mathbf{q}\cdot\mathbf{u}) \to q v_s$. Note that the emission caused by the transformation of a longitudinal fluctuation wave into a transverse one is large only in the long wavelength region.

The kinetic equation for waves we made use of for the analysis of the wave scattering and transformation in a plasma is nowadays the basis for the studies of various processes in turbulent plasmas.†

† For example we refer to the papers devoted to the theory of the modulation instability (Vedenov and Rudakov, 1965; Vedenov, Gordeev, and Rudakov, 1967; and Galeev, Sagdeev, Shapiro, and Shevchenko, 1976), and Langmuir collapse (Zakharov, 1972). An original approach to the theory of strong plasma turbulence was worked out by Dupree (1966, 1967, 1972), Weinstock (1965, 1969, 1970), and Weinstock and Bezzerides (1973).

CHAPTER 11

Wave Interaction in a Semi-bounded Plasma

11.1. Waves in a semi-bounded plasma

The non-linear equation for the potential field in a semi-bounded plasma

The distinguishing property of a spatially bounded plasma is that together with bulk oscillations there exist in it also surface waves which propagate along the boundary surface and are damped deep in the plasma. The structure of the surface waves depends in an essential way on the shape of the surface and the nature of the boundary conditions. It is clear that the latter are determined by the character of particle interactions in a plasma with a bounding surface. The surface waves are described in the simplest way in the case of the so-called specular reflection model, when one assumes all charged particles incident on the surface to be specularly reflected from it (Landau, 1946; Reuter and Sondheimer, 1948; Shafranov, 1958; Silin and Fetisov, 1962). A number of authors (Romanov, 1965; Stepanov, 1966; Kondratenko, 1965; Guernsey, 1969) have studied in detail the properties of various surface waves for the simplest case of a semi-bounded plasma. The first indication that surface Langmuir waves exist in a semi-bounded plasma is in the papers by Schumann (1950) and Fainberg (1956). The excitation of surface Langmuir waves in a semi-bounded plasma during the motion of a charged particle along the boundary surface has been considered by Sitenko and Tkalich (1960). Romanov (1965) has developed a kinetic theory of high-frequency Langmuir and low-frequency ion-acoustic surface waves in a semi-bounded plasma. Thermal fluctuations due to bulk and surface eigenoscillations excited in a plasma half-space were studied by Sitenko and Yakimenko (1974). Surface fluctuations in and spontaneous radiation from a semi-bounded non-equilibrium plasma were studied by Yakimenko and Zagorodny (1974, 1976). It was shown by Popov and Yakimenko (1976) that surface excitations can cause critical fluctuations. Fuse and Ichimaru (1975) considered the instability arising in a semi-bounded plasma due to the growth of the ion-acoustic surface waves in the presence of relative electron-ion motion.

It is clear that when the intensities of the surface waves are very large there come into play effects associated with non-linear wave interactions. A number of authors (Kondratenko and Shaptala, 1969; Karplyuk, Kolesnichenko, and Oraevsky, 1970) studied non-linear wave interactions in a semi-bounded plasma in the hydrodynamic approximation;

in particular, Karplyuk and Oraevskii (1969) described the decay instability of surface waves. Sitenko, Pavlenko, and Zasenko (1975a, b, 1976) used the kinetic approach to show how non-linear interactions of waves in a semi-bounded plasma can give rise to echo surface oscillations.

It is convenient for the investigation of surface and bulk eigenoscillations and non-linear interaction of those in a semi-bounded plasma, as in the case of an unbounded plasma, to start from the non-linear equation for the field which one can derive using the kinetic equations for the particle distribution functions and the equation for the self-consistent field. We shall assume that the plasma fills the half-space $z > 0$ and is spatially homogeneous and in a stationary state. The half-space $z < 0$ will be assumed to be filled by a dielectric characterized by a dielectric constant ε_0. The considerations will be restricted to electrostatic interactions between charged particles (in that case the self-consistent field is a potential field).

The kinetic equations for the electron and ion distribution functions, and also the equation for the self-consistent field in the $z > 0$ region may be written in the form

$$\frac{\partial f}{\partial t}+\left(\mathbf{v}\cdot\frac{\partial f}{\partial \mathbf{r}}\right)+\frac{e}{m}\left(\mathbf{E}\cdot\frac{\partial}{\partial \mathbf{v}}\right)(f_0+f) = 0, \tag{11.1}$$

$$\operatorname{div}\mathbf{E} = 4\pi\left(\sum e\int d\mathbf{v}\, f+\varrho^0\right), \tag{11.2}$$

where f is the deviation of the electron or ion distribution function from the unperturbed distribution function f_0 (we must use for f_0 the Maxwell distribution function for an equilibrium plasma), $\mathbf{E}$ is the self-consistent electric field, and ϱ^0 the density of the external charges. The sum $\sum$ in (11.2) indicates summation over the electron and ion components. In the region of space $z < 0$ the electric field $\mathbf{E}$ satisfies the equation

$$\operatorname{div}\mathbf{E} = 0. \tag{11.3}$$

(We assume that there are no external charges in the region of space outside the plasma).

We must supplement the kinetic equation (11.1) with boundary conditions which are imposed on the distribution function deviations f at the boundary surface $z = 0$. If we assume that the particles are specularly reflected from the boundary surface, these conditions may be written in the form

$$f(x, y, z = 0;\ v_x, v_y, v_z) = f(x, y, z = 0;\ v_x, v_y, -v_z). \tag{11.4}$$

The electric field at the boundary surface $z = 0$ must satisfy the usual boundary conditions which are reduced to the requirement that the tangential components of the electric field and the normal components of the electric induction are continuous.

To find the solution of the set of (11.1) and (11.2) given in the region of space $z > 0$ we use the following formal method. We continue into the region of space $z < 0$ the components $E_\perp$ of the electric field so that they are even functions and the component E_z so that it is an odd function (we denote the electric field continued in this way by E^+) and we assume that the kinetic equations determine the distribution function in the whole of space (we denote these distribution functions by f^+):

$$\frac{\partial f^+}{\partial t}+\left(\mathbf{v}\cdot\frac{\partial f^+}{\partial \mathbf{r}}\right)+\frac{e}{m}\left(\mathbf{E}^+\cdot\frac{\partial}{\partial \mathbf{v}}\right)(f_0+f^+) = 0. \tag{11.5}$$

As the differential operator occurring in these equations for the given continuation of the electric field is invariant under the substitution $z, v_z \to -z, -v_z$, the solutions of the equations will also be invariant under that substitution

$$f(x, y, z; v_x, v_y, v_z) = f(x, y, -z; v_x, v_y, -v_z). \tag{11.6}$$

(We assume the unperturbed distributions f_0 to be even functions of v_z.) It is clear that in the region of space $z > 0$ the distribution functions f and f^+ must be the same as at $z = 0$. The boundary conditions (11.4) follow directly from (11.6).

The electric field $\mathbf{E}^+$ satisfies the equation

$$\operatorname{div} \mathbf{E}^+ = 4\pi\left(e \int d\mathbf{v} f^+ + \varrho^0\right) + 2E_z^+(x\, y, z = 0)\, \delta(z), \tag{11.7}$$

which differs from (11.2) by taking into account an additional surface charge which guarantees the jump in the normal component of the field at the boundary surface. (The density of external charges ϱ^0 is assumed to be continued into the region $z < 0$ in an even fashion.) The electric field determined in the region of space $z < 0$ by (11.3) is similarly continued into the region of space $z > 0$. This field which we shall denote by $\mathbf{E}^-$ satisfies the equation

$$\operatorname{div} \mathbf{E}^- = -2E_z^-(x, y, z = 0)\, \delta(z). \tag{11.8}$$

It is clear that the solution of (11.7) in the region $z > 0$ describes the electric field in the plasma ($\mathbf{E}^+(\mathbf{r}) = \mathbf{E}(\mathbf{r})$) and the solution of (11.8) describes in the region the field outside the plasma ($\mathbf{E}^-(\mathbf{r}) = \mathbf{E}(\mathbf{r})$). At the surface of the plasma the following boundary conditions must be satisfied:

$$\mathbf{E}_\perp^+(x, y, z = 0) = \mathbf{E}_\perp^-(x, y, z = 0), \tag{11.9}$$

$$E_z^+(x, y, z = 0) = \varepsilon_0 E_z^-(x, y, z = 0). \tag{11.10}$$

It is convenient to use the spatial Fourier transforms to write (11.7) in the form

$$ikE_\mathbf{k}^+ = 4\pi\left(e \int d\mathbf{v} f_\mathbf{k}^+ + \varrho_\mathbf{k}^0\right) + 2E_{z\mathbf{k}_\perp}^+(0), \tag{11.11}$$

and then, because of the longitudinal nature of the field,

$$E_{z\mathbf{k}_\perp}^+(0) \equiv \frac{1}{2\pi} \int d\mathbf{k}_z \frac{k_z}{k} E_\mathbf{k}^+. \tag{11.12}$$

Similarly we write (11.8) in the form

$$ikE_\mathbf{k}^- = -2E_{z\mathbf{k}_\perp}^-(0). \tag{11.13}$$

The quantity $E_{z\mathbf{k}_\perp}^+$, which occurs in (11.11), may be found from the boundary conditions and we can rewrite (11.11) in the form

$$ikE_\mathbf{k}^+ = 4\pi\left(e \int d\mathbf{v} f_\mathbf{k}^+ + \varrho_\mathbf{k}^0\right) - 2i\varepsilon_0 E_{\perp\mathbf{k}_\perp}^+(0), \tag{11.14}$$

where

$$E_{\perp\mathbf{k}_\perp}^+(0) \equiv \frac{k_\perp}{2\pi} \int dk_z \frac{E_k^+}{k}. \tag{11.15}$$

For the sake of simplicity we shall in what follows drop the index of **E**, i.e. we shall denote $\mathbf{E}^+$ by **E**.

Applying a space–time Fourier transformation to the kinetic equation (11.5) and solving the equation obtained by the method of successive approximations, we write the deviation of the distribution function $f^+_{\mathbf{k}\omega}$ as a power series in the field intensity $E_{\mathbf{k}\omega}$. Substituting this expansion into (11.14) we get the following non-linear equation which completely determines the electric field in the region of space occupied by the plasma:

$$\varepsilon(\omega, \mathbf{k})E_{\mathbf{k}\omega}+\sum_{\substack{\omega_1+\omega_2=\omega\\ \mathbf{k}_1+\mathbf{k}_2=\mathbf{k}}}\varkappa^{(2)}(\omega_1, \mathbf{k}_1;\, \omega_2, \mathbf{k}_2)E_{\mathbf{k}_1\omega_1}E_{\mathbf{k}_2\omega_2}$$
$$+\sum_{\substack{\omega_1+\omega_2+\omega_3=\omega\\ \mathbf{k}_1+\mathbf{k}_2+\mathbf{k}_3=\mathbf{k}}}\varkappa^{(3)}(\omega_1, \mathbf{k}_1;\, \omega_2, \mathbf{k}_2;\, \omega_3, \mathbf{k}_3)E_{\mathbf{k}_1\omega_1}E_{\mathbf{k}_2\omega_2}E_{\mathbf{k}_3\omega_3}+\dots$$
$$+2\varepsilon_0\frac{k_\perp}{k}\sum_{k'_z}\frac{E_{\mathbf{k}'\omega}}{k'} = -\frac{4\pi i}{k}\overset{0}{\varrho}_{\mathbf{k}\omega}, \tag{11.16}$$

where $\varepsilon(\omega, \mathbf{k})$ is the permittivity, and $\varkappa^{(2)}(\omega_1, \mathbf{k}_1; \omega_2, \mathbf{k}_2)$ and $\varkappa^{(3)}(\omega_1, \mathbf{k}_1; \omega_2, \mathbf{k}_2; \omega_3, \mathbf{k}_3)$ are the non-linear susceptibilities of the unbounded plasma.

The electric field in the region outside the plasma is, according to (11.13) and (11.15), directly expressed in terms of the solution of (11.16):

$$E^-_{\mathbf{k}\omega} = 2\frac{k_\perp}{k}\sum_{k'_z}\frac{E_{\mathbf{k}'\omega}}{k'}. \tag{11.17}$$

The complete solution of the problem of finding the field of a plasma half-space reduces thus to solving the non-linear equation (11.16).

One can use the non-linear equation (11.16) to study a number of effects in a semi-bounded plasma: three-wave decay interactions; four-wave interactions that cause eigen-frequency shifts and non-linear Landau damping; non-resonance echo effects connected with the undamped nature of the oscillations of the distribution function; induced scattering of waves by particles; and so on. Even if one can restrict oneself in the analysis of the decay interactions of waves to the hydrodynamic approximation (although in that case there remains an arbitrariness in the determination of the coefficients) one needs the kinetic approach for a description of the other effects listed here.

Dispersion equations for bulk and surface waves (linear approximation)

Neglecting the non-linear terms in (11.16) and putting the external charge density equal to zero, we get the basic equation of the linear approximation which describes the eigen-oscillations in a semi-bounded plasma:

$$\varepsilon(\omega, \mathbf{k})E_{\mathbf{k}\omega}+2\frac{\varepsilon_0}{k}E_{\mathbf{k}_\perp\omega} = 0. \tag{11.18}$$

Here and henceforth we introduce, for the sake of simplicity, the notation

$$E_{\mathbf{k}_\perp\omega} \equiv E_{\perp\mathbf{k}_\perp\omega}(0) = \frac{k_\perp}{2\pi}\int dk_z\frac{E_{\mathbf{k}\omega}}{k}, \tag{11.19}$$

i.e. we understand by the quantity $E_{\mathbf{k}_\perp\omega}$ the value of the tangential component of the field intensity at the surface of the plasma.

One easily finds from (11.18) the following equation for the quantity $E_{\mathbf{k}_\perp\omega}$:

$$\zeta(\omega, \mathbf{k}_\perp)E_{\mathbf{k}_\perp\omega} = 0, \tag{11.20}$$

where

$$\zeta(\omega, \mathbf{k}_\perp) \equiv 1+\frac{\varepsilon_0 k_\perp}{\pi}\int dk_z \frac{1}{k^2\varepsilon(\omega, \mathbf{k})}. \tag{11.21}$$

According to (11.18) there are possible two kinds of eigenoscillations in a semi-bounded plasma—bulk and surface oscillations.

The dispersion equation for the bulk eigenwaves $\left(E_{\mathbf{k}\omega} \neq 0, E_{\mathbf{k}_\perp\omega} = 0\right)$ is determined by the same condition as in the case of an unbounded plasma

$$\varepsilon(\omega, \mathbf{k}) = 0. \tag{11.22}$$

We shall denote the eigenfrequencies which are the solutions of (11.22) for fixed value of $\mathbf{k}$ by $\omega_\mathbf{k}$ and we write the field of the eigenoscillations in the form

$$E_{\mathbf{k}\omega} = \pi E_\mathbf{k}\{\exp(-i\phi_\mathbf{k})\,\delta(\omega-\omega_\mathbf{k})+\exp(i\phi_\mathbf{k})\,\delta(\omega+\omega_\mathbf{k})\}, \tag{11.23}$$

where $E_\mathbf{k}$ and $\phi_\mathbf{k}$ are the initial amplitude and phase. Making use of the boundary conditions (11.9) and (11.10) one can easily show that

$$E_{\mathbf{k}_\perp\omega} = \frac{i}{\varepsilon_0}\frac{1}{2\pi}\int dk_z \frac{k_z E_{\mathbf{k}\omega}}{k},$$

and as $E_\mathbf{k}$, $\phi_\mathbf{k}$, and $\omega_\mathbf{k}$ are even functions of k_z the condition $E_{\mathbf{k}_\perp\omega} = 0$ is satisfied for the bulk oscillations.

The dispersion equation for the surface eigenwaves $\left(E_{\mathbf{k}_\perp\omega} \neq 0\right)$ is given by the condition

$$\zeta(\omega, \mathbf{k}_\perp) = 0. \tag{11.24}$$

We denote the eigenfrequencies of the surface oscillations by $\omega_{\mathbf{k}_\perp}$ and write the field of the surface oscillations in the form

$$E_{\mathbf{k}_\perp\omega} = \pi E_{\mathbf{k}_\perp}\left\{\exp(-i\phi_{\mathbf{k}_\perp})\,\delta(\omega-\omega_{\mathbf{k}_\perp})+\exp(i\phi_{\mathbf{k}_\perp})\,\delta(\omega+\omega_{\mathbf{k}_\perp})\right\}. \tag{11.25}$$

Since $\varepsilon(\omega, \mathbf{k}) \neq 0$ in the case of the surface oscillations, we easily obtain from (11.18) the total spatial field component for the surface oscillations

$$E_{\mathbf{k}\omega} = -\frac{2\varepsilon_0}{k}\frac{E_{\mathbf{k}_\perp\omega}}{\varepsilon(\omega, \mathbf{k})}. \tag{11.26}$$

One can show that the field of the surface oscillations decreases exponentially when one goes away from the boundary surface.

Surface waves

Let us consider now various types of surface waves in a semi-bounded plasma, which are given by the dispersion equation (11.24). In the high-frequency region the dispersion of the surface waves is determined, just as for bulk waves, by the electron component of the plasma. In the long-wavelength limit $a^2k^2 \ll 1$ (a is the Debye radius) the eigenfrequency and the damping coefficient of the high-frequency surface waves are described by the formulas

$$\left.\begin{aligned} \omega_{\mathbf{k}_\perp} &= \frac{\Omega}{(1+\varepsilon_0)^{1/2}}\left(1+\frac{(3\varepsilon_0)^{1/2}}{2}a\,|\mathbf{k}_\perp|+\ldots\right), \\ \gamma_{\mathbf{k}_\perp} &= \left(\frac{2}{3\pi}\right)^{1/2}|\mathbf{k}_\perp|\,s, \quad s^2=\frac{3T}{m}. \end{aligned}\right\} \tag{11.27}$$

In the low-frequency region the dispersion of the surface waves depends in an essential way on both the electrons and the ions. We assume the electron temperature to be appreciably higher than that of the ions $T_e \gg T_i$ (strongly non-isothermal plasma), and we consider the frequency band satisfying the condition $s \gg \omega/k_\perp \gg s_i$. In such a case we can use for the dielectric permittivity the approximate expression

$$\varepsilon(\omega, \mathbf{k}) = 1+\frac{1}{a^2k^2}-\frac{\Omega_i^2}{\omega^2},$$

and making use of this we find easily

$$\zeta(\omega, \mathbf{k}_\perp) = 1+\frac{\varepsilon_0}{\varepsilon_i}\left(1+\frac{1}{a^2k_\perp^2\varepsilon_i}\right)^{-1/2}, \quad \varepsilon_i \equiv 1-\frac{\Omega_i^2}{\omega^2}. \tag{11.28}$$

Putting then the quantity $\zeta(\omega, \mathbf{k}_\perp)$ equal to zero we obtain for the eigenfrequencies and damping coefficients of the surface waves in the long- and short-wavelength limits, respectively: when $a^2k_\perp^2 \ll 1$

$$\left.\begin{aligned} \omega_{\mathbf{k}_\perp} &= k_\perp v_s(1-\tfrac{1}{2}a^2k_\perp^2), \\ \gamma &= \left(\frac{\pi}{8}\right)^{1/2}\left(\frac{m_e}{m_i}\right)^{1/2}\Omega_i a k_\perp\left[1+\left(\frac{m_i}{m_e}\right)^{1/2}\left(\frac{T_e}{T_i}\right)^{3/2}\exp\left(-\frac{T_e}{2T_i}\right)\right], \end{aligned}\right\} \tag{11.29a}$$

and when $a^2k_\perp^2 \gg 1$,

$$\omega_{\mathbf{k}_\perp} = \frac{\Omega_i}{(1+\varepsilon_0)^{1/2}}\left(1-\frac{1}{2a^2k_\perp^2(1+\varepsilon_0)}\right), \quad \gamma \simeq \left(\frac{2}{\pi}\right)^{1/2} k_\perp s_i. \tag{11.29b}$$

The high-frequency electron and low-frequency ion-sound and ion-surface waves considered here are characterized by positive energies. It is well known that in an unbounded non-equilibrium plasma there are possible not only waves with positive energy, but also negative energy waves (Kadomtsev, Mihailovskii, and Timofeev, 1965). In a non-equilibrium semi-bounded plasma surface waves with negative energy are also possible. As an example we consider a plasma through which a compensated low-density ion beam ($n_0' \ll n_0$) propagates with velocity $\mathbf{u}$ (the vector $\mathbf{u}$ is parallel to the boundary surface). Neglecting the thermal ion motion we can write the permittivity of the plasma in the form

$$\varepsilon(\omega, \mathbf{k}) = 1+\frac{1}{a^2k^2}-\frac{\Omega_i^2}{\omega^2}\left[1+\eta\frac{\omega^2}{[\omega-(\mathbf{k}_\perp\cdot\mathbf{u})]^2}\right], \quad \eta\equiv\frac{n_0'}{n_0}\ll 1. \tag{11.30}$$

(This expression is applicable in the frequency range $\omega < k_\perp s$.) Substituting (11.30) into the general equation (11.21) and integrating over k_z we obtain (11.28) for the quantity $\zeta(\omega, \mathbf{k}_\perp)$ in which we must understand by

$$\varepsilon_i = 1-\frac{\Omega_i^2}{\omega^2}\left[1+\eta\frac{\omega^2}{[\omega-(\mathbf{k}_\perp\cdot\mathbf{u})]^2}\right].$$

After that, putting $\zeta(\omega, \mathbf{k}_\perp)$ equal to zero, and restricting the consideration to the long-wavelength limit $a^2k_\perp^2 \ll 1$, we can write the dispersion equation in the form

$$1+\frac{1}{a^2k_\perp^2}+a^2k_\perp^2\varepsilon_0^2 = \frac{\Omega_i^2}{\omega^2}\left[1+\eta\frac{\omega^2}{[\omega-(\mathbf{k}_\perp\cdot\mathbf{u})]^2}\right]. \tag{11.31}$$

Assuming the beam density to be sufficiently small ($\eta \ll 1$), we can easily find the roots of that equation which correspond to the eigenfrequencies of the surface waves:

$$\left.\begin{aligned}\omega_{\mathbf{k}_\perp}^{(1)} &= k_\perp v_s/(1+a^2k^2)^{1/2},\\ \omega_{\mathbf{k}_\perp}^{(2,3)} &= \mathbf{k}_\perp\mathbf{u}\mp\eta^{1/2}\Omega_i[1+1/a^2k_\perp^2-\Omega_i^2/(\mathbf{k}_\perp\cdot\mathbf{u})^2]^{-1/2}.\end{aligned}\right\} \tag{11.32}$$

The waves corresponding to the eigenfrequencies $\omega_{\mathbf{k}_\perp}^{(1)}$ and $\omega_{\mathbf{k}_\perp}^{(3)}$ are characterized by positive energies, while the wave corresponding to the eigenfrequency $\omega_{\mathbf{k}_\perp}^{(2)}$ is characterized by a negative energy.

The non-linear equation for the field (general case)

We consider now the non-linear equation for the electric field in a semi-bounded plasma making no assumption of potentiality of the field. In such a case, besides the self-consistent electric field, the self-consistent magnetic field also enters the kinetic equation. We continue the self-consistent electric and magnetic fields into the space outside the plasma (the components $\mathbf{E}_\perp$ and B_z, respectively, to be even, and E_z and $\mathbf{B}_\perp$ to be odd functions under the substitution $z \to -z$) and supplement the definition of the distribution function f as in the case of the potential electric field. The induced current density in a semi-bounded plasma is expressed in terms of the linear and non-linear susceptibilities of an unbounded homogeneous plasma, and the space limitation leads to an additional surface current in the field equation. The non-linear equation for the electric field in a semi-bounded plasma (in the absence of external sources) may be written as

$$\Lambda_{ij}(\omega, \mathbf{k})E_{j\mathbf{k}\omega}+\sum_{1+2=\omega,\mathbf{k}}\varkappa_{ijk}^{(2)}(1, 2)E_{j1}E_{k2}$$
$$+\sum_{1+2+3=\omega,\mathbf{k}}\varkappa_{ijkl}^{(3)}(1, 2, 3)E_{j1}E_{k2}E_{l3}+\ldots-2i\frac{c}{\omega}\varepsilon_{zij}B_{j\mathbf{k}_\perp\omega}(0) = 0, \tag{11.33}$$

where $B_{j\mathbf{k}_\perp\omega}$ is the magnetic field on the boundary

$$B_{j\mathbf{k}_\perp\omega}(0) \equiv \frac{1}{2\pi}\int dk_z B_{j\mathbf{k}\omega}. \tag{11.34}$$

Assuming that there is a dielectric with dielectric permittivity ε_0 outside the plasma and using the boundary conditions for the fields, one can easily show that

$$\varepsilon_{zij} B_{j\mathbf{k}_\perp\omega}(0) = i\alpha_{ij}(\omega, \mathbf{k}_\perp) E_{j\mathbf{k}_\perp\omega}(0), \tag{11.35}$$

where

$$\alpha_{ij} = \begin{pmatrix} \alpha_1 & 0 & 0 \\ 0 & \alpha_2 & 0 \\ 0 & 0 & 0 \end{pmatrix}, \quad \alpha_1 = \frac{\varepsilon_0}{\sqrt{\eta_\perp^2 - \varepsilon_0}}, \quad \alpha_2 = -\sqrt{\eta^2 - \varepsilon_0}.$$

Now we can rewrite the basic non-linear equation for the electric field in a semi-bounded plasma in a closed form:

$$\begin{aligned} \Lambda_{ij}(\omega, \mathbf{k}) E_{j\mathbf{k}\omega} + & \sum_{1+2=\omega, \mathbf{k}} \varkappa^{(2)}_{ijk}(1, 2) E_{j1} E_{k2} \\ & + \sum_{1+2+3=\omega, \mathbf{k}} \varkappa^{(3)}_{ijkl}(1, 2, 3) E_{j1} E_{k2} E_{l3} + \dots \\ & + \frac{2c}{\omega} \alpha_{ij}(\omega, \mathbf{k}_\perp) \sum_{k_z} E_{j\mathbf{k}\omega} = 0. \end{aligned} \tag{11.36}$$

In the linear approximation the equation for the field in a semi-bounded plasma is of the form

$$\Lambda_{ij}(\omega, \mathbf{k}) E_{j\mathbf{k}\omega} + \frac{2c}{\omega} \alpha_{ij}(\omega, \mathbf{k}_\perp) E_{j\mathbf{k}_\perp\omega} = 0, \tag{11.37}$$

where $E_{j\mathbf{k}_\perp\omega}$ is the electric field on the boundary

$$E_{j\mathbf{k}_\perp\omega} \equiv E_{j\mathbf{k}_\perp\omega}(0) = \frac{1}{2\pi}\int dk_z E_{j\mathbf{k}\omega}. \tag{11.38}$$

It follows from (11.37) that eigenoscillations of two types can exist in a semi-bounded plasma—the bulk and surface oscillations.

The dispersion equation for the bulk eigenoscillations $\left(E_{j\mathbf{k}\omega} \neq 0,\ E_{j\mathbf{k}_\perp\omega} = 0\right)$ is the same as for the unbounded plasma

$$|\Lambda_{ij}(\omega, \mathbf{k})| = 0.$$

The dispersion equation for the surface eigenoscillations $\left(E_{j\mathbf{k}_\perp\omega} \neq 0\right)$ follows from the condition

$$\left| \delta_{ik} + \frac{c}{\pi\omega} \int dk_z \Lambda^{-1}_{ij}(\omega, \mathbf{k})\, \alpha_{jk}(\omega, \mathbf{k}_\perp) \right| = 0. \tag{11.39}$$

When the particle distribution is isotropic, the dispersion equation decomposes in two independent equations:

$$1-\frac{c}{\pi\omega}\sqrt{\eta_\perp^2-\varepsilon_0}\int dk_z\frac{1}{\varepsilon_t(\omega,\mathbf{k})-\eta^2}=0, \tag{11.40}$$

$$1+\frac{c}{\pi\omega}\frac{\varepsilon_0}{\sqrt{\eta_\perp^2-\varepsilon_0}}\int\frac{dk_z}{k^2}\left\{\frac{k_\perp^2}{\varepsilon_l(\omega,\mathbf{k})}+\frac{k^2}{\varepsilon_t(\omega,\mathbf{k})-\eta^2}\right\}=0, \tag{11.41}$$

which describe the surface oscillations with different polarizations of the electric field (s and p are the polarizations). In the case of an s-polarized surface wave the vector of the electric field intensity is perpendicular to the direction of wave propagation; that of a p-polarized surface wave is directed along the wave propagation.

The dispersion equation (11.40) has no solutions, which indicates that there cannot exist s-polarized surface waves in a plasma. As it follows from (11.41) the p-polarized surface wave is neither longitudinal nor transverse. Since the phase velocity of such a wave $v_{ph} < c/\sqrt{\varepsilon_0}$, it cannot be radiated in the form of an electromagnetic wave. In the $c\to\infty$ limiting case, (11.41) reduces to the dispersion equation for potential surface waves (11.24).

The non-linear equation (11.36) is suitable for an investigation of three-wave decays in a semi-bounded plasma in the general case and is basic for the derivation of the kinetic equation for waves which describes also the transverse electromagnetic waves.

11.2. Non-linear Interaction of Surface Waves

Resonant wave interaction

Non-linear interactions of bulk and surface waves in a semi-bounded plasma are described by the general non-linear equation (11.16). Assuming there to be no external charges we rewrite this equation in the form

$$\begin{aligned}\varepsilon(\omega,\mathbf{k})E_{\mathbf{k}\omega}&+\sum_{\substack{\omega_1+\omega_2=\omega\\ \mathbf{k}_1+\mathbf{k}_2=\mathbf{k}}}\varkappa^{(2)}(\omega_1,\mathbf{k}_1;\omega_2,\mathbf{k}_2)E_{\mathbf{k}_1\omega_1}E_{\mathbf{k}_2\omega_2}\\ &+\sum_{\substack{\omega_1+\omega_2+\omega_3=\omega\\ \mathbf{k}_1+\mathbf{k}_2+\mathbf{k}_3=\mathbf{k}}}\varkappa^{(3)}(\omega_1,\mathbf{k}_1;\omega_2,\mathbf{k}_2;\omega_3,\mathbf{k}_3)E_{\mathbf{k}_1\mathbf{k}_1}E_{\mathbf{k}_2\omega_2}E_{\mathbf{k}_3\omega_3}+\dots\\ &+\frac{2\varepsilon_0}{c}E_{\mathbf{k}_\perp\omega}=0.\end{aligned} \tag{11.42}$$

Using a multiple-time-scale expansion method one can obtain from this equation a hierarchy of equations which determine the time-dependence of the amplitudes caused by the non-linear resonant interactions of the waves.

The simplest example of a non-linear resonant wave interaction is the three-wave resonance which occurs in the case when the frequencies of the interacting waves satisfy the condition

$$\omega_{\mathbf{k}_1}+\omega_{\mathbf{k}_2}=\omega_{\mathbf{k}}. \tag{11.43}$$

It is clear that three-wave resonance is possible in a semi-bounded plasma in the case of the interaction of three bulk waves, two bulk and one surface wave (two cases are possible: as the result of the interaction of two bulk waves a surface wave is formed, and the interaction of a bulk wave with a surface wave leads to the formation of a bulk wave), two surface waves with a bulk wave, and of three surface waves. In the case when there are no three-wave resonances, the four-wave resonant interaction, which occurs when the following resonance condition between the frequencies holds,

$$\omega_{\mathbf{k}_1}+\omega_{\mathbf{k}_2}+\omega_{\mathbf{k}_3}=\omega_{\mathbf{k}} \tag{11.44}$$

turns out to be the most important one.

We restrict ourselves to a more detailed discussion of the resonant interactions of surface waves. We multiply (11.42) by $k_\perp/k\varepsilon(\omega, \mathbf{k})$ and integrate over k_z and then, having in mind the surface nature of the interacting waves, we use (11.26) to express the fields $E_{\mathbf{k}_1\omega_1}$, $E_{\mathbf{k}_2\omega_2}$, ... in terms of the surface components $E_{\mathbf{k}_{1\perp}\omega_1}$, $E_{\mathbf{k}_{2\perp}\omega_2}$. ... As a result, the basic equation describing the interactions of surface waves in a semi-bounded plasma may be written in the form

$$\zeta(\omega, \mathbf{k}_\perp)E_{\mathbf{k}_\perp\omega}+\sum_{\substack{\omega_1+\omega_2=\omega\\ \mathbf{k}_{1\perp}+\mathbf{k}_{2\perp}=\mathbf{k}_\perp}}\tilde{\varkappa}^{(2)}(\omega_1, \mathbf{k}_{1\perp};\,\omega_2, \mathbf{k}_{2\perp})E_{\mathbf{k}_{1\perp}\omega_1}E_{\mathbf{k}_{2\perp}\omega_2}$$

$$+\sum_{\substack{\omega_1+\omega_2+\omega_3=\omega\\ \mathbf{k}_{1\perp}+\mathbf{k}_{2\perp}+\mathbf{k}_{3\perp}=\mathbf{k}_\perp}}\tilde{\varkappa}^{(3)}(\omega_1, \mathbf{k}_{1\perp};\,\omega_2, \mathbf{k}_{2\perp};\,\omega_3, \mathbf{k}_{3\perp})E_{\mathbf{k}_{1\perp}\omega_1}E_{\mathbf{k}_{2\perp}\omega_2}E_{\mathbf{k}_{3\perp}\omega_3}+\ldots=0, \tag{11.45}$$

where $\tilde{\varkappa}^{(2)}$ and $\tilde{\varkappa}^{(3)}$ are the non-linear surface susceptibilities of the plasma, determined by equations

$$\tilde{\varkappa}^{(2)}(\omega_1, \mathbf{k}_{1\perp}, \omega_2, \mathbf{k}_{2\perp})=\frac{\varepsilon_0^2 k_\perp}{\pi^2}\int dk_{1z}\int dk_{2z}\,\frac{\varkappa^{(2)}(\omega_1, \mathbf{k}_1;\,\omega_2, \mathbf{k}_2)}{k_1k_2k\varepsilon(\omega_1, \mathbf{k}_1)\,\varepsilon(\omega_2, \mathbf{k}_2)\,\varepsilon(\omega, \mathbf{k})}, \tag{11.46}$$

$$\begin{aligned}
&\tilde{\varkappa}^{(3)}(\omega_1, \mathbf{k}_{1\perp};\,\omega_2, \mathbf{k}_{2\perp};\,\omega_3, \mathbf{k}_{3\perp})\\
&\quad=-\frac{\varepsilon_0^3k_\perp}{\pi^3}\int dk_{1z}\int dk_{2z}\int dk_{3z}\,\frac{1}{k_1k_2k_3k\varepsilon(\omega_1, \mathbf{k}_1)\,\varepsilon(\omega_2, \mathbf{k}_2)\,\varepsilon(\omega_3, \mathbf{k}_3)\,\varepsilon(\omega, \mathbf{k})}\\
&\quad\times\Bigg\{\varkappa^{(3)}(\omega_1, \mathbf{k}_1;\,\omega_2, \mathbf{k}_2;\,\omega_3, \mathbf{k}_3)-\frac{2}{3}\Bigg[\frac{\varkappa^{(2)}(\omega_1, \mathbf{k}_1;\,\omega_2+\omega_3, \mathbf{k}_2+\mathbf{k}_3)\,\varkappa^{(2)}(\omega_2, \mathbf{k}_2;\,\omega_3, \mathbf{k}_3)}{\varepsilon(\omega_2+\omega_3, \mathbf{k}_2+\mathbf{k}_3)}\\
&\quad+\frac{\varkappa^{(2)}(\omega_2, \mathbf{k}_2;\,\omega_1+\omega_3, \mathbf{k}_1+\mathbf{k}_3)\,\varkappa^{(2)}(\omega_1, \mathbf{k}_1;\,\omega_3, \mathbf{k}_3)}{\varepsilon(\omega_1+\omega_3, \mathbf{k}_1+\mathbf{k}_3)}\\
&\quad+\frac{\varkappa^{(2)}(\omega_3, \mathbf{k}_3;\,\omega_1+\omega_2, \mathbf{k}_1+\mathbf{k}_2)\,\varkappa^{(2)}(\omega_1, \mathbf{k}_1;\,\omega_2, \mathbf{k}_2)}{\varepsilon(\omega_1+\omega_2;\,\mathbf{k}_1+\mathbf{k}_2)}\Bigg]\Bigg\}.
\end{aligned} \tag{11.47}$$

We apply to the non-linear equation (11.47) a multiple-time-scale expansion method. In the first approximation the fields of the surface waves are, as before, given by (11.25), but if the non-linear interactions of the waves are taken into account the amplitudes $E_{\mathbf{k}_\perp}$ and phases $\phi_{\mathbf{k}_\perp}$ must be assumed to be slowly varying functions of the time. The equations determining the time-dependence of the amplitudes $E_{\mathbf{k}_\perp}$ and the phases $\phi_{\mathbf{k}_\perp}$ may be found from the conditions that the secular parts of the higher approximations of (11.45) must vanish. Under the conditions of three-wave resonance for surface waves:

$$\omega_{\mathbf{k}_{1\perp}}+\omega_{\mathbf{k}_{2\perp}} = \omega_{\mathbf{k}_\perp}, \tag{11.48}$$

the equation determining the time-dependence of the amplitude of the surface waves in the linear approximation is of the form

$$\frac{\partial}{\partial t} E_{\mathbf{k}_\perp} \exp(-i\phi_{\mathbf{k}_\perp}) = \frac{i}{2}\left(\frac{\partial\zeta(\omega_{\mathbf{k}_\perp}, \mathbf{k}_\perp)}{\partial\omega_{\mathbf{k}_\perp}}\right)^{-1} \times \sum_{\mathbf{k}_{1\perp}+\mathbf{k}_{2\perp}=\mathbf{k}_\perp} \tilde{\varkappa}^{(2)}(\omega_{\mathbf{k}_{1\perp}}, \mathbf{k}_{1\perp}; \omega_{\mathbf{k}_{2\perp}}, \mathbf{k}_{2\perp}) E_{\mathbf{k}_{1\perp}} E_{\mathbf{k}_{2\perp}} \exp[-i(\phi_{\mathbf{k}_{1\perp}}+\phi_{\mathbf{k}_{2\perp}})]. \tag{11.49}$$

This equation describes three-wave processes, i.e. processes in which two waves fuse to form one wave and processes in which one wave decays into two others. The decay interaction appears also when we neglect the thermal motion of the particles, i.e. it may be described also on the basis of hydrodynamical considerations.

If condition (11.48) is not satisfied, the correction to the field is in the second approximation expressed in terms of the field in the first approximation as follows:

$$E^{(2)}_{\mathbf{k}_\perp\omega} = -\frac{1}{\zeta(\omega, \mathbf{k}_\perp)} \sum_{\substack{\omega_1+\omega_2=\omega \\ \mathbf{k}_{1\perp}+\mathbf{k}_{2\perp}=\mathbf{k}_\perp}} \tilde{\varkappa}^{(2)}(\omega_1, \mathbf{k}_{1\perp}; \omega_2, \mathbf{k}_{2\perp}) E_{\mathbf{k}_{1\perp}\omega_1} E_{\mathbf{k}_{2\perp}\omega_2}. \tag{11.50}$$

The time-dependence of the amplitude and phase may in this case be found from the condition for removing the secularity in the equation of the third approximation. Resonant interactions occur if the following relation holds between the frequencies:

$$\omega_{\mathbf{k}_{1\perp}}+\omega_{\mathbf{k}_{2\perp}}+\omega_{\mathbf{k}_{3\perp}} = \omega_{\mathbf{k}_\perp}. \tag{11.51}$$

In the four-wave resonance case the equation determining the time-dependence of the field of the surface wave is of the form

$$\begin{aligned} \frac{\partial}{\partial t} E_{\mathbf{k}_\perp} \exp(-i\phi_{\mathbf{k}_\perp}) &= -\frac{i}{4}\left(\frac{\partial\zeta(\omega_{\mathbf{k}_\perp}, \mathbf{k}_\perp)}{\partial\omega_{\mathbf{k}_\perp}}\right)^{-1} \\ &\times \sum'_{\mathbf{k}_{1\perp}+\mathbf{k}_{2\perp}+\mathbf{k}_{3\perp}=\mathbf{k}_\perp} \Bigg\{\frac{2}{\zeta(\omega_{\mathbf{k}_{2\perp}}+\omega_{\mathbf{k}_{3\perp}}, \mathbf{k}_{2\perp}+\mathbf{k}_{3\perp})} \tilde{\varkappa}^{(2)}(\omega_{\mathbf{k}_{1\perp}}, \mathbf{k}_{1\perp}; \omega_{\mathbf{k}_{2\perp}}+\omega_{\mathbf{k}_{3\perp}}, \mathbf{k}_{2\perp}+\mathbf{k}_{3\perp}) \\ &\times \tilde{\varkappa}^{(2)}(\omega_{\mathbf{k}_{2\perp}}, \mathbf{k}_{2\perp}; \omega_{\mathbf{k}_{3\perp}}, \mathbf{k}_{3\perp}) - \tilde{\varkappa}^{(3)}(\omega_{\mathbf{k}_{1\perp}}, \mathbf{k}_{1\perp}; \omega_{\mathbf{k}_{2\perp}}, \mathbf{k}_{2\perp}; \omega_{\mathbf{k}_{3\perp}}, \mathbf{k}_{3\perp}\Bigg\} \\ &\times E_{\mathbf{k}_{1\perp}} E_{\mathbf{k}_{2\perp}} E_{\mathbf{k}_{3\perp}} \exp[-i(\phi_{\mathbf{k}_{1\perp}}+\phi_{\mathbf{k}_{2\perp}}+\phi_{\mathbf{k}_{3\perp}})]. \end{aligned} \tag{11.52}$$

The prime on the summation symbol on the right-hand side of (11.52) indicates the necessity to take all possible wave combinations which are in accordance with the condition (11.51) for various signs in front of the frequencies into account.

Equation (11.52) describes induced wave decay processes and processes of induced scattering of waves by particles which lead to the appearance of a non-linear shift in the eigenfrequencies and the non-linear Landau damping. These effects, like the linear damping, can only be described in the framework of a kinetic theory.

Three-wave surface wave decays

We consider the resonant interaction of three surface waves with frequencies $\omega_{\mathbf{k}_\perp}$, $\omega_{\mathbf{k}_{1\perp}}$, and $\omega_{\mathbf{k}_{2\perp}}$, and fixed values of the wave vectors $\mathbf{k}_\perp$, $\mathbf{k}_{1\perp}$, and $\mathbf{k}_{2\perp}$ for which the resonance conditions

$$\omega_{\mathbf{k}_{1\perp}}+\omega_{\mathbf{k}_{2\perp}} = \omega_{\mathbf{k}_\perp}, \quad \mathbf{k}_{1\perp}+\mathbf{k}_{2\perp} = \mathbf{k}_\perp \tag{11.53}$$

are satisfied. Each of the interacting waves is characterized by an energy

$$W_{\mathbf{k}_\perp} = -\frac{\varepsilon_0}{8\pi}\frac{\omega_{\mathbf{k}_\perp}}{k_\perp}\frac{\partial\zeta(\omega_{\mathbf{k}_\perp}, \mathbf{k}_\perp)}{\partial\omega_{\mathbf{k}_\perp}}|E_{\mathbf{k}_\perp}|^2. \tag{11.54}$$

The energies of the separate waves can be either positive or negative (the nature of the wave energy is determined by the sign of the derivative $\zeta'_{\mathbf{k}_\perp} \equiv \partial\zeta(\omega_{\mathbf{k}_\perp}, \mathbf{k}_\perp)/\partial\omega_{\mathbf{k}_\perp}$). For the sake of convenience we introduce the amplitudes $A_{\mathbf{k}_\perp}$ and sign factors $s_{\mathbf{k}_\perp}$ defining them by means of equations

$$A_{\mathbf{k}_\perp} \equiv [\varepsilon_0|\zeta'_{\mathbf{k}_\perp}|/8\pi\mathbf{k}_\perp]^{1/2}E_{\mathbf{k}_\perp}\exp(-i\phi_{\mathbf{k}_\perp}), \quad s_{\mathbf{k}_\perp} \equiv -\operatorname{sgn}\zeta'_{\mathbf{k}_\perp}. \tag{11.55}$$

The expressions for the energy and momentum of the waves then take the form

$$W_{\mathbf{k}_\perp} = s_{\mathbf{k}_\perp}\omega_{\mathbf{k}_\perp}|A_{\mathbf{k}_\perp}|^2, \quad P_{\mathbf{k}_\perp} = s_{\mathbf{k}_\perp}\mathbf{k}_\perp|A_{\mathbf{k}_\perp}|^2. \tag{11.56}$$

Making use of the definitions (11.55) we can write the basic equation (11.49) in the form of a Schrödinger equation in the interaction representation

$$i\frac{\partial A_{\mathbf{k}_\perp}}{\partial t} = s_{\mathbf{k}_\perp}V_{\mathbf{k}_\perp;\mathbf{k}_{1\perp},\mathbf{k}_{2\perp}}A_{\mathbf{k}_{1\perp}}A_{\mathbf{k}_{2\perp}}, \tag{11.57}$$

where $V_{\mathbf{k}_\perp;\mathbf{k}_{1\perp},\mathbf{k}_{2\perp}}$ is an interaction matrix element defined by means of

$$V_{\mathbf{k}_\perp;\mathbf{k}_{1\perp},\mathbf{k}_{2\perp}} \equiv -\left(\frac{2\pi k_{1\perp}k_{2\perp}}{\varepsilon_0 k_\perp}\right)^{1/2}\frac{\tilde{\varkappa}^{(2)}(\omega_{\mathbf{k}_{1\perp}}, \mathbf{k}_{1\perp};\, \omega_{\mathbf{k}_{2\perp}}, \mathbf{k}_{2\perp})}{|\zeta'_{\mathbf{k}_\perp}\zeta'_{\mathbf{k}_{1\perp}}\zeta'_{\mathbf{k}_{2\perp}}|^{1/2}}. \tag{11.58}$$

Within the context of the symmetry properties of the non-linear plasma susceptibility $\tilde{\varkappa}^{(2)}(\omega_{\mathbf{k}_{1\perp}}, \mathbf{k}_{1\perp};\, \omega_{\mathbf{k}_{2\perp}}, \mathbf{k}_{2\perp})$ one can easily show that the time-dependence of the amplitudes of surface waves with frequencies $\omega_{\mathbf{k}_{1\perp}}$ and $\omega_{\mathbf{k}_{2\perp}}$ is described by equations

$$\left.\begin{aligned} i\frac{\partial A_{\mathbf{k}_1-}}{\partial t} &= s_{\mathbf{k}_{1\perp}}V^*_{\mathbf{k}_\perp;\mathbf{k}_{1\perp},\mathbf{k}_{2\perp}}A_{\mathbf{k}_\perp}A^*_{\mathbf{k}_{2\perp}},\\ i\frac{\partial A_{\mathbf{k}_{2\perp}}}{\partial t} &= s_{\mathbf{k}_{2\perp}}V^*_{\mathbf{k}_\perp;\mathbf{k}_{1\perp},\mathbf{k}_{2\perp}}A_{\mathbf{k}_\perp}A_{\mathbf{k}_{1\perp}}, \end{aligned}\right\} \tag{11.59}$$

in which the same interaction matrix element occurs as in (11.57). The set of coupled equations (11.57) and (11.59) completely describe the dynamics of three interacting surface waves (Sitenko and Pavlenko, 1978).

In the case of the interaction of three surface waves with energies of the same (positive or negative) sign, i.e. when the condition

$$s_{\mathbf{k}_{1\perp}} = s_{\mathbf{k}_{2\perp}} = s_{\mathbf{k}_\perp} \tag{11.60}$$

is satisfied, a decay instability arises in the system. Let a wave of frequency $\omega_{\mathbf{k}_\perp}$ initially, at time $t=0$, be characterized by a large amplitude $|A_{\mathbf{k}_\perp}|^2 \gg |A_{\mathbf{k}_{1\perp}}|^2$ and $|A_{\mathbf{k}_\perp}|^2 \gg |A_{\mathbf{k}_{2\perp}}|^2$. As a result of resonant interactions the amplitude $A_{\mathbf{k}_\perp}$ in the first stage of the temporal evolution will change insignificantly, while the amplitudes $A_{\mathbf{k}_{1\perp}}$ and $A_{\mathbf{k}_{2\perp}}$ will exponentially increase with time. The growth rate of the waves with frequencies $\omega_{\mathbf{k}_{1\perp}}$ and $\omega_{\mathbf{k}_{2\perp}}$ is determined by the intensity of the wave of frequency $\omega_{\mathbf{k}_\perp}$:

$$|\gamma| = |V_{\mathbf{k}_\perp;\mathbf{k}_{1\perp},\mathbf{k}_{2\perp}} A_{\mathbf{k}_\perp}|. \tag{11.61}$$

The reciprocal of (11.61) determines the decay time. As an example of a decay interaction of surface waves in a semi-bounded plasma we consider the decay of a Langmuir surface wave into a Langmuir and an ion-sound surface wave, the dispersion of which is given by (11.27) and (11.29) ($ak_\perp \ll 1$). The frequencies and wave vectors of the interacting waves satisfy the decay conditions

$$\left.\begin{aligned} &\frac{\Omega}{(1+\varepsilon_0)^{1/2}}+\frac{1}{2}\left(\frac{3}{1+\varepsilon_0}\right)^{1/2}\Omega a\,|\mathbf{k}_\perp| \\ &\quad = \frac{\Omega}{(1+\varepsilon_0)^{1/2}}+\frac{1}{2}\left(\frac{3}{1+\varepsilon_0}\right)^{1/2}\Omega a\,|\mathbf{k}_{1\perp}|+|\mathbf{k}_{2\perp}|\,v_s, \\ &k_{2\perp}^2 = k_\perp^2+k_{1\perp}^2-2\,|\mathbf{k}_\perp|\,|\mathbf{k}_{1\perp}|\cos\alpha, \end{aligned}\right\} \tag{11.62}$$

where α is the angle between $\mathbf{k}_\perp$ and $\mathbf{k}_{1\perp}$. Putting $|\mathbf{k}_\perp| = |\mathbf{k}_{1\perp}| = \mathbf{k}_0$ we have $k_{2\perp} = 2k_0 \sin\frac{\alpha}{2}$ and the decay conditions are then satisfied if

$$\sin\frac{\alpha}{2} > 2\left(\frac{1+\varepsilon_0}{3}\right)^{1/2}\left(\frac{m}{M}\right)^{1/2}.$$

Making use of (11.58) and (11.46) we find the growth rate of the surface oscillations ($E_{\mathbf{k}_\perp} = E_0$):

$$\gamma^2 = \frac{1}{9}\left(\frac{eE_0}{m}\right)^2 \frac{\varepsilon_0^5}{(1+\varepsilon_0)^{5/2}} \frac{a^2k_0^5v_s}{\Omega\Omega_i^2}\sin^3\frac{\alpha}{2}. \tag{11.63}$$

The expression (11.63) for the non-linear growth rate was obtained assuming that the linear damping rates of the interacting surface waves were small compared to the corresponding eigenfrequencies. It is clear that the growth of the Langmuir and ion-sound surface waves will occur if the non-linear growth rate given by (11.63) turns out to be larger than the linear damping rates given by (11.27) and (11.29), i.e. when the conditions

$$\gamma^2 > \frac{4}{3\pi}k_0^2a^2\Omega^2,$$

$$\gamma^2 > \frac{\pi}{2}\left(\frac{m}{M}\right)^2\left[1+\left(\frac{M}{m}\right)^{1/2}\left(\frac{T_e}{T_i}\right)^{3/2}\exp\left(-\frac{T_e}{2T_i}\right)\right]^2 k_0^2a^2\Omega^2\sin^2\frac{\alpha}{2}$$

are satisfied. However, in that case it is necessary for the validity of the considerations that non-linear growth rate does not exceed in magnitude the eigenfrequencies of the growing surface waves, i.e. that the conditions

$$\gamma^2 < \Omega^2/(1+\varepsilon_0), \quad \gamma^2 < 4k_0^2 v_s^2 \sin^2 \frac{\alpha}{2}$$

are satisfied.

In the case of the decay of a Langmuir bulk wave into Langmuir and ion-sound bulk waves the growth rate is given by the expression (Oraevskii and Sagdeev, 1963)

$$\bar{\gamma}^2 = \frac{1}{8}\left(\frac{eE_0}{m}\right)^2 \frac{k_0^3 v_s}{\Omega \Omega_i^2} \sin^3 \frac{\alpha}{2} \cos^2 \alpha. \tag{11.64}$$

We note that the growth rate of the decay instability of surface waves is appreciably larger than the growth rate of the decay instability of bulk waves ($\gamma/\bar{\gamma} \sim (1/ak_0) \gg 1$).

In the case of a resonant interaction of three surface waves with different signs of the energy, for instance, when the condition

$$s_{\mathbf{k}_{1\perp}} = s_{\mathbf{k}_{2\perp}} = -s_{\mathbf{k}_\perp} \tag{11.65}$$

holds, an explosive instability occurs in the system in which the amplitudes of the interacting waves turn to infinity at some finite time t_∞. The negative-energy wave gives energy to the positive-energy waves (or the negative-energy waves give energy to positive-energy waves) and the amplitudes of the interacting waves increase without bounds, notwithstanding the conservation of the total energy of the system. By an appropriate choice of the initial conditions one can achieve that the amplitude of the (initially) strongest wave changes with time according to

$$A_{\mathbf{k}_\perp}(t) = \frac{A_{\mathbf{k}_\perp}(0)}{(1-t/t_\infty)}\,. \tag{11.66}$$

The explosive time t_∞ will be in such a case determined by the initial value of the amplitude and the matrix element of the non-linear interaction

$$t_\infty^{-1} = |V_{\mathbf{k}_\perp;\mathbf{k}_{1\perp},\mathbf{k}_{2\perp}} A_{\mathbf{k}_\perp}(0)|. \tag{11.67}$$

An explosive instability is possible in a semi-bounded plasma, when a compensated beam of ions passes through it, as the result of resonant interactions of three surface waves, the dispersion of which is given by (11.42). Putting $|\mathbf{k}_\perp| = |\mathbf{k}_{1\perp}| = k_0$ we find from the decay conditions $k_{2\perp} = -2\eta^{1/2} k_0 v_s/u$.

The explosion time t_∞ for the decay of a negative-energy wave $\omega_{\mathbf{k}_\perp}^{(2)}$ into positive-energy waves $\omega_{\mathbf{k}_{1\perp}}^{(3)}$ and $\omega_{\mathbf{k}_{2\perp}}^{(1)}$ is given by the expression

$$t_\infty^{-1} = 8\varepsilon_0^{5/2} \frac{eE_0}{mu} \eta \left(\frac{M}{m}\right)^2 \left(\frac{v_s}{u}\right)^{1/2} (k_0 a)^3. \tag{11.68}$$

We note that in the case of the explosive instability caused by the interaction of bulk waves the explosion time equals

$$\bar{t}_\infty^{-1} = \frac{1}{8^{1/2}} \frac{eE_0}{mu} \eta^{1/2} \left(\frac{M}{m}\right)^2 \left(\frac{v_s}{u}\right)^{5/2}. \tag{11.69}$$

The ratio of the explosion time for bulk waves to that for surface waves is of the order of

$$\bar{t}_\infty / t_\infty \sim \eta^{1/2}(k_0 a)^3 (u/v_s)^2. \tag{11.70}$$

It is to be observed that the characteristics (growth rate and explosion time) of the decay and explosive instabilities depend on the nature of the boundary conditions at the surface bounding the plasma. For the applicability of the specular reflection model it is necessary that all characteristic dimensions of the problem (wavelengths, Debye radii, and so on) are considerably larger than the size of the inhomogeneity of the plasma density near the boundary. As we have studied three-wave interactions of just the long-wavelength surface oscillations the choice of such a model is qualitatively justified (Ginzburg and Rukhadze, 1975; Abu-Asali, Alterkop, and Rukhadze, 1975).

Non-linear interactions of bulk and surface waves

Let us consider the decay of an s-polarized transverse bulk wave ($E_y \neq 0, E_x = E_z = 0$) into an s-polarized transverse bulk wave and a potential surface wave ($t \rightarrow t+l$ process) (Sitenko, Pavlenko, and Revenchuk, 1979). The wave vectors and phases of the interacting waves will be taken to be fixed; let the wave vectors be directed as

$$\mathbf{k} = (k_x, 0, k_0), \quad \mathbf{k}_1 = (k_{1x}, 0, k_0), \quad \mathbf{k}_{2\perp} = (k_{2x}, 0, 0)$$

The frequencies $\omega_{\mathbf{k}}$, $\omega_{\mathbf{k}_1}$, $\omega_{\mathbf{k}_{2\perp}}$ and the wave vector components along the boundary $\mathbf{k}_\perp$, $\mathbf{k}_{1\perp}$, $\mathbf{k}_{2\perp}$ of the interacting waves satisfy the decay conditions

$$\sqrt{\Omega^2+(k_x^2+k_0^2)c^2} = \sqrt{\Omega^2+(k_{1x}^2+k_0^2)c^2}+\frac{\Omega}{\sqrt{1+\varepsilon_0}}\left(1+\frac{\sqrt{3\varepsilon_0}}{2}k_{2x}a\right), \quad k_x = k_{1x}+k_{2x}. \tag{11.71}$$

Let the electric fields of the interacting waves be of the form

$$E(\mathbf{r}, t) = E_{\mathbf{k}} \cos((\mathbf{k}_\perp \cdot \mathbf{r}_\perp) - \omega_{\mathbf{k}} t - i\varphi_{\mathbf{k}}) \cos k_0 z, \tag{11.72}$$

where we must put $k_0 = 0$ for the surface waves.

We begin with the consideration of the initial stage of the decay instability development, when the amplitude of the decaying wave with wave vector $\mathbf{k}$ and frequency $\omega_{\mathbf{k}}$ is much greater than the amplitudes of the growing waves with wave vectors $\mathbf{k}_1$, $\mathbf{k}_{2\perp}$, and frequencies $\omega_{\mathbf{k}_1}$, $\omega_{\mathbf{k}_{2\perp}}$. The equation for the amplitude of the transverse wave that grows as the decay instability progresses, may be obtained from (11.36) by means of the multiple-time-scale expansion within the context of the explicit form of the fields of the interacting waves (11.72):

$$i\left(\frac{\partial}{\partial t}+\gamma_1\right)A_{\mathbf{k}_1} = s_{\mathbf{k}_1} V^*_{\mathbf{k},\mathbf{k}_1,\mathbf{k}_{2\perp}} A_{\mathbf{k}} A^*_{\mathbf{k}_{2\perp}}, \tag{11.73}$$

where the amplitude $A_{\mathbf{k}}$ and the sign factor $s_{\mathbf{k}}$ of the transverse bulk wave are determined by the equalities

$$A_{\mathbf{k}} = \frac{1}{\sqrt{16\pi}}\left[\frac{\partial}{\partial \omega_{\mathbf{k}}}(\varepsilon_t(\omega_{\mathbf{k}}, \mathbf{k})-\eta^2)\right]^{1/2} E_{\mathbf{k}} e^{-i\varphi_{\mathbf{k}}}, \quad s_{\mathbf{k}} = \operatorname{sgn}\frac{\partial}{\partial \omega_{\mathbf{k}}}[\varepsilon_t(\omega_{\mathbf{k}}, \mathbf{k})-\eta^2], \tag{11.74}$$

γ_1 is the damping rate of the collisional damping of the transverse wave with wave vector $\mathbf{k}_1$. The interaction matrix element is the following:

$$V_{\mathbf{k},\mathbf{k}_1,\mathbf{k}_{2\perp}} = \sqrt{\frac{2\varepsilon_0 k_{2\perp}^3}{\pi L}} \int dk_{1z} \int dk_{2z} \frac{\varkappa_{yyx}^{(2)}(\omega_{\mathbf{k}_1}, \mathbf{k}_1; \omega_{\mathbf{k}_{2\perp}}, \mathbf{k}_2)}{k_2^2 \varepsilon(\omega_{\mathbf{k}_{2\perp}}, \mathbf{k}_2)} \delta(k_{1z}+k_0), \tag{11.75}$$

$$L = |\Lambda'_{yy}(\omega_{\mathbf{k}}, \mathbf{k})\, \Lambda'_{yy}(\omega_{\mathbf{k}_1}, \mathbf{k}_1)\, \zeta'(\omega_{\mathbf{k}_{2\perp}}, \mathbf{k}_{2\perp})|,$$

$$\Lambda'_{ij}(\omega_{\mathbf{k}}, \mathbf{k}) \equiv \frac{\partial}{\partial \omega_{\mathbf{k}}} \Lambda_{ij}(\omega_{\mathbf{k}}, \mathbf{k}), \quad \zeta'(\omega_{\mathbf{k}_\perp}, \mathbf{k}_\perp) \equiv \frac{\partial}{\partial \omega_{\mathbf{k}_\perp}} \zeta(\omega_{\mathbf{k}_\perp}, \mathbf{k}_\perp).$$

Equation (11.72) yields also an equation for the time-dependence of the potential surface wave amplitude:

$$i\left(\frac{\partial}{\partial t}+\gamma_{\mathbf{k}_{2\perp}}\right) A_{\mathbf{k}_{2\perp}} = s_{\mathbf{k}_{2\perp}} V'^{*}_{\mathbf{k},\mathbf{k}_1,\mathbf{k}_{2\perp}} A_{\mathbf{k}} A^{*}_{\mathbf{k}_1}, \tag{11.76}$$

where $\gamma_{\mathbf{k}_{2\perp}}$ is the damping rate of the surface wave resonant damping; the amplitude $A_{\mathbf{k}_\perp}$ and the sign factor $s_{\mathbf{k}_\perp}$ of the potential surface wave are given by

$$A_{\mathbf{k}_\perp} = \sqrt{\frac{\varepsilon_0 |\zeta'(\omega_{\mathbf{k}_\perp}, \mathbf{k}_\perp)|}{8\pi k_\perp}} E_{\mathbf{k}_\perp} \exp(-i\varphi_{\mathbf{k}_\perp}), \quad s_{\mathbf{k}_\perp} = -\operatorname{sgn} \zeta'(\omega_{\mathbf{k}_\perp}, \mathbf{k}_\perp). \tag{11.77}$$

The interaction matrix element may be written as

$$V'_{\mathbf{k},\mathbf{k}_1,\mathbf{k}_{2\perp}} = \sqrt{\frac{\varepsilon_0 \pi}{2k_{2\perp} L}} \int dk_{1z} \frac{\varkappa_{xyy}^{(2)}(\omega_{\mathbf{k}_1}, \mathbf{k}_1; \omega_{\mathbf{k}_{2\perp}}, \mathbf{k}_{2\perp})}{\varepsilon(\omega_{\mathbf{k}_{2\perp}}, \mathbf{k}_{2\perp})} \delta(k_{1z}+k_0). \tag{11.78}$$

Since the interaction of high-frequency waves is considered, we may use for the nonlinear susceptibility $\varkappa_{ijl}^{(2)}(\omega_{\mathbf{k}_1}, \mathbf{k}_1; \omega_{\mathbf{k}_2}, \mathbf{k}_2)$ the cold plasma limiting value (2.30).

Having solved the obtained set of equations we come at $s_{\mathbf{k}} = s_{\mathbf{k}_2} = s_{\mathbf{k}_{2\perp}}$ to the following decay instability growth rate:

$$\gamma = -\frac{\gamma_1+\gamma_{\mathbf{k}_{2\perp}}}{2} + \sqrt{\frac{(\gamma_1+\gamma_{\mathbf{k}_{2\perp}})^2}{4} + \gamma_0^2 - \gamma_1 \gamma_{\mathbf{k}_{2\perp}}}, \tag{11.79}$$

where

$$\gamma_0^2 = V_{\mathbf{k},\mathbf{k}_1,\mathbf{k}_{2\perp}} V'^{*}_{\mathbf{k},\mathbf{k}_1,\mathbf{k}_{2\perp}} |A_{\mathbf{k}}|^2 = \left(\frac{eE_{\mathbf{k}}}{m\omega_{\mathbf{k}}}\right)^2 \frac{\sqrt{1+\varepsilon_0}}{256} \frac{k_{2x}^2 \Omega}{\omega_{\mathbf{k}_1}}. \tag{11.80}$$

It follows from (11.79) that the decay instability threshold is determined by the relation

$$\gamma_0^2 = \gamma_1 \gamma_{\mathbf{k}_{2\perp}}, \tag{11.81}$$

i.e. it is governed by the collisional damping of the transverse bulk wave.

In a collisionless plasma $\gamma_1 = 0$ and (11.79) within the context of the condition $\gamma_0 \ll \gamma_{\mathbf{k}_{2\perp}}$ reduces to the following:

$$\gamma \simeq \gamma_0^2/\gamma_{\mathbf{k}_{2\perp}}. \tag{11.82}$$

The transverse bulk waves involved in the three-wave decay process may be radiated into vacuum. Since their frequencies differ by the surface wave frequency and their wave vectors by that of the surface wave, the experimental study of such decays may be useful for the investigation of the surface-wave dispersion.

Let us consider a decay of an s-polarized transverse bulk wave of frequency $\omega_{\mathbf{k}}$ into two surface Langmuir waves with frequencies $\omega_{\mathbf{k}_{1\perp}}$ and $\omega_{\mathbf{k}_{2\perp}}(t \to l+l)$. Wave vectors and phases of all waves are fixed, the former being directed as

$$\mathbf{k} = \{k_x, 0, k_0\}, \quad \mathbf{k}_{1\perp} = \{k_{1x}, k_{1y}, 0\}, \quad \mathbf{k}_{2\perp} = \{k_{2x}, k_{2y}, 0\}, \tag{11.83}$$

where

$$|k_{1x}| = |k_{2x}|, \quad |k_{1y}| = |k_{2y}|.$$

The frequencies and the wave vector components of the interacting waves parallel to the boundary satisfy the decay conditions

$$\sqrt{\Omega^2+(k_x^2+k_0^2)c^2} = \frac{\Omega}{\sqrt{1+\varepsilon_0}}\left(1+\frac{\sqrt{3\varepsilon_0}}{2}k_{1\perp}a\right)+\frac{\Omega}{\sqrt{1+\varepsilon_0}}\left(1+\frac{\sqrt{3\varepsilon_0}}{2}k_{2\perp}a\right),$$
$$\mathbf{k}_\perp = \mathbf{k}_{1\perp}+\mathbf{k}_{2\perp}. \tag{11.84}$$

One can easily verify that these conditions are satisfied provided $k_\perp a \sim s/c$.

Let the electric fields of the interacting waves be of the form (11.72). Equations (11.36) yield the following equations for potential surface waves which grow by virtue of the decay instability under consideration:

$$i\left(\frac{\partial}{\partial t}+\gamma_{\mathbf{k}_{1,2\perp}}\right)A_{\mathbf{k}_{1,2\perp}} = s_{\mathbf{k}_{1,2\perp}}W^*_{\mathbf{k},\mathbf{k}_{1\perp},\mathbf{k}_{2\perp}}A_{\mathbf{k}}A^*_{\mathbf{k}_{2,1\perp}}, \tag{11.85}$$

where the amplitude and the sign factor of the transverse bulk wave are given by formulas (11.74); those of the surface Langmuir waves are determined by (11.77). The interaction matrix element is equal to

$$W_{\mathbf{k},\mathbf{k}_{1\perp},\mathbf{k}_{2\perp}} = \frac{\varepsilon_0}{\sqrt{\pi}}k_{1\perp}L_1^{-\frac{1}{2}}\sum_{i,j=x,y}k_{1i}k_{2j}\int dk_{1z}\int dk_{2z}\frac{\varkappa^{(2)}_{iyj}(\omega_{\mathbf{k}_{1\perp}},\mathbf{k}_1;\omega_{\mathbf{k}_{2\perp}},\mathbf{k}_2)}{k_1^2k_2^2\varepsilon(\omega_{\mathbf{k}_{1\perp}},\mathbf{k}_1)\,\varepsilon(\omega_{\mathbf{k}_{2\perp}},\mathbf{k}_2)}, \tag{11.86}$$

where

$$L_1 = |\Lambda'_{yy}(\omega_{\mathbf{k}},\mathbf{k})\,\zeta'(\omega_{\mathbf{k}_{1\perp}},\mathbf{k}_{1\perp})\,\zeta'(\omega_{\mathbf{k}_{2\perp}},\mathbf{k}_{2\perp})|.$$

The solution of the set (11.85) at $s_{\mathbf{k}} = s_{\mathbf{k}_{1\perp}} = s_{\mathbf{k}_{2\perp}}$ yields the growth rate of the instability under consideration in the form (11.79), where $\gamma_1 \to \gamma_{\mathbf{k}_{1\perp}}$ and γ_0 is given by the following expression:

$$\gamma_0^2 = \left(\frac{eE_{\mathbf{k}}}{m\omega_{\mathbf{k}}}\right)^2\frac{1+\varepsilon_0}{64}\frac{\omega_{\mathbf{k}}^2}{\Omega^2}\frac{k_{1\perp}^4k_{1y}^2}{(k_0^2+4k_{1\perp}^2)^2}. \tag{11.87}$$

The instability threshold is determined by the equality

$$\gamma_0^2 = \gamma_{\mathbf{k}_{1\perp}}\gamma_{\mathbf{k}_{2\perp}}. \tag{11.88}$$

It follows from (11.27) and (11.87) that condition (11.88) can be satisfied only if $eE_{\mathbf{k}}/(m\omega_{\mathbf{k}}s) \gtrsim 1$, i.e. the threshold of this instability is much higher than that of the $t \to t+l$ process.

11.3. Fluctuations and Kinetic Equations for Surface Waves

Fluctuations

When considering fluctuations in a non-equilibrium semi-bounded plasma it is convenient, as in the case of an unbounded plasma, to make use of the non-linear equation for the field including the fluctuation sources, which may be derived from the Maxwell equations and the equation for the microscopic density describing the particle motion in the plasma. We restrict ourselves for the sake of simplicity to a consideration of a potential field and assume the plasma in the half-space to be homogeneous and stationary. We separate the fluctuating part δf of the microscopic density and present it as a sum of the microscopic density fluctuation part in the absence of particle interaction δf^0 and the difference between the exact microscopic density and the microscopic density for non-interacting particles f. The quantity f and the microscopic field $\mathbf{E}$ are described by the equations which differ from (11.1) and (11.2) only by additional terms, respectively,

$$\frac{e}{m}\left(\mathbf{E}\cdot\frac{\partial}{\partial \mathbf{v}}\right)f^0 \quad \text{and} \quad 4\pi e\int d\mathbf{v}f^0.$$

Assuming the specular reflection conditions to be satisfied on the surface, and continuing the electric field outside the plasma as before, we obtain for f^+ an equation that is determined in the whole space. Presenting the solution of this equation in a series expansion form in the field amplitude and substituting it in the expression for the induced charge, we thus obtain the following non-linear equation for the fluctuation field:

$$\begin{aligned}
&\varepsilon(\omega, \mathbf{k})E_{\mathbf{k}\omega} + \sum_{\substack{\omega_1+\omega_2=\omega\\ \mathbf{k}_1+\mathbf{k}_2=\mathbf{k}}} \varkappa^{(2)}(\omega_1, \mathbf{k}_1; \omega_2, \mathbf{k}_2)E_{\mathbf{k}_1\omega_1}E_{\mathbf{k}_2\omega_2}\\
&\quad + \sum_{\substack{\omega_1+\omega_2+\omega_3=\omega\\ \mathbf{k}_1+\mathbf{k}_2+\mathbf{k}_3=\mathbf{k}}} \varkappa^{(3)}(\omega_1, \mathbf{k}_1; \omega_2, \mathbf{k}_2; \omega_3, \mathbf{k}_3)E_{\mathbf{k}_1\omega_1}E_{\mathbf{k}_2\omega_2}E_{\mathbf{k}_3\omega_3} + \ldots\\
&\quad + \sum_{\omega', \mathbf{k}'} \delta\varepsilon(\omega, \mathbf{k}; \omega', \mathbf{k}')E_{\mathbf{k}'\omega'} + \sum_{\substack{\omega_1+\omega_2=\omega\\ \mathbf{k}_1+\mathbf{k}_2=\mathbf{k}\\ \omega', \mathbf{k}'}} \delta\varkappa^{(2)}(\omega_1, \mathbf{k}_1; \omega_2, \mathbf{k}_2; \omega', \mathbf{k}')E_{\mathbf{k}_1\omega_1}E_{\mathbf{k}'\omega'} + \ldots\\
&\quad + \frac{2\varepsilon_0}{k}E_{\mathbf{k}_\perp\omega} = -\frac{4\pi i}{k}\varrho^0_{\mathbf{k}\omega},
\end{aligned} \tag{11.89}$$

where $\varrho^0_{\mathbf{k}\omega}$ is the fluctuation charge density due to the random motion of the individual charged particles:

$$\varrho^0_{\mathbf{k}\omega} = \sum e\int d\mathbf{v}\,\delta f^0_{\mathbf{k}\omega}. \tag{11.90}$$

($\varrho^0(z)$ is continued into the half-space outside the plasma in an even way); $\delta\varepsilon$ and $\delta\varkappa^{(2)}$ are the fluctuation variations of the dielectric permittivity and the non-linear susceptibility.

If the field intensity is weak and the non-linear effects are negligible, we obtain from (11.85) in the linear approximation

$$\varepsilon(\omega, \mathbf{k})E_{\mathbf{k}\omega}+\frac{2\varepsilon_0}{k}E_{\mathbf{k}_\perp\omega}=-\frac{4\pi i}{k}\varrho^0_{\mathbf{k}\omega} \tag{11.91}$$

and the surface fluctuation field $E_{\mathbf{k}_\perp\omega}$ satisfies the following equation:

$$\zeta(\omega, \mathbf{k}_\perp)E_{\mathbf{k}_\perp\omega}=-4\pi i k_\perp\sum_{k_z}\frac{\varrho^0_{\mathbf{k}\omega}}{k^2\varepsilon(\omega, \mathbf{k})}\,. \tag{11.92}$$

The relation (11.92) connects the fluctuation field in the plasma with the distribution function fluctuations in the absence of particle interactions. Making use of this relation, we can express the field correlation functions in the plasma directly in terms of the correlation functions for the system of non-interacting particles.

The spectral distribution of the fluctuations of the distribution function in neglect of particle interactions (but taking into account specular reflection from the boundary) is determined by the formula:

$$\begin{aligned}\langle\delta f^0_{k_z}(\mathbf{v})\,\delta f^0_{k'_z}(\mathbf{v}')\rangle_{\mathbf{k}_\perp\omega}=(2\pi)^2\,\delta(\mathbf{v}_\perp-\mathbf{v}'_\perp)\,\{\delta(v_z-v'_z)\,\delta(k_z-k'_z)\\+\delta(v_z+v'_z)\,\delta(k_z+k'_z)\}\,\delta(\omega-(\mathbf{k}\cdot\mathbf{v}))\,f_0(\mathbf{v}).\end{aligned} \tag{11.93}$$

The first term on the right-hand side of (11.93) describes fluctuations in an unbounded system in the absence of particle interactions; the second term is due to specular reflection of particles from the boundary. Integrating (11.93) over velocities, we find the spectral distribution of the charge density fluctuations in neglect of the interaction between particles (but taking into account reflection from the boundary) to be of the form

$$\langle\varrho_{k_z}\varrho_{k'_z}\rangle^0_{\mathbf{k}_\perp\omega}=4\pi^2\sum e^2\int d\mathbf{v}\,\delta(\omega-(\mathbf{k}\cdot\mathbf{v}))\,f_0(\mathbf{v})\,\{\delta(k_z-k'_z)+\delta(k_z+k'_z)\}. \tag{11.94}$$

Within the context of (11.92) we find the distribution of the electric field surface fluctuations:

$$\langle E^2\rangle_{\mathbf{k}_\perp\omega}=\frac{16\pi^2k_\perp^2}{|\zeta(\omega, \mathbf{k}_\perp)|^2}\sum_{k_z, k'_z}\frac{\langle\varrho_{k_z}\varrho_{k'_z}\rangle^0_{\mathbf{k}_\perp\omega}}{k^2k'^2\varepsilon(\omega, \mathbf{k})\,\varepsilon^*(\omega, \mathbf{k}')}\,. \tag{11.95}$$

Expressing the surface component of the field $E_{\mathbf{k}_\perp\omega}$ in (11.91) in terms of the fluctuation charge density $\varrho^0_{\mathbf{k}\omega}$, we obtain, as before, the total spectral distribution of the field fluctuations in a semi-bounded plasma:

$$\begin{aligned}\langle E_{k_z}E_{k'_z}\rangle_{\mathbf{k}_\perp\omega}=\frac{16\pi^2}{kk'\varepsilon(\omega, \mathbf{k})\,\varepsilon^*(\omega, \mathbf{k}')}\Bigg\{\langle\varrho_{k_z}\varrho_{k'_z}\rangle^0_{\mathbf{k}_\perp\omega}-\frac{2\varepsilon_0k_\perp}{\zeta(\omega, \mathbf{k}_\perp)}\sum_{k''_z}\frac{\langle\varrho_{k''_z}\varrho_{k'_z}\rangle^0_{\mathbf{k}_\perp\omega}}{k''^2\varepsilon(\omega, \mathbf{k}'')}\\-\frac{2\varepsilon_0k_\perp}{\zeta^*(\omega, \mathbf{k}_\perp)}\sum_{k''_z}\frac{\langle\varrho_{k_z}\varrho_{k''_z}\rangle^0_{\mathbf{k}_\perp\omega}}{k''^2\varepsilon^*(\omega, \mathbf{k}'')}+\frac{4\varepsilon_0^2k_\perp^2}{|\zeta(\omega, \mathbf{k}_\perp)|^2}\sum_{k_z, k'_z}\frac{\langle\varrho_{k_z}\varrho_{k'_z}\rangle^0_{\mathbf{k}_\perp\omega}}{k^2k'^2\varepsilon(\omega, \mathbf{k})\,\varepsilon^*(\omega, \mathbf{k}')}\Bigg\}.\end{aligned} \tag{11.96}$$

The first term on the right-hand side of (11.96) describes fluctuations due to bulk field oscillations in a semi-bounded plasma. As this term contains delta-functions of the difference or the sum of k_z and k_z', it exceeds all other terms in (11.96) in the relevant range of k_z and k_z' values (bulk fluctuation domain). Thus the spectral distribution of the bulk fluctuations of the electric field in a semi-bounded plasma may be written approximately in the form:

$$\langle E_{k_z} E_{k_z'} \rangle_{\mathbf{k}_\perp \omega} = \frac{16\pi}{kk'} \frac{\langle \varrho_{k_z} \varrho_{k_z'} \rangle^0_{\mathbf{k}_\perp \omega}}{\varepsilon(\omega, \mathbf{k})\, \varepsilon^*(\omega, \mathbf{k}')} . \tag{11.97}$$

We note that, by integrating (11.96) over k_z and k_z' components, we immediately obtain the spectral distributions of the surface fluctuations (11.95).

Expressions (11.95) and (11.97) are valid for the description of surface and bulk fluctuations of the electric field both in thermal equilibrium and in non-equilibrium (but stationary and stable) plasmas. Expressions (11.95) and (11.97) can be considerably simplified for a plasma in thermal equilibrium. Upon noting that, in an equilibrium plasma,

$$\langle \varrho_{k_z} \varrho_{k_z'} \rangle^0_{\mathbf{k}_\perp \omega} = k^2 \{\delta(k_z - k_z') + \delta(k_z + k_z')\} \frac{T}{\omega} \operatorname{Im} \varepsilon(\omega, \mathbf{k}), \tag{11.98}$$

we can present the spectral distributions of surface and bulk fluctuations in the form

$$\langle E^2 \rangle_{\mathbf{k}_\perp \omega} = \frac{8\pi k_\perp}{\varepsilon_0} \frac{T}{\omega} \frac{\operatorname{Im} \zeta^*(\omega, \mathbf{k}_\perp)}{|\zeta(\omega, \mathbf{k}_\perp)|^2} , \tag{11.99}$$

$$\langle E_{k_z} E_{k_z'} \rangle_{\mathbf{k}_\perp \omega} = 16\pi^2 \frac{T}{\omega} \frac{\operatorname{Im} \varepsilon(\omega, \mathbf{k})}{|\varepsilon(\omega, \mathbf{k})|^2} \{\delta(k_z - k_z') + \delta(k_z + k_z)\}. \tag{11.100}$$

In the spectra of surface and bulk fluctuations, in addition to a broad maximum in the low-frequency range, which is due to the random motion of charged particles, there are present also sharp maxima corresponding to the collective surface or bulk fluctuation oscillations of the field. The spectral distributions of surface and bulk fluctuations near the eigenfrequencies are determined by the following expressions:

$$\langle E^2 \rangle_{\mathbf{k}_\perp \omega} = \pi I_{\mathbf{k}_\perp} \{\delta(\omega - \omega_{\mathbf{k}_\perp}) + \delta(\omega + \omega_{\mathbf{k}_\perp})\}, \quad I_{\mathbf{k}_\perp} = \frac{8\pi}{\varepsilon_0} \frac{T}{\omega_{\mathbf{k}_\perp} \zeta'_{\mathbf{k}_\perp}} , \tag{11.101}$$

$$\langle E_{k_z} E_{k_z'} \rangle_{\mathbf{k}_\perp \omega} = 2\pi^2 I_{\mathbf{k}} \{\delta(\omega - \omega_{\mathbf{k}}) + \delta(\omega + \omega_{\mathbf{k}})\} \{\delta(k_z - k_z') + \delta(k_z + k_z')\}, \quad I_{\mathbf{k}} = 8\pi \frac{T}{\omega_{\mathbf{k}} \varepsilon'_{\mathbf{k}}} . \tag{11.102}$$

It is to be observed that the intensities of the fluctuation oscillations in a non-equilibrium plasma may considerably differ fom the thermal level. The intensities may be appreciably larger than (11.101) and (11.102) if the plasma state is close to the kinetic instability threshold. In such a case the non-linear interactions of the fluctuation oscillations must be taken into account.

We separate in (11.85) the part corresponding to the non-linear interactions of surface waves. As a result the equation becomes of the form

$$\begin{aligned}
\zeta(\omega, \mathbf{k}_\perp)E_{\mathbf{k}_\perp\omega} &+ \sum_{\substack{\omega_1+\omega_2=\omega \\ \mathbf{k}_{1\perp}+\mathbf{k}_{2\perp}=\mathbf{k}_\perp}} \tilde{\varkappa}^{(2)}(\omega_1, \mathbf{k}_{1\perp}; \omega_2, \mathbf{k}_{2\perp})E_{\mathbf{k}_{1\perp}\omega_1}E_{\mathbf{k}_{2\perp}\omega_2} \\
&+ \sum_{\substack{\omega_1+\omega_2+\omega_3=\omega \\ \mathbf{k}_{1\perp}+\mathbf{k}_{2\perp}+\mathbf{k}_{3\perp}=\mathbf{k}_\perp}} \tilde{\varkappa}^{(3)}(\omega_1, \mathbf{k}_{1\perp}; \omega_2, \mathbf{k}_{2\perp}; \omega_3, \mathbf{k}_{3\perp})E_{\mathbf{k}_{1\perp}\omega_1}E_{\mathbf{k}_{2\perp}\omega_2}E_{\mathbf{k}_{3\perp}\omega_3} + \dots \\
&+ \sum_{\omega', \mathbf{k}'_\perp} \delta\zeta(\omega, \mathbf{k}_\perp; \omega', \mathbf{k}'_\perp)E_{\mathbf{k}'_\perp\omega'} \\
&+ \sum_{\substack{\omega_1+\omega_2=\omega \\ \mathbf{k}_{1\perp}+\mathbf{k}_{2\perp}=\mathbf{k}_\perp \\ \omega', \mathbf{k}'_\perp}} \delta\tilde{\varkappa}^{(2)}(\omega_1, \mathbf{k}_{1\perp}; \omega_2, \mathbf{k}_{2\perp}; \omega', \mathbf{k}'_\perp)E_{\mathbf{k}_{1\perp}\omega_1}E_{\mathbf{k}'_\perp\omega'} + \dots \\
&= -4\pi i k_\perp \sum_{k'_z} \frac{\overset{0}{\varrho}_{\mathbf{k}\omega}}{k^2\varepsilon(\omega, \mathbf{k})}, \qquad (11.103)
\end{aligned}$$

where

$$\delta\zeta(\omega, \mathbf{k}_\perp; \omega', \mathbf{k}'_\perp) = -\sum_{k_z, k'_z} \frac{2\varepsilon_0 k_\perp}{kk'} \frac{\delta\varepsilon(\omega, \mathbf{k}; \omega', \mathbf{k}')}{\varepsilon(\omega, \mathbf{k})\,\varepsilon(\omega', \mathbf{k}')},$$

and so on. In order to describe the fluctuation field, taking the non-linear interaction into account, it is necessary in the general case to find, besides the quadratic correlation function also the higher-order correlators (i.e. the third and the fourth order in the accepted approximation). Multiplying successively the left- and right-hand sides of (11.103) by themselves, we can obtain a set of inhomogeneous integral equations which determine a set of correlation functions of interest. In particular, we obtain the following equation for the second-order correlation function

$$\begin{aligned}
\zeta(\omega, \mathbf{k}_\perp)\langle E^2\rangle_{\mathbf{k}_\perp\omega} &- \frac{2}{\zeta^*(\omega, \mathbf{k}_\perp)} \sum_{\omega', \mathbf{k}'_\perp} |\varkappa^{(2)}(\omega-\omega', \mathbf{k}_\perp-\mathbf{k}'_\perp; \omega', \mathbf{k}'_\perp)|^2 \langle E^2\rangle_{\mathbf{k}'_\perp\omega'}\langle E^2\rangle_{\mathbf{k}_\perp-\mathbf{k}'_\perp, \omega-\omega'} \\
&- \sum_{\omega', \mathbf{k}'} \tilde{a}(\omega, \mathbf{k}_\perp; \omega', \mathbf{k}'_\perp)\langle E^2\rangle_{\mathbf{k}'_\perp\omega'}\langle E^2\rangle_{\mathbf{k}_\perp\omega} \\
&= \frac{1}{\zeta^*(\omega, \mathbf{k}_\perp)}\left\{\sum_{\omega', \mathbf{k}'_\perp} \tilde{b}(\omega, \mathbf{k}_\perp; \omega', \mathbf{k}'_\perp)\langle E^2\rangle_{\mathbf{k}'_\perp\omega'} + \tilde{q}_{\mathbf{k}_\perp\omega}\right\}, \qquad (11.104)
\end{aligned}$$

where the quantity $\tilde{a}$ is expressed in terms of the non-linear susceptibilities

$$\begin{aligned}
\tilde{a}(\omega, \mathbf{k}_\perp; \omega', \mathbf{k}'_\perp) = 4 &\frac{\tilde{\varkappa}^{(2)}(\omega-\omega', \mathbf{k}_\perp-\mathbf{k}'_\perp; \omega', \mathbf{k}'_\perp)\,\tilde{\varkappa}^{(2)}(\omega, \mathbf{k}_\perp; -\omega', -\mathbf{k}'_\perp)}{\zeta(\omega-\omega', \mathbf{k}_\perp-\mathbf{k}'_\perp)} \\
&+2\frac{\tilde{\varkappa}^{(2)}(\omega, \mathbf{k}_\perp; 0, 0)\,\tilde{\varkappa}^{(2)}(\omega', \mathbf{k}'_\perp; -\omega', -\mathbf{k}'_\perp)}{\zeta(0, 0)} \\
&-3\tilde{\varkappa}^{(3)}(\omega, \mathbf{k}'; \omega, \mathbf{k}_\perp; -\omega', -\mathbf{k}'_\perp) \qquad (11.105)
\end{aligned}$$

and the quantities $\tilde{b}$ and $\tilde{q}$ are expressed in terms of the correlation functions for the fluctuations sources

$$\tilde{q}_{\mathbf{k}_\perp\omega} = 16\pi^2 k_\perp^2 \sum_{k_z, k'_z} \frac{\langle \varrho_{k_z}\varrho_{k'_z}\rangle^0_{\mathbf{k}_\perp\omega}}{k^2k'^2\varepsilon(\omega, \mathbf{k})\,\varepsilon(\omega, \mathbf{k}')} + \dots \,. \qquad (11.106)$$

Using (11.104) we can find the spectral distribution of the fluctuation field taking into account the non-linear interactions of surface waves. These interactions, in particular, cause additional maxima in the spectrum at the combination frequencies as well as saturation of the critical surface fluctuations in a non-equilibrium plasma by virtue of non-linear shifts of the eigenfrequencies of the surface oscillations. Equation (11.104) which determines the spectral distribution of the electric field surface fluctuations, was derived within the context of the assumption that these distributions are stationary. However, under realistic conditions, when we take into account the non-linear interactions of surface waves together with the linear damping or growth of the oscillations, a possibility may arise that the spectral distributions of the field fluctuations vary in time. If the particle distributions are stationary, the equation governing the time-evolution of the distribution of surface fluctuations may be derived from (11.104) by the substitution:

$$\zeta(\omega, \mathbf{k}_\perp)\langle E^2\rangle_{\mathbf{k}_\perp\omega} \to \left\{\zeta(\omega, \mathbf{k}_\perp)+\frac{i}{2}\,\frac{\partial\,\mathrm{Re}\,\zeta(\omega, \mathbf{k}_\perp)}{\partial\omega}\,\frac{\partial}{\partial t}\right\}\langle E^2\rangle_{\mathbf{k}_\perp\omega} \tag{11.107}$$

and taking the imaginary part. As a result we obtain

$$\begin{aligned}
\frac{1}{2}\,\frac{\partial\,\mathrm{Re}\,\zeta(\omega, \mathbf{k}_\perp)}{\partial\omega}\,\frac{\partial}{\partial t}\langle E^2\rangle_{\mathbf{k}_\perp\omega} &= -\mathrm{Im}\,\zeta(\omega, \mathbf{k}_\perp)\langle E^2\rangle_{\mathbf{k}_\perp\omega}\\
&+2\,\mathrm{Im}\,\frac{1}{\zeta^*(\omega, \mathbf{k}_\perp)}\sum_{\omega', \mathbf{k}'_\perp}|\tilde{\varkappa}^{(2)}(\omega-\omega', \mathbf{k}_\perp-\mathbf{k}'_\perp;\,\omega', \mathbf{k}'_\perp)|^2\langle E^2\rangle_{\mathbf{k}'_\perp\omega}\langle E^2\rangle_{\mathbf{k}_\perp-\mathbf{k}'_\perp,\,\omega-\omega'}\\
&+\mathrm{Im}\sum_{\omega', \mathbf{k}'_\perp}\tilde{a}(\omega, \mathbf{k}_\perp;\,\omega', \mathbf{k}'_\perp)\langle E^2\rangle_{\mathbf{k}'_\perp\omega'}\langle E^2\rangle_{\mathbf{k}_\perp\omega}\\
&+\mathrm{Im}\,\frac{1}{\zeta^*(\omega, \mathbf{k}_\perp)}\left\{\sum_{\omega', \mathbf{k}'_\perp}\tilde{b}(\omega, \mathbf{k}_\perp;\,\omega', \mathbf{k}'_\perp)\langle E^2\rangle_{\mathbf{k}'_\perp\omega'}+\tilde{q}_{\mathbf{k}_\perp\omega}\right\}.
\end{aligned} \tag{11.108}$$

This equation describes the time-dependence of the spectral density $\langle E^2\rangle_{\mathbf{k}_\perp\omega}$ due to linear dissipation and non-linear wave interactions. One can easily derive a kinetic equation for surface waves in a semi-bounded plasma on the basis of (11.108).

Kinetic equation for surface waves

The general solution of (11.108) in the linear approximation is

$$\langle E^2\rangle_{\mathbf{k}_\perp\omega} = \langle E^2\rangle^0_{\mathbf{k}_\perp\omega}+\pi I_{\mathbf{k}_\perp}(t)\{\delta(\omega-\omega_{\mathbf{k}_\perp})+\delta(\omega+\omega_{\mathbf{k}_\perp})\}, \tag{11.109}$$

where the first term, which is determined by the inhomogeneous part of (11.108), characterizes the stationary level of surface fluctuations, and the second term is associated with surface eigenoscillations of the electric field due to the initial conditions. The stationary level of surface fluctuations in an equilibrium plasma is governed by the temperature, therefore we may neglect the thermal oscillations provided the level of the induced oscillations is sufficiently high.

Substituting the spectral distribution (11.109) in (11.108) we obtain the following equation for the intensity of oscillations of a certain type $I_{\mathbf{k}_\perp}$:

$$\frac{\partial I_{\mathbf{k}_\perp}}{\partial t} = -2\gamma_{\mathbf{k}_\perp} I_{\mathbf{k}_\perp} + \frac{2\pi}{(\zeta'_{\mathbf{k}_\perp})^2} \sum_{\mathbf{k}'_\perp} |\varkappa^{(2)}(\omega_{\mathbf{k}_\perp} \mp \omega_{\mathbf{k}'_\perp}, \mathbf{k}_\perp - \mathbf{k}'_\perp; \pm\omega_{\mathbf{k}'_\perp}, \mathbf{k}'_\perp)|^2$$

$$\times \delta(\omega_{\mathbf{k}_\perp} \mp \omega_{\mathbf{k}'_\perp} \mp \omega_{\mathbf{k}_\perp - \mathbf{k}'_\perp}) I_{\mathbf{k}'_\perp} I_{\mathbf{k}_\perp - \mathbf{k}'_\perp} + \frac{1}{\zeta'_{\mathbf{k}_\perp}} \operatorname{Im} \sum_{\mathbf{k}'_\perp} \tilde{a}(\omega_{\mathbf{k}_\perp}, \mathbf{k}_\perp; \pm\omega_{\mathbf{k}'_\perp}, \mathbf{k}'_\perp) I_{\mathbf{k}'_\perp} I_{\mathbf{k}_\perp}$$

$$+ \frac{2}{\zeta'_{\mathbf{k}_\perp}} \operatorname{Im} \sum_{\omega', \mathbf{k}'_\perp} \tilde{a}(\omega_{\mathbf{k}_\perp}, \mathbf{k}_\perp; \omega', \mathbf{k}'_\perp) \langle E^2 \rangle^0_{\mathbf{k}'_\perp \omega'} I_{\mathbf{k}_\perp}$$

$$+ \frac{1}{(\zeta'_{\mathbf{k}_\perp})^2} \sum_{\mathbf{k}'_\perp} \{\tilde{b}(\omega_{\mathbf{k}_\perp}, \mathbf{k}_\perp; \pm\omega_{\mathbf{k}'_\perp}, \mathbf{k}'_\perp)$$

$$+ 4|\varkappa^{(2)}(\omega_{\mathbf{k}_\perp} \mp \omega_{\mathbf{k}'_\perp}, \mathbf{k}_\perp - \mathbf{k}'_\perp; \pm\omega_{\mathbf{k}'_\perp}, \mathbf{k}'_\perp)|^2 \langle E^2 \rangle^0_{\mathbf{k}_\perp - \mathbf{k}'_\perp, \omega_{\mathbf{k}_\perp} - \omega_{\mathbf{k}'_\perp}}\} I_{\mathbf{k}} \,. \quad (11.110)$$

The last equation describes the dynamics of surface waves taking into account the interactions both of the surface waves with each other and with surface fluctuations in the plasma. If we neglect in (11.110) the intensity of the fluctuation oscillations $\langle E^2 \rangle^0_{\mathbf{k}_\perp \omega}$ and the quantity $\tilde{b}(\omega_{\mathbf{k}_\perp}, \mathbf{k}_\perp; \omega_{\mathbf{k}'_\perp}, \mathbf{k}'_\perp)$ associated with the fluctuations of the particle distributions, we obtain a kinetic equation for surface waves:

$$\frac{\partial I_{\mathbf{k}}}{\partial t} = -2\gamma_{\mathbf{k}_\perp} I_{\mathbf{k}_\perp} + \frac{2\pi}{(\zeta'_{\mathbf{k}_\perp})^2} \sum_{\mathbf{k}'_\perp} |\tilde{\varkappa}^{(2)}(\omega_{\mathbf{k}_\perp} \mp \omega_{\mathbf{k}'_\perp}, \mathbf{k}_\perp - \mathbf{k}'_\perp; \pm\omega_{\mathbf{k}'_\perp}, \mathbf{k}'_\perp)|^2$$

$$\times \delta(\omega_{\mathbf{k}_\perp} \mp \omega_{\mathbf{k}'_\perp} \mp \omega_{\mathbf{k}_\perp - \mathbf{k}'_\perp}) I_{\mathbf{k}_\perp} I_{\mathbf{k}_\perp - \mathbf{k}'_\perp} + \frac{1}{\zeta'_{\mathbf{k}_\perp}} \operatorname{Im} \sum_{\mathbf{k}'_\perp} \tilde{a}(\omega_{\mathbf{k}_\perp}, \mathbf{k}_\perp; \pm\omega_{\mathbf{k}'_\perp}, \mathbf{k}'_\perp) I_{\mathbf{k}'_\perp} I_{\mathbf{k}_\perp}. \quad (11.111)$$

The kinetic equation (11.111) describes the changes in the spectral density of surface waves $I_{\mathbf{k}_\perp}$ caused by linear dissipation and non-linear wave–wave and wave-particle interactions. Three-wave decay processes, i.e. the transformation of two waves into one and the decay of a wave into two others, are also taken into account in (11.111). Besides, induced scattering of waves by particles is described, which causes an additional damping of waves—non-linear Landau damping.

As already mentioned, the kinetic equation (11.111) is adequate only when the wave intensities are large enough and the fluctuation oscillations in the plasma may be neglected. That is why wave scattering and transformation due to interactions with the fluctuation fields are not described by (11.111). In contrast to (11.111), the kinetic equation (11.110) does take into account interactions between the waves and the fluctuation fields and therefore may be used for the description of wave scattering and transformation by the fluctuations in a semi-bounded plasma. Equation (11.108) is also suitable for the investigation of the scattering of charged particles by electric field fluctuations accompanied by bremsstrahlung. That is why, using the extended equation (11.108), taking into account both potential and vorticity electric fields, we can study spontaneous emission from the plasma into the surrounding medium. This radiation may be very different from thermal emission in a non-equilibrium plasma.

References[†]

ABU-ASALI, E. A., ALTERKOP, B. A., and RUKHADZE, A. A. (1975) Nonlinear ion-acoustic oscillations of a nonisothermal current carrying plasma close to the instability threshold, *Plasma Phys.* **17,** 189.

AKHIEZER, A. I., AKHIEZER, I. A., POLOVIN, R. V., SITENKO, A. G., and STEPANOV, K. N. (1967) *Collective Oscillations in a Plasma*, Pergamon Press, Oxford.

AKHIEZER, A. I., AKHIEZER, I. A., POLOVIN, R. V., SITENKO A. G., and STEPANOV, K. N. (1975) *Plasma Electrodynamics*, Pergamon Press, Oxford.

AKHIEZER, A. I., AKHIEZER, I. A., and SITENKO, A. G. (1961) On the theory of fluctuations in a plasma, *Soviet Phys. JETP* **14,** 462.

AKHIEZER, A. I. and LYUBARSKII, G. YA. (1951) Nonlinear theory of oscillations of an electron plasma, *Dokl. Akad. Nauk SSSR* **80,** 193.

AKHIEZER, A. I. and POLOVIN, R. V. (1955) On relativistic plasma oscillations, *Dokl. Akad. Nauk SSSR* **102,** 919.

AKHIEZER, A. I., PROKHODA, I. G., and SITENKO, A. G. (1958) Electromagnetic wave scattering in plasma, *Soviet Phys. JETP* **6,** 576.

AKHIEZER, I. A. and BOLOTIN, YU. L. (1963) Interaction of waves and particles with a nonequilibrium plasma, *Nucl. Fusion* **3,** 271.

AKHIEZER, I. A., DANELIYA, I. A., and TSYNTSADZE, N. L. (1964) On the theory of scattering and transformation of electromagnetic waves in non-equilibrium plasma, *Soviet Phys. JETP* **19,** 208.

AKOPIAN, A. V. and TSYTOVICH, V. N. (1975) Bremsstrahlung in a nonequilibrium plasma, *Sov. J. Plasma Physics* **1,** 371.

ALLIS, W. P., BUCHSBAUM, S. J., and BERS, A. (1963) *Waves in Anisotropic Plasmas*, MIT Press, Cambridge, Mass.

AL'TSHUL, L. M. and KARPMAN, V. I. (1965) Kinetics of waves in a weakly turbulent plasma, *Soviet Phys, JETP* **20,** 1043.

ANDREEV, N. E., KIRII, A. YU., and SILIN, V. P. (1970) Parametric excitation of longitudinal echo oscillations in a plasma by a weak highfrequency electric field, *Soviet Phys. JETP* **30,** 866.

ARMSTRONG, J. A., BLOEMBERGEN, N., DUCUING, J., and PERSHAN, P. S. (1962) Interactions between light waves in a nonlinear dielectric, *Phys. Rev.* **127,** 1918.

BACHMANN, P., SAUER, K., and WALLIS, G. (1972) Zeitliche Phasenmischungsechos, *Fortschritte der Physik* **20,** 147.

BALESCU, R. (1960) Irreversible processes in ionized gases, *Phys. Fluids.* **3,** 52.

BALESCU , R. (1963) *Statistical Mechanics of Charged Particles*, Interscience, New York.

BEKEFI, G. (1966) *Radiation Processes in Plasmas*, Wiley, New York.

BEKEFI, G., HIRSHFIELD, J. L., and BROWN, S. C. (1961) Kirchhoff's radiation law for plasmas with non-Maxwellian distributions, *Phys. Fluids* **4,** 173.

BERNARD, W. and CALLEN, H. B. (1960) Irreversible thermodynamics of a nonlinear R–C system, *Phys. Rev.* **118,** 1466.

BERNSTEIN, I. B. (1958) Waves in a plasma in a magnetic field, *Phys. Rev.* **109,** 10.

BERNSTEIN, I. B., GREENE, J. M., and KRUSKAL M. D. (1957) Exact nonlinear plasma oscillations, *Phys. Rev.* **108,** 546.

BIRMINGHAM, T., DAWSON, J., and OBERMAN, C. (1965) Radiation processes in plasmas, *Phys. Fluids* **8,** 297.

BOGDANKEVICH, L. S., RUKHADZE, A. A., and SILIN, V. P. (1962) On electromagnetic field fluctuations in nonequilibrium plasma, *Radiofizika* **5,** 1093.

[†] The originals of many papers are written in Russian. The English translations are given where known—(*Translator*).

BOGOLIUBOV, N. N. (1962) Problems of a dynamical theory in statistical physics, in *Studies in Statistical Mechanics*, vol. 1 (ed. J. de Boer and G. E. Uhlenbeck), Wiley, New York.

BOLTZMANN, L. (1964) *Lectures on Gas Theory*, University of California Press, Berkeley.

BRIGGS, R. J. (1964) *Electron-stream Interaction with Plasmas*, MIT Press, Cambridge.

BUNKIN, F. V. (1962) Phenomenological theory of thermal fluctuations in macroscopic systems, *Radiofizika* **5,** 83.

BYCHENKOV, V. YU., PUSTOVALOV, V. V., SILIN, V. P., and TIKHONCHUK, V. T. (1976) Saturation of plasma turbulence by nonlinear frequency shifts, *Sov. J. Plasma Physics* **2,** 251.

CALLEN, H. B. and WELTON, T. A. (1951) Irreversibility and generalized noise, *Phys. Rev.* **83,** 34.

CANUTO, V. and CHIU, H. (1970) Nonrelativistic electron bremsstrahlung in a strongly magnetized plasma, *Phys. Rev.* **A2,** 518.

CHANDRASEKHAR, S. (1943) Stochastic problems in physics and astronomy, *Rev. Mod. Phys.* **15,** 1.

COHEN, R., SPITZER, L., and ROUTLY, P. (1950) The electrical conductivity of an ionized gas, *Phys. Rev.* **80,** 230.

COOK, I. and TAYLOR, J. B. (1973) Electric field fluctuations in turbulent plasmas, *J. Plasma Phys.* **9,** 131.

COPPI, B., ROSENBLUTH, M. N., and SUDAN, R. N. (1969) Nonlinear interactions of positive and negative energy modes in rarefied plasmas, 1, 2. *Annals Phys.* **55,** 207; 248.

DAVIDSON, R. C. (1968) Resonant four-wave interaction of electron-plasma oscillations, *Phys. Rev.* **176,** 344.

DAVIDSON, R. C. (1972) *Methods in Nonlinear Plasma Theory*, Academic Press, New York.

DAWSON, J. M. (1968) Radiation from a plasma, in *Advances in Plasma Physics*, vol. 1 (ed. by A. Simon and W. B. Tompson), Wiley, New York.

DAWSON, J. and OBERMAN, C. (1962) High-frequency conductivity and the emission and absorption coefficients of a fully ionized plasma, *Phys. Fluids* **5,** 517.

DAWSON, J. and OBERMAN, C. (1963) Effect of ion correlations on high-frequency conductivity, *Phys. Fluids* **6,** 394.

DIKASOV, V. M., RUDAKOV, L. I., and RYUTOV, D. D. (1965) On the interaction of negative energy waves in a weakly turbulent plasma, *Soviet Phys. JETP* **21,** 608.

DNESTROVSKII, YU. N., and KOSTOMAROV, D. P. (1961) On the dispersion equation for an ordinary wave propagating in a plasma transverse to the external magnetic field, *Soviet Phys. JETP* **13,** 986.

DOUGHERTY, J. P. and FARLEY, D. T. (1960) A theory of incoherent scattering of radio waves by a plasma, *Proc. Roy. Soc.* **A259,** 79.

DRAKE, J. F. KAW, P. K., LEE, Y. C., SCHMIDT, G., LIU, C. S., and ROSENBLUTH, M. N. (1974) Parametric instabilities of electromagnetic waves in plasmas, *Phys. Fluids* **17,** 778.

DRUMMOND, W. E. (1958) Basic microwave properties of a hot magnetoplasma, *Phys. Rev.* **110,** 293.

DRUMMOND, W. E. and PINES, D. (1962) Non-linear stability of plasma oscillations, *Nucl. Fusion* **3,** 1049.

DRUMMOND, W. E. and PINES, D. (1964) Nonlinear plasma oscillations, *Annals Phys.* **28,** 478.

DUBOIS, D. F. and GOLDMAN, M. V. (1967) Parametrically excited plasma fluctuations, *Phys. Rev.* **164,** 207.

DUPREE, T. H. (1963) Kinetic theory of plasma and electromagnetic field, *Phys. Fluids* **6,** 1714.

DUPREE, T. H. (1966) A perturbation theory for strong plasma turbulence, *Phys. Fluids* **9,** 1773.

DUPREE, T. H. (1967) Nonlinear theory of drift-wave turbulence and enhanced diffusion, *Phys. Fluids* **10,** 1049.

DUPREE, T. H. (1972) Theory of phase space density fluctuations in plasma, *Phys. Fluids* **75,** 334.

ECKER, G. (1972) *Theory of Fully Ionized Plasmas*, Academic Press, New York.

ELEONSKII, V. M., ZYRIANOV, P. S., and SILIN, V. P. (1962) Collision integral of charged particles in a magnetic field, *Soviet Phys. JETP* **15,** 619.

FAINBERG, YA. B. (1956) The use of plasma waveguides as accelerating structures in linear accelerators, *Proc. Symposium CERN* **1,** 84.

FRIEMAN, E. A. (1963) On a new method in the theory of irreversible processes, *J. Math. Phys.* **4,** 410.

FUSE, M. and ICHIMARU, S. (1975) Theory of surface collective modes in semi-infinite plasmas, *J.Phys. Soc. Japan* **38,** 559.

GALEEV, A. A. and KARPMAN, V. I. (1963) Turbulent theory of a weakly nonequilibrium rarefied plasma and structure of shock waves, *Soviet Phys. JETP* **17,** 403.

GALEEV, A. A., KARPMAN, V. I., and SAGDEEV, R. Z. (1965) Many-particle aspects of turbulent plasma theory, *Nucl. Fusion* **5,** 20.

GALEEV, A. A. and SAGDEEV, R. Z. (1973) Non-linear plasma theory, *Rev. Plasma Phys.* **7,** 1.

GALEEV, A. A., SAGDEEV, R. Z., SHAPIRO, V. D., and SHEVCHENKO, V. I. (1976) Non-linear theory of modulation instability of Langmuir waves, in *Problems of Plasma Theory*, p. 154, Naukova Dumka, Kiev.

GAPONOV, A. V. and MILLER, M. A. (1958) Potential wells for charged particles in a high-frequency electromagnetic field, *Soviet Phys. JETP* **7**, 185.

GIBBS, J. W. (1902) *Elementary Principles in Statistical Mechanics*, Scribner, New York.

GINZBURG, V. L. (1970) *Propagation of Electromagnetic Waves in a Plasma*, Pergamon Press, Oxford.

GINZBURG, V. L. (1975) *Theoretical Physics and Astrophysics*, Pergamon Press, Oxford.

GINZBURG, V. L. and RUKHADZE, A. A. (1970) Handb. Phys. **4914**, 395.

GOLANT, V. E. (1968) *Super-high-frequency Methods in Plasma Investigations*, Nauka, Moscow.

GOLDEN, K. I., KALMAN, G. and DATTA, T. (1975) Sum rules for nonlinear plasma response functions, *Phys. Rev.* **A11**, 2147.

GOLDEN, K. I., KALMAN, G., and SILEVICH, M. B. (1972) Nonlinear fluctuation-dissipation theorem, *J. Stat. Phys.* **6**, 87.

GORBUNOV, L. M. (1969) Perturbation of a medium by a powerful electromagnetic wave, *Soviet Phys. JETP* **20**, 2037.

GORBUNOV, L. M., PUSTOVALOV, V. V., and SILIN, V. P. (1965) Nonlinear interaction of electromagnetic waves in a plasma, *Soviet Phys. JETP* **20**, 967.

GORBUNOV, L. M. and SILIN, V. P. (1966) Scattering of waves in a plasma, *Soviet Phys. JETP* **23**, 895.

GORBUNOV, L. M. and TIMBERBULATOV, A. M. (1968) Dispersion law and nonlinear interaction of Langmuir waves in a weakly turbulent plasma, *Soviet Phys. JETP* **26**, 861.

GOULD, R. W., O'NEIL, T. M., and MALMBERG, J. H. (1967) Plasma wave echo, *Phys. Rev. Lett.* **19**, 219.

GUERNSEY, R. L. (1969) Surface waves in hot plasmas, *Phys. Fluids* **12**, 1852.

GUROV, K. P. (1966) *Foundations of the Kinetic Theory*, Nauka, Moscow.

HAGFORS, T. and BROCKELMAN, R. (1971) A theory of collision dominated electron density fluctuations in a plasma with application to incoherent scattering, *Phys. Fluids* **14**, 1143.

HAMASAKI, S. and KRALL, N. A. (1971) Self-consistent calculation of an explosive instability, *Phys. Fluids* **14**, 1441.

HEALD, M. and WHARTON, S. (1965) *Plasma Diagnostics with Microwaves*, J. Wiley, New York.

HINTON, F. L. and OBERMAN, C. (1968) Test-particle propagator and its application to the decay of plasma wave echoes, *Phys. Fluids* **11**,1982.

HIRSHFIELD, J. L. (1973) Wave–wave contribution to the high-frequency resistivity of nonequilibrium plasma, *Phys. Fluids* **16**, 1460.

HUBBARD, J. (1961a) The friction and diffusion coefficients of the Fokker–Planck equation in a plasma, 1, *Proc. Roy. Soc.* **A260**, 114.

HUBBARD, J. (1961b) The friction and diffusion coefficients of the Fokker–Planck equation in a plasma, 2, *Proc. Roy. Soc.* **A261**, 371.

ICHIMARU, S. (1962) Theory of fluctuations in a plasma, *Annals Phys.* **20**, 78.

ICHIMARU, S. (1970) Theory of strong turbulence in plasmas, *Phys. Fluids* **13**, 1560.

ICHIMARU, S. (1973) *Basic Principles of Plasma Physics*, Benjamin, London.

ICHIMARU, S., PINES, D., and ROSTOKER, N. (1962) Observation of critical fluctuations associated with plasma-wave instabilities, *Phys. Rev. Lett.* **8**, 231.

ICHIMARU, S. and ROSENBLUTH, M. N. (1970) Relaxation processes in plasmas with magnetic field, *Phys. Fluids* **13**, 2778.

ICHIMARU, S. and STARR, S. (1970) Electromagnetic radiation from anisotropic turbulent plasma, *Phys. Rev.* **A2**, 821.

JACKSON, E. A. (1960) Nonlinear oscillations in a cold plasma, *Phys. Fluids* **3**, 831.

KADOMTSEV, B. B. (1957) On fluctuations in a gas, *Soviet Phys. JETP* **5**, 771.

KADOMTSEV, B. B. (1965) *Plasma Turbulence*, Academic Press, New York.

KADOMTSEV, B. B., MIKHAILOVSKII, A. B., and TIMOFEEV, A. V. (1965) Negative energy waves in dispersive media, *Soviet Phys. JETP* **20**, 1517.

KADOMTSEV, B. B. and PETVIASHVILI, V. I. (1963) Weakly turbulent plasma in a magnetic field, *Soviet Phys. JETP* **16**, 1578.

KARPLYUK, K. S. KOLESNICHENKO, YA. I., and ORAEVSKY, V. N. (1970) Interaction of magnetohydrodynamic waves in a bounded plasma, *Nucl. Fusion* **10**, 3.

KARPLYUK, K. S. and ORAEVSKII, V. N. (1969) Nonlinear interaction of bulk and surface waves in a bounded plasma, *Soviet Phys. Tech. Phys.* **13**, 1003.

KARPMAN, V. I. (1975) *Non-linear Waves in Dispersive Media*, Pergamon Press, Oxford.

KLIMONTOVICH, YU. L. (1965) Nonlinear interaction of waves in plasma, *Soviet Phys. JETP* **21**, 326.

KLIMONTOVICH, YU. L. (1967) *Statistical Theory of Non-equilibrium Processes in a Plasma*, Pergamon Press, Oxford.

KLIMONTOVICH, YU. L. (1975) *Kinetic Theory of Nonideal Gas and Nonideal Plasma*, Nauka, Moscow; Pergamon Press, Oxford (1981).

KLIMONTOVICH, YU. L. and SILIN, V. P. (1962) Fluctuations in a collisionless plasma, *Soviet Phys. Doklady* **7**, 698.

KOCHERGA, O. D. (1975) Correlation functions for currents and nonlinear susceptibilities of a nonequilibrium isotropic plasma, *Ukr. Fiz. Zh.* **20,** 148.

KONDRATENKO, A. N. (1965) Kinetic theory of electromagnetic waves in a limited plasma, *Nucl. Fusion* **5,** 267.

KONDRATENKO, A. N. and SHAPTALA, V. G. (1969) On non-linear interaction of surface waves, *Ukr. Fiz. Zh.* **14,** 1092.

KOVRIZHNYCH, L. M. (1966) Theory of nonlinear wave interaction in a plasma, in. *Proceedings of the P. N. Lebedev Physical Institute* **32,** 173, Nauka, Moscow.

KRYLOV, N. M. and BOGOLIUBOV, N. N. (1947) *Introduction to Nonlinear Mechanics*, Princeton University Press, Princeton.

KUBO, R. J. (1957) Statistical-mechanical theory of irreversible processes: 1, General theory and simple applications to magnetic and conduction problem, *J. Phys. Soc. Japan* **12,** 570.

LANDAU, L. D. (1937) The kinetic equation for the case of Coulomb interaction, *Zh. Eksp. Teor. Fiz.* **7,** 203 (*Collected Papers*, Pergamon Press, Oxford, 1965, p. 163).

LANDAU, L. D. (1946) On the oscillations of an electron plasma, *Zh. Eksp. Teor. Fiz.* **16,** 574 (*Collected Papers*, Pergamon Press, Oxford, 1965, p. 445).

LANDAU, L. D. and LIFSHITZ, E. M. (1960) *Electrodynamics of Continuous Media*, Pergamon Press, Oxford.

LANDAU, L. D. and LIFSHITZ, E. M. (1969) *Statistical Physics*, Pergamon Press, Oxford.

LÉNARD, A. (1960) On Bogoliubov's kinetic equation for a spatially homogeneous plasma, *Annals Phys.* **10,** 390.

LEONTOVICH, M. A. and RYTOV, S. M. (1952) On a differential law for the intensity of electric fluctuations and the influence of the skin-effect, *Zh. Eksp. Teor. Fiz.* **23,** 246.

LEVIN, M. L. and RYTOV, S. M. (1967) *Theory of Equilibrium Thermal Fluctuations in Electrodynamics*, Nauka, Moscow.

LINHART, J. G. (1969) *Plasma Physics*, EAEC, Brussels.

LOMINADZE, D. G. (1975) *Cyclotron Waves in a Plasma*, Metsniereba, Tbilisi; Pergamon Press, Oxford (1981).

MIKHAILOVSKII, A. B. (1974) *Plasma Instabilities*, Plenum Press, New York.

MONTGOMERY, D. C. and TIDMAN, D. A. (1964) *Plasma Kinetic Theory*, McGraw-Hill, New York.

NISHIKAWA, K. (1967) Instability of a weakly ionized plasma induced by an alternating electric field, *Progr. Theor. Phys.* **37,** 769.

NYQUIST, H. (1928) Thermal agitation of electric charge in conductors, *Phys. Rev.* **32,** 110.

O'NEIL, T. M. (1968) Effect of Coulomb collisions and microturbulence on the plasma wave echo, *Phys. Fluids* **11,** 2420.

O'NEIL, T. M. and GOULD, R. W. (1968) Temporal and spatial plasma wave echoes, *Phys. Fluids* **11,** 134.

ORAEVSKI, V. N. and SAGDEEV, R. Z. (1963) Stability of steady-state Langmuir plasma oscillations, *Soviet Phys. Tech. Phys.* **7,** 955.

ORAEVSKII, V. N. WILHELMSSON, H., KOGAN, E. YA., and PAVLENKO, V. P. (1973) On the stabilization of explosive instabilities by nonlinear frequency shifts, *Physica Scripta* **7,** 217.

POLOVIN, R. V. (1957) On the nonlinear theory of longitudinal plasma oscillations, *Soviet Phys. JETP* **4,** 290.

POPOV, V. S. and YAKIMENKO, I. P. (1976) Electromagnetic fluctuations in a half-space of nonequilibrium plasma, *Soviet Phys. Tech. Phys.* **20,** 138.

PUSTOVALOV, V. V., ROMANOV, A. B. SAVCHENKO, M. A. SILIN, V. P., and CHERNIKOV, A. A. (1975) The functional mean approach in the kinetic plasma theory, Preprint FIAN-156, Moscow.

PUSTOVALOV, V. V. ROMANOV, A. B. SAVCHENKO, M. A., SILIN, V. P., and CHERNIKOV, A. A. (1976) On one method of solving the kinetic Vlasov equation, *Short Comm. in Physics* **12,** 28., Moscow.

PUSTOVALOV, V. V. and SILIN, V. P. (1972a) Nonlinear theory of wave interaction in a plasma, *Proceedings of the P. N. Lebedev Physical Institute* **61,** 42, Nauka, Moscow.

PUSTOVALOV, V. V. and SILIN, V. P. (1972b) On the stationary turbulence of a parametrically unstable plasma, *JETP Lett.* **16,** 217.

PUSTOPALOV, V. V., SILIN, V. P., and TIKHONCHUK, V. T. (1974) Nonlinear transformation of radiation into plasma waves, *Soviet Phys. JETP* **38,** 1880.

REUTER, C. E. H. and SONDHEIMER, E. H. (1948) The theory of the anomalous skin effect in metals, *Proc. Roy. Soc.* **A195,** 336.

ROMANOV, YU. A. (1965) Low-frequency surface waves in a two-component plasma, *Soviet Phys. Solid State*, **7,** 782.

ROMANOV, YU. A. and FILIPPOV, G. F. (1961) Interaction of fast electron fluxes with longitudinal plasma waves, *Soviet Phys. JETP* **13,** 87.

ROSENBLUTH, M. N. and ROSTOKER, N. (1962) Scattering of electromagnetic waves by a nonequilibrium plasma, *Phys. Fluids* **5,** 776.

ROSTOKER, N. (1960) Kinetic equation with a constant magnetic field, *Phys. Fluids* **3,** 922.

ROSTOKER, N. (1961) Fluctuations of a plasma, *Nucl. Fusion* **1,** 101.

ROSTOKER, N. and ROSENBLUTH, M. N. (1960) Test particles in a completely ionized plasma, *Phys. Fluids* **3,** 1.

RYTOV, S. M. (1953) *Theory of Electrical Fluctuations and Thermal Radiation*, Izd. Akad. Nauk SSSR, Moscow.

SAGDEEV, R. Z. (1958) On plasma confinement by the pressure of a standing electromagnetic wave, in *Plasma Physics and Controlled Thermonuclear Fusion* **3,** 346, Moscow.

SAGDEEV, R. Z. and GALEEV, A. A. (1969) *Nonlinear Plasma Theory*, Benjamin, New York.

SALPETER, E. E. (1960) Electron density fluctuations in a plasma, *Phys. Rev.* **120,** 1528.

SALPETER, E. E. (1961) Plasma density fluctuations in a magnetic field, *Phys. Rev.* **122,** 1663.

SANDRI, G. (1963) The foundations of nonequilibrium statistical mechanics, 1, 2, *Annals Phys.* **24,** 332; 380.

SANDRI, G. (1965) A new method of expansion in mathematical physics, *Nuovo Cimento* **36,** 67.

SCHUMANN, W. O. (1950) Über elektrische Wellen längs eines dielektrischen Zylinders in einer dielektrischen Umgebung, wobei eines oder beide der Medien Plasmen sind, *Z. Naturforsch.* **5a,** 181.

SHAFRANOV, V. D. (1958) Electromagnetic wave propagation in a medium with spatial dispersion, *Soviet Phys. JETP* **7,** 1019.

SHAFRANOV, V. D. (1963) Electromagnetic waves in a plasma, *Rev. Plasma Phys.* **3,** 1.

SHEFFIELD, J. (1975) *Plasma Scattering of Electromagnetic Radiation*, Academic Press, New York.

SILIN, V. P. (1959) Electromagnetic fluctuations in media with spatial dispersion, *Radiofizika* **2,** 198.

SILIN, V. P. (1962) On the theory of electromagnetic fluctuations in a plasma, *Soviet Phys. JETP* **14,** 689.

SILIN, V. P. (1964a) One possible plasma instability, *Soviet Phys. JETP* **18,** 559.

SILIN, V. P. (1964b) On the kinetic theory of interaction of plasma waves, *Prikl. Mat. Teor. Fiz.* **1,** 31.

SILIN, V. P. (1965) Parametric resonance in a plasma, *Soviet Phys. JETP* **21,** 1127.

SILIN, V. P. (1971) *Introduction to the Kinetic Theory of Gases*, Nauka, Moscow.

SILIN, V. P. (1973) *Parametric Effect of Powerful Radiation on a Plasma*, Nauka, Moscow.

SILIN, V. P. and FETISOV, E. P. (1962) On the electromagnetic properties of a relativistic plasma, *Soviet Phys. JETP* **14,** 115.

SILIN, V. P. and RUKHADZE, A. A. (1961) *Electromagnetic Properties* of *Plasmas and Plasma-like Media*, Atomizdat, Moscow.

SITENKO, A. G. (1966) On the fluctuation-dissipation theorem for nonequilibrium systems, *Ukr. Fiz. Zh.* **11,** 1161.

SITENKO, A. G. (1967) *Electromagnetic Fluctuations in Plasma*, Academic Press, New York.

SITENKO, A. G. (1973a) Fluctuations in plasma and nonlinear susceptibilities, *Physica Scripta* **7,** 190.

SITENKO, A. G. (1973b) Nonlinear wave interaction and fluctuations in plasma, *Physica Scripta* **7,** 193.

SITENKO, A. G. (1973c) Connection between nonlinear susceptibilities and fluctuations in a plasma, *Soviet Phys. Doklady* **18,** 483.

SITENKO, A. G. (1975) Fluctuations in a turbulent plasma, *Plasma Physics (USSR)* **1,** 45.

SITENKO, A. G. (1976) Plasma fluctuations and the statistical description of a plasma, in *Problems of Plasma Theory*, p. 5, Naukova Dumka, Kiev.

SITENKO, A. G. (1978) Kinetic theory of surface waves in semibounded plasmas, in *Theoretical and Computational Physics*, p. 145. IAEA, Vienna.

SITENKO, A. G. (1979) On the fluctuation–dissipation theorem in nonlinear electrodynamics, *Soviet Phys. JETP* **48,** 103.

SITENKO, A. G. and GURIN, A. A. (1966) Effect of particle collisions on plasma fluctuations, *Soviet Phys. JETP* **22,** 1089.

SITENKO, A. G. and KIROCHKIN, YU. A. (1966) Scattering and transformation of waves in a magnetoactive plasma, *Soviet Phys. Uspekhi* **9,** 430.

SITENKO, A. G. and KOCHERGA, O. D. (1977a) Nonlinear electromagnetic fluctuations in nonequilibrium plasmas, *Ukr. Fiz. Zh.* **22,** 1790.

SITENKO, A. G. and KOCHERGA, O. D. (1977b) Nonlinear wave interaction and stationary fluctuation spectrum in nonequilibrium plasmas. *Ukr. Fiz. Zh.* **22,** 1800.

SITENKO, A. G. and KOCHERGA, O. D. (1977c) Kinetic equation for waves and wave scattering in nonequilibrium plasmas, *Ukr. Fiz. Zh.* **22,** 1808.

SITENKO, A. G. and NGUEN VAN CHONG (1978) Nonlinear wave interaction in a semibounded plasma, *Ukr. Fiz. Zh.* **23,** 999.

SITENKO, A. G., NGUEN VAN CHONG and PAVLENKO, V. N. (1970a) Theory of plasma echoes: three-pulse echo oscillations, *Soviet Phys. JETP* **31,** 1165.

SITENKO, A. G., NGUEN VAN CHONG, and PAVLENKO, V. N. (1970b) Contribution to the theory of echo phenomena in a plasma, *Nucl. Fusion* **10,** 259.

SITENKO, A. G., NGUEN VAN CHONG, and PAVLENKO, V. N. (1970c) Spatial echoes in a plasma, *Ukr. Fiz. Zh.* **15,** 1372.

SITENKO, A. G. and ORAEVSKII, V. N. (1968) Fluctuations in solid state plasmas, *J. Phys. Chem. Solids* **29,** 1783.

SITENKO, A. G. and PAVLENKO, V. N. (1978) Kinetic theory of the nonlinear wave interaction in a semibounded plasma, *Soviet Phys. JETP* **47,** 65.

SITENKO, A. G., PAVLENKO, V. N., and REVENCHUK, S. M. (1979) Non-linear interaction of bulk and surface waves in a semibounded plasma, *Sov. J. Plasma Physics (USSR)* (to be published).

SITENKO, A. G. PAVLENKO, V. N., and ZASENKO, V. I. (1975a) Echoes in semibounded plasmas, *Ukr. Fiz. Zh.* **20,** 323.

SITENKO, A. G., PAVLENKO, V. N., and ZASENKO, V. I. (1975b) Echo in a half-space plasma, *Phys. Lett.* **53A,** 259.

SITENKO, A. G. PAVLENKO, V. N. and ZASENKO, V. I. (1976) Echo surface waves in a plasma, *Sov. J. Plasma Physics* **2,** 448.

SITENKO, A. G. and RADZIEVSKII, V. N. (1966) Fluctuations in a nonequilibrium magnetoactive plasma, *Soviet Phys. Tech. Phys.* **10,** 903.

SITENKO, A. G. and STEPANOV, K. N. (1957) On oscillations of an electron plasma in a magnetic field, *Soviet Phys. JETP* **4,** 512.

SITENKO, A. G. and TKALICH, V. S. (1960) On the Cherenkov effect for the charge motion along the boundary between two media, *Soviet Phys. Tech. Phys.* **4,** 981.

SITENKO, A. G. and TSZYAN'YU-TAI (1963) On dynamic friction and diffusion coefficients in a plasma, *Soviet Phys. Tech. Phys.* **7,** 978.

SITENKO, A. G. and YAKIMENKO, I. P. (1974) Method of inversion of the fluctuation–dissipation ratio in plasma theory, in *Advances in Plasma Physics* **5,** 15 (ed. A. Simon and W. B. Thompson), Wiley, New York.

SITENKO, A. G. and ZASENKO, V. I. (1978a) Saturation of the fluctuation level in a magnetoactive plasma with an anisotropic velocity distribution, *Ukr. Fiz. Zh.* **23,**715.

SITENKO, A. G. and ZASENKO, V. I. (1978b) On the nonlinear eigenfrequency shift in a plasma, *Ukr. Fiz. Zh.* **23,** 1277.

SIZONENKO, V. L. and STEPANOV, K. N. (1966) Quasilinear relaxation of longitudinal oscillations of a plasma, *Soviet Phys. JETP* **22,** 832.

SOMMERFELD, A. (1956) *Thermodynamics and Statistical Mechanics*, Academic Press, New York.

SPITZER, L. (1956) *Physics of a Fully Ionized Gas*, Interscience, New York.

STENFLO, L. (1970) Effects of collisions on resonant three-wave coupling, *Plasma Phys.* **12,** 509.

STEPANOV, K. N. (1966) On the surface wave propagation in a magnetoactive plasma, *Soviet Phys. Tech. Phys.* **10,** 1048.

STIX, T. (1962) *Theory of Plasma Waves*, McGraw-Hill, New York.

STURROCK, P. A. (1957) Nonlinear effects in electron plasmas, *Proc. Roy. Soc.* **A242,** 277.

STURROCK, P. A. (1964) Nonlinear theory of electrostatic waves in plasmas, in *Proceedings of the International School of Physics Enrico Fermi*, p. 180, Academic Press, New York.

THOMPSON, W. B. (1962) *An Introduction to Plasma Physics*, Pergamon Press, Oxford.

THOMPSON, W. B. and HUBBARD, J. (1960) Long-range forces and the diffusion coefficients of a plasma, *Rev. Mod. Phys.* **32,** 714.

TIDMAN, D. A. and DUPREE, T. H. (1965) Enhanced bremsstrahlung from plasmas containing nonthermal electrons, *Phys. Fluids* **8,** 1860.

TSYTOVICH, V. N. (1970) *Non-linear Effects in a Plasma*, Plenum Press, New York.

TSYTOVICH, V. N. (1973) *An Introduction to the Theory of Plasma Turbulence*, Pergamon Press Oxford.

TSYTOVICH, V. N. and SHAPIRO, V. D. (1965) Nonlinear stabilization of beam plasma instabilities, *Nucl. Fusion* **5,** 228.

UHLENBECK, G. E. and FORD, G. W. (1963) *Lectures in Statistical Mechanics*, Providence.

VANDENPLAS, P. E. (1968) *Electron Waves and Resonances in Bounded Plasmas*, Wiley, New York.

VEDENOV, A. A. (1965) *Theory of a Turbulent Plasma*, Israel Program for Scientific Translation, Jerusalem.

VEDENOV, A. A., GORDEEV, A. V., and RUDAKOV, L. I. (1967) Oscillations and instability of a weakly turbulent plasma, *Plasma Phys.* **9,** 719.

VEDENOV, A. A. and RUDAKOV, L. I. (1965) Interaction of waves in continuous media, *Soviet Phys. Doklady* **9,** 1073.

VEDENOV, A. A., VELIKHOV, E. P., and SAGDEEV, R. Z. (1961) Nonlinear oscillations of a rarefied plasma, *Nucl. Fusion* **1,** 82.

VEDENOV, A. A., VELIKHOV, E. P., and SAGDEEV, R. Z. (1962) Quasilinear theory of plasma oscillations, *Nucl. Fusion Suppl.* **2,** 465.

VEKSLER, V. I. and KOVRIZHNYKH, L. M. (1959) Cyclic acceleration of particles in a high-frequency fields, *Soviet Phys. JETP* **8,** 947.

VLASOV, A. A. (1938) On the vibration properties of an electron gas, *Zh. Eksp. Teor. Fiz.* **8,** 291.

VLASOV, A. A. (1950) *Many-particle Theory*, Gostekhizdat, Moscow.

VOLKOV, T. F. (1958) Influence of a high-frequency electromagnetic field on plasma oscillations, in *Plasma Physics and Controlled Thermonuclear Fusion*, vol. 4, Moscow.

WEINSTOCK, J. (1965) New approach to the theory of fluctuations in a plasma, *Phys. Rev.* **139A,** 388.

WEINSTOCK, J. (1969) Formulation of a statistical theory of strong plasma turbulence, *Phys. Fluids* **12,** 1045.

WEINSTOCK, J. (1970) Turbulent plasma in a magnetic field, *Phys. Fluids* **13,** 2308.

WEINSTOCK, J. and BEZZERIDES, B. (1973) Nonlinear saturation of parametric instabilities: spectrum of turbulence and enhanced collision frequency, *Phys. Fluids* **16,** 2287.

WILHELMSSON, H. (1961) Stationary nonlinear plasma oscillations, *Phys. Fluids* **4,** 335.

WILHELMSSON, H., STENFLO, L., and ENGELMANN, F. (1970) Explosive instabilities in the well-defined phase description, *J. Math. Phys.* **11,** 1738.

WILHELMSSON, H. and YAKIMENKO, I. P. (1978) Fluctuations in active molecular plasmas, *Physica Scripta* **17,** 523.

YAKIMENKO, I. P. (1976) Statistical theory of electromagnetic processes in bounded plasmas, in *Problems of Plasma Theory*, p. 80, Naukova Dumka, Kiev.

YAKIMENKO, I. P. and ZAGORODNY, A. G. (1974) Transition probabilities, correlation functions and dielectric permittivity tensor for semi-bounded and bounded plasmas, *Physica Scripta* **10,** 244.

YAKIMENKO, I. P. and ZAGORODNY, A. G. (1976) Spontaneous emission of electromagnetic waves by a semi-bounded non-equilibrium plasma, *Physica Scripta* **14,** 242.

YEFREMOV, G. F. (1969) Fluctuation–dissipation theorem for nonlinear media, *Soviet Phys. JETP* **28,** 2322.

YEH, K. C. and LIU, C. H. (1972) *Theory of Ionospheric Waves*, Academic Press, New York.

ZAGORODNY, A. G., TEGEBACK, R., USENKO, A. S., and YAKIMENKO, I. P. (1978) Enhanced fluctuations in semi-infinite plasmas, *Physica Scripta* **18,** 182.

ZAKHAROV, V. E. (1972) Collapse of Langmuir waves, *Soviet Phys, JETP* **35,** 908.

ZHELEZNYAKOV, V. V. (1970) *Radio Emission of Sun and Planets*, Pergamon Press, Oxford.

Index

18

OTHER TITLES IN THE SERIES IN NATURAL PHILOSOPHY

Vol. 1. Davydov—Quantum Mechanics
Vol. 2. Fokker—Time and Space, Weight and Inertia
Vol. 3. Kaplan—Interstellar Gas Dynamics
Vol. 4. Abrikosov, Gor'kov and Dzyaloshinskii—Quantum Field Theoretical Methods in Statistical Physics
Vol. 5. Okun'—Weak Interaction of Elementary Particles
Vol. 6. Shklovskii—Physics of the Solar Corona
Vol. 7. Akhiezer *et al.*—Collective Oscillations in a Plasma
Vol. 8. Kirzhnits—Field Theoretical Methods in Many-body Systems
Vol. 9. Klimontovich—The Statistical Theory of Nonequilibrium Processes in a Plasma
Vol. 10. Kurth—Introduction to Stellar Statistics
Vol. 11. Chalmers—Atmospheric Electricity (2nd Edition)
Vol. 12. Renner—Current Algebras and their Applications
Vol. 13. Fain and Khanin—Quantum Electronics, Volume 1—Basic Theory
Vol. 14. Fain and Khanin—Quantum Electronics, Volume 2—Maser Amplifiers and Oscillators
Vol. 15. March—Liquid Metals
Vol. 16. Hori—Spectral Properties of Disordered Chains and Lattices
Vol. 17. Saint James, Thomas and Sarma—Type II Superconductivity
Vol. 18. Margenau and Kestner—Theory of Intermolecular Forces (2nd Edition)
Vol. 19. Jancel—Foundations of Classical and Quantum Statistical Mechanics
Vol. 20. Takahashi—An Introduction to Field Quantization
Vol. 21. Yvon—Correlations and Entropy in Classical Statistical Mechanics
Vol. 22. Penrose—Foundations of Statistical Mechanics
Vol. 23. Visconti—Quantum Field Theory, Volume 1
Vol. 24. Furth—Fundamental Principles of Theoretical Physics
Vol. 25. Zheleznyakov—Radioemission of the Sun and Planets
Vol. 26. Grindlay—An Introduction to the Phenomenological Theory of Ferroelectricity
Vol. 27. Unger—Introduction to Quantum Electronics
Vol. 28. Koga—Introduction to Kinetic Theory: Stochastic Processes in Gaseous Systems
Vol. 29. Galasiewicz—Superconductivity and Quantum Fluids
Vol. 30. Constantinescu and Magyari—Problems in Quantum Mechanics
Vol. 31. Kotkin and Serbo—Collection of Problems in Classical Machanics
Vol. 32. Panchev—Random Functions and Turbulence
Vol. 33. Taipe—Theory of Experiments in Paramagnetic Resonance
Vol. 34. Ter Haar—Elements of Hamiltonian Mechanics (2nd Edition)
Vol. 35. Clarke and Grainger—Polarized Light and Optical Measurement
Vol. 36. Haug—Theoretical Solid State Physics, Volume 1
Vol. 37. Jordan and Beer—The Expanding Earth
Vol. 38. Todorov—Analytical Properties of Feynman Diagrams in Quantum Field Theory
Vol. 39. Sitenko—Lectures in Scattering Theory
Vol. 40. Sobel'man—Introduction to the Theory of Atomic Spectra
Vol. 41. Armstrong and Nicholls—Emission, Absorption and Transfer of Radiation in Heated Atmospheres
Vol. 42. Brush—Kinetic Theory, Volume 3
Vol. 43. Bogolyubov—A Method for Studying Model Hamiltonians
Vol. 44. Tsytovich—An Introduction to the Theory of Plasma Turbulence
Vol. 45. Pathria—Statistical Mechanics
Vol. 46. Haug—Theoretical Solid State Physics, Volume 2
Vol. 47. Nieto—The Titius–Bode Law of Planetary Distances: Its History and Theory
Vol. 48. Wagner—Introduction to the Theory of Magnetism
Vol. 49. Irvine—Nuclear Structure Theory
Vol. 50. Strohmeier—Variable Stars
Vol. 51. Batten—Binary and Multiple Systems of Stars